Combinatorial, Computational, and Applied Algebraic Geometry: A Tribute to Bernd Sturmfels

Proceedings of Symposia in
PURE MATHEMATICS

Volume 111

Combinatorial, Computational, and Applied Algebraic Geometry: A Tribute to Bernd Sturmfels

Serkan Hoşten
Diane Maclagan
Frank Sottile
Editors

AMERICAN MATHEMATICAL SOCIETY
Providence, Rhode Island

2020 *Mathematics Subject Classification.* Primary 12D10, 13P10, 13P15, 13P25, 14M25, 14Q20, 14Txx, 52Bxx, 62R01, 65H14.

For additional information and updates on this book, visit
www.ams.org/bookpages/series-volume

Library of Congress Cataloging-in-Publication Data

DOI: https://doi.org/10.1090/pspum/111

∞ The paper used in this book is acid-free and falls within the guidelines
established to ensure permanence and durability.
Visit the AMS home page at `https://www.ams.org/`

10 9 8 7 6 5 4 3 2 1   30 29 28 27 26 25

Contents

Preface

Bernd Sturmfels turned sixty on March 28, 2022. To celebrate his birthday and to honor his immense contributions to mathematics, over 150 participants converged to Seattle, Washington, to attend *Combinatorial, Computational, and Applied Algebraic Geometry* (CCAAGS'22) during June 27–July 1, 2022. The conference was sponsored by Georgia Tech, MIT, NSA, NSF, PIMS, SFSU, UC Davis, and the University of Washington. Seven of Bernd's former students composed the organizing committee and the local organizers: Jesús De Loera, Serkan Hoşten, Diane Maclagan, Caroline Uhler, Josephine Yu, Rekha Thomas, and Cynthia Vinzant.

Over five days, 21 speakers gave hour-long talks on a diverse range of topics: from commutative algebra, tensors, algebraic statistics, and computational biology to tropical geometry, Hodge theory, spectral theory, and numerical algebraic geometry. A lively poster session was held where work was presented on equally diverse topics. There was cake cutting. And there was a session where Bernd was "roasted" with personal stories presented by his current and former students and postdocs, colleagues, and friends. It was towards the end of the COVID times, but it was still risky to travel and hold a conference. By and large, things were almost uneventful on that front, with one big exception: after the first day of the meeting, we woke up to the news that Bernd had tested positive! It could have been a huge damper, but Bernd himself was the first person to make the best out of the situation. He participated in all of the talks and other events via Zoom, asked questions, and made comments. He even gave a talk when, in characteristic fashion, he gave short presentations of recent work of his current students. We joked that, had he been healthy, Bernd probably would have organized an evening session for young participants to discuss mathematics. The conference dinner was a great time to catch up with those one had not seen for a while. Bernd had a finger in creating many of these friendships through his mathematical matchmaking capacity.

We felt that, parallel to this wonderful celebration, something more permanent should be created to mark Bernd's place in the combinatorial, computational, and applied algebraic geometry world, and beyond. This volume is a result of this effort. We solicited, from experts in their fields, articles that would highlight Bernd's unforgettable contributions. We received fourteen such articles over two years. With more time and effort, it could have been double the number, since the mathematics that Bernd helped to get created is vast. It is just impossible to stop and give a snapshot or a summary of a project that has been going on since 1986.

In addition, as the first article of this volume, we included something from Bernd. He is known not only for his extraordinary energy initiating and doing new mathematics, but equally for his amazing talent in mentoring students, postdocs, and other early career mathematicians (63 former and 9 current students as of

January 2025 and counting). To highlight this side of him we chose to reprint an early career advice column Bernd wrote for the AMS *Notices*. It is wonderful to hear from him what he thinks about his own idiosyncratic ways of mentoring graduate students.

We thank all of the authors in this volume who answered our call to write about Bernd and his mathematics. We also thank many others who anonymously helped us to review the articles. And last but not least, a huge thank you goes to Bernd Sturmfels who helped build the mathematical and human community to which we all belong.

Serkan Hoşten
Diane Maclagan
Frank Sottile

Proceedings of Symposia in Pure Mathematics
Volume **111**, 2025
https://doi.org/10.1090/pspum/111/02036

Adventures in mentoring

Bernd Sturmfels

ABSTRACT. This is reprinted from an early career advice column of *Notices of the American Mathematical Society.*

I started mentoring students and postdocs thirty years ago, as a young assistant professor at Cornell. This continued after I moved to Berkeley in 1995, and I now enjoy working with young scholars in Germany. I always thought of mentoring as one of the best parts of my job. With these notes, I'd like to share some anecdotes and experiences. These might be useful for colleagues who are starting their first tenure-track jobs, and will perhaps be entertaining and thought-provoking for all readers. Before beginning, I would like to use this opportunity to thank all my students and postdocs. I learned a lot from you. Please continue to teach me.

Doing the opposite. My first PhD student was older and more experienced. He came to my office once a week, for an entire afternoon, asking me for my advice on his research. I eagerly offered my advice. After leaving he did precisely the opposite of what I had suggested. The same happened the following week. I recommended A, he did not-A. The same events, week after week. I was puzzled. Is that what it meant to be an advisor? The answer is yes, for this particular student. Having already made up his mind on what to do, he just needed me to test his ideas. And to debate. He became a successful professor at a top institution.

A threat. It did not go so well with another student. One day, he messed up badly in presenting some material in a graduate seminar. I told him what I thought, right then and there. He was extremely upset because he felt humiliated in front of his peers. He threatened to shoot me if I ever did this again. I learned my lesson and decided to be more respectful.

Finding good students. Don't ever compare students at your current institution with yourself, from back when you were a grad student. Most students are capable and motivated. Don't worry so much whether they are "good" yet. It is your job to help them become good.

2020 *Mathematics Subject Classification.* 97B99.

Published in the September 2019 issue of *Notices of the American Mathematical Society* https://www.ams.org/cgi-bin/notices/nxgnotices.pl?fm=main&current=201908.

Being an imposter. When I was a postdoc, I could not believe my luck. Someone paid me for what I loved doing. But how come they missed that I was not qualified for the job? That worried me. I was sitting in seminars and did not understand. I went to the library, and the math books were incomprehensible. But everyone else seemed on top of it. I decided to wait and see. Surely, some day they'll discover that I am a fraud. But this has not happened yet.

Fordern und fördern. This is a phrase in German that my father liked to say. It is composed of two verbs, and translates roughly into "demand and encourage." I have been awfully demanding towards my mentees, but I also strongly encourage them. Try to read that German phrase aloud. Pay attention to the umlaut. It sounds pretty nice, doesn't it?

Follow the child. When my children entered preschool, I learned about Maria Montessori's famous line "Follow the Child." She was absolutely right. You have to trust your children. They will show you what they need. You should likewise trust your students and postdocs.

Questions. Here are questions I ask mentees when we first meet. What do you excel at? Do you enjoy teaching? What have you written lately? What is the difference between pure and applied mathematics? Do you like to compute? Where will you be ten years from now?

A path forward. Inexperienced professors like to say "I'll just give it to a student" when referring to loose ends from their past research publications. This almost never works. You'd better solve your own problems. Your students are either unable to do it, or they will choose their own problems. Instead, show your students a path forward. Their own path, not yours.

Excellence through diversity. Some people believe that there is a conflict between fostering diversity and striving for academic excellence. I found the opposite to be true. True excellence is inclusive. If you wish to train students and postdocs for academic careers at top research schools, then my advice is to build a diverse, inclusive, and supportive team.

Gender. Another lesson I learned from Montessori preschool: "everyone has their gifts and challenges." This applies to all learners, including you and me. Sometimes people ask how I became so successful in training outstanding female mathematicians. I already revealed two of the secret ingredients: Respect and encouragement. These two go a long way. Try to learn who your mentees really are. Gender, gender identity, and sexual orientation matter.

Phase transition. The research seminar or group meeting can be a tough place for young scholars. Tone and atmosphere are tremendously important. I found that active participation by women and members of other underrepresented groups makes a huge difference. There is a threshold in their participation rate, around 30% in my empirical experience, at which a phase transition occurs. Above that threshold, seminars are fun and productive for everyone.

On writing well. I often recommend William Zinsser's classic book with that title. But there are also good texts on mathematical writing. Writing a dissertation is writing. Writing research papers is writing. Writing job applications is writing. Writing grant proposals is writing. The importance of writing skills cannot be overemphasized. The mentor makes a huge difference. Lead by example. And, make them read aloud the writings they produce.

On lecturing well. Almost all mathematicians end up in careers where they must give clear and persuasive presentations. This is obvious for teachers and professors, but it also applies to those working in industry or government. Nobody is born a good lecturer. Practice makes all the difference. I discovered that some of the best-prepared incoming PhD students were on the debate team back in their high school. We all need to train with the debate team.

On coding well. To me, our cell phones are supercomputers. All of us spend most of the day performing heavy computations with these machines. But many young mathematicians are still afraid of using software for their mathematics research. Help your mentees to overcome that reluctance, and to make friends with tools like `Julia`, `Macaulay2`, `Magma`, `Maple`, `Mathematica`, `Matlab`, `OSCAR`, `R`, `SAGE`, etc. Do admire the output of their computations.

Listening. Prospective graduate students recently visited Berkeley. Before the visit day, my current students were asking me "What should we tell them?" I reply, "Nothing, just listen to them." At all stages of mentoring, listening is often more beneficial than telling things.

Being a scholar. There are many ways to become a scholar. Here are two points I recommend to my mentees. First, learn about mathematics. Here is a good exercise: for each of the two-digit numbers xx in the Mathematics Subject Classification, from "03 Logic" to "94 Information and Communications," state in your own words what xx is about. Name one person you know who works in xx. Second, find out about the history of your speciality. Read papers that were written more than 100 years ago. Don't fear foreign language sources.

Matchmaking. After we got married in 1990, my wife and I traveled Korea by car. On a country road we picked up a hitchhiker, an old lady who was a professional matchmaker. When she learned that my wife and I had not used a matchmaker, she was shocked: how was it possible for us to meet in Seattle, without the services of her profession? That old lady had a good point. Matchmakers are useful. That is why I am a matchmaker for doing mathematics. I connect my mentees to collaborators. Many good matches have come from this.

Language acquisition. I believe the phrase "Good things happen to those who show up." Encounters can be planned or circumstance. But if you don't show up then you will miss out. Students are hesitant to attend colloquia because they don't understand anything (like all imposters). But think of this as a foreign language immersion class. Just keep going. A few months later you will understand and speak. One colleague recommended uttering unfamiliar words over and over again. One learns their meaning by gauging the reaction of the listeners. Like a toddler. He said he tested this when he was a PhD student. He wanted to know what "holomorphic vector bundle" means. So, he went to the math lounge and repeated those three words over and over again. Now he is an expert in complex geometry.

Waking up to email. Every mentor has a different style. So does every mentee. A while back, PhD students at Berkeley complained that they do not get enough attention from their advisors. My students think that they receive too much attention from their advisor. A friend once referred to my style as "the samurai method." It involves a barrage of emails, usually sent between 4am and 5am. While this is fine for some, it does not work for others.

In the bullets above, I shared some perspectives and anecdotes on mentoring. It is most definitely not meant as a "how to" manual. There are many ways for you to be a terrific mentor. Make yourself available to young scholars. Help them in finding their perfect match.

Max Planck Institute for Mathematics in Sciences, Leipzig, Germany
Email address: bernd@mis.mpg.de

Proceedings of Symposia in Pure Mathematics
Volume **111**, 2025
https://doi.org/10.1090/pspum/111/02037

Invariant theory: A third lease of life

Gregor Kemper

ABSTRACT. In 1993, just about a century after the epoch of classical invariant theory and almost thirty years after Mumford's seminal book on geometric invariant theory, Bernd Sturmfels approached the subject from a new, algorithmic perspective in his book *Algorithms in invariant theory*. This article aims to highlight some of the developments that followed the book. Inspired by Bernd's style of teaching mathematics, the goal is neither comprehensiveness nor maximal generality, but to emphasize the main ideas and to convey the beauty of the subject. The article is intended as an invitation to invariant theory, and in particular to Bernd's work and the ideas inspired by it.

Introduction

When Bernd Sturmfels's book on *Algorithms in invariant theory* [**43**] appeared in 1993, I immediately devoured it, since at the time I was busy developing a package for computing invariant rings in the Maple computer algebra system. A major part of my PhD thesis was concerned with computational invariant theory, and Bernd's book served as an indication that this would be the part with the best chances of getting people interested. As it turned out, Bernd's book gave the subject a significant boost, especially by its overarching theme of wedding invariant theory to Gröbner basis methods. This is the idea behind the "third lease of life" in the title of this article, with the previous "leases" thought of as classical invariant theory and geometric invariant theory. Bernd's book came at a time when there was a confluence of new developments in computational invariant theory and modular invariant theory, with Harm Derksen's algorithm emerging shortly after the book's publication.

The book also got me in contact with Bernd, who later invited me as a contributor to a course he gave on the book in Eindhoven—my first international mathematical gig. Years after that, in 1999, he invited Harm Derksen and me to Berkeley at a time when the two of us were embarking on a joint book project in computational invariant theory [**10**]. Let me quote from the preface: "*... it was an invitation by Bernd Sturmfels to spend two weeks together in Berkeley that really got us started on this book project. We thank Bernd for his strong encouragement and very helpful advice. During the stay at Berkeley, we started outlining the book, making decisions about notation, etc.*"

2020 *Mathematics Subject Classification.* Primary 13A50, 13P10.

After writing his invariant theory book, Bernd, as he so often does, rather quickly moved on to other topics. But invariant theory, having received a new impetus from his book, also moved on, and it is the purpose of this article to cast some light on what has happened since 1993. However, this is not meant to be a typical survey article, since it will not aim for comprehensiveness and will contain quite a lot of proofs. Readers should not only learn what is true, but also, when space permits, why it is true. In this way, I will try to mimic what I understand to be Bernd's style of teaching mathematics. For the selection of material to be presented here, one criterion is that it should be at least loosely related to computation. Another criterion is that the results and the proofs, when they are included, should be nice, or even beautiful and elegant. So there is definitely some cherry-picking going on. (And, inevitably, my personal taste will favor some cherries over others.) To make the material as accessible as possible and to keep things simple for readers, attaining maximal generality will not be a paramount goal. This, I believe, is also in keeping with Bernd's style of teaching.

Keeping things simple for readers also means to avoid hopping between different situations and changing notation. In this article, we will always consider a group G acting on a polynomial ring $K[x_1, \dots, x_n] =: K[\mathbf{x}]$ over a field by linear transformations of the variables x_i. First and foremost, the object of our interest will be the invariant ring $K[\mathbf{x}]^G$. Here is another convention used throughout: when I say "Bernd's book", I will always be referring to his book *Algorithms in invariant theory* [**43**].

The first three sections will look at the case where G is finite, distinguishing between the **nonmodular** case (where the characteristic of K does not divide $|G|$, including $\operatorname{char}(K) = 0$) and the **modular** case. Specifically, Section 1 is devoted to nonmodular invariant theory and will present Noether's degree bound and King's algorithm for computing $K[\mathbf{x}]^G$. In Section 2 on modular invariant theory of finite groups, we address Symonds's degree bound (without proof) and look at results about the Cohen–Macaulay property (giving a full proof for one of them). Section 3 **??** deals with separating invariants. We show that they always satisfy Noether's degree bound, and give a self-contained proof of a result by Emilie Dufresne about separating invariants and reflection groups.

The last three sections are about invariants of infinite groups. They are significantly shorter than the first three sections, since there are fewer new results to report. But this does not mean that they are any less important. Maybe even to the contrary, as exemplified by Derksen's algorithm, which is the topic of Section 4. In Section 5 **??** we show how invariant fields of potentially nonreductive groups can be computed using the same ideal that occurs prominently in Derksen's algorithm, but in a very different way. The section also explains how to compute invariant rings of reductive groups in positive characteristic. For this, the stepping stone is the computation of separating invariants, which are the topic of Section 6. Here we present a proof of the (folklore) statement that there is always a separating set of size $\leq 2n + 1$, independently of the group.

1. Nonmodular invariants of finite groups

In nonmodular invariant theory of finite groups (more generally, in invariant theory of linearly reductive groups) the supreme ruler is the **Reynolds operator**

$$\mathcal{R}: \ K[\mathbf{x}] \to K[\mathbf{x}]^G, \ f \mapsto \frac{1}{|G|} \sum_{\sigma \in G} \sigma(f),$$

which averages over the group. It makes an appearance in the proof of Proposition 1.1, which, in some form, was already known to Hilbert and Noether. (More than a hundred years later, I learned it from Bernd's book.) The proposition tells us how the ideal

$$H := K[\mathbf{x}] \cdot K[\mathbf{x}]_+^G$$

generated by all nonconstant homogeneous invariants plays a central role. (Here $K[\mathbf{x}]_+^G$ stands for the set of all invariants with constant coefficient equal to zero, and the product $K[\mathbf{x}] \cdot K[\mathbf{x}]_+^G$ consists of all finite sums of products of an element from $K[\mathbf{x}]$ and one from $K[\mathbf{x}]_+^G$.) H has come to be known as the **Hilbert ideal**, and it is an ideal in $K[\mathbf{x}]$.

PROPOSITION 1.1. *For nonconstant homogeneous invariants $f_1, \dots, f_m \in K[\mathbf{x}]^G$, it is equivalent that they generate the Hilbert ideal H (as an ideal) and that they generate the invariant ring $K[\mathbf{x}]^G$ (as an algebra).*

Another equivalent condition is that (the classes of) the f_i generate the quotient space $H/(K[\mathbf{x}]_+ \cdot H)$ as a vector space over K. This fact is not needed here, and its proof is left as an exercise.

PROOF OF PROPOSITION 1.1. If the f_i generate $K[\mathbf{x}]^G$, then every invariant $h \in K[\mathbf{x}]_+^G$ can be written as $h = F(f_1, \dots, f_m)$ with F a polynomial in m variables having zero constant coefficient. Since every product of powers of some of the f_i lies in the $K[\mathbf{x}]$-ideal $(f_1, \dots, f_m)_{K[\mathbf{x}]}$ generated by the f_i, so does h. We conclude that $K[\mathbf{x}]_+^G \subseteq (f_1, \dots, f_m)_{K[\mathbf{x}]}$ and thus $H \subseteq (f_1, \dots, f_m)_{K[\mathbf{x}]}$. The reverse inclusion follows from $f_i \in H$.

For the converse, assume that $H = (f_1, \dots, f_m)_{K[\mathbf{x}]}$ and let f be a homogeneous invariant of degree $d > 0$. Then $f \in H$, so $f = \sum_{i=1}^m g_i f_i$ with $g_i \in K[\mathbf{x}]$. Applying the Reynolds operator yields

$$f = \mathcal{R}(f) = \sum_{i=1}^{m} \mathcal{R}(g_i) f_i.$$

By considering the homogeneous part of degree d of this equation, we may assume that all summands are homogeneous of degree d, so the $\mathcal{R}(g_i)$ have degree $< d$. So, by induction on d we may assume that they lie in the algebra $K[f_1, \dots, f_m]$ generated by the f_i. Therefore, the same is true for f. □

Together with Hilbert's basis theorem, which by historical accounts was proved for this very purpose, the proposition already shows that if there is a Reynolds operator, then the invariant ring is finitely generated. It also follows that any degree bound for generators of the Hilbert ideal is automatically a degree bound for the invariant ring.

This brings us to Noether's degree bound, proved by Emmy Noether [**39**] in 1916, which says that the invariant ring is generated by homogeneous invariants of degree at most the group order $|G|$. Actually, she gave two proofs. However, the first proof requires that $|G|!$ is invertible in K, so the characteristic must be 0 or bigger than $|G|$, and the second one, which can be found in Bernd's book [**43**, Theorem 2.1.4], only works in characteristic 0. A proof that extends to the full nonmodular case—people were speaking of the "Noether gap"—remained elusive for quite some time, until it was finally found independently by Fleischmann [**18**] and Fogarty [**19**]. We give a version that has gone through various stages of simplification. In fact, Noether's bound is an almost immediate consequence of the following lemma, which makes a surprising observation and is proved by a nice elementary trick.

LEMMA 1.2. *The Hilbert ideal H contains every monomial in the x_i of degree $|G|$.*

PROOF. We will show a bit more: that H contains every product of $|G|$ (not necessarily distinct) polynomials of $K[\mathbf{x}]_+$. Such polynomials may as well be indexed by the elements of G, so we claim that any product $\prod_{\sigma \in G} f_\sigma$, with $f_\sigma \in K[\mathbf{x}]_+$, lies in H. For every $\tau \in G$ we have

$$\prod_{\sigma \in G} \left(f_\sigma - (\tau\sigma)(f_\sigma)\right) = 0.$$

Fully expanding the product and summing over all $\tau \in G$ yields

$$\sum_{M \subseteq G} (-1)^{|M|} \left(\prod_{\sigma \in G \setminus M} f_\sigma \right) \cdot \left(\sum_{\tau \in G} \prod_{\sigma \in M} (\tau\sigma)(f_\sigma) \right) = 0.$$

The summand for $M = \emptyset$ is $|G| \cdot \prod_{\sigma \in G} f_\sigma$, and all other summands lie in H. So dividing by $|G|$ proves the claim. □

THEOREM 1.3 (Noether's degree bound). *If $|G|$ is invertible in K, then the invariant ring $K[\mathbf{x}]^G$ is generated as an algebra by homogeneous invariants of degree $\leq |G|$.*

PROOF. Let $I \subseteq K[\mathbf{x}]$ be the ideal generated by all nonconstant homogeneous invariants of degree $\leq |G|$. By Lemma 1.2, I contains every monomial of degree $|G|$, and therefore also every monomial of degree $\geq |G|$. So I contains every homogeneous polynomial of degree $\geq |G|$. This implies that I contains every nonconstant homogeneous invariant. Thus the Hilbert ideal H is generated in degree $\leq |G|$, and by Proposition 1.1 the same is true for $K[\mathbf{x}]^G$. □

In summary, we have presented a self-contained proof of Noether's bound that, after removal of comments and other luxuries, fits on a single page and also closes the Noether gap.

With a degree bound, we automatically have an algorithm for computing generators of the invariant ring: simply apply the Reynolds operator to all monomials of degree $\leq |G|$. Of course, this is extremely inefficient. In Bernd's book two algorithms are given for computing nonmodular invariant rings of finite groups. The first one [**43**, Algorithm 2.2.5] is driven by knowledge of the Hilbert series of the invariant ring, and the second [**43**, Algorithm 2.5.8] computes a so-called Hironaka decomposition. This subdivides the generating invariants into primary invariants, which are algebraically independent; and secondary invariants, which generate the invariant ring as a module over the subalgebra generated by the primary invariants. The second algorithm from Bernd's book has gone through various optimizations (for example, [**28**] improves the finding of primary invariants), but remained the state of the art for about twenty years until Simon King [**35**] proposed a new one in 2013. Although more elementary than Sturmfels's algorithms, King's algorithm is very clever and therefore makes good material for this article. Unlike the algorithms given by Sturmfels, King's algorithm requires only *truncated* Gröbner bases (see, for example, [**36**, Section 4.5.B]). For this reason, intuition says that it should be more effective and, more importantly, experience confirms this. That is why King's algorithm is now the standard in `Magma` [**5**] and `Macaulay2` [**17**], qualifying it to be called today's state of the art. At the time of writing this, it is also becoming part of the brand new computer algebra system OSCAR.[1]

ALGORITHM 1.4 (King's algorithm [**35**]).

Input: A finite group G acting linearly on the variables x_i, with $|G|$ not a multiple of $\mathrm{char}(K)$.

Output: A minimal generating set of the invariant ring $K[\mathbf{x}]^G$ as an algebra.

(1) Set $S := \emptyset$. Choose a monomial ordering on $K[\mathbf{x}]$.
(2) For $d = 1, 2, \ldots$ execute the following steps:
  (3) Let $\mathcal{G}$ be a d-truncated Gröbner basis of the ideal $(S) \subseteq K[\mathbf{x}]$ generated by S. This can be computed by Buchberger's algorithm, but considering only s-polynomials of degree $\leq d$.
  (4) Let M be the set of all monomials in $K[\mathbf{x}]$ of degree d that are not divisible by any leading monomial $\mathrm{LM}(g)$, $g \in \mathcal{G}$.
  (5) For $t \in M$ execute the following two steps:
    (6) Compute $f := \mathcal{R}(t)$, with $\mathcal{R}$ the Reynolds operator, and $h := \mathrm{NF}_{\mathcal{G}}(f)$, the *normal form* of f with respect to $\mathcal{G}$. By this we mean the unique polynomial h with $f - h \in (\mathcal{G})$ such that no monomial occurring in h is divisible by any $\mathrm{LM}(g)$, $g \in \mathcal{G}$.
    (7) If $h \neq 0$, then adjoin f to S and h to $\mathcal{G}$, and set $M := M \setminus \{\mathrm{LM}(h)\}$.
  (8) If $M = \emptyset$, then terminate the algorithm and return S.

I find it hard to understand immediately what is going on in the algorithm and why it works, and readers might feel the same way. Especially the consecutive application of the Reynolds operator and the normal form in step (6) and the business of removing monomials from M during the second loop may seem a bit obscure. To me, the only way to understand the algorithm is through a proof of correctness, which is presented here.

PROOF OF CORRECTNESS OF ALGORITHM 1.4. The proof is subdivided into five claims.

CLAIM 1. *During the inner loop initiated by step* (5), $\mathcal{G}$ *remains a d-truncated Gröbner basis of* (S) *and M remains the set of monomials of degree d not divisible by any* $\mathrm{LM}(g)$, $g \in \mathcal{G}$.

Indeed, since each h that is adjoined to $\mathcal{G}$ is in normal form with respect to the elements already in $\mathcal{G}$, no new s-polynomials of degree $\leq d$ occur after adjoining h, so the "new" $\mathcal{G}$ is again a d-truncated Gröbner basis. It also generates the same ideal as the "new" S. Moreover, since $\mathrm{LM}(h)$ has degree d, removing it from M amounts to removing all monomials divisible by it.

[1] `https://oscar.computeralgebra.de/`

Claim 2. *After d passes through the outer loop initiated by step (2), all homogeneous invariants of degree $\le d$ lie in the algebra generated by S: $K[\mathbf{x}]^G_{\le d} \subseteq K[S]$.*

The proof proceeds by induction on d, so we assume $K[\mathbf{x}]^G_{\le d-1} \subseteq K[S]$ at the start of the dth pass. In the following, let S, $\mathcal{G}$, and M denote the sets as they are *after* the dth pass. Let $\widetilde{f}$ be a homogeneous invariant of degree d. Using Claim 1, we get $\widetilde{f} - \mathrm{NF}_{\mathcal{G}}(\widetilde{f}) \in (\mathcal{G}) = (S)$, so $\widetilde{f} - \mathrm{NF}_{\mathcal{G}}(\widetilde{f}) = \sum_{i=1}^r c_i f_i$ with $f_i \in S$ and $c_i \in K[\mathbf{x}]$. Again by Claim 1, $\mathrm{NF}_{\mathcal{G}}(\widetilde{f})$ is a linear combination of monomials $t_1, \dots, t_s$ from M, so

$$\widetilde{f} = \sum_{i=1}^r c_i f_i + \sum_{i=1}^s a_i t_i$$

with $a_i \in K$. Applying the Reynolds operator yields

$$\widetilde{f} = \sum_{i=1}^r \mathcal{R}(c_i) f_i + \sum_{i=1}^s a_i \mathcal{R}(t_i).$$

By considering only the degree-d part we may assume the $\mathcal{R}(c_i)$ to be homogeneous, so their degrees are $< d$, and thus $\mathcal{R}(c_i) \in K[S]$ by induction. So Claim 2 is proved if we can show $\mathcal{R}(t_i) \in K[S]$ for all i. Since the t_i lie in the set M as it is after the dth pass through the outer loop, they have *not* been removed in step (7). So for $t = t_i$ and $f = \mathcal{R}(t)$ we have $h = \mathrm{NF}_{\mathcal{G}}(f) = 0$. Hence it is enough to prove Claim 3.

Claim 3. *For f and h in step (6), we have the equivalence*

$$h = 0 \iff f \in K[S],$$

where S denotes the set as it is in step (6).

In fact, Claim 1 shows that $h = 0$ is equivalent to $f \in (S)$. If that happens, we have $f = \sum_{i=1}^r q_i f_i$ with $f_i \in S$ and $q_i \in K[\mathbf{x}]$. Again, we can apply the Reynolds operator, assume homogeneity, and use induction to show that the q_i may be taken from $K[S]$, so $f \in K[S]$. Conversely, if $f \in K[S]$, then clearly $f \in (S)$ and hence $h = 0$. So Claims 2 and 3 are proved. Let us also note that the equivalence in Claim 3 means that an invariant f is adjoined to S only if this enlarges $K[S]$. This implies that throughout the algorithm, S is a *minimal* generating set. Now we aim to show that the algorithm terminates, and that upon termination, S generates $K[\mathbf{x}]^G$.

Claim 4. *The algorithm terminates after $|G|$ passes through the outer loop or earlier.*

Indeed, Claim 2 implies that after d passes, (S) contains all homogeneous elements of the Hilbert ideal that have degree $\le d$. So if $d = |G|$, Lemma 1.2 tells us that (S) contains all monomials of degree d. Now it follows from Claim 1 that all these monomials are divisible by $\mathrm{LM}(g)$ for some $g \in \mathcal{G}$. So using Claim 1 again, we see that after the $|G|$th pass, M will be empty, triggering termination.

Claim 5. *Upon termination, $K[\mathbf{x}]^G = K[S]$.*

Indeed, if every monomial of degree d is divisible by some $\mathrm{LM}(g)$, then the same is true for every monomial of degree $> d$. This remains true if the truncated Gröbner basis is continued to a higher degree. So if the algorithm were left to keep running, it would henceforth always produce $M = \emptyset$ in step (4) and would thus never adjoin a new invariant to S. So by Claim 2, homogeneous invariants of any degree lie in $K[S]$, which proves the claim. □

Readers who take a look at the paper [**35**] of King will notice some differences. In fact, Algorithm 1.4 is a modification the original one. For one thing, it is simpler, and for another, the original algorithm does in fact require the computation of a full Gröbner basis at some point, so the previous statement that only truncated Gröbner bases are used really applied to the modification given here. Specifically, in the original algorithm the computation of a full Gröbner basis of the ideal (S) is triggered if for some d no invariants of that degree have been added. If (S) has dimension 0 at that point, the Gröbner basis is used to determine a bound for termination. If it does not, a full Gröbner basis is computed again at a later point. The variant presented here as Algorithm 1.4 does not need such heuristics and terminates at the same d as the original one, or earlier.

EXAMPLE 1.5. Let us run Algorithm 1.4 on a group taken from Bernd's book [**43**, Proposition 2.2.10]. This is the symmetry group $G = D_8$ of a regular octagon, generated by a reflection

$$\tau = \begin{pmatrix} 1 & 0 \\ 0 & -1 \end{pmatrix}$$

and a rotation

$$\sigma = \frac{1}{\sqrt{2}} \begin{pmatrix} 1 & -1 \\ 1 & 1 \end{pmatrix}$$

by an angle of 45°. G has order 16, and plays an interesting role in coding theory, as explained in Bernd's book.

Notice that G contains the scalar matrix

$$\sigma^4 = \begin{pmatrix} -1 & 0 \\ 0 & -1 \end{pmatrix},$$

which has the consequence that no homogeneous invariant of odd degree exists. This means that for odd d the invariants produced by the algorithm will all be 0, so no additions to the generating set S are made and we might as well skip those d. We write x and y for the variables, and choose a monomial ordering with $x > y$. One further comment before we delve into the workings of the algorithm: in the inner loop over the monomials $t \in M$, it is clever to start with the least monomial because that gives a better chance that in step (7) a monomial will be taken out of M that has not been treated yet. Now here is what the algorithm does, rendered in telegraphic style, with the computations not shown in detail but left to the computer.

$\mathbf{d = 2}$**:** As initialized, $S = \mathcal{G} = \emptyset$, so $M := \{x^2, xy, y^2\}$.
  $\mathbf{t = y^2}$**:** $f := \mathcal{R}(t) = \frac{1}{2}(x^2 + y^2) =: f_2$, $h := f$. So we update $S := \{f_2\}$, $\mathcal{G} := \{f_2\}$, and $M := \{xy, y^2\}$.
  $\mathbf{t = xy}$**:** $f := \mathcal{R}(t) = 0$.
$\mathbf{d = 4}$**:** As before, $S = \mathcal{G} = \{f_2\}$ with leading monomial x^2, so $M := \{xy^3, y^4\}$.
  $\mathbf{t = y^4}$**:** $f := \mathcal{R}(t)$ is divisible by f_2, so has normal form $h := 0$.
  $\mathbf{t = xy^3}$**:** $f := \mathcal{R}(t) = 0$.
$\mathbf{d = 6}$**:** As before, $S = \mathcal{G} = \{f_2\}$, so $M := \{xy^5, y^6\}$.
  $\mathbf{t = y^6}$**:** $f := \mathcal{R}(t)$ is divisible by f_2, so has normal form $h := 0$.
  $\mathbf{t = xy^5}$**:** $f := \mathcal{R}(t) = 0$.
$\mathbf{d = 8}$**:** As before, $S = \mathcal{G} = \{f_2\}$, so $M := \{xy^7, y^8\}$.

t $= \mathbf{y^8}$**:** $f := \mathcal{R}(t) = \frac{1}{32}(9x^8 + 28x^6y^2 + 70x^4y^4 + 28x^2y^6 + 9y^8) =: f_8$ has normal form $h := y^8$. So we update $S := \{f_2, f_8\}$, $\mathcal{G} := \{f_2, y^8\}$.

t $= \mathbf{xy^7}$**:** $f := \mathcal{R}(t) = 0$.

d $=$ **9:** (Even though we have until now skipped odd degrees, we do look at $d = 9$ since the computer, oblivious of our above argument about odd degrees, would do so too.) $\mathcal{G} = \{f_2, y^8\}$ is a Gröbner basis, and therefore also a 9-truncated Gröbner basis, with leading monomials x^2 and y^8, so now $M := \emptyset$. This means that the algorithm terminates here.

In this example, the algorithm terminates at $d = 9$, one degree after the last invariant has been added but before reaching $d = |G|$. It has found generating invariants f_2 and f_8, which may be replaced by

$$g_2 = 2f_2 = x^2 + y^2 \quad \text{and} \quad g_8 := 18f_2^4 - 4f_8 = x^2y^2(x^2 - y^2)^2.$$

These are exactly the generating invariants given in Bernd's book. The invariant g_8 happens to be 4 times the product over the orbit of x. ◁

2. Modular invariants of finite groups

In this section we assume that the characteristic $p = \mathrm{char}(K)$ divides the group order $|G|$. Thus we no longer have a Reynolds operator. So chaos is to be expected, and indeed many pleasant features of nonmodular invariant theory break down.

The first victim is Noether's degree bound. Perhaps the easiest example where it fails is the cyclic group C_2 of order 2 acting on the polynomial ring

$$\mathbb{F}_2[x_1, x_2, x_3, y_1, y_2, y_3]$$

by interchanging the x_i and y_i. An explicit calculation showing that the invariant ring is not generated in degrees ≤ 2 is given in [**12**, Example 3.3.1], so we direct interested readers there. Worse, Richman [**40**] showed that in the modular case there cannot exist any degree bound that only depends on $|G|$. Reasonable degree bounds for the modular case were long elusive (an unreasonably large bound appeared in [**10**, Theorem 3.9.11]), until Peter Symonds [**44**] proved in 2011 a bound that had been conjectured for a while. His paper tells us that if $n \ge 2$ is the number of variables and $|G| \ge 2$, then $K[\mathbf{x}]^G$ is generated by homogeneous invariants of degrees $\le n\big(|G| - 1\big)$. The proof makes heavy use of the paper by Karagueuzian and Symonds [**27**], which establishes the fascinating result that the polynomial ring $K[\mathbf{x}]$—viewed as a direct sum of infinitely many indecomposable KG-modules—only contains a finite number of isomorphism types of such modules. As the title of [**44**] suggests, it not only provides a bound on the generators of $K[\mathbf{x}]^G$ but also on the algebraic relations between them. For an (even) better appreciation of Symonds's bound, and for understanding why it qualifies as "reasonable", it should be pointed out that it is really a bound on secondary invariants.

This brings us to the topic of primary and secondary invariants, which we already mentioned in passing before presenting King's algorithm. As they are covered in Bernd's book, we will just briefly recall them here. **Primary invariants**, which owe their existence to Noether normalization, are homogeneous invariants $f_1, \dots, f_n$ (again with n the number of variables) such that $K[\mathbf{x}]^G$ is finitely generated as a module over the subalgebra $A := K[f_1, \dots, f_n]$ generated by them. After having chosen primary invariants, we call homogeneous generators $g_1, \dots, g_m$ of $K[\mathbf{x}]^G$ as A-module **secondary invariants**. Together, the primary and secondary invariants form a generating system of $K[\mathbf{x}]^G$ as an algebra. Now, what Symonds's bound really says is that if the secondary invariants are chosen to *minimally* generate $K[\mathbf{x}]^G$ as an A-module, then

$$\deg(g_i) \leq \sum_{i=1}^{n}(\deg(f_i) - 1). \tag{1}$$

A typical example is the invariants of the alternating group A_n acting naturally by permuting the x_i, which are treated in Bernd's book [**43**, Proposition 1.1.3]. In this case, the elementary symmetric polynomials can be taken as primary invariants and, in characteristic $\neq 2$, secondary invariants are given by $g_1 = 1$ and $g_2 = \prod_{i<j}(x_i - x_j)$. So here Symonds's bound for secondary invariants is sharp, as it quite often is. It is, however, seldom sharp as a bound for algebra generators, since the secondary invariant(s) of top degree can typically be chosen as a product of lower-degree secondary invariants. But precisely because it is a bound on secondary invariants, it is extremely useful for computations, since in the modular case the state-of-the-art algorithms for computing $K[\mathbf{x}]^G$ do proceed by consecutively computing primary and secondary invariants.

How does this lead to the bound $n(|G| - 1)$ stated above? To see this, we need to recall how primary invariants are characterized geometrically: homogeneous invariants $f_1, \dots, f_n$ are primary invariants if and only if the affine variety $\mathcal{V}_{\overline{K}^n}(f_1, \dots, f_n)$ over an algebraic closure of K defined by them consists of the point $(0, \dots, 0)$ alone (see, for example, [**42**, Proposition 5.3.7]). We will often use the less context-specific term **homogeneous system of parameters** (**hsop**) for the set $\{f_1, \dots, f_n\}$. Also notice that an equivalent condition to the above criterion is that $\mathcal{V}_{\overline{K}^n}(f_1, \dots, f_n)$ is zero-dimensional. This can be refined: a set $\{f_1, \dots, f_k\}$ of homogeneous polynomials, with $k \leq n$, can be extended to an hsop if and only if the variety it defines has dimension $n - k$, the least dimension possible. In this case we speak of a **partial homogeneous system of parameters** (**phsop**). It is these geometric tools that are at the heart of algorithms for constructing primary invariants, as those given in Bernd's book [**43**, Section 2.5] or the refinement [**28**], which still seems to be the state of the art.

In particular, at least if the field K has enough elements, we can form primary invariants by taking orbit products of linear forms, if only these forms are chosen "in general position." This method is referred to as *Dade's algorithm*, and more details can be found in Bernd's book [**43**, Subroutine 2.5.12]. This brings us back to Symonds's bound. The primary invariants constructed by Dade's algorithm have degree $\leq |G|$, so the bound $n(|G| - 1)$ follows directly from (1). What about the assumption that K has enough elements? This is actually a without-loss assumption, since it can be shown that the maximal degree of a minimal generating system of $K[\mathbf{x}]^G$ does not change when the ground field K is extended.

Having reported progress on degree bounds in the modular case, let us now address a different issue, the Cohen–Macaulay property. In the case of invariant rings, this property can be defined in a very simple way, which also makes it immediately clear that it is a property worth having: $K[\mathbf{x}]^G$ is Cohen–Macaulay if it is free as a module over the subalgebra generated by some (or, equivalently, every) hsop. Now one of the prominent results in nonmodular invariant theory is that in the nonmodular case, $K[\mathbf{x}]^G$ always is Cohen–Macaulay (see Bernd's book [**43**, Theorem 2.3.5]). And once again, this often breaks down in the modular case. At the time of writing Bernd's book, results on the Cohen–Macaulay property in modular invariant theory were rather haphazard: there were just a few examples known where the Cohen–Macaulay property failed, and a few others where it held. The only notable (and, indeed, very remarkable) result was the paper by Ellingsrud and Skjelbred [**16**], which answers the question completely in the case of cyclic p-groups. Later, Campbell, Geramita, Hughes, Shank, and Wehlau [**7**] considered the case of *vector invariants*, which means that a given linear action of a group G is replicated on several sets of variables; so if $\sigma(x_i) = a_{1,i}x_1 + \cdots + a_{n,i}x_n$, then the action on the new variables $x_{i,j}$ is by $\sigma(x_{i,j}) = a_{1,i}x_{1,j} + \cdots + a_{n,i}x_{n,j}$, for j between 1 and some k. An example with $k = 3$ is the action of C_2 at the beginning of the section. The result in [**7**] says that if G is a p-group acting nontrivially and if $k \geq 3$, then the ring of vector invariants is not Cohen–Macaulay. In fact, the C_2-action just recalled is not only the most accessible example where Noether's bound fails, but also where the Cohen-Macaulay property fails. The following result came out later in the same year as [**7**].

THEOREM 2.1 (Kemper [**29**]). *If $K[\mathbf{x}]^G$ is Cohen–Macaulay, then G is generated by p'-elements (i.e., by elements of order not divisible by p) and by bireflections (i.e., by elements that fix a subspace of codimension 2).*

The term bireflection is modeled after the concept of a reflection, which is a linear transformation fixing a subspace of codimension 1. Notice that "fixing a subspace of codimension 2" allows that the fixed space actually has smaller codimension; so, in particular, every reflection is also a bireflection. On the other hand, if we consider vector invariants of $k \geq 3$ sets of variables (the case of [**7**]), then the only bireflection is the identity; so [**7**] is contained in Theorem 2.1. It should be mentioned that, unfortunately, the converse of Theorem 2.1 is not true, and the quest for a group or representation theoretic if-and-only-if criterion for the Cohen–Macaulay property is still one of the holy grails in modular invariant theory.

The paper by Kemper [**29**] proves Theorem 2.1 within a broader framework, so it may be worthwhile to present a proof here that is shorter and more streamlined, and thus better suited to make the arguments explicit. As we go along in the proof, we will introduce the relative trace map and recall the concept of a regular sequence. Both are interesting in themselves.

This is an outline of the proof: We assume that G is *not* generated by p'-elements and bireflections, and derive the existence of a certain normal subgroup N in Lemma 2.2. Lemma 2.3 then reveals the variety defined by the image I of the relative trace with respect to N, and Lemma 2.4 tells us that I contains a phsop (see on page 13) of length 3. Finally, in Lemma 2.6, where the endgame of the proof plays out, we show that such a phsop is not a regular sequence, implying that $K[\mathbf{x}]^G$ is not Cohen–Macaulay. The first lemma is purely group theoretic and holds, like everything else, under the assumption that G is not generated by p'-elements and bireflections.

LEMMA 2.2. *There is a normal subgroup $N \subset G$ of index p that contains all bireflections.*

PROOF. The subgroup $N_0 \subseteq G$ generated by all p'-elements and bireflections is normal. Since the order of every element from G/N_0 is a power of p, G/N_0 is a p-group. By assumption, it is nontrivial, so it has a subgroup N/N_0 of index p. But a subgroup of index p in a p-group is always normal (see [**25**, Kapitel I, Satz 8.9]). $\square$

Since in modular invariant theory we cannot average over the group, what is left to do is just summing. The resulting map $K[\mathbf{x}] \to K[\mathbf{x}]^G$ is called the **trace map** or **transfer**. It has a relative version: for $H \subseteq G$ a subgroup, the **relative trace** (or **relative transfer**) is the map

$$\mathrm{Tr}_{G/H} : K[\mathbf{x}]^H \to K[\mathbf{x}]^G, \ f \mapsto \sum_{\sigma \in G/H} \sigma(f),$$

where the summation is over a set of left coset representatives. This is surjective if $p \nmid [G : H]$, but how big is its image (which is an ideal in $K[\mathbf{x}]^G$) otherwise? An answer in geometric terms (i.e., determining the radical ideal of the image) can be found in [**37**, Lemma 1.1]. What we need here is the special case $H = N$ with N from Lemma 2.2. For simplicity (and since it is an assumption that can be made without loss of generality) we will assume K to be algebraically closed. The following lemma determines the variety in $V := K^n$ defined by the ideal

$$I_{G/N} := \mathrm{Tr}_{G/N}\big(K[\mathbf{x}]^N\big) \subseteq K[\mathbf{x}]^G.$$

LEMMA 2.3. $\mathcal{V}_V(I_{G/N}) = \bigcup_{\sigma \in G \setminus N} V^\sigma$. *In particular, since all bireflections are contained in N, the variety has dimension $\leq n - 3$.*

PROOF. For the inclusion "$\supseteq$", assume a vector $v \in V$ is fixed by some $\sigma \in G \setminus N$. The class of σ generates G/N, so for $f \in K[\mathbf{x}]^N$ we obtain

$$\big(\mathrm{Tr}_{G/N}(f)\big)(v) = \sum_{i=0}^{p-1} \big(\sigma^i(f)\big)(v) = \sum_{i=0}^{p-1} f\big(\sigma^{-i}(v)\big) = \sum_{i=0}^{p-1} f(v) = 0,$$

since K has characteristic p. For the reverse inclusion, assume $\sigma(v) \neq v$ for all $\sigma \in G \setminus N$, so the sets $\{\sigma(v) \mid \sigma \in G \setminus N\}$ and $\{\tau(v) \mid \tau \in N\}$ are disjoint. By interpolation, we can find a polynomial $f \in K[\mathbf{x}]$ that is constantly zero on the first set but constantly 1 on the second. Replacing f by $\prod_{\tau \in N} \tau(f)$, we may assume $f \in K[\mathbf{x}]^N$. Then $\big(\mathrm{Tr}_{G/N}(f)\big)(v) = 1$, so $v \notin \mathcal{V}_V(I_{G/N})$. □

LEMMA 2.4. *$I_{G/N}$ contains a phsop of length 3.*

PROOF. We will prove the following more general statement using induction on k: if $I \subseteq K[\mathbf{x}]^G$ is a homogeneous ideal such that $\mathcal{V}_V(I)$ has dimension $\leq n - k$, then I contains a phsop of length k. There is nothing to show for $k = 0$, so we can assume $k \geq 1$ and, by induction, that we have already found a phsop $f_1, \ldots, f_{k-1} \in I$. This means that the prime ideals $Q_1, \ldots, Q_m$ in $K[\mathbf{x}]$ that minimally lie over $(f_1, \ldots, f_{k-1})$ all have dimension $n - k + 1$ (some readers may prefer to say that the height is $k - 1$). So, by hypothesis, the ideal $J := K[\mathbf{x}] \cdot I$ generated by I is contained in none of the Q_i; so by prime avoidance (see [**15**, Lemma 3.3]), there is a homogeneous $g \in J$ with $g \notin Q_i$ for all i. Since the G-action permutes the Q_i, this is also true for all $\sigma(g)$, so the product $f_k := \prod_{\sigma \in G} \sigma(g)$ also avoids all Q_i. This implies that $f_1, \ldots, f_k$ are a phsop.

So what is left is to show $f_k \in I$. Since $g \in J$ we have $g = \sum_{i=1}^r h_i g_i$ with $g_i \in I$ and $h_i \in K[\mathbf{x}]$. With $t_1, \ldots, t_r$ new variables, we have a G-equivariant map φ: $K[\mathbf{x}][t_1, \ldots, t_r] \to K[\mathbf{x}]$ of $K[\mathbf{x}]$-algebras sending t_i to g_i (with trivial G-action on the t_i). So $\varphi(h_1 t_1 + \cdots + h_r t_r) = g$ and hence

$$\varphi\left(\prod_{\sigma \in G} \sigma(h_1 t_1 + \cdots + h_r t_r)\right) = \prod_{\sigma \in G} \sigma(g) = f_k.$$

But the product of the $\sigma(h_1t_1 + \cdots + h_rt_r)$ lies in

$$\big(K[\mathbf{x}][t_1,\ldots,t_r]\big)^G = K[\mathbf{x}]^G[t_1,\ldots,t_r],$$

so applying φ to it yields an element of I since $g_i \in I$. This completes the proof. □

REMARK 2.5. In the above proof we really showed the following statement: every phsop in a homogeneous ideal $I \subseteq K[\mathbf{x}]^G$ with $\dim\big(\mathcal{V}_V(I)\big) \le n-k$ can be extended to a phsop in I of length k. ◁

Before presenting the final step in the proof of Theorem 2.1 we need to recall the concept of a regular sequence. In our context this can be defined as a sequence $f_1,\ldots,f_k \in K[\mathbf{x}]^G$ of nonconstant homogeneous invariants such that for each i, the map

$$K[\mathbf{x}]^G/(f_1,\ldots,f_{i-1}) \to K[\mathbf{x}]^G/(f_1,\ldots,f_{i-1})$$

given by multiplication with f_i is injective. It is easily seen that if $K[\mathbf{x}]^G$ is Cohen–Macaulay, then every phsop is a regular sequence. (The converse also holds, but we do not need it here.) In particular, every phsop is an $K[\mathbf{x}]$-regular sequence, i.e., regular when regarded as a sequence in $K[\mathbf{x}]$. It is interesting that for proving the following lemma, the $K[\mathbf{x}]$-regularity is actually used to show that $K[\mathbf{x}]^G$-regularity fails.

LEMMA 2.6. *A phsop $f_1, f_2, f_3 \in I_{G/N}$ of length 3 is not a regular sequence. Therefore $K[\mathbf{x}]^G$ is not Cohen–Macaulay.*

PROOF. Choose an element $\sigma \in G \setminus H$. By the definition of $I_{G/N}$ we have $f_i = \sum_{j=0}^{p-1} \sigma^j(h_i)$ with $h_i \in K[\mathbf{x}]^N$. Since N is a normal subgroup we have $\sigma^j(h_i) \in K[\mathbf{x}]^N$. In characteristic p, we have the polynomial identity $1+t+\cdots+t^{p-1} = (1-t)\big(1+2t+\cdots+(p-1)t^{p-2}\big)$. So, setting $g_i := -\sum_{j=1}^{p-1} j\sigma^{j-1}(h_i)$ yields

$$f_i = \sigma(g_i) - g_i \quad \text{with} \quad g_i \in K[\mathbf{x}]^N \qquad (i = 1,2,3). \tag{2}$$

A consequence is that for $1 \le i < j \le 3$ the $u_{i,j} := f_ig_j - f_jg_i \in K[\mathbf{x}]^N$ satisfy

$$\sigma(u_{i,j}) = f_i\sigma(g_j) - f_j\sigma(g_i) \underset{(2)}{=} f_i(f_i + g_i) - f_j(f_i + g_i) = u_{i,j},$$

so, in fact, $u_{i,j} \in K[\mathbf{x}]^G$. The claim that the sequence f_1, f_2, f_3 is not $K[\mathbf{x}]^G$-regular will follow from the relation

$$f_1u_{2,3} - f_2u_{1,3} + f_3u_{1,2} = \det\begin{pmatrix} f_1 & f_2 & f_3 \\ f_1 & f_2 & f_3 \\ g_1 & g_2 & g_3 \end{pmatrix} = 0.$$

Indeed, assuming regularity would yield $u_{1,2} = (f_1, f_2)$, i.e., $u_{1,2} = f_1\widetilde{g}_2 - f_2\widetilde{g}_1$ with $\widetilde{g}_i \in K[\mathbf{x}]^G$. With the definition of $u_{1,2}$ this gives $f_1(g_2 - \widetilde{g}_2) = f_2(g_1 - \widetilde{g}_1)$. Since f_1, f_2 is a $K[\mathbf{x}]$-regular sequence, we obtain $g_1 - \widetilde{g}_1 \in (f_1)$, so $g_1 - \widetilde{g}_1 = f_1d$ with $d \in K[\mathbf{x}]$. By (2) and since $\widetilde{g}_1 \in K[\mathbf{x}]^G$, this implies

$$f_1\big(\sigma(d) - d\big) = \sigma(g_1 - \widetilde{g}_1) - (g_1 - \widetilde{g}_1) = f_1.$$

Now, because f_1 is $K[\mathbf{x}]$-regular, we obtain $\sigma(d) - d = 1$, which is a contradiction since for every polynomial the homogeneous part of degree 0 is an invariant. □

With this, the proof of Theorem 2.1 is complete. The proof can be framed in homological terms: there is a nonzero element in the first cohomology $H^1(G/N, R^N)$, and this cohomology group is annihilated when multiplied by elements from the relative trace ideal $I_{G/N}$. Now, from the fact that a phsop of length 3 annihilates a nonzero class in $H^1(G, R)$, it can be concluded that $K[\mathbf{x}]^G$ fails to be Cohen–Macaulay. This approach, using higher cohomology as well, has brought forth some further results, such as:

- If G acts by the regular representation, then $K[\mathbf{x}]^G$ is Cohen–Macaulay if and only if $G \cong C_2$, $G \cong C_3$, or $G \cong C_2 \times C_2$ (see [**29**]).
- If G acts simultaneously on k sets of variables (the case of *vector invariants* that was also treated in [**7**]), then $K[\mathbf{x}]^G$ is not Cohen–Macaulay for k large enough (see [**29**]). In fact, with rising k, the *Cohen–Macaulay defect* $\dim\big(K[\mathbf{x}]^G\big) - \operatorname{depth}\big(K[\mathbf{x}]^G\big)$ tends to infinity (see [**21**]).
- If G acts as a permutation group and $|G|$ is not divisible by p^2 (the *mildly modular* case), the exact depth of $K[\mathbf{x}]^G$ was determined in [**30**]. For example, if $G \not\cong C_2$ acts by the regular representation,
$$\operatorname{depth}\big(K[\mathbf{x}]^G\big) = \frac{|G|}{p} + 2,$$
consistent with the result mentioned above. As another example, the Cohen–Macaulay defect of vector invariants of S_n, with $p \le n < 2p$ acting on $k \ge 2$ sets of variables, is $(k-2)(p-1)$.

In the above listing we have always assumed that the characteristic p divides $|G|$ and G acts faithfully.

Much more recently, Ben Blum-Smith and Sophie Marques [**4**] scrutinized the case of permutation actions further and achieved the following result: for a permutation group G (meaning that G permutes the variables x_i) the invariant rings in all characteristics are Cohen–Macaulay if and only if G is generated by bireflections. (Notice that in the case of permutation actions, bireflections are transpositions, double-transpositions, and 3-cycles.) This is remarkable not only because of its simplicity, but also since it reaches the gold standard of directly translating a group or representation theoretic property of the action to an algebraic property of the invariant ring.

So far, we have looked at two nice features of nonmodular invariant theory that are missing in the modular case: Noether's degree bound and the Cohen–Macaulay property. A further feature, which is especially useful for computations, is Molien's formula, presented in Bernd's book [**43**, Theorem 2.2.1]. It affords the computation of the Hilbert series of $K[\mathbf{x}]^G$ without computing a single invariant. But again, in the modular case Molien's formula is absent. What was present at the time of Bernd's book was a paper by Almkvist and Fossum [**1**] that provides formulas for the Hilbert series in the case of an indecomposable action of the cyclic group C_p of order p. In a paper with Ian Hughes [**24**], we picked up this thread and were surprised to find that the methods can be carried further to the *mildly modular* case, i.e., the case where $|G|$ is divisible by p but not by p^2. The paper [**24**] gives a recipe for computing the Hilbert series of $K[\mathbf{x}]^G$ which, like Molien's formula, does the job without considering any invariant. Also in terms of computational cost, the recipe is very similar to Molien's formula, it just takes longer to explain (or implement) it. The book by Derksen and Kemper [**12**, Section 3.4.2] presents an outline, so we direct interested readers there for details. We hit multiple roadblocks, however, when trying to move anywhere beyond the mildly modular case. Perhaps a better understanding of Karagueuzian and Symonds's methods in [**27**] will open new pathways to the a priori computation of Hilbert series in modular invariant theory.

3. Separating invariants of finite groups

One of the main purposes of invariants, perhaps even their raison d'être, is to separate groups orbits. The theme permeates Bernd's book, and it prompted the development of geometric invariant theory. So it seems natural to consider the separation properties of given invariants, which leads to the concept of separating invariants. Let us recall the definition. A subset $S \subseteq K[\mathbf{x}]^G$ is called **separating** if for any two points $v, w \in V := K^n$ we have: if there is an invariant $f \in K[\mathbf{x}]^G$ such that $f(v) \neq f(w)$, then there is a $g \in S$ with $g(v) \neq g(w)$.

Here G need not be finite, although we are considering the finite case in this section. Clearly, every set of generating invariants is also separating. Perhaps the easiest example which reveals that a separating set can really be smaller than a generating one is the action of a cyclic group C_n on $\mathbb{C}^2$ by the scalar matrices

$$e^{2k\pi i/n} \cdot \begin{pmatrix} 1 & 0 \\ 0 & 1 \end{pmatrix}.$$

A minimal *generating* set consists of all $n+1$ monomials of degree n; however, $f_1 := x_1^n$, $f_2 := x_1^{n-1}x_2$, and $f_3 := x_2^n$ form a separating set. This follows from the formula

$$x_1^{n-k}x_2^k = \frac{f_2^k}{f_1^{k-1}}.$$

As we are dealing with a weakening of the concept of generating the invariant ring, this offers some hope that the bad behavior of generating sets in the modular case might be mitigated by considering separating sets instead. And, indeed, Noether's degree bound is reinstated by the following elementary result.

THEOREM 3.1 (Noether's degree bound for separating invariants). *Let G be a finite group acting on $K[\mathbf{x}]$ by linear transformations of the x_i. With additional variables t and y, form the polynomial*

$$F(t, y) := \prod_{\sigma \in G} \left(y - \sum_{i=1}^{n} \sigma(x_i) t^{i-1} \right) \in K[\mathbf{x}]^G[t, y].$$

If all the coefficients of F agree on two points $v, w \in K^n$, then v and w are in the same G-orbit. In particular, the coefficients of F form a separating set consisting of homogeneous invariants of degree $\leq |G|$, and all G-orbits can be separated by invariants.

PROOF. The hypothesis gives

$$\prod_{\sigma \in G} \left(y - \sum_{i=1}^{n} (\sigma(x_i))(v) t^{i-1} \right) = \prod_{\sigma \in G} \left(y - \sum_{i=1}^{n} (\sigma(x_i))(w) t^{i-1} \right).$$

So, there is a $\sigma \in G$ such that $\sum_{i=1}^n x_i(w) t^{i-1} = \sum_{i=1}^n \big(\sigma(x_i)\big)(v) t^{i-1}$. Of course x_i is nothing but the ith coordinate functional on K^n. For all i we obtain

$$x_i(w) = \big(\sigma(x_i)\big)(v) = x_i\big(\sigma^{-1}(v)\big),$$

which implies $w = \sigma^{-1}(v)$. So the first statement is proved, and the other statements are now clear. □

Of course, the theorem contains a simple method for constructing separating invariants that does not require any Gröbner basis computation at all. However, even for moderate-sized groups, the computational cost of expanding the product and extracting coefficients is enormous, as is the number of invariants that the method produces.

By now, we have reported progress on almost all topics treated in the chapter on finite group actions in Bernd's book, with one notable exception: reflection groups. These are groups generated by reflections, where a reflection is an element of G that fixes a codimension-1 subspace of K^n, including reflections of orders other than 2 (if the ground field K allows them) and even elements acting trivially. Also very well known is the result that in the nonmodular case, the invariant ring $K[\mathbf{x}]^G$ can be generated by n invariants (equivalently, is isomorphic to a polynomial ring) if and only if G is a reflection group (see Bernd's book [**43**, Theorem 2.4.1]). As readers will be expecting by now, this breaks down in the modular case. But not completely: while there are many examples of modular reflection groups whose invariant ring is *not* a polynomial ring, it is still true that for $K[\mathbf{x}]^G$ to be polynomial, G has to be a reflection group. This implication is often attributed to Serre, but in [**41**] he says that it was proved in 1955 by Chevalley. Meanwhile, Chevalley's article [**8**] does not appear to contain the result, and Benson [**3**] gives the citation [**6**, Chapitre 5, § 5, Exercice 7] where there is indeed a blueprint of a proof which uses purity of the branch locus.

So far, this has nothing to do with separating invariants. But since generating invariants are always separating, a stronger result would be if there exist n separating invariants, then G has to be a reflection group. And this is precisely what Emilie Dufresne proved in her amazing paper [**13**]. It is striking that Dufresne's result has a rather short but elegant proof. But before turning to that, let us present the precise statement.

THEOREM 3.2 (Dufresne [**13**]). *Assume that K is algebraically closed and that there is a separating set of n invariants, with n the number of variables. Then G is a reflection group.*

Dufresne's paper [**13**] also offers an example of an invariant ring that has n separating invariants, but does *not* have n generating invariants. What about the hypothesis that K is algebraically closed? The following example shows that the theorem fails without that hypothesis. It also shows that when it comes to separating invariants, the ground field K makes a difference.

EXAMPLE 3.3. The group $G := \langle A \rangle \subset \mathrm{GL}_2(\mathbb{R})$ generated by $A := \left(\begin{smallmatrix} 0 & 1 \\ -1 & -1 \end{smallmatrix}\right)$ (the companion matrix of the third cyclotomic polynomial) is not a reflection group. The transformation given by A maps x_1 to $-x_2$ and x_2 to $x_1 - x_2$, and thus, with $\omega := e^{2\pi i/3}$ and $y_j := x_1 + \omega^j x_2 \in \mathbb{C}[\mathbf{x}]$ $(j = 1, 2)$, it maps y_j to $\omega^j y_j$. Therefore,

$$f_1 := y_1^3 + y_2^3 \quad \text{and} \quad f_2 := \sqrt{-3}\left(y_1^3 - y_2^3\right)$$

are invariant under G, and also under complex conjugation. So $f_1, f_2 \in \mathbb{R}[\mathbf{x}]^G$. We claim that the f_i are separating invariants. Observe that they are also invariant under the transformation given by $D := \left(\begin{smallmatrix} \omega & 0 \\ 0 & \omega \end{smallmatrix}\right)$, so $f_i \in \mathbb{C}[\mathbf{x}]^{\widetilde{G}}$ with $\widetilde{G} := \langle A, D \rangle$. In fact, $\widetilde{G}$ is generated by the transformations

$$\begin{pmatrix} y_1 \\ y_2 \end{pmatrix} \mapsto \begin{pmatrix} \omega y_1 \\ y_2 \end{pmatrix}$$

and

$$\begin{pmatrix} y_1 \\ y_2 \end{pmatrix} \mapsto \begin{pmatrix} y_1 \\ \omega y_2 \end{pmatrix}.$$

So it is a reflection group, and

$$\mathbb{C}[\mathbf{x}]^{\widetilde{G}} = \mathbb{C}[y_1, y_2]^{\widetilde{G}} = \mathbb{C}[y_1^3, y_2^3] = \mathbb{C}[f_1, f_2].$$

Being generating invariants, the f_i are also separating invariants in $\mathbb{C}[\mathbf{x}]^{\widetilde{G}}$ and by Theorem 3.1, they separate $\widetilde{G}$-orbits in $\mathbb{C}^2$. But our claim was that they are separating invariants in $\mathbb{R}[\mathbf{x}]^G$. To prove this, let $v, w \in \mathbb{R}^2$ such that $f_i(v) = f_i(w)$ for $i = 1, 2$. Then by what we have just seen there exists $\widetilde{\sigma} \in \widetilde{G}$ such that $w = \widetilde{\sigma}(v)$. We can write $\widetilde{\sigma} = \omega^i \cdot \sigma$ with $\sigma \in G$ and $i \in \mathbb{Z}$, so $w = \omega^i \sigma(v)$. But w and $\sigma(v)$ have components in $\mathbb{R}$, so $\omega^i = 1$ or $v = w = 0$. In both cases $G(v) = G(w)$, and our claim is proved. Of course, the f_i are *not* separating invariants in $\mathbb{C}[\mathbf{x}]^G$.

To sum up, this example provides a finite group $G \subset \mathrm{GL}_2(\mathbb{R})$ which is not a reflection group, but there exists a separating set of two invariants. $\triangleleft$

Over finite ground fields, there can even exist separating sets with fewer than n elements (see [**34**, Theorem 1.1]).

Even though Theorem 3.2 requires K to be algebraically closed, the implication "$K[\mathbf{x}]^G$ polynomial $\Rightarrow$ G reflection group" follows from it in full generality, i.e., without the algebraic closedness hypothesis. In fact, if there are are n invariants generating $K[\mathbf{x}]^G$, they also generate $\overline{K}[\mathbf{x}]^G$ and are therefore separating for the G-action on $\overline{K}^n$. From this, Theorem 3.2 yields that G is a reflection group.

In a nutshell, Dufresne's proof of Theorem 3.2 consists of the crucial observation that the graph of the action is connected in codimension 1 if and only if G is a reflection group (see Proposition 3.4), combined with Hartshorne's connectedness theorem. Let us explain. By the **graph of the action** we mean the set $\Gamma := \{(v, \sigma(v)) \mid v \in V,\ \sigma \in G\} \subset V \times V$, where $V := K^n$. Since G is finite, this is (Zariski-)closed in $V \times V$, and, by Theorem 3.1, a pair (v, w) of points lies in Γ if and only if all invariants take the same value on v and w. Furthermore, a variety is called **connected in codimension** k if it is connected even after removing a closed subset of codimension $> k$. (A bit of care must be taken to get the definition of codimension right if the variety has irreducible components of different dimensions, but this is not the case for the graph of the action.) For example, two planes meeting in a line are connected in codimension 1, but two planes meeting in a point are not. *Not* to be connected in codimension k means that the variety can be written as a union of two closed subsets, none contained in the other, whose intersection has codimension $> k$. For want of a better term, let us call such a pair of closed subsets a *witness of disconnectedness*.

PROPOSITION 3.4 (Dufresne [**13**]). *The graph of the action is connected in codimension* 1 *if and only if G is a reflection group. If this is not the case, there is a witness (X, Y) of disconnectedness with X and Y given by homogeneous ideals.*

PROOF. The irreducible components of Γ, the graph of the action, are the sets

$$Z_\sigma := \{(v, \sigma(v)) \mid v \in V\} \cong V \quad \text{for } \sigma \in G,$$

and their intersections are

$$Z_\sigma \cap Z_\tau = \{(v, \sigma(v)) \mid v \in V,\ \sigma(v) = \tau(v)\} \cong V^{\sigma^{-1}\tau}. \tag{3}$$

If Γ is not connected in codimension 1, then there is a witness (X, Y) of disconnectedness. Each irreducible component Z_σ is contained in X or in Y, but not in both. So writing $\sigma \sim \tau$ if Z_σ and Z_τ are both contained in X or both contained in Y defines an equivalence relation on G. If $\sigma \not\sim \tau$, then $\operatorname{codim}(Z_\sigma \cap Z_\tau) > 1$, which, by (3), means that $\sigma^{-1}\tau$ is *not* a reflection. Using transitivity, we conclude that if σ and τ lie in the same left coset of the subgroup $H \subseteq G$ generated by all reflections, then $\sigma \sim \tau$. But not all group elements can be equivalent, since otherwise $\Gamma = X$ or $\Gamma = Y$. it follows that $H \subsetneqq G$, so G is not a reflection group.

This also works backward: If $H \subsetneqq G$, form $X := \bigcup_{\sigma \in H} Z_\sigma$ and $Y := \bigcup_{\tau \in G \setminus H} Z_\tau$ (or lump irreducible components corresponding to cosets of H together in any other way). Then (X, Y) is a witness for disconnectedness. Since every Z_σ is given by homogeneous linear equations, X and Y are given by homogeneous ideals. □

REMARK 3.5. Of course, by the same argument we see that Γ is connected in codimension 2 if and only if G is generated by bireflections, and so on. ◁

Hartshorne's connectedness theorem is also about connectedness in codimension 1, but how can we use it? Assume there are n separating invariants $f_1, \ldots, f_n$. Taking additional variables $y_1, \ldots, y_n$ and setting $\Delta f_i := f_i(\mathbf{x}) - f_i(\mathbf{y})$, which can be evaluated at points from $V \times V$, we see that the variety in $V \times V$ given by the Δf_i is precisely Γ. Since $\dim(\Gamma) = n$, this implies that Γ is a complete intersection. Therefore, $R := K[\mathbf{x}, \mathbf{y}]/(\Delta f_1, \ldots, \Delta f_n)$ is Cohen–Macaulay (see [**15**, Proposition 18.13]), and now the connectedness theorem [**22**, Corollary 2.4 and Remark 1.3.2] tells us that Γ (more precisely, $\operatorname{Spec}(R)$, a potentially nonreduced version of Γ) is connected in codimension 1. So applying Proposition 3.4 gives us Theorem 3.2.

If Serre or Hartshorne had been aware of Proposition 3.4, it seems likely that they would have made the connection to the connectedness theorem (no pun intended) and come up with this simple idea to prove the implication "$K[\mathbf{x}]^G$ polynomial $\Rightarrow$ G reflection group." Quitting this idle speculation, let us ask how hard or easy Defresne's proof really is by presenting a textbook-style variant proof, which, instead of citing Hartshorne's connectedness theorem, only uses knowledge from Bernd's book. We will restrict our attention to the case that the n separating invariants are homogeneous. This restriction is a mild for two reasons: (1) most invariant theorists (almost) never use inhomogeneous invariants in the first place, and (2) if there are n possibly inhomogeneous *generating* invariants, then there are also n homogeneous generating invariants (it is left as an exercise to prove this); thus in order to show the implication "$K[\mathbf{x}]^G$ polynomial $\Rightarrow$ G reflection group," one may assume homogeneity without loss of generality.

PROOF OF THEOREM 3.2. As explained above, this proof only treats the case of *homogeneous* separating invariants, and it does not quote any "heavier machinery" such as Hartshorne [**22**, Corollary 2.4 and Remark 1.3.2].

By hypothesis there are homogeneous invariants $f_1, \ldots, f_n$ such that $\Gamma = \mathcal{V}_{V \times V}(D)$ with $D := (\Delta f_1, \ldots, \Delta f_n)$. Since $\dim(\Gamma) = n$, the Δf_i form a phsop in $K[\mathbf{x}, \mathbf{y}]$. With Proposition 3.4 in mind, we assume that $\Gamma = X \cup Y$ with X and Y closed, given by homogeneous ideals, such that $\operatorname{codim}_\Gamma(X \cap Y) \geq 2$. So if we can show that $X \subseteq Y$ or $Y \subseteq X$, we are done.

We have $X = \mathcal{V}_{V\times V}(g_1,\dots,g_l)$ and $Y = \mathcal{V}_{V\times V}(h_1,\dots,h_m)$ with g_i and h_j homogeneous. Each product $g_i h_j$ lies in the vanishing ideal of $X \cup Y = \Gamma$, which is $\sqrt{D}$, so there is k such that $g_i^k h_j^k \in D$. The ideals $I := (g_1^k,\dots,g_l^k) + D$ and $J := (h_1^k,\dots,h_m^k) + D$ are homogeneous, satisfy $I \cdot J \subseteq D \subseteq I \cap J$, and define the algebraic sets X and Y. So it suffices to show $I \subseteq J$ or $J \subseteq I$.

The ideal $I + J$ defines $X \cap Y$, which has dimension $\le n - 2 = 2n - (n+2)$. So by Remark 2.5, the phsop formed by the Δf_i can be extended by two further polynomials, both lying in $I + J$. Since $K[\mathbf{x},\mathbf{y}]$ is Cohen–Macaulay, our phsop is a regular sequence of length $n+2$. Forming the ring $R := K[\mathbf{x},\mathbf{y}]/D$ and considering the ideals I/D and J/D, we have the situation of Lemma 3.6. This tells us that $I/D \subseteq J/D$ or $J/D \subseteq I/D$, and the desired inclusion follows. □

The following lemma, which may be called the graded version of Hartshorne's connectedness theorem in ideal-theoretic form, was used in the above proof.

LEMMA 3.6. *Let R be a graded ring with $K := R_0$ a field, and let $I, J \subseteq R$ be homogeneous ideals such that $I \cdot J = \{0\}$. If $I + J$ contains a regular sequence of length 2, then $I \subseteq J$ or $J \subseteq I$.*

PROOF. We have a regular sequence $a_1, a_2 \in I + J$, so

$$a_1 \cdot (I \cap J) \subseteq (I + J) \cdot (I \cap J) \subseteq I \cdot J = \{0\}.$$

Therefore, $I \cap J = \{0\}$ since a_1 is not a zero divisor. So the map φ: $R \to R/I \oplus R/J$, $x \mapsto (x + I, x + J)$ is injective. With ψ: $R/I \oplus R/J \to R/(I + J)$, $(x + I, y + J) \mapsto (x - y) + (I + J)$, the sequence

$$\{0\} \longrightarrow R \xrightarrow{\varphi} R/I \oplus R/J \xrightarrow{\psi} R/(I + J) \longrightarrow \{0\}$$

is easily checked to be exact. So Lemma 3.7 yields $x, y \in R$ such that $(x - y) + (I + J) = 1 + (I + J)$ and $(I + J) \cdot x \subseteq I$ and $(I + J) \cdot y \in J$. The equality implies that we are in at least one of the following cases: (1) I or J contains an element with nonzero component in degree 0, (2) x has nonzero component in degree 0, or (3) y has nonzero component in degree 0. In case (1), I or J is equal to R, and the assertion follows. In case (2), $(I + J) \cdot x \subseteq I$ implies that every homogeneous element of J lies in I, so $J \subseteq I$, and case (3) works analogously. □

Some readers may recognize the formula $\operatorname{Ext}_R^1(R/I, R) = 0$ in the assertion of the next lemma, which was used in the proof of Lemma 3.6.

LEMMA 3.7. *Let $I \subset R$ be an ideal in a ring such that I contains a regular sequence of length 2. Then for every exact sequence*

$$\{0\} \longrightarrow R \xrightarrow{\varphi} M \xrightarrow{\psi} R/I \longrightarrow \{0\}$$

of R-modules, there exists $m \in M$ with $\psi(m) = 1 + I$ and $I \cdot m = \{0\}$. (Equivalently, the sequence splits.)

PROOF. We have a regular sequence $a_1, a_2 \in I$. Choose $m' \in M$ with $\psi(m') = 1 + I$. For $x \in I$ we have $\psi(xm') = x\psi(m') = 0$. So

$$I \cdot m' \subseteq \operatorname{im}(\varphi). \tag{4}$$

In particular, $a_i m' = \varphi(b_i)$ with $b_i \in R$, so

$$\varphi(a_2 b_1) = a_2 \varphi(b_1) = a_2 a_1 m' = a_1 \varphi(b_2) = \varphi(a_1 b_2).$$

Thus $a_2 b_1 = a_1 b_2$, and the regularity of a_1, a_2 implies $b_1 = a_1 y$ with $y \in R$. Set $m := m' - \varphi(y)$. Then $\psi(m) = \psi(m') = 1 + I$. Now let $x \in I$. By (4) there is $z \in R$ with $xm' = \varphi(z)$, so $\varphi(a_1 z) = a_1 x m' = x\varphi(b_1) = \varphi(x b_1)$, and we obtain $a_1 z = x b_1 = a_1 x y$. By the regularity of a_1, this implies $z = xy$, so

$$xm = xm' - x\varphi(y) = xm' - \varphi(xy) = \varphi(z) - \varphi(z) = 0.$$

This concludes the proof. □

Theorem 3.2 was extended a few years later, which led to the following beautiful result.

THEOREM 3.8 (Emilie Dufresne and Jack Jeffries [**14**]). *Assume that K is algebraically closed and there is a separating set of $n + k - 1$ invariants. Then G is generated by k-reflections, i.e., by elements fixing a subspace of codimension k.*

Even the corollary that if there are $n + k - 1$ *generating* invariants, then G is a k-reflection group, is new. The proof is bafflingly simple: just replace Hartshorne's connectedness theorem by Grothendieck's connectedness theorem and combine this with Remark 3.5. (But some care should be taken since Grothendieck's result is actually about complete local rings.)

The topic of separating invariants has attracted some sustained interest. At the time of writing this, searching articles with "separating invariants" in the title on MathSciNet produces twenty-five hits, the latest one from 2022. Since comprehensiveness is not among the main goals of this article, let us desist from giving accounts of them and move on to the topic of infinite group invariants.

4. Invariants of linearly reductive groups

I have to report fewer new developments in algorithmic invariant theory of infinite groups than of finite groups. This is why the remaining sections of the paper are shorter than the previous ones. But arguably the developments relating to infinite groups are more important, since in classical invariant theory and in geometric invariant theory the focus has always been on infinite groups.

It should be safe to say that the most dramatic development in this area has been Harm Derksen's discovery [**9**] of an algorithm for computing generating invariants for linearly reductive groups. To shift the narrative from the mathematical to the personal: I was told that Bernd Sturmfels became completely ecstatic when he learned about the algorithm for the first time. And for good reason, because Derken's algorithm ties in perfectly with the philosophy of Bernd's book. To quote from Frank Grosshans's MathSciNet review of Bernd's book: "The efficacy of the Gröbner basis algorithms presented in this book and their place among the tools of invariant theory is yet to be decided." In this context, Derksen's algorithm establishes the place of Gröbner basis methods in invariant theory as front and center, once again showing Bernd's remarkable knack for foreseeing where and how things will happen.

Going back to mathematics, let us explain Derksen's algorithm. We will skip the proofs, directing readers to the presentation in [**12**, Section 4.1]. The algorithm connects nicely to the last section since the graph of the action $\Gamma := \{(v, \sigma(v)) \mid v \in V,\ \sigma \in G\} \subset V \times V$ (with $V := K^n$ as before) plays an important role again. Its vanishing ideal $D := \mathcal{I}(\Gamma) \subseteq K[\mathbf{x}, \mathbf{y}]$ has come to be known as the **Derksen ideal**, and we will discuss how it can be computed in a moment. The following result paves the way for Derksen's algorithm. Before we state it, recall that a linear algebraic group G is called **linearly reductive** if there is a Reynolds operator $\mathcal{R}$: $K[\mathbf{x}] \to K[\mathbf{x}]^G$ (i.e., a projection that is constant on orbits) for every morphic G-action by linear transformations of the x_i. Throughout, we assume K to be algebraically closed.

THEOREM 4.1 (Derksen [**9**]). *Assume G is linearly reductive and let $f_1, \dots, f_m$ be homogeneous generators of the Derksen ideal D. Then the $f_i(\mathbf{x}, \mathbf{0}) \in K[\mathbf{x}]$, obtained by setting $y_j = 0$ for all j, generate the Hilbert ideal $H := K[\mathbf{x}] \cdot K[\mathbf{x}]_+^G$ (see the beginning of Section 1).*

So, from homogeneous generators of D we obtain homogeneous generators of H, and by a slight extension of Proposition 1.1, applying the Reynolds operator to them yields generators of the invariant ring. Since the Reynolds operator is not always known or easily implementable, a less elegant but more effective variant consists of throwing away the generators f_i and only remembering their degrees d_i, and then computing bases of the spaces $K[\mathbf{x}]_{d_i}^G$ of homogeneous invariants of degrees d_i from scratch, which is a rather simple matter (see [**12**, Algorithm 4.5.1]).

With this, everything comes down to the question of how the Derksen ideal can be computed. To answer it, we must first specify how the action of G is given. For this, we make "a morphic action of a linear algebraic group by linear transformations of the x_i" explicit. So G is given as an affine variety in K^r by its vanishing ideal $I_G = (g_1, \dots, g_l) \subseteq K[z_1, \dots, z_r]$, with the z_i new variables. Moreover, the action is given by $\sigma(x_i) = f_i(\sigma)$ for $\sigma \in G = \mathcal{V}(I_G) \subseteq K^r$, where $f_i = \sum_{j=1}^n a_{i,j} x_j$ with $a_{i,j} \in K[\mathbf{z}]$, and $f_i(\sigma) := \sum_{j=1}^n a_{i,j}(\sigma) x_j$. Now it is easy to see that the ideal

$$\widehat{D} := \big(g_1, \dots, g_l, f_1 - y_1, \dots, f_n - y_n\big) \subseteq K[\mathbf{x}, \mathbf{y}, \mathbf{z}] \tag{5}$$

is the vanishing ideal of

$$\widehat{\Gamma} := \big\{(v, w, \sigma) \in V \times V \times K^r \mid \sigma \in G,\ w = \sigma(v)\big\}.$$

This implies that the elimination ideal $K[\mathbf{x}, \mathbf{y}] \cap \widehat{D}$ is the vanishing ideal $D = \mathcal{I}(\Gamma)$ of Γ. So the Derksen ideal can be computed as an elimination ideal of $\widehat{D}$, where $\widehat{D}$ can be formed immediately from the data defining the group and the action. It is the computation of this elimination ideal where Gröbner basis methods come in, and where the bulk of the work of Derksen's algorithm lies.

Now we are ready to present the algorithm.

ALGORITHM 4.2 (Derksen's algorithm [**9**]).

Input: A linearly reductive group G, given by its vanishing ideal $I_G = (g_1, \dots, g_l) \subseteq K[z_1, \dots, z_r]$, and a morphic G-action given by polynomials $f_1, \dots, f_n \in K[x_1, \dots, x_n]$ as described above.

Output: A generating set of the invariant ring $K[\mathbf{x}]^G$ as an algebra.

(1) Form the ideal $\widehat{D} \subseteq K[\mathbf{x}, \mathbf{y}, \mathbf{z}]$ as in Equation (5).
(2) Using Gröbner basis methods, compute the elimination ideal $D := K[\mathbf{x}, \mathbf{y}] \cap \widehat{D}$. Let $f_1, \dots, f_m$ be homogeneous generators of D.
(3) For $i = 1, \dots, m$, set $d_i := \deg(f_i)$ and calculate a basis B_i of the space $K[\mathbf{x}]_{d_i}^G$ of homogeneous invariants of degree d_i. For this, [**12**, Algorithm 4.5.1] may be used. Alternatively, if the Reynolds operator has been implemented, set $B_i := \big\{\mathcal{R}\big(f_i(\mathbf{x}, \mathbf{0})\big)\big\}$.
(4) Now $B_1 \cup \dots \cup B_m$ is a (usually not minimal) generating set of $K[\mathbf{x}]^G$.

Recall that in characteristic 0, all reductive groups are linearly reductive, which includes the classical groups. So Derksen's algorithm has brought much of classical invariant theory into the realm of algorithmic computability. But practically, its reach is not unlimited, since it requires a huge Gröbner basis computation. For example, the more advanced known results about invariants of binary forms (see the account in [**12**, Example 2.1.2]) are out of reach for Derksen's algorithm. Still, Derksen's algorithm is a great tool if you need to know a particular invariant ring that cannot be found in the literature, or if the algorithm happens to perform faster than your search of the literature.

5. Other infinite groups

In this section we take a brief look at infinite groups that are not linearly reductive. These fall in two categories: reductive groups in positive characteristic, and nonreductive groups. Typical examples for the first case are the classical groups, and for the second case the additive group. The section could also be titled "other ways to use the Derksen ideal," since this ideal turns up in different contexts.

The Derksen ideal is defined as the ideal of polynomials in $K[\mathbf{x},\mathbf{y}]$ vanishing on all $\bigl(v,\sigma(v)\bigr) \in V \times V$ with $\sigma \in G$, $v \in V = K^n$, so

$$D = \bigcap_{\sigma \in G} \bigl(y_1 - \sigma(x_1), \ldots, y_n - \sigma(x_n)\bigr).$$

This formula is better suited for generalizations than the original geometric definition of D. For example, if a group G acts on a finitely generated (but not necessarily finite) field extension $L = K(a_1, \ldots, a_n)$ of K, we define the Derksen ideal as

$$D := \bigcap_{\sigma \in G} \bigl(y_1 - \sigma(a_1), \ldots, y_n - \sigma(a_n)\bigr) \subseteq L[y_1, \ldots, y_n].$$

Of course, this depends on the choice of the generators a_i. An important special case is that the a_i are algebraically independent, so L is a rational function field. If G is infinite, this definition does not provide a way to compute D. So let us assume that G, as in the previous section, is a linear algebraic group given as a closed subset of K^r by its vanishing ideal $\mathcal{I}(G) = (g_1, \ldots, g_l) \subseteq K[z_1, \ldots, z_r]$. Moreover, assume that the G-action is by K-automorphisms and given by polynomials $f_1, \ldots, f_n \in L[\mathbf{z}]$ such that

$$\sigma(a_i) = f_i(\sigma) \tag{6}$$

(where $f_i(\sigma)$ means specializing the z_j to the coordinates of $\sigma \in K^r$). These are reasonable assumptions, and they are good enough to make the Derksen ideal computable. In fact, we have virtually the same result as in the last section.

PROPOSITION 5.1 (Computing the Derksen ideal). *In the above situation, set*

$$\widehat{D} := (g_1, \ldots, g_l, f_1 - y_1, \ldots, f_n - y_n) \subseteq L[\mathbf{y}, \mathbf{z}].$$

Then $D = L[\mathbf{y}] \cap \widehat{D}$.

PROOF. We start by taking $f \in D$. Set $\widetilde{f} := f(f_1, \ldots, f_n) \in L[\mathbf{z}]$. Then $f - \widetilde{f} \in \widehat{D}$, so for the inclusion "$\subseteq$" we need to show $\widetilde{f} \in \widehat{D}$. Let $\sigma \in G \subseteq K^r$. Then

$$\widetilde{f}(\sigma) = f\bigl(f_1(\sigma), \ldots, f_n(\sigma)\bigr) = f\bigl(\sigma(a_1), \ldots, \sigma(a_n)\bigr) = 0,$$

where the last equality comes from $f \in (y_1 - \sigma(a_1), \ldots, y_n - \sigma(a_n))$. Let B be a basis of L as a vector space over K. Then we can write $\widetilde{f} = \sum_{b \in B} h_b \cdot b$ with $h_b \in K[\mathbf{z}]$, so $\sum_{b \in B} h_b(\sigma) \cdot b = \widetilde{f}(\sigma) = 0$, which implies $h_b(\sigma) = 0$ for all b. Since this holds for every $\sigma \in G$, we conclude $h_b \in (g_1, \ldots, g_l)$ (as an ideal in $K[\mathbf{z}]$), so $\widetilde{f} \in (g_1, \ldots, g_l)$ (as an ideal in $L[\mathbf{z}]$). This implies $\widetilde{f} \in \widehat{D}$.

For the reverse inclusion, take $f \in L[\mathbf{y}] \cap \widehat{D}$. Then $f = \sum h_i g_i + \sum h_i'(f_i - y_i)$ with $h_i, h_i' \in L[\mathbf{y}, \mathbf{z}]$. Let $\sigma \in G$. Viewing f as a polynomial in $L[\mathbf{y}, \mathbf{z}]$ we can specialize the z_i to the coordinates of σ, which does not change f. So

$$\begin{aligned} f &= f(\sigma) \\ &= \sum h_i(\sigma) g_i(\sigma) + \sum h_i'(\sigma)\big(f_i(\sigma) - y_i\big) \\ &= \sum h_i'(\sigma)\big(\sigma(a_i) - y_i\big) \in \big(y_1 - \sigma(a_1), \ldots, y_n - \sigma(a_n)\big). \end{aligned}$$

Since this holds for every $\sigma \in G$, we obtain $f \in D$. □

The following result is amazing because it uses the Derksen ideal in a way that is very different from Theorem 4.1 but still arrives at computing generating invariants. It is also remarkable that there is no hypothesis such as reductivity required for the group. The theorem probably goes back to Müller-Quade and Beth [**38**], but their result was modified, simplified, generalized and extended by quite a few authors; see [**23, 26, 33**].

THEOREM 5.2 (Computing invariant fields). *In the above situation, let $\mathcal{G} \subseteq L[\mathbf{y}]$ be a reduced Gröbner basis, with respect to any monomial order, of the Derksen ideal. Then the invariant field L^G is generated, as an extension of K, by the coefficients of all polynomials in $\mathcal{G}$.*

PROOF. With G acting coefficient-wise on $L[\mathbf{y}]$, the action preserves the set of monomials in a polynomial. So for $\sigma \in G$, $\sigma(\mathcal{G})$ is a reduced Gröbner basis of $\sigma(D) = D$. The uniqueness of reduced Gröbner bases (see [**2**, Theorem 5.43]) yields $\sigma(\mathcal{G}) = \mathcal{G}$, so σ fixes every polynomial in $\mathcal{G}$. Therefore the field L' generated by the coefficients of all polynomials in $\mathcal{G}$ is contained in L^G.

For the reverse inclusion, let $b \in L^G$, which we can write as

$$b = \frac{f(a_1, \ldots, a_n)}{g(a_1, \ldots, a_n)}$$

with $f, g \in K[\mathbf{y}]$. Setting $h := f - bg \in L[\mathbf{y}]$ and taking $\sigma \in G$, we have

$$h\big(\sigma(a_1), \ldots, \sigma(a_n)\big) = \sigma\big(f(a_1, \ldots, a_n) - bg(a_1, \ldots, a_n)\big) = 0,$$

so $h \in D$ and therefore the normal form $\mathrm{NF}_{\mathcal{G}}(h)$ is zero. So $\mathrm{NF}_{\mathcal{G}}(f) - b\,\mathrm{NF}_{\mathcal{G}}(g) = 0$ by the L-linearity of the normal form. Since $g(a_1, \ldots, a_n) \neq 0$, we have $g \notin D$, so $\mathrm{NF}_{\mathcal{G}}(g) \neq 0$. This gives $b = \mathrm{NF}_{\mathcal{G}}(f)/\mathrm{NF}_{\mathcal{G}}(g)$. But computing the normal forms of f and g only involves polynomials from $L'[\mathbf{y}]$, so $b \in L'(\mathbf{y}) \cap L = L'$. □

EXAMPLE 5.3. As a toy example, consider the action of the multiplicative group $G = \mathbb{G}_m$ on two variables x_1 and x_2 with weight $(1, -1)$. With the notation of Proposition 5.1 we have

$$\begin{aligned}\widehat{D} &= (z_1z_2 - 1, z_1x_1 - y_1, z_2x_2 - y_2)\\ &= \left(z_1z_2 - 1, z_1 - \frac{1}{x_1}y_1, z_2 - \frac{1}{x_2}y_2\right)\\ &= \left(y_1y_2 - x_1x_2, z_1 - \frac{1}{x_1}y_1, z_2 - \frac{1}{x_2}y_2\right),\end{aligned}$$

where the last displayed generating set is the reduced Gröbner basis w.r.t. any monomial order with $z_i > y_j$. So $y_1y_2 - x_1x_2$ is the reduced Gröbner basis of D, and $K(x_1, x_2)^G = K(x_1 \cdot x_2)$. Of course, we have known this all along. ◁

As mentioned above, four papers were cited for a result as simple as Theorem 5.2 because it can be generalized and extended in various ways. For example, for a group action on an irreducible affine variety X, it is often possible to find a G-invariant common denominator f for the coefficients of $\mathcal{G}$, and then one gets a generating set of the localization $K[X]^G_f$. This does not require G to be reductive, or even $K[X]^G$ to be finitely generated. In fact, there is a semi-algorithm for computing $K[X]^G$ from $K[X]^G_f$, which terminates if and only if the desired invariant ring is finitely generated. (But, alas, there is no algorithm known for determining finite generation.)

Another way of using the Derksen ideal forms a bridge to the next section about separating invariants. Let G be a reductive group in positive characteristic, so it is not linearly reductive unless it is a torus with a nonmodular finite group on top of it. Then, as for any reductive group, we know that all invariants agree on two points $v, w \in V$ if and only if the orbit closures intersect:

$$f(v) = f(w) \quad \text{for all} \quad f \in K[\mathbf{x}]^G \quad \Longleftrightarrow \quad \overline{G(v)} \cap \overline{G(w)} \neq \emptyset. \tag{7}$$

It is not hard to derive from this how to calculate the **separating variety** $\mathcal{S}$, defined to contain all pairs of points $(v, w) \in V \times V$ where all invariants agree. In fact, $\mathcal{S}$ is given by the elimination ideal $K[\mathbf{x}, \mathbf{y}] \cap (D_{\mathbf{x},\mathbf{z}}, D_{\mathbf{y},\mathbf{z}})$, where $D_{\mathbf{x},\mathbf{z}}$ and $D_{\mathbf{y},\mathbf{z}}$ stand for the Derksen ideal, as first defined in Section 4, with the y_i or x_i, respectively, replaced by new variables z_i. With this, one can also compute separating invariants: just add homogeneous invariants f_i of rising degrees until the $f_i(\mathbf{x}) - f_i(\mathbf{y})$ define the variety $\mathcal{S}$. So we can compute a separating homogeneous subalgebra $A \subseteq K[\mathbf{x}]^G$. Now it is well known that (as we are in characteristic $p > 0$) the entire invariant ring is the purely inseparable closure of A, i.e.,

$$K[\mathbf{x}]^G = \{f \in K[\mathbf{x}] \mid f^q \in A \text{ for some } p\text{-power } q\}.$$

And for the calculation of the purely inseparable closure there is also an algorithm (using Gröbner basis methods, as readers are probably expecting). In summary, we obtain an algorithm for computing invariant rings of reductive groups in positive characteristic, which complements Derksen's algorithm for reductive groups in characteristic 0. For more details, see [**31**], [**12**, Section 4.9], and [**11**], where the methods are extended to reductive group actions on affine varieties instead of only vector spaces.

6. Separating invariants of infinite groups

In Section 5 we considered separating invariants of reductive group, so the focus here is on nonreductive groups. For such groups, the invariant ring may not be finitely generated, and the criterion (7) may fail. Nevertheless, there always exists a finite set of separating invariants. This may seem like an amazing, deep fact, until one realizes how simple it is to prove: the ideal in $K[\mathbf{x}, \mathbf{y}]$ generated by all $\Delta f := f(\mathbf{x}) - f(\mathbf{y})$ with $f \in K[\mathbf{x}]^G$ can, by Hilbert's basis theorem, be generated by the deltas of finitely many invariants, which then form a separating set.

This argument is similar in spirit to Hilbert's first proof of finite generation of $K[\mathbf{x}]^G$ for linearly reductive groups. After Hilbert's proof, it took about 100 years until a constructive version emerged in the form of Derksen's algorithm. For separating invariants, we still seem to be in the 100-year waiting period: at the time of writing this, no algorithm has been found for computing a finite separating set in the case of nonreductive groups.

We can, however, go further in another direction and quantify our finiteness statement. The following result gives an upper bound on the minimal number of separating invariants that is, surprisingly, independent of the group. It is hard to pin down who first proved the result, which has been folklore for quite a while.

THEOREM 6.1 (The number of separating invariants). *If K is an infinite field, there is a set of separating invariants of size $\leq 2n + 1$, with n the number of variables.*

Since the proof is nice and short, we present it here.

PROOF. We know from the above observation that there is a finite set of separating invariants $f_1, \ldots, f_k$. Assume $k > 2n + 1$. Then with t an additional variable, the

$$g_i := t \cdot \big(f_i(\mathbf{x}) - f_i(\mathbf{y})\big) \in K[\mathbf{x}, \mathbf{y}, t] \quad (i = 1, \ldots, k)$$

are algebraically dependent, so we have a nonzero polynomial H in k variables such that $H(g_1, \ldots, g_k) = 0$. We can choose $\alpha_1, \ldots, \alpha_k \in K$ with $H(\alpha_1, \ldots, \alpha_k) \neq 0$ and $\alpha_1 \neq 0$. Now we claim that the invariants

$$\widetilde{f}_i := \alpha_1 f_i - \alpha_i f_1 \quad (i = 2, \ldots, k)$$

form a separating set. If that is proved, we can repeat this until reaching a separating set of size $2n + 1$. To prove the claim, assume that the $\widetilde{f}_i$ do not form a separating set, so there are points $v, w \in K^n$ such that all $\widetilde{f}_i$ agree on v and w, but not all f_i. Then from the definition of the $\widetilde{f}_i$ we see that

$$\alpha_1\big(f_i(v) - f_i(w)\big) = \alpha_i\big(f_1(v) - f_1(w)\big),$$

so $f_1(v) \neq f_1(w)$. Now let $\Phi\colon K[\mathbf{x}, \mathbf{y}, t] \to K$ be the map given by evaluating a polynomial at the point (v, w, η) with $\eta := \frac{\alpha_1}{f_1(v) - f_1(w)}$. Then

$$\Phi(g_i) = \eta \cdot \big(f_i(v) - f_i(w)\big) = \alpha_i \quad (i = 1, \ldots, k),$$

from which the contradiction

$$H(\alpha_1, \ldots, \alpha_k) = H\big(\Phi(g_1), \ldots, \Phi(g_k)\big) = \Phi\big(H(g_1, \ldots, g_k)\big) = \Phi(0) = 0$$

ensues. □

REMARK 6.2. The above proof really establishes the upper bound

$$2\dim\bigl(K[\mathbf{x}]^G\bigr)+1,$$

because for the invariant ring, as for any subalgebra of a finitely generated algebra, the Krull dimension is equal to the transcendence degree (see [**32**, Exercise 5.3] or [**20**, Proposition 2.3]). Also notice that the proof never uses the group action, so the theorem holds in a broader context (see [**26**, Theorem 5.3]). The proof is constructive, so once a finite separating set is known, it can be boiled down to one of size $\leq 2n+1$. But a disadvantage is that starting out with a homogeneous separating set will produce a smaller separating set that is almost certainly inhomogeneous, since the smaller set consists of linear combinations of the larger set. ◁

Even though the focus in this section lies on nonreductive groups, Theorem 6.1 tells us something new also in the case of reductive and even finite groups. In fact, there are lots of examples where the minimum number of *generating* invariants vastly exceeds the bound $2n+1$ (see [**26**, Section 5]).

Acknowledgments

The author would like to thank Serkan Hoşten, Diane Maclagan, and Frank Sottile for organizing this volume and, of course, for inviting me to make a contribution. It was a pleasure to do so. Also, many thanks to the anonymous reviewers for their careful reading of the manuscript and for numerous helpful suggestions, corrections and comments.

References

[1] G. Almkvist and R. Fossum, *Decomposition of exterior and symmetric powers of indecomposable* $\mathbf{Z}/p\mathbf{Z}$*-modules in characteristic* p *and relations to invariants*, Séminaire d'Algèbre Paul Dubreil, 30ème année (Paris, 1976), Lecture Notes in Math., Vol. 641, Springer, Berlin-New York, 1978, pp. 1–111. MR499459

[2] T. Becker and V. Weispfenning, *Gröbner bases: A computational approach to commutative algebra*, Graduate Texts in Mathematics, vol. 141, Springer-Verlag, New York, 1993. In cooperation with Heinz Kredel, DOI 10.1007/978-1-4612-0913-3. MR1213453

[3] D. J. Benson, *Polynomial invariants of finite groups*, London Mathematical Society Lecture Note Series, vol. 190, Cambridge University Press, Cambridge, 1993, DOI 10.1017/CBO9780511565809. MR1249931

[4] B. Blum-Smith and S. Marques, *When are permutation invariants Cohen-Macaulay over all fields?*, Algebra Number Theory **12** (2018), no. 7, 1787–1821, DOI 10.2140/ant.2018.12.1787. MR3871511

[5] W. Bosma, J. Cannon, and C. Playoust, *The Magma algebra system. I. The user language*, J. Symbolic Comput. **24** (1997), no. 3-4, 235–265, DOI 10.1006/jsco.1996.0125. Computational algebra and number theory (London, 1993). MR1484478

[6] N. Bourbaki, *Éléments de mathématique: Groupes et algèbres de Lie* (French), Masson, Paris, 1982. Chapitres 4–6. MR682756

[7] H. E. A. Campbell, A. V. Geramita, I. P. Hughes, R. J. Shank, and D. L. Wehlau, *Non-Cohen-Macaulay vector invariants and a Noether bound for a Gorenstein ring of invariants*, Canad. Math. Bull. **42** (1999), no. 2, 155–161, DOI 10.4153/CMB-1999-018-4. MR1692004

[8] C. Chevalley, *Invariants of finite groups generated by reflections*, Amer. J. Math. **77** (1955), 778–782, DOI 10.2307/2372597. MR72877

[9] H. Derksen, *Computation of invariants for reductive groups*, Adv. Math. **141** (1999), no. 2, 366–384, DOI 10.1006/aima.1998.1787. MR1671758

[10] H. Derksen and G. Kemper, *Computational invariant theory*, Invariant Theory and Algebraic Transformation Groups, vol. I, Springer-Verlag, Berlin, 2002. Encyclopaedia of Mathematical Sciences, 130, DOI 10.1007/978-3-662-04958-7. MR1918599

[11] H. Derksen and G. Kemper, *Computing invariants of algebraic groups in arbitrary characteristic*, Adv. Math. **217** (2008), no. 5, 2089–2129, DOI 10.1016/j.aim.2007.08.016. MR2388087

[12] H. Derksen and G. Kemper, *Computational invariant theory*, Second enlarged edition, Encyclopaedia of Mathematical Sciences, vol. 130, Springer, Heidelberg, 2015. With two appendices by Vladimir L. Popov, and an addendum by Norbert A'Campo and Popov; Invariant Theory and Algebraic Transformation Groups, VIII, DOI 10.1007/978-3-662-48422-7. MR3445218

[13] E. Dufresne, *Separating invariants and finite reflection groups*, Adv. Math. **221** (2009), no. 6, 1979–1989, DOI 10.1016/j.aim.2009.03.013. MR2522833

[14] E. Dufresne and J. Jeffries, *Separating invariants and local cohomology*, Adv. Math. **270** (2015), 565–581, DOI 10.1016/j.aim.2014.11.003. MR3286543

[15] D. Eisenbud, *Commutative algebra: With a view toward algebraic geometry*, Graduate Texts in Mathematics, vol. 150, Springer-Verlag, New York, 1995, DOI 10.1007/978-1-4612-5350-1. MR1322960

[16] G. Ellingsrud and T. Skjelbred, *Profondeur d'anneaux d'invariants en caractéristique p* (French), Compositio Math. **41** (1980), no. 2, 233–244. MR581583

[17] L. Ferraro, F. Galetto, F. Gandini, H. Huang, M. Mastroeni, and X. Ni, *The InvariantRing package for Macaulay2*, J. Softw. Algebra Geom. **14** (2024), no. 1, 5–11, DOI 10.2140/jsag.2024.14.5. MR4722167

[18] P. Fleischmann, *The Noether bound in invariant theory of finite groups*, Adv. Math. **156** (2000), no. 1, 23–32, DOI 10.1006/aima.2000.1952. MR1800251

[19] J. Fogarty, *On Noether's bound for polynomial invariants of a finite group*, Electron. Res. Announc. Amer. Math. Soc. **7** (2001), 5–7, DOI 10.1090/S1079-6762-01-00088-9. MR1826990

[20] J. M. Giral, *Krull dimension, transcendence degree and subalgebras of finitely generated algebras*, Arch. Math. (Basel) **36** (1981), no. 4, 305–312, DOI 10.1007/BF01223706. MR623141

[21] N. Gordeev and G. Kemper, *On the branch locus of quotients by finite groups and the depth of the algebra of invariants*, J. Algebra **268** (2003), no. 1, 22–38, DOI 10.1016/S0021-8693(02)00572-0. MR2004478

[22] R. Hartshorne, *Complete intersections and connectedness*, Amer. J. Math. **84** (1962), 497–508, DOI 10.2307/2372986. MR142547

[23] E. Hubert and I. A. Kogan, *Rational invariants of a group action. Construction and rewriting*, J. Symbolic Comput. **42** (2007), no. 1-2, 203–217, DOI 10.1016/j.jsc.2006.03.005. MR2284293

[24] I. Hughes and G. Kemper, *Symmetric powers of modular representations for groups with a Sylow subgroup of prime order*, J. Algebra **241** (2001), no. 2, 759–788, DOI 10.1006/jabr.2000.8710. MR1843324

[25] B. Huppert, *Endliche Gruppen. I* (German), Die Grundlehren der mathematischen Wissenschaften, Band 134, Springer-Verlag, Berlin-New York, 1967. MR224703

[26] T. Kamke and G. Kemper, *Algorithmic invariant theory of nonreductive groups*, Qual. Theory Dyn. Syst. **11** (2012), no. 1, 79–110, DOI 10.1007/s12346-011-0059-4. MR2902726

[27] D. B. Karagueuzian and P. Symonds, *The module structure of a group action on a polynomial ring: A finiteness theorem*, J. Amer. Math. Soc. **20** (2007), no. 4, 931–967, DOI 10.1090/S0894-0347-07-00563-2. MR2328711

[28] G. Kemper, *An algorithm to calculate optimal homogeneous systems of parameters*, J. Symbolic Comput. **27** (1999), no. 2, 171–184, DOI 10.1006/jsco.1998.0247. MR1672128

[29] G. Kemper, *On the Cohen-Macaulay property of modular invariant rings*, J. Algebra **215** (1999), no. 1, 330–351, DOI 10.1006/jabr.1998.7716. MR1684150

[30] G. Kemper, *The depth of invariant rings and cohomology*, J. Algebra **245** (2001), no. 2, 463–531, DOI 10.1006/jabr.2001.8840. With an appendix by Kay Magaard. MR1863889

[31] G. Kemper, *Computing invariants of reductive groups in positive characteristic*, Transform. Groups **8** (2003), no. 2, 159–176, DOI 10.1007/s00031-003-0305-1. MR1976458

[32] G. Kemper, *A course in commutative algebra*, Graduate Texts in Mathematics, vol. 256, Springer, Heidelberg, 2011, DOI 10.1007/978-3-642-03545-6. MR2766370

[33] G. Kemper, *Using extended Derksen ideals in computational invariant theory*, J. Symbolic Comput. **72** (2016), 161–181, DOI 10.1016/j.jsc.2015.02.004. MR3369055

[34] G. Kemper, A. Lopatin, and F. Reimers, *Separating invariants over finite fields*, J. Pure Appl. Algebra **226** (2022), no. 4, Paper No. 106904, 18, DOI 10.1016/j.jpaa.2021.106904. MR4310052

[35] S. A. King, *Minimal generating sets of non-modular invariant rings of finite groups*, J. Symbolic Comput. **48** (2013), 101–109, DOI 10.1016/j.jsc.2012.05.002. MR2980468

[36] M. Kreuzer and L. Robbiano, *Computational commutative algebra. 2*, Springer-Verlag, Berlin, 2005. MR2159476

[37] M. Lorenz and J. Pathak, *On Cohen-Macaulay rings of invariants*, J. Algebra **245** (2001), no. 1, 247–264, DOI 10.1006/jabr.2001.8900. MR1868191

[38] J. Müller-Quade and T. Beth, *Calculating generators for invariant fields of linear algebraic groups*, Applied algebra, algebraic algorithms and error-correcting codes (Honolulu, HI, 1999), Lecture Notes in Comput. Sci., vol. 1719, Springer, Berlin, 1999, pp. 392–403, DOI 10.1007/3-540-46796-3_37. MR1846513

[39] E. Noether, *Der Endlichkeitssatz der Invarianten endlicher Gruppen* (German), Math. Ann. **77** (1915), no. 1, 89–92, DOI 10.1007/BF01456821. MR1511848

[40] D. R. Richman, *Invariants of finite groups over fields of characteristic p*, Adv. Math. **124** (1996), no. 1, 25–48, DOI 10.1006/aima.1996.0076. MR1423197

[41] J.-P. Serre, *Groupes finis d'automorphismes d'anneaux locaux réguliers* (French), Colloque d'Algèbre (Paris, 1967), École normale supérieure de jeunes filles, Secrétariat mathématique, Paris, 1968, pp. Exp. 8, 11. MR234953

[42] L. Smith, *Polynomial invariants of finite groups*, Research Notes in Mathematics, vol. 6, A K Peters, Ltd., Wellesley, MA, 1995. MR1328644

[43] B. Sturmfels, *Algorithms in invariant theory*, Texts and Monographs in Symbolic Computation, Springer-Verlag, Vienna, 1993, DOI 10.1007/978-3-7091-4368-1. MR1255980

[44] P. Symonds, *On the Castelnuovo-Mumford regularity of rings of polynomial invariants*, Ann. of Math. (2) **174** (2011), no. 1, 499–517, DOI 10.4007/annals.2011.174.1.14. MR2811606

Department of Mathematics, Technische Universiät München, Boltzmannstrasse 3, 85748 Garching, Germany

Email address: kemper@ma.tum.de

Proceedings of Symposia in Pure Mathematics
Volume **111**, 2025
https://doi.org/10.1090/pspum/111/02038

A universal tale of determinants and Gröbner bases

Aldo Conca

Dedicated to Bernd Sturmfels
on the occasion of his sixtieth birthday

Abstract. In 1965 Buchberger defined Gröbner bases and an algorithm to compute them. Despite a slow start, in the 1980s Gröbner bases had already become the main device for symbolic computations involving polynomials as well as a theoretical tool for the investigation of ideals and varieties via the so-called Gröbner deformation techniques. Rings and algebraic varieties defined by means of determinants are among the most classical objects in commutative algebra, algebraic geometry, and invariant theory. By the end of the 1980s the time was ripe for the computation of Gröbner bases of determinantal ideals. We will tell the tale of how (universal) Gröbner bases of determinantal ideals were identified and the key role played by Bernd Sturmfels and his collaborators in this enterprise.

1. Generalities on Gröbner bases and deformations

Let K be a field and S be the polynomial ring $K[x_1, \dots, x_n]$. As a K-vector space S has very natural basis: the set of the monomials $\{x^a : a \in \mathbb{N}^n\}$. A term order on S is a total order $\leq$ on the set of monomials of S such that $x^a \leq x^b$ implies $x^a x^c \leq x^b x^c$ and $1 = x^0 \leq x^c$ for all $c \in \mathbb{N}^n$. Given a term order $\leq$, any nonzero polynomial $f \in S$ has an initial monomial $\text{in}(f)$, i.e., the largest monomial that appears in f. Furthermore, any ideal I of S has an initial ideal $\text{in}(I) = (\text{in}(f) : f \in I \setminus \{0\})$. A subset $G = \{f_1, \dots, f_s\}$ of nonzero elements of I is a Gröbner basis of I with respect to $\leq$ if $\text{in}(f_1), \dots, \text{in}(f_s)$ generates $\text{in}(I)$ as an ideal. Equivalently, G is a Gröbner basis of I if for every nonzero $f \in I$ there exists $f_i \in G$ such that $\text{in}(f_i)$ divides $\text{in}(f)$. The Noetherianity of S implies that I has a finite Gröbner basis and the Buchberger algorithm [**BCRV**, Chapter 1] computes it starting from any finite set of generators of I. A universal Gröbner basis of I is a set polynomials of I that is a Gröbner basis with respect to all the possible term orders. There are infinitely many term orders but, essentially, only finitely many different Gröbner bases of I so that I has a finite universal Gröbner basis.

2020 *Mathematics Subject Classification.* Primary 13C13.

The author was partially supported by PRIN 2020355B8Y "Squarefree Gröbner degenerations, special varieties and related topics" and by GNSAGA-INdAM. The author thanks the referees for the careful reading of the paper and for the improvements of the exposition suggested.

The relationship between S/I to $S/\text{in}(I)$ can be understood as the passage from the generic fiber to the special one of a 1-dimensional flat family. An important consequence is that most of the algebraic properties can be transferred from $S/\text{in}(I)$ to S/I; see [**BC3**, **BCRV**] for details. For example, S/I is Cohen–Macaulay whenever $S/\text{in}(I)$ is Cohen–Macaulay. By the very definition, $\text{in}(I)$ is a monomial ideal and hence an object that can be investigated by means of combinatorial methods. The special case when $\text{in}(I)$ is a radical ideal is very important for at least two reasons. The first is that a radical monomial ideal corresponds to a simplicial complex in the Stanley–Reisner duality and hence an entire toolkit of new ideas and resources can be employed in the investigation of the properties of $S/\text{in}(I)$. The second is a recent result [**CV**] asserting that when $\text{in}(I)$ is radical the relationship between $S/\text{in}(I)$ to S/I is tighter, for example, $S/\text{in}(I)$ is Cohen–Macaulay if and only if S/I is Cohen–Macaulay.

2. Generalities on determinantal ideals

Let A be a matrix with entries in a Noetherian ring S. The ideal generated by all the minors (i.e., subdeterminants) of A of a given size, say, t, is denoted by $I_t(A)$. It is called the determinantal ideal associated to A. The classical Cauchy–Binet formula actually asserts that $I_t(AB) \subseteq I_t(A)I_t(B)$ and this easily implies that $I_t(CAB) = I_t(A)$ whenever C, B are invertible matrices. Said otherwise, if $f : F \to G$ is a S-linear map and F and G are finitely generated free S-modules, the ideal $I_t(f)$ can be defined as the ideal of t-minors of any matrix representing f.

Determinantal ideals are ubiquitous in algebra and geometry. Indeed, many rings and algebraic varieties that are classically studied are, in one way or another, defined by determinantal conditions. Among them a central role is played by the "generic determinantal ideals" that are defined as follows. Let K be a field and m, n be positive integers. Let $S = K[X] = K[x_{ij} : (i,j) \in [m] \times [n]]$ be the polynomial ring in mn variables and X be $m \times n$ matrix whose (i,j)th entry is equal to x_{ij}. We call X the generic $m \times n$ matrix (over K) and $I_t(X)$ the generic determinantal ideal. The algebraic variety defined by $I_t(X)$ is the set of linear maps $f : K^n \to K^m$ of rank $< t$. For simplicity we will assume that $m \leq n$ which is not a restrictive assumption because ideals of minors do not change under transposition.

The ideal $I_t(X)$ shows up in classical invariant theory. Let us recall how. Consider the polynomial ring $K[Y, Z]$ with Y, Z generic matrices over K of size $m \times (t-1)$ and $(t-1) \times n$. The group $G = \text{GL}_{t-1}(K)$ acts on $K[Y, Z]$ by $Y \to YA$ and $Z \to A^{-1}Z$ for $A \in \text{GL}_{t-1}(K)$. Under mild assumption on K (see [**BV**, Chapter 7]) the ring of invariants $K[Y, Z]^G$ is generated by the entries of the product matrix YZ. Mapping $X \to YZ$ we get a surjective map $\phi : K[X] \to K[Y, Z]^G$. By rank reasons its kernel clearly contains $I_t(X)$ and a result of classical invariant theory asserts that $\ker \phi = I_t(X)$. Hence,

$$K[X]/I_t(X) \simeq K[YZ] = K[Y, Z]^G. \tag{2.1}$$

In the early 1970s Hochster conjectured that the ring of the invariants of any reasonably good group (i.e., linearly reductive) acting on a polynomial ring over a field is Cohen–Macaulay. The conjecture was proved in full generality by Hochster and Roberts [**HR**] and re-proved later on by Hochster and Huneke [**HH**] using characteristic p methods, in particular, tight closure. Other approaches to the Cohen–Macaulay property of $K[X]/I_t(X)$ have been developed more or less in the same years. One of them, which is important for our story, is based on the so-called standard monomial theory.

3. Standard monomial theory

The polynomial ring $K[X] = K[x_{ij} : (i,j) \in [m] \times [n]]$ has a very natural K-basis: the set $\mathfrak{M}$ of the monomials in the x_{ij}'s. This basis is closed under multiplication and hence monomial ideals are (relatively) easy to deal with. However, to study the determinantal ideals $I_t(X)$, it would be better to have a K-basis $\mathfrak{S}$ of $K[X]$ such that $I_t(X)$ is generated, as a vector space, by a subset of the $\mathfrak{S}$. Miraculously, such a basis $\mathfrak{S}$ was discovered in the 1970s by Doubilet, Rota, and Stein [**BCRV**, Theorem 3.2.1] and its elements are certain products of minors of various sizes called standard monomials. The price to pay to deal with the standard monomial basis $\mathfrak{S}$ is that it is not closed under product. But there is a replacement for this drawback: the so-called straightening law that allows one to keep the rewriting of the product of two elements in $\mathfrak{S}$ under control as a K-linear combination of elements of $\mathfrak{S}$. The entire package (the basis and rewriting rule) goes under the name of algebras with straightening law and it has been developed as an abstract theory in the 1970s and 1980s; see [**BV**, **E**, **DEP**, **MuS**]. We will not enter into full details of this side of the story but we will recall the definition of Doubilet, Rota, and Stein's standard monomials. A single t-minor δ of X is denoted by $[r_1,\dots,r_t|c_1,\dots,c_t]$ where the r_i's and the c_i's are the row and column indices given in increasing order. The set of minors is partially ordered by

$$[r_1,\dots,r_t|c_1,\dots,c_t] \le [s_1,\dots,s_p|d_1,\dots,d_p]$$

$$\text{if and only if } \begin{cases} t \ge p, \\ r_i \le s_i \text{ for } i = 1,\dots p \\ c_i \le d_i \text{ for } i = 1,\dots p. \end{cases}$$

A standard monomial is a product of minors $\delta_1\delta_2\cdots\delta_v$ such that $\delta_1 \le \delta_2 \le \cdots \le \delta_v$. Writing the row indices as rows of a Young tableau and similar for the column indices, a standard monomial gets identified with a Young bitableau which is standard (i.e., with increasing rows from left to right and weakly increasing columns from top to bottom). For example, the standard monomial

$$[1,3,4,5|1,2,3,6] \cdot [2,3,5|1,4,5] \cdot [4|2]$$

is represented by the bitableau

$$\begin{array}{|c|c|c|c|} \hline 1 & 3 & 4 & 5 \\ \hline 2 & 3 & 5 \\ \cline{1-3} 4 \\ \cline{1-1} \end{array} \qquad \begin{array}{|c|c|c|c|} \hline 1 & 2 & 3 & 6 \\ \hline 1 & 4 & 5 \\ \cline{1-3} 2 \\ \cline{1-1} \end{array} \tag{3.1}$$

4. Gröbner bases of determinantal ideals via the RSK correspondence

A term order on $S = K[X] = K[x_{ij} : (i,j) \in [m] \times [n]]$ is called diagonal if the initial term of every minor δ is the product of the entries in the main diagonal, i.e.,

$$\mathrm{in}([r_1,r_2,\dots,r_t|c_1,c_2,\dots,c_t]) = x_{r_1c_1}x_{r_2c_2}\cdots x_{r_tc_t}$$

and antidiagonal if

$$\mathrm{in}([r_1,r_2,\dots,r_t|c_1,c_2,\dots,c_t]) = x_{r_1c_t}x_{r_2c_{t-1}}\cdots x_{r_tc_1}.$$

There are plenty of diagonal or antidiagonal term orders. For example, the lexicographic term order induced by

$$x_{11} > x_{12} > \cdots > x_{1n} > x_{21} > x_{22} > \cdots > x_{mn}$$

is diagonal and reverse lexicographic term order induced by the same total order of the variables is antidiagonal.

THEOREM 1. *The set of t-minors of X is a Gröbner basis of $I_t(X)$ with respect to every diagonal or antidiagonal order.*

Actually, since what really matters are the initial terms of the minors and since diagonals and antidiagonals coincide up to an obvious symmetry, it is enough to prove the statement for a single diagonal or antidiagonal term order.

This theorem has been proved, more or less simultaneously, by Sturmfels [**S**] and Caniglia, Guccione, and Guccione [**CGG**] at the end of the 1980s. A version of it can be deduced from an earlier paper of Narashiman [**N**], which is unfortunately written using an unusual terminology and with no mention of Gröbner bases. The proof given by Sturmfels is a priceless jewel of beauty. We sketch it.

The polynomial ring $K[X] = K[x_{ij} : (i,j) \in [m] \times [n]]$ has hence two very natural K-bases: the set $\mathfrak{M}$ of the monomials in the x_{ij}'s and the set of standard monomials $\mathfrak{S}$. Indeed, they both consist of homogeneous elements. Sturmfels observed that a version of the classical Robinson–Schensted–Knuth correspondence (RSK for short, denoted by KRS in some papers) can be reinterpreted as an explicit, combinatorially defined, and degree preserving bijection between $\mathfrak{S}$ and $\mathfrak{M}$. Hence it can be extended to a degree preserving K-linear map:

$$\text{RSK} : K[X] \to K[X],$$

sending each standard monomial to an ordinary monomial. Now since $I_t(X)$ has a K-basis of standard monomials its image $\text{RSK}(I_t(X))$ has the following two properties:

(a) $\text{RSK}(I_t(X))$ is generated by monomials,
(b) $\text{RSK}(I_t(X))$ and $I_t(X)$ have the same Hilbert function.

Note that (a) and (b) are typical properties of an initial ideal of $I_t(X)$. However, by the very definition, $\text{RSK}(I_t(X))$ is only a K-vector subspace of $K[X]$ and not (yet) an ideal. On the other hand, consider the ideal J_t generated by the main diagonals of the t-minors of X and observe that if

$$\text{RSK}(I_t(X)) \subseteq J_t \tag{4.1}$$

then we would have:

$$\text{RSK}(I_t(X)) \subseteq J_t \subseteq \text{in}(I_t(X)) \tag{4.2}$$

for every diagonal order. Now in (4.2) the leftmost and rightmost vectors spaces are graded and have the same Hilbert function. Hence in (4.2) one has equality throughout. In particular, $J_t = \text{in}(I_t(X))$, that is, Theorem 1 holds and one has a bonus:

$$\text{RSK}(I_t(X)) = \text{in}(I_t(X)). \tag{4.3}$$

The final observation of Sturmfels is that the inclusion (4.1) holds by a result of Schensted [**Sc**] (again suitably reinterpreted) and this completes the proof of Theorem 1.

Another important remark of Sturmfels is that, for a diagonal term order, the initial ideal $\text{in}(I_t(X))$ is the Stanley–Reisner ideal associated with a simplicial complex that can be identified with the $(t-1)$-order complex of a planar distributive lattice. Björner proved in [**Bj**] that such a simplicial complex is shellable and hence Cohen–Macaulay. Summing up, one establishes the Cohen–Macaulay property of the determinantal ring $K[X]/I_t(X)$ via Gröbner deformation.

Sturmfels's approach to Gröbner bases of determinantal ideals and related ideas were later extended to other types of determinantal ideals and to their ordinary and symbolic powers; see [**BC1**, **BC2**, **BC3**, **C94**, **C98**, **HT**]. In particular, it turned out that (4.3) holds for a very large class of determinantal ideals.

Back then it was very surprising, and still is, to discover that a combinatorial map, the RSK, that has nothing to do with the selection of monomials in the support of polynomials actually "computes" initial ideals and Gröbner bases of determinantal ideals. It is another example of the unreasonable effectiveness of mathematical ideas.

5. Universal Gröbner bases

A universal Gröbner basis of an ideal I in a polynomial ring is a subset of I that is a Gröbner basis with respect to every term order. There are infinitely many term orders but, as a single polynomial has only finitely many initial terms, one can prove that every ideal I has only a finite number initial ideals. This implies that I has a finite universal Gröbner basis.

5.1. Universal Gröbner bases for maximal minors. For the ideal of maximal minors $I_m(X)$ of the generic matrix X the result is the best possible.

THEOREM 2. *The set of maximal minors of X is a universal Gröbner basis of $I_m(X)$.*

The original proof of Theorem 2, obtained by combining work of Sturmfels and Zelevinsky [**SZ**] and of Bernstein and Zelevinsky [**BZ**], is based on a detailed analysis of the Newton polytope associated to the product of the maximal minors and on the notion of matching fields. Indeed, in [**SZ**] the proof of Theorem 2 is reduced to a conjectural assertion on matching fields later proved in [**BZ**]. Another proof of Theorem 2 is given by Kalinin in [**K**]. A simpler and shorter proof of Theorem 2 was later found by Conca, De Negri, and Gorla [**CDG1**]. It is based on the observation that under the substitution $x_{ij} \to a_{ij}x_{1j}$ with $a_{ij} \in K$ generic elements both $I_m(X)$ and the ideal generated by the initial terms of the maximal minors of X equals the ideal of all the square-free monomials of degree m in the variables $x_{11}, x_{12}, \dots, x_{1n}$. This proof and the main results of [**C07**, **CS**] were the starting points of a development that led to the identification and study of Cartwright–Sturmfels ideals, i.e., multigraded ideals whose multigraded generic initial ideal is radical; see [**CCC**, **CDG2**, **CDG3**, **CDG4**, **CDG5**, **CDG6**, **CW**]. A corollary of the proof of Theorem 2 given in [**CDG1**] is the following.

COROLLARY 3. *Let D be any initial ideal of $I_m(X)$. Then:*

(1) *D is radical,*
(2) *S/D is Cohen–Macaulay, and*
(3) *$\beta_{ij}(S/D) = \beta_{ij}(S/I_m(X))$ for all i, j.*

Corollary 3(2) and (3) were first proved by Boocher [**Bo**] in a more general setting and follows also from the recent result of Conca and Varbaro [**CV**] and the Cohen–Macaulayness of $K[X]/I_m(X)$.

5.2. Universal Gröbner bases for 2-minors. For $t = 2$ the isomorphism (2.1) takes the form:

$$K[X]/I_2(X) \simeq K[y_i z_j : (i,j) \in [m] \times [n]] \text{ where } x_{ij} \to y_i z_j. \tag{5.1}$$

In other words, $I_2(X)$ is the toric ideal, i.e., a prime ideal generated by binomials, associated with the classical Segre variety. Special techniques have been developed to study toric ideals and varieties; see the recent volumes [**BG, CLS**] for an overview. In particular, Sturmfels wrote a fundamental monograph on the Gröbner bases for toric ideals [**S2**]. Given a finite set of vectors in $\mathcal{A} = \{a_1, \dots, a_n\} \subset \mathbb{Z}^d$ one considers the associated toric ideal $P_{\mathcal{A}}$, i.e., the kernel of the map

$$K[x_1, \dots, x_n] \to K[t_1^{\pm 1}, \dots t_d^{\pm 1}] \quad x_i \to t^{a_i}.$$

In [**S2**, Chapter 4] Sturmfels defined:

(a) the set $U_{\mathcal{A}}$ of all the binomials that belong to some reduced Gröbner basis of $P_{\mathcal{A}}$,
(b) the set of circuits $C_{\mathcal{A}}$ where a circuit of $P_{\mathcal{A}}$ is a binomial in $P_{\mathcal{A}}$ that is the difference of coprime monomials and has a minimal support,
(c) the Graver basis $G_{\mathcal{A}}$ of $P_{\mathcal{A}}$, i.e., the set of the binomials $x^a - x^b$ in $P_{\mathcal{A}}$ such that there is no other binomial $x^c - x^d$ in $P_{\mathcal{A}}$ such that x^c divides x^a and x^d divides x^b,

and observed that $U_{\mathcal{A}}$ is a universal Gröbner basis of $P_{\mathcal{A}}$. He also proved in [**S2**, Proposition 4.11]:

$$C_{\mathcal{A}} \subseteq U_{\mathcal{A}} \subseteq G_{\mathcal{A}}. \tag{5.2}$$

For special configurations $\mathcal{A}$, called unimodular, Sturmfels then proved in [**S2**, Proposition 8.11] that the equality

$$C_{\mathcal{A}} = G_{\mathcal{A}} \tag{5.3}$$

holds. The conclusion is that $C_{\mathcal{A}}$ is a universal Gröbner basis of $P_{\mathcal{A}}$ if $\mathcal{A}$ is unimodular.

Returning to $I_2(X)$, taking the canonical bases $e_1, \dots, e_m$ and $f_1, \dots, f_n$ of $\mathbb{Z}^m$ and $\mathbb{Z}^n$ and

$$\mathcal{A} = \{e_i + f_j : (i,j) \in [m] \times [n]\} \subset \mathbb{Z}^m \oplus \mathbb{Z}^n,$$

the corresponding toric ideal $P_{\mathcal{A}}$ is exactly $I_2(X)$. It turns out that $\mathcal{A}$ is indeed unimodular. Hence one has a description of a universal Gröbner basis of $I_2(X)$ as the set of the circuits of $\mathcal{A}$. The circuits, in this case, are described as cycles in the complete bipartite graph $K_{m,n}$. In practice a circuit is described as follows. For a pair (I, J) of sequences of distinct integers, say, $I = i_1, \dots i_s$, $J = j_1, \dots, j_s$ with $2 \leq s \leq m$ and $i_k \in [m]$ and $j_k \in [n]$ for all k one considers the binomial

$$F_{I,J} = x_{i_1 j_1} x_{i_2 j_2} \cdots x_{i_s j_s} - x_{i_1 j_2} x_{i_2 j_3} \cdots x_{i_{s-1} j_s} x_{i_s j_1}.$$

It turns out that the set of the $F_{I,J}$ is indeed the set of the circuits of $I_2(X)$. Summing up, combining [**S2**, Proposiiton 4.11] with [**S2**, Proposition 8.11], one obtains the following characterization of a universal Gröbner basis of $I_2(X)$.

THEOREM 4. *The set of binomials $F_{I,J}$ is a universal Gröbner basis of the ideal $I_2(X)$.*

Theorem 4 is also a special case of a theorem proved by Villarreal [**V**, 10.1.11] that describes a universal Gröbner basis of the toric ideal associated to the edge algebra of a graph.

REMARK 5. Obviously different pairs of sequences I, J's can give the same polynomial $F_{I,J}$ up to sign. A canonical choice can be made as follows: let C be the set of polynomials $F_{I,J}$ where $I = i_1, \dots, i_s \in$ and $J = j_1, \dots, j_s$ are sequences of distinct elements such that

$$i_1 < i_k \text{ for } k = 2, \dots, s \quad \text{and} \quad j_1 < j_2.$$

The set C is indeed a universal Gröbner basis of $I_2(X)$ and it is minimal, that is, no proper subset is a universal Gröbner basis of $I_2(X)$.

EXAMPLE 6. For a 3×3 matrix X the minimal universal Gröbner basis of $I_2(X)$ described in Remark 5 consists of the 9 minors of size 2 and by 6 binomials of degree 3:

$$\begin{aligned}
F_{123,123} &= x_{11}x_{22}x_{33} - x_{12}x_{23}x_{31}, & F_{123,132} &= x_{11}x_{23}x_{32} - x_{13}x_{22}x_{31},\\
F_{123,231} &= x_{12}x_{23}x_{31} - x_{13}x_{21}x_{32}, & F_{132,123} &= x_{11}x_{32}x_{23} - x_{12}x_{33}x_{21},\\
F_{132,132} &= x_{11}x_{33}x_{22} - x_{13}x_{32}x_{21}, & F_{132,231} &= x_{12}x_{33}x_{21} - x_{13}x_{31}x_{22}.
\end{aligned}$$

An important consequence is the following.

COROLLARY 7. *Let D be any initial ideal of $I_2(X)$. Then:*

(1) *D is radical, and*
(2) *S/D is Cohen–Macaulay.*

Assertion (1) follows immediately from Theorem 4. Statement (2) can be seen either as a special case of the main result of [**CV**] or derived from the following general result of Sturmfels [**S2**, Chapter 8].

THEOREM 8. *Let $<$ be a term order and I a toric ideal. Then $\sqrt{\mathrm{in}_<(I)}$ is the Stanley–Reisner ideal of a regular triangulation of a ball and hence it defines a Cohen–Macaulay ring.*

REMARK 9. The Betti numbers of the initial ideals of $I_2(X)$ are in general strictly larger than those of $I_2(X)$. For example, if $m, n > 2$ and $<$ a suitable term order, then the corresponding initial ideal D of $I_2(X)$ has a generator in degree 3 (the initial term of one of the binomials in Example 6) so that $\beta_{13}(S/D) > \beta_{13}(S/I_2(X)) = 0$. Already for $m = n = 4$ there is no initial ideal D of $I_2(X)$ with $\beta_{2,j}(R/D) = \beta_{2,j}(R/I_2(X))$ for all j; see Conca, Hoşten, and Thomas [**CHT**, Example 1.4]. This behavior, (i.e., no initial ideals of $I_2(X)$ have the Betti numbers of $I_2(X)$) can, at least for large m and n, be explained by a very interesting result of Dao, Huneke, and Schweig in [**DHS**]. Indeed, if $I_2(X)$ has an initial ideal D with $\beta_{2,j}(S/D) = \beta_{2,j}(S/I_2(X))$ for all j, then D is generated in degree 2 and has only linear first syzygies (because $I_2(X)$ has only linear first syzygies). For a monomial ideal D generated in degree 2 and with only linear syzygies in [**DHS**, Theorem 4.1] it is proved that the regularity $\mathrm{reg}D$ of D is bounded above by $\log_\alpha(f(mn)) + \beta$

where the numbers $\alpha > 1$ and β and the linear function f are given explicitly. It would then follow that

$$\text{reg}I_2(X) \leq \text{reg}D \leq \log_\alpha(f(mn)) + \beta,$$

which is a contradiction since $\text{reg}I_2(X) = m$ and $\log_\alpha(f(mn)) + \beta$ grows much slower than that.

The 2-minors do not form a universal Gröbner basis of $I_2(X)$ in general; nevertheless one has the following proposition.

PROPOSITION 10. *The 2-minors of X are a universal revlex Gröbner basis, i.e., they are a Gröbner basis for all the reverse lexicographic orders.*

This assertion, originally due to Restuccia and Rinaldo [**RR**], can be proved by a careful analysis of the S-pairs reduction. It can be proved also with the help of a computer algebra system because, by the Buchberger algorithm, it is enough to prove it for a 3×3 matrix where that are "only" 9! reverse lexicographic orders and this number can be drastically reduced by taking into consideration the obvious symmetries.

5.3. Universal Gröbner bases for t-minors with $2 < t < m$. A universal Gröbner basis of $I_t(X)$ is unknown for $2 < t < m$. Experiments with small values of m, n, t show that $I_t(X)$ has initial ideals that are not radical and do not define Cohen–Macaulay rings. In particular:

(1) For $t = 3$ and $m = n = 4$ the ideal $I_t(X)$ has a lexicographic initial ideal which is not radical.
(2) For $t = 3$ and $m = 4$, $n = 5$ the ideal $I_t(X)$ has a lexicographic initial ideal J such that J is not radical and both S/J and $S/\sqrt{J}$ are not Cohen–Macaulay.
(3) For $t = 4$ and $m = n = 5$ the ideal $I_t(X)$ has a reverse lexicographic initial ideal which is not radical.
(4) For $t = 4$ and $m = 5$, $n = 6$ the ideal $I_t(X)$ has a reverse lexicographic initial ideal J such that J is not radical and both S/J and $S/\sqrt{J}$ are not Cohen–Macaulay.

REMARK 11. For $t = 3$ all the reverse lexicographic initial ideals we have computed are radical. For example, for $m = n = 4$ we have found 69 distinct revlex initial ideals up to the natural $S_4 \times S_4$ and $\mathbb{Z}/2\mathbb{Z}$ symmetry and they are all square-free. It is hence possible that the ideal $I_3(X)$ has indeed a square-free universal revlex Gröbner basis. This would indeed explain, from a different perspective, a result proved by Conca and Welker [**CW**] asserting that, at least in characteristic 0, the ideal of 3-minors of the $m \times n$ matrix of variables remains radical after setting to 0 an arbitrary subset of the variables.

6. Gröbner bases of determinantal ideals via secants

Sturmfels and Sullivant published a paper [**SS**] in 2006 proposing a new and simpler approach to the study of Gröbner bases of determinantal ideals. The new insight is to use secant ideals and varieties. Let us sketch briefly the main steps. The arguments work over any base field but, in order to avoid technical complications, we assume right away that the base field is $\mathbb{C}$.

Let I, J be ideals in a polynomial ring $S = \mathbb{C}[x_1, \dots, x_n]$. The join $I * J$ of I and J is by definition the kernel of the $\mathbb{C}$-algebra map

$$S \to S/I \otimes_{\mathbb{C}} S/J$$

that sends x_i to $y_i \otimes 1 + 1 \otimes y_i$ for $i = 1, \dots, n$. Here y_i denotes the coset of x_i in the corresponding quotient ring. More concretely, $I * J$ is obtained by eliminating the variables $y = y_1, \dots, y_n$ and $z = z_1, \dots, z_n$ from the ideal

$$I(y) + J(z) + (x_i - y_i - z_i : i = 1, \dots, n)$$

where $I(y)$ is the ideal obtained from I by replacing the x's with the y's and similarly for $J(z)$.

If I and J are homogeneous ideals defining projective varieties X and Y in $\mathbb{P}^{n-1}$ then the join (ideal) $I * J$ defines the join variety $X * Y$ of X and Y, i.e. the Zariski closure of the union of the lines through the points p and q with $p \in X$ and $q \in Y$. Furthermore, $I * J$ is radical if I and J are radical and prime if I and J are prime.

More generally, given ideals $I_1, \dots, I_r$ of S one defines the join ideal $I_1 * I_2 * \cdots * I_r$ as the kernel of the map

$$S \to S/I_1 \otimes_{\mathbb{C}} S/I_2 \otimes_{\mathbb{C}} \cdots \otimes_{\mathbb{C}} S/I_r,$$

sending x_i to

$$y_i \otimes 1 \otimes \cdots \otimes 1 + 1 \otimes y_i \otimes \cdots \otimes 1 + \cdots + 1 \otimes 1 \otimes \cdots \otimes y_i.$$

The rth secant ideal of an ideal I is the join $I * \cdots * I$ (with r factors) and it will be denoted by $I^{\{r\}}$. Geometrically it defines the so-called r-secant variety, that is, the closure of the union of the linear span of r points $p_1, \dots, p_r$ that belong to the variety defined by I. The following two results, proved in [**SS**] and [**SU**], are crucial for the application to determinantal ideals.

LEMMA 12. *Let $<$ be a term order on S and I be an ideal. Then*

$$\mathrm{in}_<(I^{\{r\}}) \subseteq \mathrm{in}_<(I)^{\{r\}}$$

for all r.

LEMMA 13. *If J is the Stanley–Reisner ideal of a simplicial complex Δ then $J^{\{r\}}$ is the Stanley–Reisner ideal of the simplicial complex $\Delta^{\{r\}} = \{\cup_{i=1}^r F_i : F_i \in \Delta\}$.*

Back to the determinantal ideals $I_t(X)$ of the generic matrix $X = (x_{ij})$ and $S = \mathbb{C}[x_{ij}]$ equipped with an antidiagonal term order. Let J_t be the ideal generated by the antidiagonals of the t-minors of X. By construction we have

$$J_t \subseteq \mathrm{in}(I_t(X)). \tag{6.1}$$

Furthermore, one has

$$I_t(X) = I_2(X)^{\{t-1\}} \tag{6.2}$$

for $2 \le t \le m$. This equality can be proved by purely algebraic arguments but follows also from the well-known fact that any matrix of rank $\le t-1$ is the sum of at most $t-1$ matrices of rank ≤ 1.

Let $P = [m] \times [n]$ with poset structure defined by $(i, j) \le (h, k)$ if $i \le h$ and $j \le k$.

By construction J_t is generated by the antichains (pairwise incomparable elements) of P of cardinality t. In particular, J_2 is the Stanley–Reisner ideal of the order complex of P, i.e., the simplicial complex Δ_2 whose elements are the chains (totally ordered subsets) in P. Denote by Δ_t the simplicial complex associated to J_t. Since the union of $(t-1)$ chains cannot contain an antichain of cardinality t and, because of Lemma 13, we have $\Delta_2^{\{t-1\}} \subseteq \Delta_t$. On the other hand, in this case it is easy to see that a subset of P that does not contain an antichain of cardinality t is actually the union of at most $t-1$ chains. This is indeed what Dilworth's theorem states in general for posets: the size of the largest antichain of any poset equals the minimal number of chains needed to partition the poset. So we have $\Delta_2^{\{t-1\}} = \Delta_t$ or, equivalently,

$$J_t = J_2^{\{t-1\}}. \tag{6.3}$$

The final observation is that for an antidiagonal term order the 2-minors of X form a Gröbner basis, i.e.,

$$\text{in}(I_2(X)) = J_2. \tag{6.4}$$

This can be checked easily using the Buchberger algorithm or using the toric presentation or (for the lazy) using a computer algebra system as the "action takes place" in a 3×3 matrix at most.

Summing up, if $<$ is an antidiagonal term oder on S, we have

$$J_t \underset{(6.1)}{\subseteq} \text{in}(I_t(X)) \underset{(6.2)}{=} \text{in}(I_2(X)^{\{t-1\}}) \underset{12}{\subseteq} \text{in}(I_2(X))^{\{t-1\}} \underset{(6.4)}{=} J_2^{\{t-1\}} \underset{(6.3)}{=} J_t.$$

Therefore one has equality throughout and, in particular, in (6.1). This proves Theorem 1 for an antidiagonal order.

7. Gröbner bases via degrees and multidegrees

Another approach to Theorem 1 is presented in the book of Miller and Sturmfels [**MS**] and it is based on ideas developed in the paper of Knutson and Miller [**KM**]. To explain the main ideas, let us first recall the following well-known lemma. Let $S = K[x_1, \dots, x_n]$ and I be an homogeneous ideal. We denote by $\dim S/I$ the Krull dimension of S/I, by $\deg S/I$ its degree and by $P_{S/I}(x)$ its Hilbert polynomial. With $e = \deg S/I$ and $d = \dim S/I$ one has

$$P_{S/I}(x) = \frac{e}{(d-1)!} x^{d-1} + \text{ lower degree terms.}$$

Hence the knowledge of $\dim S/I$ and $\deg S/I$ is equivalent to the knowledge of the largest degree component of $P_{S/I}(x)$.

LEMMA 14. *Let I and J be homogeneous ideals in S. Assume that $J \subseteq I$, $\dim S/J = \dim S/I$, $\deg S/J = \deg S/I$ and J is pure (i.e., all its associated primes have the same dimension). Then $I = J$.*

Said otherwise, given an inclusion of homogeneous ideals $J \subseteq I$ to prove equality one can try to prove the two ideals have the same Hilbert functions. But if J is pure it is enough to prove that the highest degree component of the Hilbert polynomials of S/J and S/I agree. Lemma 14 follows by simple considerations on the primary decomposition and the additivity of the degree.

REMARK 15. Now suppose we have a homogeneous ideal I and we want to prove that $f_1, \dots, f_s \in I$ are indeed a Gröbner basis of I with respect to a term order. We look at $J = (\text{in}(f_1), \dots, \text{in}(f_s))$. We have inclusion $J \subseteq \text{in}(I)$ and we want to prove it is an equality. In view of Lemma 14 if J is pure we just need to prove that:

(1) $\dim S/J = \dim S/I$ and
(2) $\deg S/J = \deg S/I$

because then Lemma 14 applied to J and $\text{in}(I)$ implies that $J = \text{in}(I)$. Here we have used the fact that $\dim S/I = \dim S/\text{in}(I)$ and $\deg S/I = \deg S/\text{in}(I)$.

In the determinantal setting, with $S = K[x_{ij} : (i,j) \in [m] \times [n]]$ we consider the ideal of minors $I_t(X)$ and a diagonal or antidiagonal term order. We apply the scheme of reasoning of Remark 15 to the ideal $I_t(X)$ to prove Theorem 1. We take J_t to be the ideal generated by the diagonals (or antidiagonals) of length t in X. The facets of the simplicial complex Δ_t associated to J_t can be described as families of nonintersecting paths in the grid $[m] \times [n]$; see [**HT**] for details. From this description it follows that J_t is pure and $\dim S/J_t = (m+n-t+1)(t-1)$. Furthermore, the degree of S/J_t equals the number of facets of Δ_t. The latter can be computed using the Gessel–Viennot determinantal formula and one obtains

$$\deg S/J_t = \det\left(\binom{m-i+n-j}{m-i}\right)_{i,j=1,\dots,t-1}. \tag{7.1}$$

Now the dimension of $S/I_t(X)$ is easily seen to be $(m+n-t+1)(t-1)$ and its degree was computed by various authors (not using Gröbner bases); see, for example, [**HaT**, **ACGH**]. Therefore, one can check that degree and dimension of S/J_t and $S/I_t(X)$ actually agree. Some efforts are needed to check that the formulas given in [**HaT**, **ACGH**] agree with (7.1) but usually one gets along by using Laplace expansions and Vandermonde determinants. Summing up, one obtains a proof of Theorem 1 using Remark 15.

A critic to the argument just given is that it uses information about determinantal ideals (the formula for the degree) to compute a Gröbner basis, while, in many cases, e.g., [**HT**], the goal is the opposite, i.e., to compute Gröbner bases in order to derive information about the determinantal ideals as, for example, their degree and Cohen–Macaulayness.

The approach proposed by Knutson and Miller [**KM**] and Miller and Sturmfels [**MS**] to Theorem 1 does not use already known results on the determinantal rings and has the advantage to work simultaneously for a large class of determinantal ideals. Another important point is that a larger class is more suitable to inductive processes. The same type of strategy is used, under the name of principal radical system, by Hochster and Eagon [**HE**] to prove the Cohen–Macaulay property of determinantal rings.

In practice one identifies a carefully chosen class of ideals, say, $\{I_w : w \in \Omega\}$ where Ω is a set (or a poset), such that I_w is generated by minors of X of possibly different sizes (that depend on w). Furthermore, $I_t(X)$ is among the I_w's. For each $w \in \Omega$ one identifies a somehow canonical set of generators of I_w whose initial terms with respect to an appropriate term order generates a monomial ideal J_w. To prove that canonical set of generators of I_w is a Gröbner basis using Remark 15 one has to prove the following.

LEMMA 16. *For every $w \in \Omega$ the ideal J_w is pure and*

(1) $\dim S/J_w = \dim S/I_w$,
(2) $\deg S/J_w = \deg S/I_w$.

As the minors of the matrix X are naturally $\mathbb{Z}^m \times \mathbb{Z}^n$ multigraded both S/I_w and S/J_w acquire a multigraded structure and hence have multigraded Hilbert polynomials and multidegrees. Indeed, each finitely generated $\mathbb{Z}^m \times \mathbb{Z}^n$-module M has an associated multidegree, denoted by $\mathrm{Deg}(M)$, which itself a homogeneous polynomial whose degree is the codimension of M, see [**MS**, 8.5] for details. The polynomial $\mathrm{Deg}(M)$ captures both the codimension and, in finer form, the degrees of M in the various directions. Summing up, to prove Lemma 16 it is enough to prove the following lemma.

LEMMA 17. *For every $w \in \Omega$ the ideal J_w is pure and*

$$\mathrm{Deg}(S/J_w) = \mathrm{Deg}(S/I_w).$$

The ideal J_w is a generated by square-free monomials and the associated simplicial complex Δ_w has facets that can be described. In this way one checks that J_w is pure. Finally, Lemma 17 is proved, roughly speaking, by showing that both $\mathrm{Deg}(S/J_w)$ and $\mathrm{Deg}(S/I_w)$ satisfy the same recursion involving multidegrees of ideals of the same type for "smaller" values of w. This is usually a hard task that requires mastery in identifying and describing the "right" combinatorial gadgets (e.g., Schubert polynomials and pipe dreams in [**MS**]). The final step is to check the statement for the initial values of w which is usually easy.

Yet another approach to Theorem 1, similar in spirit and inductive in nature, makes use of the notion of linkage and has been applied by Gorla, Nagel, and Migliore to various classes of determinantal rings in [**GMN**].

8. Conclusion

The search of Gröbner bases of determinantal ideals has been a very active research field in the last thirty years. Many important results have been achieved, many new techniques and ideas have been developed. In this process the contribution given by Bernd Sturmfels has been fundamental.

References

[ACGH] E. Arbarello, M. Cornalba, P. A. Griffiths, and J. Harris, *Geometry of algebraic curves. Vol. I*, Grundlehren der mathematischen Wissenschaften [Fundamental Principles of Mathematical Sciences], vol. 267, Springer-Verlag, New York, 1985, DOI 10.1007/978-1-4757-5323-3. MR770932

[BZ] D. Bernstein and A. Zelevinsky, *Combinatorics of maximal minors*, J. Algebraic Combin. **2** (1993), no. 2, 111–121, DOI 10.1023/A:1022492222930. MR1229427

[Bj] A. Björner, *Shellable and Cohen-Macaulay partially ordered sets*, Trans. Amer. Math. Soc. **260** (1980), no. 1, 159–183, DOI 10.2307/1999881. MR570784

[Bo] A. Boocher, *Free resolutions and sparse determinantal ideals*, Math. Res. Lett. **19** (2012), no. 4, 805–821, DOI 10.4310/MRL.2012.v19.n4.a6. MR3008416

[BC1] W. Bruns and A. Conca, *KRS and powers of determinantal ideals*, Compositio Math. **111** (1998), no. 1, 111–122, DOI 10.1023/A:1000287107308. MR1611055

[BC2] W. Bruns and A. Conca, *KRS and determinantal ideals*, Geometric and combinatorial aspects of commutative algebra (Messina, 1999), Lecture Notes in Pure and Appl. Math., vol. 217, Dekker, New York, 2001, pp. 67–87. MR1824218

[BC3] W. Bruns and A. Conca, *Gröbner bases and determinantal ideals*, Commutative algebra, singularities and computer algebra (Sinaia, 2002), NATO Sci. Ser. II Math. Phys. Chem., vol. 115, Kluwer Acad. Publ., Dordrecht, 2003, pp. 9–66. MR2030262

[BCRV] W. Bruns, A. Conca, C. Raicu, and M. Varbaro, *Determinants, Gröbner bases and cohomology*, Springer Monographs in Mathematics, Springer, Cham, [2022] ©2022, DOI 10.1007/978-3-031-05480-8. MR4627943

[BG] W. Bruns and J. Gubeladze, *Polytopes, rings, and K-theory*, Springer Monographs in Mathematics, Springer, Dordrecht, 2009, DOI 10.1007/b105283. MR2508056

[BV] W. Bruns and U. Vetter, *Determinantal rings*, Lecture Notes in Mathematics, vol. 1327, Springer-Verlag, Berlin, 1988, DOI 10.1007/BFb0080378. MR953963

[CCC] A. Caminata, Y. Cid-Ruiz, and A. Conca, *Multidegrees, prime ideals, and non-standard gradings*. part A, Adv. Math. **435** (2023), no. part A, Paper No. 109361, 46, DOI 10.1016/j.aim.2023.109361. MR4658828

[CS] D. Cartwright and B. Sturmfels, *The Hilbert scheme of the diagonal in a product of projective spaces*, Int. Math. Res. Not. IMRN **9** (2010), 1741–1771, DOI 10.1093/imrn/rnp201. MR2643580

[CGG] L. Caniglia, J. A. Guccione, and J. J. Guccione, *Ideals of generic minors*, Comm. Algebra **18** (1990), no. 8, 2633–2640, DOI 10.1080/00927879008824043. MR1074248

[C94] A. Conca, *Gröbner bases of ideals of minors of a symmetric matrix*, J. Algebra **166** (1994), no. 2, 406–421, DOI 10.1006/jabr.1994.1160. MR1279266

[C98] A. Conca, *Straightening law and powers of determinantal ideals of Hankel matrices*, Adv. Math. **138** (1998), no. 2, 263–292, DOI 10.1006/aima.1998.1740. MR1645574

[C07] A. Conca, *Linear spaces, transversal polymatroids and ASL domains*, J. Algebraic Combin. **25** (2007), no. 1, 25–41, DOI 10.1007/s10801-006-0026-3. MR2284146

[CHT] A. Conca, S. Hoşten, and R. R. Thomas, *Nice initial complexes of some classical ideals*, Algebraic and geometric combinatorics, Contemp. Math., vol. 423, Amer. Math. Soc., Providence, RI, 2006, pp. 11–42, DOI 10.1090/conm/423/08073. MR2298753

[CDG1] A. Conca, E. De Negri, and E. Gorla, *Universal Gröbner bases for maximal minors*, Int. Math. Res. Not. IMRN **11** (2015), 3245–3262, DOI 10.1093/imrn/rnu032. MR3373050

[CDG2] A. Conca, E. De Negri, and E. Gorla, *Multigraded generic initial ideals of determinantal ideals*, Homological and computational methods in commutative algebra, Springer INdAM Ser., vol. 20, Springer, Cham, 2017, pp. 81–96. MR3751880

[CDG3] A. Conca, E. De Negri, and E. Gorla, *Cartwright-Sturmfels ideals associated to graphs and linear spaces*, J. Comb. Algebra **2** (2018), no. 3, 231–257, DOI 10.4171/JCA/2-3-2. MR3845718

[CDG4] A. Conca, E. De Negri, and E. Gorla, *Universal Gröbner bases and Cartwright-Sturmfels ideals*, Int. Math. Res. Not. IMRN **7** (2020), 1979–1991, DOI 10.1093/imrn/rny075. MR4089441

[CDG5] A. Conca, E. De Negri, and E. Gorla, *Radical generic initial ideals*, Vietnam J. Math. **50** (2022), no. 3, 807–827, DOI 10.1007/s10013-022-00551-w. MR4447407

[CDG6] A. Conca, E. De Negri, and E. Gorla, *Radical support for multigraded ideals*, São Paulo J. Math. Sci. **17** (2023), no. 1, 64–75, DOI 10.1007/s40863-022-00318-8. MR4603899

[CW] A. Conca and V. Welker, *Lovász-Saks-Schrijver ideals and coordinate sections of determinantal varieties*, Algebra Number Theory **13** (2019), no. 2, 455–484, DOI 10.2140/ant.2019.13.455. MR3927052

[CV] A. Conca and M. Varbaro, *Square-free Gröbner degenerations*, Invent. Math. **221** (2020), no. 3, 713–730, DOI 10.1007/s00222-020-00958-7. MR4132955

[CLS] D. A. Cox, J. B. Little, and H. K. Schenck, *Toric varieties*, Graduate Studies in Mathematics, vol. 124, American Mathematical Society, Providence, RI, 2011, DOI 10.1090/gsm/124. MR2810322

[DHS] H. Dao, C. Huneke, and J. Schweig, *Bounds on the regularity and projective dimension of ideals associated to graphs*, J. Algebraic Combin. **38** (2013), no. 1, 37–55, DOI 10.1007/s10801-012-0391-z. MR3070118

[DEP] C. De Concini, D. Eisenbud, and C. Procesi, *Hodge algebras*, Astérisque, vol. 91, Société Mathématique de France, Paris, 1982. With a French summary. MR680936

[E] D. Eisenbud, *Introduction to algebras with straightening laws*, Ring theory and algebra, III (Proc. Third Conf., Univ. Oklahoma, Norman, Okla., 1979), Lect. Notes Pure Appl. Math., vol. 55, Dekker, New York, 1980, pp. 243–268. MR584614

[GMN] E. Gorla, J. C. Migliore, and U. Nagel, *Gröbner bases via linkage*, J. Algebra **384** (2013), 110–134, DOI 10.1016/j.jalgebra.2013.02.017. MR3045153

[HaT] J. Harris and L. W. Tu, *On symmetric and skew-symmetric determinantal varieties*, Topology **23** (1984), no. 1, 71–84, DOI 10.1016/0040-9383(84)90026-0. MR721453

[HT] J. Herzog and N. V. Trung, *Gröbner bases and multiplicity of determinantal and Pfaffian ideals*, Adv. Math. **96** (1992), no. 1, 1–37, DOI 10.1016/0001-8708(92)90050-U. MR1185786

[HE] M. Hochster and J. A. Eagon, *Cohen-Macaulay rings, invariant theory, and the generic perfection of determinantal loci*, Amer. J. Math. **93** (1971), 1020–1058, DOI 10.2307/2373744. MR302643

[HR] M. Hochster and J. L. Roberts, *Rings of invariants of reductive groups acting on regular rings are Cohen-Macaulay*, Advances in Math. **13** (1974), 115–175, DOI 10.1016/0001-8708(74)90067-X. MR347810

[HH] M. Hochster and C. Huneke, *Tight closure, invariant theory, and the Briançon-Skoda theorem*, J. Amer. Math. Soc. **3** (1990), no. 1, 31–116, DOI 10.2307/1990984. MR1017784

[K] M. Yu. Kalinin, *Universal and comprehensive Gröbner bases of the classical determinantal ideal* (Russian, with English and Russian summaries), Zap. Nauchn. Sem. S.-Peterburg. Otdel. Mat. Inst. Steklov. (POMI) **373** (2009), no. Teoriya Predstavleniĭ, Dinamicheskie Sistemy, Kombinatornye Metody. XVII, 134–143, 348, DOI 10.1007/s10958-010-9990-1; English transl., J. Math. Sci. (N.Y.) **168** (2010), no. 3, 385–389. MR2749259

[KM] A. Knutson and E. Miller, *Gröbner geometry of Schubert polynomials*, Ann. of Math. (2) **161** (2005), no. 3, 1245–1318, DOI 10.4007/annals.2005.161.1245. MR2180402

[MS] E. Miller and B. Sturmfels, *Combinatorial commutative algebra*, Graduate Texts in Mathematics, vol. 227, Springer-Verlag, New York, 2005. MR2110098

[MuS] C. Musili and C. S. Seshadri, *Standard monomial theory*, Paul Dubreil and Marie-Paule Malliavin Algebra Seminar, 33rd Year (Paris, 1980), Lecture Notes in Math., vol. 867, Springer, Berlin, 1981, pp. 441–476. MR633532

[RR] G. Restuccia and G. Rinaldo, *On certain classes of degree reverse lexicographic Gröbner bases*, Int. Math. Forum **2** (2007), no. 21-24, 1053–1068, DOI 10.12988/imf.2007.07092. MR2334020

[N] H. Narasimhan, *The irreducibility of ladder determinantal varieties*, J. Algebra **102** (1986), no. 1, 162–185, DOI 10.1016/0021-8693(86)90134-1. MR853237

[Sc] C. Schensted, *Longest increasing and decreasing subsequences*, Canadian J. Math. **13** (1961), 179–191, DOI 10.4153/CJM-1961-015-3. MR121305

[SU] A. Simis and B. Ulrich, *On the ideal of an embedded join*, J. Algebra **226** (2000), no. 1, 1–14, DOI 10.1006/jabr.1999.8091. MR1749874

[S] B. Sturmfels, *Gröbner bases and Stanley decompositions of determinantal rings*, Math. Z. **205** (1990), no. 1, 137–144, DOI 10.1007/BF02571229. MR1069489

[S2] B. Sturmfels, *Gröbner bases and convex polytopes*, University Lecture Series, vol. 8, American Mathematical Society, Providence, RI, 1996, DOI 10.1090/ulect/008. MR1363949

[SZ] B. Sturmfels and A. Zelevinsky, *Maximal minors and their leading terms*, Adv. Math. **98** (1993), no. 1, 65–112, DOI 10.1006/aima.1993.1013. MR1212627

[SS] B. Sturmfels and S. Sullivant, *Combinatorial secant varieties*, Pure Appl. Math. Q. **2** (2006), no. 3, Special Issue: In honor of Robert D. MacPherson., 867–891, DOI 10.4310/PAMQ.2006.v2.n3.a12. MR2252121

[V] R. H. Villarreal, *Monomial algebras*, Monographs and Textbooks in Pure and Applied Mathematics, vol. 238, Marcel Dekker, Inc., New York, 2001. MR1800904

Dipartimento di Matematica, Dipartimento di Eccellenza 2023-2027, Universitá di Genova, Italy

Email address: `conca@dima.unige.it`

Proceedings of Symposia in Pure Mathematics
Volume **111**, 2025
https://doi.org/10.1090/pspum/111/02039

An early history of toric ideals

Serkan Hoşten

ABSTRACT. Toric ideals are everywhere. They have been in the commutative algebra lexicon since about 1990 when Bernd Sturmfels used the term. The early days of toric ideals and their Gröbner bases were full of new results and promising developments in their applications. Bernd has been consistently their biggest promoter through his own work and that of his collaborators and students. This article is a personal and subjective recalling of the first decade of toric ideals when Bernd played a central role.

1. Introduction

I arrived to the United States as a graduate student at Cornell University in August 1992. I came to study operations research, in particular, discrete optimization such as integer linear programming and network flow algorithms. As the typical first year in the PhD program goes, I took foundational courses in optimization (including a course taught by Mike Todd in interior point methods for linear programming that was then the rage) and other core courses in stochastic processes and statistics. At the time, each first year student in the OR Department had to thoroughly read and then present to a committee of three faculty members a published research article of their choice. I chose Hendrik Lenstra's paper on solving integer programs in polynomial time in fixed dimension [**35**]. I was certain that I would work on algorithms and computation for integer programs (cutting planes, branch-and-bound, and all that) with Les Trotter.

At Cornell, besides the major subject, one had to also take courses in two minor subjects. I knew one of my minor subjects would be computer science. What should have the other one been? In my first semester, I took a course in advanced linear algebra from the math department that was taught by Keith Dennis. I almost flunked it but I really liked Keith's teaching style, so when I found out he would be teaching an honors course in abstract algebra (groups, rings, etc.) I decided to take this course. Since my undergrad education was in engineering, I had no idea what the course would be about but I signed up anyway. And I *loved* it (and did very well). The natural choice for the second minor topic was emerging to be mathematics. What other courses would have been good to take then? I knew I

2020 *Mathematics Subject Classification.* Primary 14M25, 14Q15, 13P10, 13P25.
Key words and phrases. toric ideals, Gröbner bases, initial ideals, integer programming, regular triangulation, Gröbner fan.

would do the year-long graduate algebra sequence (to be taught by Mark Gross), but what else was out there?

Rekha Thomas was also a graduate student in the OR Department who was couple years ahead of me. My memory is dark about this point, but maybe I heard Rekha giving a talk on her work about Gröbner bases of toric ideals and solving integer programs or I simply struck up a conversation with her about the same topic. We had a long and fascinating discussion. She told me in detail what she was doing and it looked beautiful to me. It looked like one can do both optimization and algebra at the same time. A perfect fit! She also told me that in Fall 1993 her advisor, a certain Bernd Sturmfels, would be teaching a course on Gröbner bases. It was clear that I had to take this course and also learn about toric ideals.

2. Toric ideals and their Gröbner bases

Bernd used the first edition of *Ideals, Varieties, and Algorithms* [**8**] that was published just the previous year. Agnes Szanto was also in the course and Hal Schenck was the grader. In fact, Bernd was also teaching a course on toric varieties during the same semester. I knew zero algebraic geometry so I stuck with Gröbner bases.

Bernd had been thinking and publishing about Gröbner bases since 1989. His paper co-authored with Neil White on invariant theory and Gröbner bases [**47**] is seminal. Equally important is his solo paper from 1990 where he proved the following famous result. See also the article by Aldo Conca in this volume.

THEOREM 2.1 ([**42**, Theorem 1]). *The set of r-minors of a generic $m \times n$ matrix (x_{ij}) is a reduced Gröbner basis of the ideal they generate with respect to the lexicographic order induced by*

$$x_{1,n} \succ x_{1,n-1} \succ \cdots \succ x_{1,1} \succ x_{2,n} \succ x_{2,n-1} \succ \cdots \succ x_{2,1} \\ \succ \cdots \succ x_{m,n} \succ x_{m,n-1} \succ \cdots \succ x_{m,1}.$$

This brings us to the paper that started all regarding toric ideals and their Gröbner bases, namely *Gröbner bases of toric varieties* [**43**]. Although published in 1991, it was submitted in March 1990. This means that Bernd must have known the contents of the article sometime in late 1989 or early 1990. The only reason this fundamental article has only 38 citations on the MathSciNet is because everything in it and a lot more appears in Bernd's book *Gröbner bases and convex polytopes* [**44**] (949 citations in January 2025).

DEFINITION 2.2. Let $A = [a_1\, a_2\, \cdots\, a_n]$ be a $d \times n$ matrix with integer entries and $\text{rank}(A) = d$. Consider the $\mathbb{C}$-algebra homomorphism

$$\phi_A \colon \mathbb{C}[x_1, \ldots, x_n] \longrightarrow \mathbb{C}[t_1^{\pm}, t_2^{\pm}, \ldots, t_d^{\pm}] \qquad \phi_A(x_i) = t_1^{a_{1i}} t_2^{a_{2i}} \cdots t_d^{a_{di}}.$$

The toric ideal I_A is the kernel of ϕ_A.

REMARK 2.3. The above is the most general definition of a toric ideal. Typically, we want cone(A), the rational polyhedral cone generated by $a_1, \ldots, a_n$, to be pointed. This guarantees that the toric ideal I_A does not contain binomials of the form $x^u - 1$. Furthermore, if we assume that the monoid cone$(A) \cap \mathbb{N}^d$ is generated by $a_1, \ldots, a_n$, i.e., it is equal to $\mathbb{N}(A) := \{u_1 a_1 + \cdots + u_n a_n \colon u_1, \ldots, u_n \in \mathbb{N}\}$, then I_A is the defining ideal of the affine (normal) toric variety X_A . Finally, if we stipulate that the first row of A is $[1\, 1\, \cdots\, 1]$, or more generally, the row span of A

contains this vector, then I_A is a homogeneous ideal defining the projective toric variety X_A in $\mathbb{P}^{n-1}$; see [**9, 44**]. In the rest of this article I will stay in this last setting as Bernd did in [**43**].

THEOREM 2.4 ([**43**, Lemmas 2.2 and 2.5]). *I_A is a prime ideal generated by the binomials*

$$\{x^u - x^v : u - v \in \ker(A) \cap \mathbb{Z}^n\}.$$

Any reduced Gröbner basis of I_A also consists of binomials.

EXAMPLE 2.5. Let

$$A = [e_i \oplus e_j \oplus e_k : 1 \leq i, j, k \leq 3],$$

where e_1, e_2, e_3 form the standard unit basis of $\mathbb{R}^3$. The corresponding monomial map is

$$\phi_A : \mathbb{C}[x_{111}, \ldots, x_{333}] \longrightarrow \mathbb{C}[a_1, a_2, a_3, b_1, b_2, b_3, c_1, c_2, c_3]$$

where $\phi_A(x_{ijk}) = a_i b_j c_k$, $i, j, k = 1, 2, 3$. The toric ideal I_A is the defining ideal of the Segre embedding of $\mathbb{P}^2 \times \mathbb{P}^2 \times \mathbb{P}^2$ in $\mathbb{P}^{26}$. It is minimally generated by 162 quadratic binomials. They form the reduced Gröbner of I_A with respect to the degree reverse lexicographic order induced by

$$x_{111} \succ x_{112} \succ x_{113} \succ \cdots \succ x_{133} \succ x_{211} \succ x_{212} \succ x_{213} \succ \cdots \succ x_{233}$$
$$\succ x_{311} \succ x_{312} \succ x_{313} \succ \cdots \succ x_{333}.$$

They are the union of the 2-minors of the following two matrices:

$$\begin{bmatrix} x_{111} & x_{112} & x_{113} & x_{121} & x_{122} & x_{123} & x_{131} & x_{132} & x_{133} \\ x_{211} & x_{212} & x_{213} & x_{221} & x_{222} & x_{223} & x_{231} & x_{232} & x_{233} \\ x_{311} & x_{312} & x_{313} & x_{321} & x_{322} & x_{323} & x_{331} & x_{332} & x_{133} \end{bmatrix},$$

$$\begin{bmatrix} x_{111} & x_{121} & x_{131} & x_{211} & x_{221} & x_{231} & x_{311} & x_{321} & x_{331} \\ x_{112} & x_{122} & x_{132} & x_{212} & x_{222} & x_{232} & x_{312} & x_{322} & x_{332} \\ x_{113} & x_{123} & x_{133} & x_{213} & x_{223} & x_{233} & x_{313} & x_{323} & x_{333} \end{bmatrix}.$$

We note that these matrices are two particular flattenings of the generic $3 \times 3 \times 3$ tensor (x_{ijk}).

Next, Bernd proved a bound on the degree of any reduced Gröbner basis element of I_A. This bound is singly exponential in the input data, namely, in d, n, and the size of entries of A. It contrasts with a previous famous result which had shown that a binomial (although not toric) ideal could have reduced Gröbner basis elements with degrees doubly exponential in the degrees of the input binomials [**37**]. To state and prove the result, I introduce a few things which will be useful later on as well. For each sign pattern $\rho \in \{+1, -1\}^n$ we let $\mathbb{R}^n_\rho := \{x \in \mathbb{R}^n : \rho_i x_i \geq 0,\ i = 1, \ldots, n\}$ and $H_\rho := \mathbb{R}^n_\rho \cap \ker_{\mathbb{Z}}(A)$. The latter is a pointed rational polyhedral cone, and it is generated by circuits of A that are in H_ρ. A circuit of A is a nonzero vector $u \in \ker_{\mathbb{Z}}(A)$ with relatively prime entries and with minimal support. The cardinality of the support of any circuit is at most $d+1$. If $\{i_1, \ldots, i_{d+1}\} \subset [n]$ contains the support of a circuit u then

$$u_{i_j} = \pm \frac{1}{c} \det(a_{i_1}, \ldots, a_{i_{j-1}}, a_{i_{j+1}}, \ldots, a_{i_{d+1}}) \tag{2.1}$$

where c is the gcd of all $(d+1)$ determinants appearing in the above formula. The following degree bound theorem is a slight improvement of the one that is in [**43**] and appears in [**44**].

THEOREM 2.6 ([**43**, Theorem 2.3] and [**44**, Theorem 4.7]). *Let $D(A)$ be the maximum of the absolute value of all d-minors of A. The degree of any reduced Gröbner basis element of I_A is at most $(n-d)(d+1)D(A)$.*

PROOF. Let $x^{v_+} - x^{v_-}$ be a reduced Gröbner basis element of I_A where $v = v_+ - v_- \in \ker_{\mathbb{Z}}(A)$. Then v is in one of H_ρ. Since the dimension of H_ρ is $n-d$, Carathéodory's theorem implies that v is a conical combination of at most $n-d$ circuits in H_ρ:

$$v = \lambda_1 u_1 + \cdots + \lambda_{n-d} u_{n-d}.$$

We claim each $\lambda_i \leq 1$. Otherwise, let $w = \sum_{i=1}^{n-d} (\lambda_i - \lfloor \lambda_i \rfloor) u_i$. We see that $v = w + (v-w)$ and observe that both w and $v-w$ are in H_ρ. But then if we write $w = w_+ - w_-$, then $x^{w_+} - x^{w_-}$ is in I_A, and x^{w_+} divides x^{v_+} and x^{w_-} divides x^{v_-}. This is a contradiction to our assumption that $x^{v_+} - x^{v_-}$ is in a reduced Gröbner basis. Since any circuit has degree at most $(d+1)D(A)$ we conclude that the degree of our binomial is at most $(n-d)(d+1)D(A)$. □

There are two other important results in [**43**] which spurred subsequent work. One is the construction of the *state polytope* of I_A. The other one is the characterization of the radical of any initial ideal of I_A. I briefly present these results.

Let $\{x^{u_1} - x^{v_1}, \ldots, x^{u_k} - x^{v_k}\}$ be a reduced Gröbner basis of I_A where we assume that the first term in each binomial is the initial term. We consider the *Gröbner cone* $\mathcal{C} = \{\omega \in \mathbb{R}^n : \omega \cdot u_i \geq \omega \cdot v_i, i = 1, \ldots, k\}$. It turns out that although there are infinitely many term orders in a polynomial ring with at least two variables, every ideal has finitely many distinct reduced Gröbner bases [**44**, Theorem 1.2]. Moreover, the collection of all Gröbner cones $\mathcal{C}$, one for each reduced Gröbner basis of I_A, is a complete polyhedral fan, called the *Gröbner fan* $\mathcal{G}_A$. The existence of Gröbner fans for any homogeneous polynomial ideal had been proved [**4, 38**]. Moreover, these fans are polytopal. In other words, there exists a polytope $\mathcal{S}_A$, called the *state polytope* of I_A whose normal fan equals the Gröbner fan $\mathcal{G}_A$ [**4**].

THEOREM 2.7 ([**43**, Proposition 2.7]). *Let E_A be the zonotope generated by all circuits of A. The state polytope $\mathcal{S}_A$ of the toric ideal I_A is a Minkowski summand of the zonotope $\sum\{[u_+, u_-] : u \in E_A \cap \mathbb{Z}^n\}$.*

For the second result, recall that a monomial ideal I is radical if and only if its minimal generators are squarefree. Moreover, there is a bijection between the radical monomial ideals in $\mathbb{C}[x_1, \ldots, x_n]$ and simplicial complexes on n vertices: given such a simplicial complex Δ we can construct the *Stanley–Reisner ideal*

$$I_\Delta := \bigcap_{\sigma \in \Delta} \langle x_i \,:\, i \notin \sigma \rangle$$

where the intersection runs over all facets of Δ. This can be reversed by computing the irredundant prime decomposition of the radical monomial ideal I. Given the matrix A, every sufficiently generic $\omega \in \mathbb{R}^n$ gives rise to a triangulation Δ_ω of $\mathrm{cone}(A)$ into simplicial cones. We think of Δ_ω as a simplicial complex on n vertices where we declare a subset $\sigma \subset [n]$ a face of Δ_ω if there is a vector $y = (y_1, \ldots, y_d) \in \mathbb{R}^d$ such that $a_i \cdot y = \omega_i$ if $i \in \sigma$ and $a_i \cdot y < \omega_i$ if $i \notin \sigma$. The triangulation Δ_ω is called the *regular triangulation* of A induced by ω; see [**13, 44**].

THEOREM 2.8 ([**43**, Theorem 3.1]). *Let ω be a vector in the interior of a Gröbner cone $\mathcal{C}$ of the toric ideal I_A. The radical of the corresponding initial ideal is the Stanley–Reisner ideal I_{Δ_ω} of the regular triangulation Δ_ω.*

It is hard not to call this paper phenomenal. In the meantime, I was continuing to learn from Cox, Little, and O'Shea [**8**] as well as from Bernd and Rekha.

3. Integer programming

I continue using the $d \times n$ integer matrix A under the assumptions of Remark 2.3. Now consider the discrete optimization problem $IP_{A,\omega}(b)$

$$\text{minimize}\, \omega \cdot x \text{ subject to } Ax = b \text{ and } x \in \mathbb{N}^n$$

where $b \in \mathbb{N}(A)$. Solving such a general integer program is NP-hard, and already by the early 1990s there had been substantial amount of work accumulated on all sorts of methods (but mostly polyhedral) to solve these hard problems; see [**41**] from those times which is still a crucial reference source. In what follows, the problem gets even harder since we will attempt to solve the above problem for all possible right-hand side vectors b while fixing ω. I will denote this family of integer programs as b varies by $IP_{A,\omega}$.

A method to accomplish this using Gröbner bases of the toric ideal I_A had been outlined in [**7**]:

(1) Form the ideal
$$J_A = \langle x_1 - t^{a_1}, x_2 - t^{a_2}, \ldots, x_n - t^{a_n} \rangle \subset \mathbb{C}[t_1, \ldots, t_d, x_1, \ldots, x_n].$$
(2) Compute the reduced Gröbner basis $\tilde{\mathcal{G}}_{A,\succ}$ of J_A with respect to the elimination term order where $t_j \succ x_i$ for every pair of $j = 1, \ldots, d$ and $i = 1, \ldots, n$, and where $x^u \succ x^v$ if $\omega \cdot u > \omega \cdot v$. In the last case if equality occurs use a lexicographic term order to break the tie $x^u \succ x^v$.
(3) Compute the normal form x^v of the monomial t^b with respect $\tilde{\mathcal{G}}_{A,\succ}$. The vector v is the optimal solution.

Note that, technically, we are not using a reduced Gröbner basis of I_A but the reduced Gröbner basis $\tilde{\mathcal{G}}_{A,\succ}$ of the (toric!) ideal J_A. However, the results of the elimination theory imply that $\mathcal{G}_{A,\succ} = \tilde{\mathcal{G}}_{A,\succ} \cap \mathbb{C}[x_1, \ldots, x_n]$ is a reduced Gröbner basis of I_A. Also note that once $\mathcal{G}_{A,\succ}$ is computed, one can determine $u \in \mathbb{N}^n$ such that $Au = b$ (still an NP-hard problem) and compute the normal form of x^u with respect to $\mathcal{G}_{A,\succ}$.

Here came Rekha's first fundamental contribution to the subject. The paper that would become [**51**] was completed in early 1994 and I knew its contents in the fall of 1993. She made two observations which tied the Gröbner basis theory of toric ideals to integer programming. First, she saw that $\mathcal{G}_{A,\succ}$ forms the unique minimal *test set* for our integer programs as b varies [**51**, Corollary 2.1.10]. A test set is any collection $\mathcal{T} \subset \ker_{\mathbb{Z}}(A)$ such that

(a) for any nonoptimal solution v for any integer program $IP_{A,\omega}(b)$ in the family $IP_{A,\omega}$ there exists $w \in \mathcal{T}$ such that $x^v \succ x^{v-w}$, and
(b) for the optimal solution u of $IP_{A,\omega}(b)$ and for all $w \in \mathcal{T}$, $u - w$ is infeasible for $IP_{A,\omega}(b)$.

Second, she observed that on the set of feasible solutions of each $IP_{A,\omega}(b)$ one can create a graph $G_{A,\omega,b}$: the vertices of this graph are the feasible solutions of $IP_{A,\omega}(b)$, and there is a directed edge from v to u if $x^{(v-u)_+} - x^{(v-u)_-} \in \mathcal{G}_{A,\succ}$.

THEOREM 3.1 ([**51**, Theorem 2.1.8]). *Each $G_{A,\omega,b}$ is a connected acyclic directed graph whose unique sink is the optimal solution to $IP_{A,\omega}(b)$.*

Finally, this paper contains a re-interpretation of the usual Buchberger's algorithm [**6**] in the case of toric ideals, stripping away the algebra of ideals and viewing a binomial $x^u - x^v$ as just $u - v$, an element of $\mathbb{Z}^n$. This was called the *geometric Buchberger's algorithm.*

By the time I graduated in 1997, Rekha published multiple papers in quick succession extending the Gröbner basis technology in integer programming; see [**29, 46, 50, 52, 53**]. The one that stands out among these is *Variations of cost functions in integer programming* [**46**], co-authored with Bernd. The contents of this article were also in the air during late 1993, and the manuscript was completed in 1994. Among other things, this paper introduced the *Graver basis* of I_A (or A).

DEFINITION 3.2. Let $\mathcal{H}_\rho$ be the Hilbert basis of the cone H_ρ. The set of binomials

$$Gr_A = \{x^u - x^v : u - v \in \bigcup_\rho \mathcal{H}_\rho\}$$

is called the Graver basis of I_A.

Recall that the union of all reduced Gröbner bases of an ideal I is called the *universal Gröbner basis* (since it is a Gröbner basis of I for any term order). We will denote the universal Gröbner basis of I_A with $\mathcal{U}_A$.

THEOREM 3.3 ([**46**, Theorem 2.7]). *The Graver basis Gr_A contains the universal Gröbner basis $\mathcal{U}_A$.*

THEOREM 3.4 ([**46**, Algorithm 4.5]). *Let*

$$\Lambda(A) = \begin{bmatrix} A & 0 \\ I_n & I_n \end{bmatrix}$$

be the Lawrence lifting of A. Let $\mathcal{G}_{\Lambda(A)}$ be the reduced Gröbner basis of $I_{\Lambda(A)}$ with respect to any term order. Then $x^u - x^v$ is in Gr_A if and only if $x^u y^v - x^v y^u$ is in $\mathcal{G}_{\Lambda(A)}$.

Along the same lines, this article also proved that every binomial corresponding to a circuit of A is in $\mathcal{U}_A$ [**46**, Lemma 2.6], and if A is a unimodular matrix then $\mathcal{U}_A$ consists of only the circuits [**46**, Theorem 2.10]. Finally, there was a close look at the Gröbner fan $\mathcal{G}_A$ and the state polytope $\mathcal{S}_A$. In particular, there was an intriguing and very concrete conjecture.

CONJECTURE 3.5. [**46**, Conjecture 6.2] *Let A be an $d \times n$ matrix where $n - d = 3$. Then every maximal cone in the Gröbner fan $\mathcal{G}_A$ has three or four facets.*

Bernd and Rekha called this the "3/4 conjecture". I liked this conjecture but I had no idea how to systematically approach it. Below, I will come back to its resolution. Rekha was determined to make progress in computing Gröbner fans of toric ideals. She came back to it in the first half of the naughts. Using a reverse search enumeration technique, she implemented a `MAPLE` code called `TiGERS` (Toric Gröbner bases Enumeration using Reverse Search) to compute Gröbner fans.

Then Birkett Huber, another student of Bernd's from Cornell, joined her in this project and TiGERS became even faster in its new implementation in C [**33**]. It could compute Gröbner fans with more than 200,000 maximal cones. Two other papers followed [**18, 19**] with Anders Jensen, Rekha's PhD student among the co-authors. Anders produced an even better implementation of reverse enumeration and Gröbner walk in his code Gfan [**34**] which is still the current state of the art. Gfan is available as a stand-alone program as well as a Macaulay 2 [**21**] package.

Eventually, I was involved in the resolution of Conjecture 3.5. Bernd had moved to UC Berkeley in 1995 and I was a postdoc at MSRI (now SLMath) during 1998/99. During that academic year and beyond, I collaborated in a project with Diane Maclagan, one of Bernd's first students from Berkeley, where we looked at a particular monomial ideal that we called the *vertex ideal* of A. While studying associated primes of vertex ideals, we constructed an interesting example associated to the corank 3 matrix $A = [15\,247\,248\,345]$. Gröbner fans were not on our mind at all. However, Diane happened to decide to compute the Gröbner fan of I_A using TiGERS; just for fun, I guess.

THEOREM 3.6 ([**26**, Theorem 3.4]). *Conjecture 3.5 is false. The Gröbner fan Gr_A of $A = [15\,247\,248\,345]$ has a maximal cone associated to $\omega = (111, 0, 341, 1)$ that has five facets.*

We even generated another example with six facets [**26**, Example 3.3]. A little later Bernd joined us to write a paper on *supernormal vector configurations* [**27**]. As a consequence of this work, examples of Gröbner cones of dimension three were found which had as many as 15 facets.

4. Computing Gröbner bases of toric ideals

As 1993 came to a close two new developments happened. First, Bernd and Rekha, jointly with Jesús De Loera (Jesús was also Bernd's student, junior to Rekha but senior to me), combinatorially described a reduced Gröbner basis of the *second hypersimplex* A_d for any d [**14**, Theorem 2.1]. Here the matrix A_d has d rows and $n = \binom{d}{2}$ columns where the columns are formed by all 0/1 vectors in $\mathbb{R}^d$ with exactly two ones. This was a first example of a family of toric ideals associated to nonunimodular matrices for which one could describe a reduced Gröbner basis.

More importantly (in hindsight!), Bernd was working on his seminal paper with Persi Diaconis where they introduced the *Markov basis* (aka minimal generating set) of a toric ideal I_A to run Monte Carlo Markov chains for sampling purposes [**16**]. This paper was not published until 1998 and it arguably marks the true beginning of the field of *algebraic statistics*. I am not going to dwell on this topic since the article in this volume by Seth Sullivant does a superb job telling that story. In any case, the explosion of the work based on [**16**] had to wait for almost another decade.

I was motivated by the optimism that was shared by Bernd, Jesús, and Rekha, that Gröbner bases of toric ideals should be part of the toolbox for solving integer programs. Of course, one had to first compute such Gröbner bases, and besides the elimination-based approach there was no other method. MAPLE had an implementation of the Buchberger's algorithm for general polynomial ideals. Macaulay, the precursor to Macaulay 2 [**21**], was already around with its earlier versions (version 3.0 for Linux came out in September 1994). It had a fast implementation of

Buchberger's, but again for general polynomial ideals. There was a need for a fast implementation of the geometric Buchberger's algorithm for toric ideals.

Before the end of the semester, Bernd invited me for a meeting to his office in White Hall. He patiently went over the advantages for a stand-alone implementation. For instance, computing the normal form of a monomial x^v with respect to a partial Gröbner basis, a crucial ingredient in the Buchberger's algorithm, could be made much more efficient: if such a basis had a binomial $x^w - x^u$ where the leading term is x^w and it divided x^v, then a one-step reduction would produce x^{v-w+u}. But maybe the same binomial could be applied multiple times for a longer chain of reductions. So instead, one could determine the largest k such that $v - k(w - u)$ did not have a negative entry and reduce x^v to $x^{v-k(w-u)}$ in one shot. Bernd demonstrated these ideas on his SUN workstation (named "Grassmann") running `UNIX`. We brainstormed similar ideas, and he convinced me that it was worth the effort to produce an implementation. I had experience in coding algorithms (mostly network flow and matching algorithms) in `C` in my undergrad days and I thought I could do it.

I convinced Lisa Fleischer, another second year student from the OR Department taking Bernd's course on Gröbner bases, to join me in the effort. We spent the entire winter break coding and refining and improving. Our first goal was to implement the geometric Buchberger's algorithm for a toric ideal I_A starting from a given generating set of I_A. Already by the end of the break Lisa was losing interest but I kept going. When I felt ready, Bernd (and Hyungsook) invited me for dinner over at their house. Rekha was there, too. Bernd asked to try the code. I dialed in into the workstation that I was working with from Bernd's home computer. It is perfectly possible that I am misremembering this, but I think he wanted to compute the Graver basis Gr_A of the ideal in Example 2.5 using Theorem 3.4. We set up the input and pushed the button. It took couple of minutes, way slow in today's standards, but it produced the correct output, something Bernd could not get `MAPLE` or `Macaulay` to do. He said something to the effect that he was impressed.

Throughout the spring of 1994 I kept refining and optimizing the code with all the bells and whistles. Mike Stillman who was the creator of `Macaulay` with Dave Bayer was also interested. He gave me more ideas. When David Eisenbud was visiting Cornell to give a talk I showed my code to him. He seemed to also like it. In the meantime, Bernd told Lisa and me that he could only take one more new student. To my relief, Lisa took herself out of the picture, and I became an official student of Bernd. So I started joining the group meetings consisting of Bernd, Jesús, and Rekha, as well as Paco Santos (who went on later to prove that toric Hilbert schemes could be disconnected and the Hirsch conjecture is false) and Anna Bigatti (of `CoCoA` fame [**2**]). Very exciting times indeed! The whole semester culminated in my first international conference and my first talk at the "Hungarian-American Workshop on Combinatorics" in Budapest in May 1994. In the audience were people like Jon Lee (combinatorial optimization) and Satish Rao (computer science). The talk was received well and people who are not so familiar with this kind of work seemed to be intrigued.

During the summer I kept working on the code, but it became clear that it reached its limit as it was. Some new idea was needed. In particular, it would have been much more desirable if one started without a generating set of I_A and not do elimination, and yet compute a Gröbner basis. Bernd pointed out the following.

PROPOSITION 4.1. [**28**, Propositions 3 and 4] *Let* $\{u^{(1)}, u^{(2)}, \ldots, u^{(n-d)}\}$ *be a lattice basis for* $\ker_{\mathbb{Z}}(A)$ *and let* $J = \langle x^{u_+^{(1)}} - x^{u_-^{(1)}}, \ldots, x^{u_+^{(n-d)}} - x^{u_-^{(n-d)}} \rangle$. *Then*

$$I_A = J\colon (x_1 \cdots x_n)^\infty = ((\cdots (J\colon x_1^\infty)\colon x_2^\infty)\cdots)\colon x_n^\infty).$$

Moreover, each saturation on the right-hand side of the formula can be computed by a single Gröbner computation with respect to a reverse lexicographic term order.

This proposition gives the desired algorithm and the only input needed is the matrix A. Computing a lattice basis of $\ker_{\mathbb{Z}}(A)$ is easy via the Hermite normal form algorithm [**41**, Section 5.3]. One can even compute a *reduced* lattice basis in the sense of the LLL-algorithm [**36**]. Then n applications of the geometric Buchberger's algorithm are needed, starting from the ideal J and going through all the intermediate binomial ideals, ending in I_A. Surprisingly, this worked and it worked really fast.

During 1994/95, Bernd was on sabbatical at NYU in New York City. I visited him there and he came back to Ithaca to check on his students. Encouraged by the success of the latest implementation we decided to make the code official. We called it `GRIN` (GRöbner bases for INteger programming). The paper we wrote (my first!) [**28**] got accepted to the "Integer Programming and Combinatorial Optimization" (IPCO) conference in May 1995. We tried to make the paper really appealing to the optimization community. Our computational experiments showed that for as large (!) as 8×16 matrices A we could solve integer programs $IP_{A,\omega}(b)$ in comparable times with the fastest integer programming solver at the time (`CPLEX`). Of course, we knew this would not scale up to solve integer programs with thousands or millions of variables. But this was a proof of concept. I have to say I was a bit disappointed with the reaction to my talk at IPCO in Copenhagen. It was mixed. Some, such as Bob Bixby, the author of `CPLEX`, saw it as a fruitful direction to pursue. Others were less impressed, mostly because they saw that the algorithm would not scale.

This disappointment affected me and the progress of `GRIN`. The plan had been to set up easy input and output formats, to work out the kinks in the algorithm, and to keep improving. I did help others do computations with `GRIN` but I never brought it to a place which was available to everyone. Maintaining a software for users looked daunting. Luckily, others picked up from there. For instance, you can compute the toric ideal I_A in `Singular` [**15**] just from A. If you issue the command `toric_ideal(A,"hs")` (the option stands for Hoşten–Sturmfels) the algorithm based on Proposition 4.1 is used.

More importantly, `4ti2` [**1**] came to the scene in 2001. Raymond Hemmecke, while doing his PhD in Duisburg, developed code for computing Graver bases solely based on computing Hilbert bases of rational cones. With the help of others, such as Ralf Hemmecke, Matthias Köppe, Peter Malkin, and Matthias Walter, `4ti2` became the go-to software for computing toric ideals, their Markov bases, Gröbner bases, Graver bases, and much more. The code helped significantly to produce results in stochastic integer programming [**3, 11, 24**], lattice point counting [**12**], and probability theory and algebraic statistics [**5, 22, 23**]. `Macaulay 2` [**21**] has an interface with `4ti2` as well. Raymond was a postdoc at UC Davis in early naughts with Jesús as `4ti2` was developed and perfected. I believe Jesús deserves a lot of credit for carrying the torch, keeping in touch with the discrete optimization community, and advocating for algebraic methods in discrete optimization. His article in this volume is a wonderful resource as well as the book he co-authored with Raymond and Matthias [**10**].

5. Gröbner bases and convex polytopes

We arrived at the peak event of all of these developments, namely the "Holiday Symposium" at New Mexico State University (NMSU), Las Cruces, that was held December 27–31, 1994. It is a bit unusual (crazy?) to have a conference between Christmas and New Year's Eve, but here was this NSF/NSA funded conference organized by Reinhard Laubenbacher who was at NMSU at the time. The whole event was scheduled around Bernd's ten one-hour talks which he delivered over five days. He wrote extensive lecture notes which included exercises. These lecture notes evolved into Bernd's blockbuster book *Gröbner bases and convex polytopes* [**44**] with the addition of four more chapters (Chapters 3, 12, 13, and 14). They were an engaging summary of the state of the art in Gröbner fans and state polytopes of general polynomial ideals, Gröbner bases of toric ideals with applications in enumeration, sampling, and integer programming, universal and Graver bases of toric ideals, and regular triangulations. The last three talks were devoted to more recent developments: the second hypersimplex, A-graded algebras, and SAGBI bases. They were meant to stimulate new ideas and research.

I got the lecture notes from Bernd probably late summer or early fall of 1994. He asked me to read the notes and give him my comments. In particular, he told me to "do the exercises!" Here is the first exercise at the end of the first lecture which also appears as exercise (1) at the end of Chapter 1 in [**44**]:

Let I be the ideal generated by the nine 2×2-minors of a 3×3-matrix of indeterminates. Find a universal Gröbner basis. How many distinct initial ideals $\mathrm{in}_{\prec}(I)$ are there?

Even today I am surprised how anyone could have answered this question after reading only the notes of the first lecture (the result that a polynomial ideal can have only finitely many initial ideals, definition of universal Gröbner bases, a few examples, and the relation between weight vectors $\omega \in \mathbb{R}^n$ and initial ideals induced by them). If one had known the contents of the lectures on toric ideals and their universal Gröbner bases, one could have developed some ideas. Knowing the contents of the lecture on regular triangulations would have also helped. Of course, all of these needed to be done without the help of the (non-existing) software such as `Gfan`. Luckily, I knew some of these things (had I really mastered them?) and I could get a start.

The ideal I is the toric ideal that is the defining ideal of $\mathbb{P}^2 \times \mathbb{P}^2$. The corresponding matrix A is a 6×9 matrix whose columns are $e_i \oplus e_j$, $1 \leq i, j \leq 3$, where e_i form the standard unit basis of $\mathbb{R}^3$. This matrix is unimodular and therefore $\mathcal{U}_A = Gr_A$, and the latter can be computed with Theorem 3.4. Or you know that $\mathcal{U}_A$ is made up of the circuits of A and they are in bijection with the cycles in the complete bipartite graph $K_{3,3}$. So there are 15 squarefree binomials in $\mathcal{U}_A$, 9 quadratics and 6 cubics: $x^{u_1} - x^{v_1}, \ldots, x^{u_{15}} - x^{v_{15}}$. In the last step one needs to check whether there is a vector $\omega \in \mathbb{R}^9$ such that

$$\omega \cdot u_1 > \omega \cdot v_1, \omega \cdot u_2 > \omega \cdot v_2, \cdots, \omega \cdot u_{15} > \omega \cdot v_{15}.$$

There are 2^{15} such linear inequality systems to check. Clearly, one needed to write some code using `MAPLE`. When I went through all of these and got the answer (108; go ahead and try yourself) I was very happy. This was hard, but it was not completely unmanagable. One needs to be willing to extend and definitely do computations. I think this is how Bernd operates, in a nutshell.

The Holiday Symposium was a huge success. There were some new faces in Las Cruces who have been contributing significantly to commutative algebra and computational algebraic geometry since then, such as Irena Peeva, Maurice Rojas, and Frank Sottile. Bernd's lectures went extremely well although he had caught some bug and therefore was sick. In characteristic fashion, Bernd announced exercise sessions and invited (!) all grad students and postdocs. One embarrassing anecdote for me happened during the first exercise session where we discussed the above exercise. At some point we landed at a place where we had to argue that circuits of a unimodular matrix have only $0, +1, -1$ entries. Bernd called me to the board. For some reason I froze. I could not reason for the life of me. He "rescued" me by basically completing the answer. On the same evening he invited me to his hotel room and we chatted about various things. He did make a point to tell me that I should have been more prepared. But the comment was very neutral. It was not angry or shaming or condescending. Also in a fine gesture, he wrote in the Introduction to *Gröbner bases and convex polytopes* that I "did a particularly great job during the last round of proof-reading". The book is to this day the starting point for someone who is learning about toric ideals, Gröbner fans, state polytopes etc. As an interesting coincidence, Bernd taught a course in Berkeley on Gröbner bases and convex polytopes from the book in spring 2022 as he turned sixty. Many students enrolled which shows that the book and the subject are still alive.

6. Degrees of Gröbner bases of toric ideals

In late spring 1995 I took my A exam at Cornell. This is an oral PhD qualification exam where I had to also pitch to my committee (Bernd, Lou Billera, Dexter Kozen, Les Trotter) my plans for my thesis. By that time, Bernd and I had settled on the question of giving a better bound on the degrees of Gröbner or Graver basis elements of toric ideals.

The starting point was Theorem 2.6. Motivated by the Eisenbud-Goto conjecture [**17**] that was still a conjecture at the time and by computational observations, we thought that the bound in this theorem should be $(d+1)D(A)$. The proof of the theorem reveals that one could first try to show that the bound is given by the maximum degree $\operatorname{cdeg}(A)$ of any circuit of A. I worked on this for a while, and together with Rekha we produced a counterexample. This counterexample is reported in [**45**] but it was already known by the time Bernd gave his lecture at the "AMS Summer Institute in Algebraic Geometry at Santa Cruz" in July 1995 on which the publication was based. Here I reproduce the counterexample as Bernd presented it.

EXAMPLE 6.1. Let

$$B = \begin{bmatrix} 1 & 3 & 4 & 6 & 0 \\ 0 & 0 & 0 & -5 & 1 \end{bmatrix},$$

and let $A = \Lambda(B)$. The ideal I_A equals

$$I_A = \langle x_2y_1^3 - x_1^3y_2, x_3y_1^4 - x_1^4y_3, x_3^3y_2^4 - x_2^4y_3^3, x_4x_5^5y_2^2 - x_2^2y_4y_5^5, x_4x_5^5y_1^6 - x_1^6y_4y_5^5,$$
$$\underline{x_4^2x_5^{10}y_3^3 - x_3^3y_4^2y_5^{10}}, x_4x_5^5y_1^2y_3 - x_1^2x_3y_4y_5^5, x_4x_5^5y_1^3y_2 - x_1^3x_2y_4y_5^5, x_3y_1y_2 - x_1x_2y_3,$$
$$x_1x_4x_5^5y_2y_3 - x_2x_3y_1y_4y_5^5, x_2^2x_4x_5^5y_3^3 - x_3^3y_2^2y_4y_5^5, x_1x_3^2y_2^3 - x_2^3y_1y_3^2, x_1^2x_3y_2^2 - x_2^2y_1^2y_3,$$
$$x_2x_4x_5^5y_1y_3^2 - x_1x_3^2y_2y_4y_5^5, x_1^2x_4x_5^5y_3^2 - x_3^2y_1^2y_4y_5^5, x_4^2x_5^{10}y_1y_2y_3^2 - x_1x_2x_3^2y_4^2y_5^{10}\rangle.$$

The generators of I_A form a Graver basis of I_A. The first six binomials including the underlined one correspond to the circuits of A. The underlined circuit attains $\mathrm{cdeg}(A) = 15$. However, the last binomial has degree $16 > 15$.

Although this example is nice it is not satisfactory since when we attempt to compute the underlined circuit using (2.1) we see that the constant $c = 2$. In other words, the "true" degree of this circuit is not 15 but 30. Therefore we define the true degree of a circuit to be the degree computed without dividing by the common factor c in (2.1). With this we can refine the conjecture.

CONJECTURE 6.2. [**45**, Conjecture 4.8] *The degree of any Graver basis element of a toric ideal I_A is at most the maximum true degree of any circuit of A.*

A small progress due to Bernd was reported in my dissertation [**25**]. Recall that $H_\rho = \mathbb{R}^n_\rho \cap \ker_{\mathbb{Z}}(A)$. This rational polyhedral cone is generated by the circuits of A conforming to the sign pattern ρ.

PROPOSITION 6.3. *If H_ρ is a simplicial cone then the degree of any Graver basis element of I_A contained in H_ρ is bounded above by the maximum true degree of any circuit in H_ρ.*

Progress on Conjecture 6.2 stayed dormant for more than a decade. A result of Sonja Petrović showed that the conjecture holds for rational normal scrolls [**39**]. Then in 2011 Christos Tatakis and Apostolos Thoma provided a counterexample.

THEOREM 6.4 ([**48**]). *There exists a toric ideal $I_{A(G)}$ associated to a graph G for which Conjecture 6.2 does not hold.*

PROOF. Given an undirected graph G on d vertices and n edges let $A(G)$ be the vertex-edge incidence matrix of G. The binomials in $I_{A(G)}$ corresponding to the circuits of $A(G)$ have been characterized in [**54**]. Similarly, Graver basis elements of $I_{A(G)}$ have been characterized in [**40**]. Now, let G be a graph consisting of a cycle of length s and s odd cycles of length ℓ, each attached to a vertex of the first cycle; see [**48**, Figure 5]. The even walk of length $\ell s + s$ traversing all edges of G corresponds to a Graver basis element of degree $\ell s + s$. Longest cycles in this graph consist of two cycles of length ℓ joined by a path of length $s - 1$. The corresponding circuits are the ones with the maximum true degree of $2(\ell + s - 1)$. Now it is easy to see that $s(\ell + 1) > 2(\ell + s - 1)$ since $s, \ell > 2$. □

It turns out one can actually create really bad counterexamples.

THEOREM 6.5 ([**49**, Theorem 4.5]). *There exist a family of graphs G_r where the maximum true degree of circuits in $I_{A(G_r)}$ grows linearly in r whereas the maximum degree of Graver basis elements of $I_{A(G_r)}$ grows exponentially in r.*

7. Initial ideals of toric ideals

I wish to conclude this brief history of toric ideals in their first decade with one last result with which I was involved. The starting point was one more time integer programming problems $IP_{A,\omega}(b)$. In order to solve these problems, one usually passes to the *linear programming relaxation* where the constraint that the solution vectors are $x \in \mathbb{N}^n$ is replaced with the easier constraint $x \geq 0$. However, in the early days of integer programming Ralph E. Gomory introduced a different and more algebraic relaxation: let σ be a subset of $\{1, \ldots, n\}$. Now replace the constraint that the solution vectors are $x \in \mathbb{N}^n$ with the constraints that $x_i \in \mathbb{N}$ for all $i \in \sigma$ and $x_i \in \mathbb{Z}$ for all $i \notin \sigma$. This is known as a *group relaxation* of the integer programs $IP_{A,\omega}(b)$ [**20**].

Bernd had earlier done some observations relating group relaxations to the structure of the monomial initial ideal $\mathrm{in}_{\succ}(I_A)$ of the toric ideal I_A with respect to the term order $\succ$ induced by the vector ω [**44**, Section 12.D]. His observations were pointing in the direction of studying minimal and embedded primes and primary decomposition of $\mathrm{in}_{\succ}(I_A)$.

One more time, Rekha and I joined forces and dug into this. Two of our papers explored the connection of these initial ideals and their *standard pair decompositions* to group relaxations [**31**, **32**]. However, a third one is my favorite since it proved a surprising result [**30**].

By the time I got my PhD, there was only one general structural result known about $\mathrm{in}_{\succ}(I_A)$.

THEOREM 7.1 ([**44**, Corollary 8.9]). *The initial ideal* $\mathrm{in}_{\succ}(I_A)$ *is square-free, i.e., it is a radical ideal, if and only if the corresponding regular triangulation of* A *is unimodular.*

Was there anything interesting if $\mathrm{in}_{\succ}(I_A)$ was not radical? We were able to characterize the associated primes of $\mathrm{in}_{\succ}(I_A)$ in terms of polytopes we could construct from the matrix A, the vector $\omega \in \mathbb{R}^n$ that gave rise to the term order $\succ$, and certain right-hand side vectors b of the family of integer programs $IP_{A,\omega}$. This characterization led to the following result that imposes strong structural restrictions on $\mathrm{in}_{\succ}(I_A)$.

THEOREM 7.2 ([**31**, Theorem 3.1]). *If* P *is an embedded prime of* $\mathrm{in}_{\succ}(I_A)$, *then* P *contains an associated prime* Q *of* $\mathrm{in}_{\succ}(I_A)$ *such that* $\dim(Q) = \dim(P) + 1$.

In other words, the associated primes of the initial ideals of toric ideals come in chains: one can start with any embedded prime and reach a minimal prime by following a chain of associated primes and hitting every dimension along the way. Oddly, there has been no work that I am aware of studying this or related phenomena since last century.

Acknowledgments

In 2016, during a happy occasion, I said that I have three families: the nuclear family of my loving wife Ingrid and my son Orhan, the family of my colleagues at the SFSU Mathematics Department who embraced me for twenty-five years, and finally my first family from the Cornell days when I did not yet have the above two. Bernd, Rekha, and Jesús, I am tremendously thankful to you for accepting me into your warm and supportive circle.

References

[1] 4ti2 team, *4ti2—A software package for algebraic, geometric, and combinatorial problems on linear spaces.*

[2] John Abbott, Anna M. Bigatti, and Lorenzo Robbiano, *CoCoA: A system for doing computations in commutative algebra*, Available at `http://cocoa.dima.unige.it`.

[3] M. Aschenbrenner and R. Hemmecke, *Finiteness theorems in stochastic integer programming*, Found. Comput. Math. **7** (2007), no. 2, 183–227, DOI 10.1007/s10208-005-0174-1. MR2324416

[4] D. Bayer and I. Morrison, *Standard bases and geometric invariant theory. I. Initial ideals and state polytopes*, J. Symbolic Comput. **6** (1988), no. 2-3, 209–217, DOI 10.1016/S0747-7171(88)80043-9. Computational aspects of commutative algebra. MR988413

[5] R. Bouckaert, R. Hemmecke, S. Lindner, and M. Studený, *Efficient algorithms for conditional independence inference*, J. Mach. Learn. Res. **11** (2010), 3453–3479. MR2756190

[6] B. Buchberger, *Ein algorithmisches Kriterium für die Lösbarkeit eines algebraischen Gleichungssystems* (German), Aequationes Math. **4** (1970), 374–383, DOI 10.1007/BF01844169. MR268178

[7] P. Conti and C. Traverso, *Buchberger algorithm and integer programming*, Applied algebra, algebraic algorithms and error-correcting codes (New Orleans, LA, 1991), Lecture Notes in Comput. Sci., vol. 539, Springer, Berlin, 1991, pp. 130–139, DOI 10.1007/3-540-54522-0_102. MR1229314

[8] D. Cox, J. Little, and D. O'Shea, *Ideals, varieties, and algorithms*, Undergraduate Texts in Mathematics, Springer-Verlag, New York, 1992. An introduction to computational algebraic geometry and commutative algebra, DOI 10.1007/978-1-4757-2181-2. MR1189133

[9] D. A. Cox, J. B. Little, and H. K. Schenck, *Toric varieties*, Graduate Studies in Mathematics, vol. 124, American Mathematical Society, Providence, RI, 2011, DOI 10.1090/gsm/124. MR2810322

[10] J. A. De Loera, R. Hemmecke, and M. Köppe, *Algebraic and geometric ideas in the theory of discrete optimization*, MOS-SIAM Series on Optimization, vol. 14, Society for Industrial and Applied Mathematics (SIAM), Philadelphia, PA; Mathematical Optimization Society, Philadelphia, PA, 2013. MR3024570

[11] J. A. De Loera, R. Hemmecke, S. Onn, and R. Weismantel, *n-fold integer programming*, Discrete Optim. **5** (2008), no. 2, 231–241, DOI 10.1016/j.disopt.2006.06.006. MR2408419

[12] J. A. De Loera, R. Hemmecke, J. Tauzer, and R. Yoshida, *Effective lattice point counting in rational convex polytopes*, J. Symbolic Comput. **38** (2004), no. 4, 1273–1302, DOI 10.1016/j.jsc.2003.04.003. MR2094541

[13] J. A. De Loera, J. Rambau, and F. Santos, *Triangulations*, Algorithms and Computation in Mathematics, vol. 25, Springer-Verlag, Berlin, 2010. Structures for algorithms and applications, DOI 10.1007/978-3-642-12971-1. MR2743368

[14] J. A. de Loera, B. Sturmfels, and R. R. Thomas, *Gröbner bases and triangulations of the second hypersimplex*, Combinatorica **15** (1995), no. 3, 409–424, DOI 10.1007/BF01299745. MR1357285

[15] W. Decker, G.-M. Greuel, G. Pfister, and H. Schönemann, SINGULAR *4-4-0 — A computer algebra system for polynomial computations*, `http://www.singular.uni-kl.de`, 2024.

[16] P. Diaconis and B. Sturmfels, *Algebraic algorithms for sampling from conditional distributions*, Ann. Statist. **26** (1998), no. 1, 363–397, DOI 10.1214/aos/1030563990. MR1608156

[17] D. Eisenbud and S. Goto, *Linear free resolutions and minimal multiplicity*, J. Algebra **88** (1984), no. 1, 89–133, DOI 10.1016/0021-8693(84)90092-9. MR741934

[18] K. Fukuda, A. N. Jensen, N. Lauritzen, and R. Thomas, *The generic Gröbner walk*, J. Symbolic Comput. **42** (2007), no. 3, 298–312, DOI 10.1016/j.jsc.2006.09.004. MR2292637

[19] K. Fukuda, A. N. Jensen, and R. R. Thomas, *Computing Gröbner fans*, Math. Comp. **76** (2007), no. 260, 2189–2212, DOI 10.1090/S0025-5718-07-01986-2. MR2336291

[20] R. E. Gomory, *On the relation between integer and noninteger solutions to linear programs*, Proc. Nat. Acad. Sci. U.S.A. **53** (1965), 260–265, DOI 10.1073/pnas.53.2.260. MR182454

[21] D. R. Grayson and M. E. Stillman, *Macaulay2, a software system for research in algebraic geometry*, Available at `http://www2.macaulay2.com`.

[22] R. Hemmecke, S. Lindner, and M. Studený, *Characteristic imsets for learning Bayesian network structure*, Internat. J. Approx. Reason. **53** (2012), no. 9, 1336–1349, DOI 10.1016/j.ijar.2012.04.001. MR2994270

[23] R. Hemmecke, J. Morton, A. Shiu, B. Sturmfels, and O. Wienand, *Three counterexamples on semi-graphoids*, Combin. Probab. Comput. **17** (2008), no. 2, 239–257, DOI 10.1017/S0963548307008838. MR2396350

[24] R. Hemmecke and R. Schultz, *Decomposition of test sets in stochastic integer programming*, Math. Program. **94** (2003), no. 2-3, Ser. B, 323–341, DOI 10.1007/s10107-002-0322-1. The Aussois 2000 Workshop in Combinatorial Optimization. MR1969115

[25] S. Hoşten, *Degrees of Gröbner bases of integer programs*, PhD thesis, Cornell University, 1997.

[26] S. Hoşten and D. Maclagan, *The vertex ideal of a lattice*, Adv. in Appl. Math. **29** (2002), no. 4, 521–538, DOI 10.1016/S0196-8858(02)00030-1. MR1943363
[27] S. Hoşten, D. Maclagan, and B. Sturmfels, *Supernormal vector configurations*, J. Algebraic Combin. **19** (2004), no. 3, 297–313, DOI 10.1023/B:JACO.0000030705.93448.ce. MR2071476
[28] S. Hoşten and B. Sturmfels, *GRIN: an implementation of Gröbner bases for integer programming*, Integer programming and combinatorial optimization (Copenhagen, 1995), Lecture Notes in Comput. Sci., vol. 920, Springer, Berlin, 1995, pp. 267–276, DOI 10.1007/3-540-59408-6_57. MR1367987
[29] S. Hoşten and R. Thomas, *Gröbner bases and integer programming*, Gröbner bases and applications (Linz, 1998), London Math. Soc. Lecture Note Ser., vol. 251, Cambridge Univ. Press, Cambridge, 1998, pp. 144–158. MR1699817
[30] S. Hoşten and R. R. Thomas, *The associated primes of initial ideals of lattice ideals*, Math. Res. Lett. **6** (1999), no. 1, 83–97, DOI 10.4310/MRL.1999.v6.n1.a6. MR1682705
[31] S. Hoşten and R. R. Thomas, *Standard pairs and group relaxations in integer programming*, J. Pure Appl. Algebra **139** (1999), no. 1-3, 133–157, DOI 10.1016/S0022-4049(99)00009-2. Effective methods in algebraic geometry (Saint-Malo, 1998). MR1700541
[32] S. Hoşten and R. R. Thomas, *Gomory integer programs*, Math. Program. **96** (2003), no. 2, Ser. B, 271–292, DOI 10.1007/s10107-003-0386-6. Algebraic and geometric methods in discrete optimization. MR1993049
[33] B. Huber and R. R. Thomas, *Computing Gröbner fans of toric ideals*, Experiment. Math. **9** (2000), no. 3, 321–331. MR1795304
[34] A. N. Jensen, *Gfan, a software system for Gröbner fans and tropical varieties*, Available at `http://home.imf.au.dk/jensen/software/gfan/gfan.html`.
[35] H. W. Lenstra Jr., *Integer programming with a fixed number of variables*, Math. Oper. Res. **8** (1983), no. 4, 538–548, DOI 10.1287/moor.8.4.538. MR727410
[36] L. Lovász, *An algorithmic theory of numbers, graphs and convexity*, CBMS-NSF Regional Conference Series in Applied Mathematics, vol. 50, Society for Industrial and Applied Mathematics (SIAM), Philadelphia, PA, 1986, DOI 10.1137/1.9781611970203. MR861822
[37] E. W. Mayr and A. R. Meyer, *The complexity of the word problems for commutative semigroups and polynomial ideals*, Adv. in Math. **46** (1982), no. 3, 305–329, DOI 10.1016/0001-8708(82)90048-2. MR683204
[38] T. Mora and L. Robbiano, *The Gröbner fan of an ideal*, J. Symbolic Comput. **6** (1988), no. 2-3, 183–208, DOI 10.1016/S0747-7171(88)80042-7. Computational aspects of commutative algebra. MR988412
[39] S. Petrović, *On the universal Gröbner bases of varieties of minimal degree*, Math. Res. Lett. **15** (2008), no. 6, 1211–1221, DOI 10.4310/MRL.2008.v15.n6.a11. MR2470395
[40] E. Reyes, C. Tatakis, and A. Thoma, *Minimal generators of toric ideals of graphs*, Adv. in Appl. Math. **48** (2012), no. 1, 64–78, DOI 10.1016/j.aam.2011.06.003. MR2845507
[41] A. Schrijver, *Theory of linear and integer programming*, Wiley-Interscience Series in Discrete Mathematics, John Wiley & Sons, Ltd., Chichester, 1986. A Wiley-Interscience Publication. MR874114
[42] B. Sturmfels, *Gröbner bases and Stanley decompositions of determinantal rings*, Math. Z. **205** (1990), no. 1, 137–144, DOI 10.1007/BF02571229. MR1069489
[43] B. Sturmfels, *Gröbner bases of toric varieties*, Tohoku Math. J. (2) **43** (1991), no. 2, 249–261, DOI 10.2748/tmj/1178227496. MR1104431
[44] B. Sturmfels, *Gröbner bases and convex polytopes*, University Lecture Series, vol. 8, American Mathematical Society, Providence, RI, 1996, DOI 10.1090/ulect/008. MR1363949
[45] B. Sturmfels, *Equations defining toric varieties*, Algebraic geometry—Santa Cruz 1995, Proc. Sympos. Pure Math., vol. 62, Amer. Math. Soc., Providence, RI, 1997, pp. 437–449, DOI 10.1090/pspum/062.2/1492542. MR1492542
[46] B. Sturmfels and R. R. Thomas, *Variation of cost functions in integer programming*, Math. Programming **77** (1997), no. 3, Ser. A, 357–387, DOI 10.1016/S0025-5610(96)00068-8. MR1447563
[47] B. Sturmfels and N. White, *Gröbner bases and invariant theory*, Adv. Math. **76** (1989), no. 2, 245–259, DOI 10.1016/0001-8708(89)90053-4. MR1013672
[48] C. Tatakis and A. Thoma, *On the universal Gröbner bases of toric ideals of graphs*, J. Combin. Theory Ser. A **118** (2011), no. 5, 1540–1548, DOI 10.1016/j.jcta.2011.01.005. MR2771599

[49] C. Tatakis and A. Thoma, *Graver degrees are not polynomially bounded by true circuit degrees*, J. Pure Appl. Algebra **219** (2015), no. 7, 2658–2665, DOI 10.1016/j.jpaa.2014.09.021. MR3313501

[50] S. R. Tayur, R. R. Thomas, and N. R. Natraj, *An algebraic geometry algorithm for scheduling in presence of setups and correlated demands*, Math. Programming **69** (1995), no. 3, Ser. A, 369–401, DOI 10.1007/BF01585566. MR1355695

[51] R. R. Thomas, *A geometric Buchberger algorithm for integer programming*, Math. Oper. Res. **20** (1995), no. 4, 864–884, DOI 10.1287/moor.20.4.864. MR1378110

[52] R. R. Thomas and R. Weismantel, *Test sets and inequalities for integer programs*, Integer programming and combinatorial optimization (Vancouver, BC, 1996), Lecture Notes in Comput. Sci., vol. 1084, Springer, Berlin, 1996, pp. 16–30, DOI 10.1007/3-540-61310-2_2. MR1441788

[53] R. R. Thomas and R. Weismantel, *Truncated Gröbner bases for integer programming*, Appl. Algebra Engrg. Comm. Comput. **8** (1997), no. 4, 241–256, DOI 10.1007/s002000050062. MR1464786

[54] R. H. Villarreal, *Rees algebras of edge ideals*, Comm. Algebra **23** (1995), no. 9, 3513–3524, DOI 10.1080/00927879508825412. MR1335312

Department of Mathematics, San Francisco State University, San Francisco, California 94132

Email address: serkan@sfsu.edu

Proceedings of Symposia in Pure Mathematics
Volume **111**, 2025
https://doi.org/10.1090/pspum/111/02040

The nonlinear algebra view of discrete mathematics

Jesús A. De Loera

Dedicated with great appreciation and gratitude to Bernd Sturmfels on his sixtieth birthday

ABSTRACT. This survey recounts how systems of nonlinear polynomials equations and inequalities help to solve many problems in discrete mathematics. In particular, it recounts the long history of contributions by Bernd Sturmfels to this method.

"The one contribution of mine that I hope will be remembered has consisted in pointing out that all sorts of problems of combinatorics can be viewed as problems of the location of the zeros of certain polynomials..."

(Gian-Carlo Rota, 1985)

"Given a set of points (or vectors, or sets) that satisfy some property, we want to say something about the size or the structure of this set. The approach is then to associate to this set a polynomial, or a collection of polynomials, and use properties of polynomials to obtain information on the size or structure of the set."

(Aart Blokhuis, 1993)

"Is it possible to modify the algebraic proofs given here so that they yield efficient ways of solving the corresponding algorithmic problems? It seems likely that such algorithms do exists."

(Noga Alon, 1999)

"Nonlinear algebra is not simply a rebranding of algebraic geometry. It is a recognition that a focus on computation and applications, and the theoretical needs that this requires, results in a body of inquiry that is complementary to the existing curriculum."

(Mateuz Michałek and Bernd Sturmfels, 2021)

2020 *Mathematics Subject Classification.* Primary 05-02, 13P25; Secondary 14Q20, 68R01.

Key words and phrases. Discrete mathematics, combinatorics, computational algebraic geometry, nonlinear algebra methods in discrete mathematics, the polynomial method, algebraic techniques in combinatorics.

The author was supported in part by NSF Grant 1818969.

1. Opening

It is evident that Bernd Sturmfels has been a major contributor to combinatorial and computational methods for algebraic geometry and commutative algebra (see, e.g., other articles in this collection as well as the many references within [**BPS21**, **BPS98**, **BJMS15**, **CDRD$^+$14**, **ES94**, **HS95b**, **HS97**, **HSYY08**, **KSZ92**, **MS15**, **MS05**, **PS93**, **Stu94a**, **Stu94b**, **Stu94c**, **Stu98**, **Stu02**, **Stu08**], [**ST08**, **STY07**, **SY08**, **SZ93**]). But since the start of his career Bernd has also cultivated and promoted the *opposite* direction of mathematical research, i.e., he promoted algebraic geometry is a useful tool to solve problems in discrete mathematics, to turn theory into algorithms, and to experiment with computers. *This survey reviews the contemporary use of methods from applied computational algebraic geometry—or nonlinear algebra as we like to call it—to answer combinatorial and computational questions.*

Clearly the working discrete mathematician benefits from knowing a diverse set of mathematical methods (see, e.g., [**Bj95**, **Alo95**] and other methods in [**GGL95**]). Soon I expect nonlinear algebra methods will become as popular as, say, linear algebra, polyhedral, or probabilistic methods that have been around longer [**AS16**, **Mat10**, **Sch95**]. Excellent surveys and books have already been written to promote the use of polynomials [**Alo99**, **Gut16**, **Tao14**], but they mostly talked about using polynomials (often only one polynomial) to prove theorems. Alon named it the *polynomial method*. Here I try to offer a larger range of possibilities than in those offered in those references. I talk about proving theorems, but I also show how to compute with systems of polynomials equations and inequalities from discrete mathematics and the associated computational complexity.

This survey is divided into two parts. First, we look at polynomials: first one single polynomial, then many, as essential elements in combinatorial proofs and then slowly move to related computational challenges. Second, we see several problems of discrete mathematics that benefited from an underlying (but sometimes hidden and unexpected) nonlinear algebra structure, an algebraic variety, or a semialgebraic set that helps with both proofs and algorithms.

This survey article is meant to be an inviting and informal presentation, a quick read. I wrote this for the nonexpert and my imagined audience is a second year graduate student curious to hear these ideas without too much effort (e.g., I will not prove any of the theorems stated), but I collected enough background and prior work in a robust bibliography for a more committed grad student or a postdoc to see the context, history, and what they need to study next in order for them to start actively producing this type of research. I apologize if I missed your favorite result. It is, of course, impossible to cover everything.

Given a collection L of N lines in R^n (which also extends to any field $\mathbb{F}$), a *joint* for L is a point p such that there are n lines in L which contain p and do not lie in a hyperplane. The *joints problem* asks: What is the maximal number of joints that can be formed from N lines? Elegant solutions came from using the polynomial method in 2008 by Guth and Katz [**GK10**], and then later by Kaplan, Sharir, and Shustin [**KSS10**], and independently by Quilodrán [**Qui10**] later proved a stronger statement.

THEOREM 1.1. *The number of joints in a collection of N lines in $\mathbb{R}^n$ is at most $C(n)N^{\frac{n}{(n-1)}}$. Where $C(n)$ is a constant that depends only in the dimension n.*

Their proof uses polynomials that vanish at the joints and then *Bezout's theorem* bounding the number of roots of a system of polynomials (see [**CLO92**] for an introduction). All known proofs involve algebraic geometry in some way.

Another notable contribution of the polynomial method is the solution of the *Erdös distinct distances problem.* Erdös asked in 1946 how many distinct distances can there be for any finite set of points in the plane. It was shown again by Guth and Katz [**GK15**] in the following theorem.

THEOREM 1.2. *There is some constant $c > 0$ such that for any set P of N points in $\mathbb{R}^2$, the set of distinct distances has cardinality*

$$|\{\|p-q\| : p, q \in P\}| \geq c\frac{N}{\log N}.$$

To prove their result, Guth and Katz reduced it to an incidence problem for points and lines in $\mathbb{R}^3$, which they then attacked with algebraic geometry: For any $\epsilon > 0$ there exist constants $D(\epsilon)$ and $K(\epsilon)$ with the following property. For any family, L, of lines in $\mathbb{R}^3$ such that no more than $|L|^{\frac{1}{2}+\epsilon}$ lines of L lie inside any algebraic surface of degree at most D, the set $P_r(L)$ of points with at least r lines from L passing through them has cardinality

$$|P_r(L)| \leq K\left(\frac{|L|^{\frac{3}{2}+\epsilon}}{r^2} + \frac{|L|}{r}\right).$$

From this, it is deduced that for any $\epsilon > 0$ there exists $c(\epsilon) > 0$ such that the bound

$$|\{\|p-q\| : p, q \in P\}| \geq cN^{1-\epsilon}$$

holds for any set P of N points in the plane. A key ingredient is to use *polynomial partitioning technique* to divide a point set P into small enough cells created by the complement of the zero set of a low-degree polynomial. Indeed, partitioning in "balanced" ways has many styles and many applications. Take, for instance, the famous *ham sandwich theorem* that guarantees that there is a hyperplane that simultaneously bisects a finite set of measures (dividing all the contents of the sandwich in half) which has been used many so times in discrete mathematics (see [**Gut16**, Chapter 10] and [**DLGMM19**]). See also [**MO22**] and the references therein for more on distinct distance problems.

It is not just discrete geometry problems that can be approached by understanding properties of polynomials! A more recent example is the *slice rank polynomial method*, a new type of polynomial method, introduced by Tao (in his blog!) generalizing earlier work by Ellenberg and Gijswijt [**EG17**] who were interested on bounding the size of subsets of $(F_p)^n$ without 3 term arithmetic progressions. Even the case $p = 3$ is hard and famous, it is called the *cap-set problem.* Several researchers have been interested on applying, improving, and extending Tao's slice rank polynomial method to tensors that are not diagonal. For example, Sauermann [**Sau21a**] studied the following natural extension of the work of Ellenberg and Gijswijt: Let p be a prime and n a positive integer. We think of p as fixed and n large. Sauermann gives new upper bounds for the size of a set $A \subseteq \mathbf{F}_p^n$ not containing distinct elements $x_1, \ldots, x_p$ with $x_1 + \cdots + x_p = 0$. The bound is $C_p(2\sqrt{p})^n$, where C_p is a constant which can be given explicitly. In [**Oma22**] Omar adapts the slice rank polynomial method to determine lower bounds on sizes of sets in F_p^n that must have solutions to a homogeneous linear equation where many but not all variables are equal. See also [**BCC$^+$17, Oma23**].

There are now many new polynomial-based proofs of discrete geometry, combinatorics, and computer science results. Examples include generalizations and variations of the Szemerédi–Trotter theorem (see [**AES20**, **SdZ18**] and references therein), sum-product estimates in combinatorial number theory (see [**IRNR18**]), and *graceful labelings* of graphs. (The Kotzig–Ringel–Rosa conjecture asserts that every tree admits a graceful labeling. Recently, using a nice new variation of the combinatorial Nullstellensatz, Gnang gave a proof of this conjecture [**Gna23**].) For more on the polynomial method, see [**Gut16**].

Note that the results we have mentioned so far use polynomials to *prove* combinatorial results, not so much to *compute* them or as the main algorithmic ingredient. Already in the conclusions of his papers [?Nogaquestion, **Alo00**] Alon asked a central philosophical question.

Alon's Question: *"Is it possible to modify the algebraic proofs given here so that they yield efficient ways of solving the corresponding algorithmic problems? It seems likely that such algorithms do exist."*

As we see next **a core mission of nonlinear algebra is to provide us with interesting answers to this question!**

Even the simplest and elementary computations with a large combinatorial polynomial can be nontrivial! Take, for instance, the following challenge: We are given a multivariate polynomial g in n variables, but the polynomial is so large that, effectively is given as a black-box, an unknown representation (perhaps is a mysterious computer program) that only allows us to evaluate the polynomial at specific points in $\mathbb{C}^n$. A natural example of a black-box polynomial is when the polynomial is the determinant of a very large symbolic matrix. With the black-box representation it would be possible to use fast numerical determinant algorithms to compute the evaluation of the polynomial at many points, but impossible to expand it as a sum of monomials. We can ask:

(A) *Is the coefficient of the monomial z^a zero? Is the coefficient positive?*
(B) *What is the exact value of the coefficient of the monomial z^a?*

Answering questions (A) and (B) for black-box polynomials is of interest in discrete mathematics and algebra. For instance, earlier we saw the graph polynomial of a graph f_G is another black-box polynomial because we cannot really expand the formula to obtain its monomials as that may involve an exponential-size set of 2^E monomials. But we still care about explicitly computing the coefficients in other situations, for example, in representation theory.

Denote by χ_μ^λ a character of the symmetric group of degree N. These are functions from the set of partitions of N into the integers. A character is completely identified by the two partitions, λ and μ, of the integer N. In this way, μ indicates the conjugacy class of group elements to which the character corresponds, and λ determines the irreducible representation from which the character emanates, as the trace of the representing matrix. The characters of the irreducible representations of finite groups, in particular, the symmetric group, play a role in many areas of mathematics [**CR06**, **Sag01**]. Explicit computation of the characters is an important and difficult task and several methods have been employed. Unfortunately, there are no general explicit formulas or algorithms for the irreducible characters of the symmetric groups. One popular method is the Murnaghan–Nakayama rule, which is a recursive method to compute the characters. Here we focus instead on the *Frobenius character formula* where one can recover the characters as the coefficients of a black-box polynomial.

To present this we introduce some notation. Let $\lambda = (\lambda_1, \lambda_2, \ldots, \lambda_k)$ and $\mu = (\mu_1, \mu_2, \ldots, \mu_t)$ be partitions of N. Consider auxiliary variables $x_1, x_2, \ldots, x_\ell$, with $\ell \geq k$. The *power symmetric function* is given by the formula

$$p_\mu(x_1, x_2, \ldots, x_\ell) = \prod_{i=1}^{t} (x_1^{\mu_i} + x_2^{\mu_i} + \cdots + x_\ell^{\mu_i}).$$

We know that the value of χ_λ^μ is equal to the coefficient of the monomial

$$\prod_{i=1}^{k} x_i^{\lambda_i + \ell - i}$$

inside the *Frobenius character polynomial*

$$F_\mu^\lambda(x_1, \ldots, x_\ell) = \left(\prod_{1 \leq i < j \leq \ell} (x_i - x_j) \right) p_\mu(x_1, \ldots, x_\ell).$$

We should note that the Frobenius character formula is equivalent to a famous identity on symmetric polynomials where s_λ is the *Schur polynomial* (there are many definitions but the Schur polynomials can be defined combinatorially as sums of monomials that encode all semi-standard Young tableaux of shape λ):

$$p_\mu = \sum_{\lambda \vdash N} \chi_\mu^\lambda s_\lambda.$$

Recently, Ikenmeyer, Pak, and Panova [**IPP22**] showed several hardness results deciding the vanishing of the character of the symmetric group as well as for deciding positivity of the characters. They also ruled out the existence of any (unsigned) compact combinatorial description for the square of the characters. This indicates the difficulty of the computational problems. The computational challenges (A) and (B) are hard in the precise sense of complexity theory [**GJ79**].

1.1. Systems of equations for proofs and computation. Now we move to survey the use of *systems* of polynomial equations to model combinatorial problems. Good introductions to methods in nonlinear algebra are available in the excellent textbooks [**BCR87, CLO05, CLO92, Mar08b, MlS21**]. Naturally we need to use more terminology about ideals and rings of polynomials from now on. Let F denote a field and when the distinction is necessary we denote its algebraic closure by $\bar{F}$. Let $F[x_1, \ldots, x_n]$ denote the ring of polynomials in n variables with coefficients over F, which will be abbreviated as $\mathbb{F}[x]$. We denote the monomials in the polynomial ring $\mathbb{F}[x]$ as $x^\alpha := x_1^{\alpha_1} x_2^{\alpha_2} \cdots x_n^{\alpha_n}$ for $\alpha \in \mathbb{N}^n$. The degree of x^α is $\deg(x^\alpha) := |\alpha| := \sum_{i=1}^n \alpha_i$. The degree of a polynomial $f = \sum_{\alpha \in \mathbb{N}^n} f_\alpha x^\alpha$, written $\deg(f)$, is the maximum degree of x^α where $f_\alpha \neq 0$ for $\alpha \in \mathbb{N}^n$. Given a set of polynomials $F \subset \mathbb{F}[x]$, we write $\deg(F)$ for the maximum degree of the polynomials in F. Given a set of polynomials $F := \{f_1, \ldots, f_m\} \subseteq \mathbb{F}[x]$, we define the *ideal* generated by F as $< F >:= \{\sum_{i=1}^m \beta_i f_i \mid \beta_i \in \mathbb{F}[x]\}$. The *variety* of F over $\mathbb{F}$, written $\mathcal{V}_\mathbb{F}(F)$, is the set of common zeros of polynomials in F in $\mathbb{F}^n$, that is, $\mathcal{V}_\mathbb{F}(F) := \{v \in \mathbb{F}^n : f(v) = 0 \; \forall f \in I\}$. Also, $\mathcal{V}_{\overline{\mathbb{F}}}(F)$, the variety of F over $\overline{\mathbb{F}}$, is the set of common zeros of F in $\overline{\mathbb{F}}^n$. Note that in combinatorial problems, the *variety* of a polynomial system typically has finitely many solutions (e.g., colorings, cuts, stable sets, etc.). For an ideal $I \subseteq \mathbb{F}[x]$, when $\mathcal{V}_{\overline{\mathbb{F}}}(I)$ is finite, the ideal is called *zero-dimensional* (this is the case for all of the applications considered here). We say that a system of polynomial equations is a *combinatorial system* when its variety encodes a combinatorial problem (e.g., zeros represent stable sets, colorings, matchings, etc.) and it is zero-dimensional. Prior work on encoding combinatorial properties using systems of polynomials touched on on colorings, flows, matchings, stable sets, and much more [**Bay82**, **DL95**, **DLHMO10**, **DLLMO09**, **Eli92**, **Fis88**, **HW08**, **LL81**, **Lov94**, **LS91**, **Onn04**]. Here are a few examples from those papers.

First, take the graph polynomial

$$f_G = \prod_{\{i,j\}\in E,\ i<j} (x_i - x_j).$$

The following theorem relating several graph polynomials was presented in [**Lov94**]. It shows the rich theory when we think beyond one single polynomial.

THEOREM 1.3 (Kleitman–Lovász). *Let $\mathcal{H}(n,k)$ be the set of all graphs with n vertices consisting of a clique of size $k+1$ and all other $n-k+1$ vertices isolated. A graph G on n vertices is not k-colorable if and only if*

$$f_G = \sum_{H\in\mathcal{H}(n,k)} \alpha_H f_H,$$

where α_H are polynomials.

One can do much more with different coloring ideals! In fact, a combinatorial problem can often be modeled nonlinearly in many different ways, and in practice choosing a "good" formulation is critical for an efficient solution. Here is a classical formulation for the k-coloring problem due to Bayer [**Bay82**].

PROPOSITION 1.4. *Let $G = (V,E)$ be a graph. For a positive integer k, consider the following polynomial system of $|V| + |E|$ polynomials equations:*

$$x_i^k - 1 = 0 \ \ \forall i \in V \ \ \text{and} \ \ \sum_{s=0}^{k-1} x_i^{k-1-s} x_j^s = 0 \ \ \forall (i,j) \in E.$$

Then the graph G is k-colorable if and only if this system has a complex solution. Furthermore, when k is odd, G is k-colorable if and only if this system has a common root over $\overline{\mathbb{F}}_2$, the algebraic closure of the finite field with two elements.

Several variations of the above ideal have been fruitful. For instance, in [**DLLMO09**] the authors investigated the problem of deciding whether a graph G has a k-colorable subgraph with R edges. This is true if and only if the following modified system of equations has a solution (and again the solutions count such subgraphs): For each vertex $i \in V(G)$ we have the equation $x_i^k = 1$. For each edge $\{i,j\} \in E(G)$ we have equations

$$y_{ij}^2 - y_{ij} = 0, \quad y_{ij}\left(x_i^{k-1} + x_i^{k-2}x_j + \cdots + x_j^{k-1}\right) = 0,$$

and finally an "edge counting equation" $\sum_{\{i,j\}\in E(G)} y_{ij} - R = 0$.

Similarly, Hillar and Windfeldt [**HW08**] also used a simple variation of Bayer's polynomial ideal to give a characterization for when a graph is uniquely k-colorable and to express the number of k-colorings as the dimension of a quotient ring.

In [**Fis88**] Fischer considered a polynomial version of *Hall's matching theorem* on bipartite graphs. Let A be a commutative ring with unity, let $x = (x_1, \ldots, x_n)$,

$y = (y_1, \ldots, y_n)$ be indeterminates, let R denote the $2n$-variable polynomial ring $A[x, y]$, and let P be the ideal of R consisting of those polynomials $p(x, y)$ for which $p(x, \sigma x) = 0$ for all permutations σ of x. Fischer showed that P is generated by $e_k(x) - e_k(y)$ $(1 \leq k \leq n)$, where e_k denotes the kth elementary symmetric function. Now let G be a bipartite graph with vertices partitioned into $\{X_1, \cdots, X_n\}$ and $\{Y_1, \cdots, Y_n\}$. Fischer pointed out that the existence of a perfect matching in G is equivalent to the condition $\prod(x_i - y_j) \notin P$, where the product is taken over all pairs (i, j) such that X_i and Y_j are nonadjacent (which is the graph polynomial for a complete bipartite graph!). He then deduced Hall's matching theorem via the above characterization of P and an application of the very important Hilbert's Nullstellensatz. Next, we recall that fundamental theorem in algebraic geometry.

THEOREM 1.5 (Hilbert's Nullstellensatz). *Let* $F := \{f_1, \ldots, f_m\} \subseteq \mathbb{F}[x]$. *Then,*

$$\nexists x \in \overline{\mathbb{F}}^n \text{ s.t. } f_1(x) = 0, \ldots, f_s(x) = 0 \Leftrightarrow \exists \beta_1, \ldots, \beta_m \in \mathbb{F}[x] \text{ such that } 1 = \sum_{i=1}^{m} \beta_i f_i.$$

Thus we have polynomial identity, the *Nullstellensatz certificate* of infeasibility of the system. No solutions! This theorem is a classic and can be found in any algebraic geometry book (see a simple proof in [**Arr06**]). We define the degree of a Nullstellensatz certificate $\sum_{i=1}^{m} \beta_i f_i = 1$ as $\max_i\{\deg(\beta_i f_i)\}$. Upper bounds on the degree of the coefficients β_i has been an extensive area of work (see [**Bro87, Kol88, Laz77**]).

Unfortunately, certifying a combinatorial statement is often complicated. Consider the interesting situation of coloring. Given $G = (V, E)$ a graph and a positive integer k, we wish to decide whether G is k-colorable. In other words, can we assign k colors to the vertices of G so that adjacent vertices receive a different color? Convincing someone that a graph is k-colorable is easy; you just need to show an explicit k-coloring, but suppose you wish to convince someone that a given graph is **not** k-colorable. What would you do? How can you convince your boss to buy more colors to color the graph? One could show one by one how each k-coloring fails, but there are way too many k-colorings, it is not a very compact certification. Its size is often much larger than the size of the graph G itself. But polynomial ideals offer a magnificent alternative to certify combinatorial statements. The exciting fact is that using Hilbert's Nullstellensatz (Theorem 1.5) one can certify many combinatorial facts, including that a graph is not k-colorable using the Hilbert Nullstellensatz and the system of polynomials of Proposition 1.4. Let us look closely at those certificates.

The Nullstellensatz certificate can also be used to find combinatorial characterizations. For example, if the graph G is not 3-colorable and it has a Nullstellensatz certificate of degree one. In the case when the coefficients are all linear polynomials there is a way to characterize which graphs are non-3-colorable of Nullstellensatz degree one (note the similarity with Theorem 1.14). Let A denote the set of all possible directed edges or arcs in the graph G. Consider two types of substructures of the graph G: oriented partial-3-cycles and oriented chordless 4-cycles An **oriented partial-3-cycle** is a set of two arcs of a 3-cycle, that is, a set $\{(i, j), (j, k)\}$ also denoted (i, j, k) where $(i, j), (j, k), (k, i) \in A$. An **oriented chordless 4-cycle** is a set of four arcs $\{(i, j), (j, l), (l, k), (k, i)\}$ also denoted (i, j, k, l) where $(i, j), (j, l), (l, k), (k, i) \in A$ and $(j, k), (i, l) \notin A$.

Now, we state a sufficient condition for non-3-colorability from [**DLHMO10**].

THEOREM 1.6. *The graph G is not 3-colorable if there exists a set C of oriented partial 3-cycles and oriented chordless 4-cycles such that*

(1) $|C_{(i,j)}| + |C_{(j,i)}| \equiv 0 \pmod 2$ *for all* $(i,j) \in E$ *and*
(2) $\sum_{(i,j)\in A, i<j} |C_{(i,j)}| \equiv 1 \pmod 2$

where $|C_{(i,j)}|$ denotes the number of cycles in C (either 3-cycles or 4-cycles) in which the arc $(i,j) \in A$ appears.

This sufficient condition for non-3-colorability is equivalent to the combinatorial system encoding 3-coloring has a Nullstellensatz certificate of degree one.

Condition (1) in Theorem 1.6 means that every undirected edge of G is covered by an even number of directed edges from cycles in C (ignoring orientation). Condition (2) in Theorem 1.6 means that, given any orientation of G, the total number of times the arcs in that orientation appear in the cycles of C is odd. Using Theorem 1.6 one can easily prove that graphs containing odd wheels (e.g., 4-cliques) are not 3-colorable (see [**DLHMO10**]). A stronger Theorem 1.6 without orientations is in [**LLO16**].

Here is another example of how Nullstellensatz certificates provide structural combinatorial descriptions: Ramsey numbers are some of the most interesting and mysterious combinatorial numbers [**GRS90**]. Recall, given graphs $G_1, \dots, G_k$ the Ramsey number $R(G_1, G_2, \dots, G_k)$ is the smallest positive integer n such that every k-coloring of the edges of K_n contains a monochromatic copy of some G_i. If $G_i = K_{r_i}$ for all i, we simply write $R(r_1, r_2, \dots, r_k)$. All of these numbers are finite by Ramsey's theorem. In [**DLW22**] we looked at the Ramsey numbers as the number of variables in certain system of polynomials. The solutions over the field $\mathbb{F}_2$ are indicator vectors that yield all *Ramsey graphs*, namely the subgraphs that do not contain monochromatic copies of any G_i.

THEOREM 1.7. *The Ramsey number $R(G_1, \dots, G_k)$ is at most n if and only if there is no solution to the following system over $\overline{\mathbb{F}_2}$, where $K_n = (V,E)$ is the complete graph on n vertices. Moreover, when the system has solutions, the number of solutions to this system is equal to the number of graphs of order n that avoid copies of G_i in color i.*

$$p_{H,i} := \prod_{e\in E(H)} x_{i,e} = 0 \qquad \forall i, 1 \le i \le k, \quad \forall H \subseteq K_n,\ H \cong G_i, \tag{1.1}$$

$$q_e := 1 + \sum_{i=1}^{k} x_{i,e} = 0 \qquad \forall e \in E, \tag{1.2}$$

$$u_{i,j,e} := x_{i,e}x_{j,e} = 0 \qquad \forall e \in E,\ \forall i,j,\ i \neq j. \tag{1.3}$$

A key consequence of Theorem 1.7 is that the very first value of n for which the system of equations has no solution is equal to the Ramsey number. In Corollary 1.8 we give degree bounds for Nullstellensatz certificates of Ramsey numbers. We use the polynomial encoding from Theorem 1.7. Assuming the number of vertices of a Ramsey ideal is n and $n \ge R(G_1, \dots, G_k)$, we give upper bounds for the degrees of the Nullstellensatz certificates. Surprisingly, our construction of Nullstellensatz certificates for the Ramsey ideals yields certificate degrees described by the *restricted online Ramsey numbers* $\tilde{R}(G_1, \dots, G_k; n)$. These numbers, first introduced by Conlon et al. [**CFGH19**], are defined using the *Builder-Painter game*: Each turn Builder selects one edge e of K_n, and Painter selects a color in $[k]$ and colors

e with this color. Builder wins the game once a monochromatic G_i in color i is constructed for some i. The number $\tilde{R}(G_1, \ldots, G_k; n)$ is the minimum number of edges Builder needs to guarantee a victory.

COROLLARY 1.8. *If* $n \geq R(G_1, \ldots, G_k)$*, then there is a Nullstellensatz certificate of degree* $\tilde{R}(G_1, \ldots, G_k; n) - 1$ *that the statement* $R(G_1, \ldots, G_k) > n$ *is false using the system of equation in Theorem* 1.7.

It is trivial that $\tilde{R}(G_1, \ldots, G_k) \leq \binom{R(G_1,\ldots,G_k)}{2}$, but it was shown in [**GHZ19**] that it can be improved. Curiously, this is an improvement over the upper bounds provided by [**BP96, Bro87, Laz77, DLLMO09**] where the degree of the Nullstellensatz certificates grows linearly in the number of variables of the system.

At the same time, tight exponential upper bounds on the degrees have been derived for general ideals (see, e.g., [**Kol88**], [**GV02**], and references therein). It is important to stress that for a fixed degree of the certificates, the calculation of a Nullstellensatz certificate can be done in polynomial time because it reduces to solving a system of *linear* equations of polynomial size. Moreover, often for combinatorial problems, the associated polynomial system contains finitely many solutions (as in the coloring problem). Combinatorial (binary) polynomial systems have been explored as algebraic proof systems [**BP96, CEI96, BIK$^+$96, IPS99, Raz98, GV02**] and Buss and Pitassi [**BP96**] first showed that a polynomial system loosely based on the "*pigeon-hole principle*" has a growing Nullstellensatz degree certification without any complexity assumption. One can prove that for combinatorial systems the degree growth is no worse than linear in the number of variables [**Laz77**]. This proves the fundamental fact that the degrees grow in the worse case as the number of variables.

There are interesting connections of computational complexity and the degree and size of a Nullstellensatz certificates. When the system of polynomial equations $f_1, \ldots, f_s$ models an NP-complete problem, such as for the k-colorability problem, the degree d must grow at least linearly with the size of the underlying NP-complete instance [**DLLMO09**]. In other words, as long as P $\neq$ NP, the certificates should be hard to find (i.e., the size of the linear systems involved should be exponential in the size of the underlying instance), and as long as NP $\neq$ coNP, the certificates should be hard to verify (i.e., the certificates should contain an exponential number of monomials). Note that [**DLLMO09**] explicitly shows growth without the $P \neq NP$ assumption in the case of stable set problems.

Regarding the k-coloring problem, the degrees of the Nullstellensatz certificates were naturally expected to grow at least linearly leading to exponential growth in the encoding size of the certificates, simply because the NP-hardness of the problem, but it took some time to construct such a family. After much work Lauria and Nordström proved in [**LN17**] that there are indeed non-k-colorable graphs whose Nullstellensatz requires arbitrarily large degrees, more precisely in the following.

THEOREM 1.9. *For any integer* $k > 3$ *there is an efficiently constructible family of graphs with* $O(k^4 n)$ *vertices of degree* $O(k^2)$ *that do not possess* k*-colorings, but for which the corresponding system of polynomial equations we saw earlier requires degree* $O(n)$ *Nullstellensatz certificates.*

We should point out that the proof of the theorem uses again a reduction to a well-known system of polynomial equations arising from the pigeon-hole principle and, in fact, gives degree lower bounds not only for Nullstellensatz certificates but also for the much stronger setting of Gröbner basis computations (a Nullstellensatz can always be recovered from such a computation). However, Lauria and Nordström's result did not give much insight into which combinatorial properties of a graph G are required to obtain very high degree Nullstellensatz certificates. Recently, in his PhD thesis, Romero-Barbosa [**Rom21**] showed that in fact non-3-colorable graphs with large enough girth, i.e., the length of the shortest cycle in the graph, need large Nullstellensatz certificates too. Tuncel and Romero-Barbosa proved the following remarkable result.

THEOREM 1.10. *Let $G = (V, E)$ be a graph with chromatic number $\chi(G) = k+1$ and girth $g > 2k$. Suppose that $d + k + \log(d + k - 1) < \frac{g}{2k}$, then G has no Nullstellensatz certificate for its non-k-colorability of degree d.*

The same kind of analysis can be done for other NP-complete problems. For stable sets, [**DLLMO09**] showed that a Nullstellensatz certificate of degree $\alpha(G)$ is needed, where $\alpha(G)$ is the stability number of G, namely, the size of the largest stable set of the input graph G. Another case where we understand the growth of Nullstellensatz degree is for the *sum partition problem* [**GJ79**], which is the question of deciding whether or not a given set of integers $W = \{w_1, \ldots, w_n\}$ can be broken into two sets, I and $W \setminus I$, such that the sums of the two sets are equal or that $\sum_{w\in I} w = \sum_{w\in W\setminus I} w$. Since it is widely believed that NP $\neq$ coNP, it is interesting to study Nullstellensatz certificates for the *non*existence of a partition in a set W.

Just as before, we can describe a system of polynomial equations that models this question, and discuss the degree and monomials in an associated minimum-degree Nullstellensatz certificate.

PROPOSITION 1.11. *Given a set of integers $W = \{w_1, \ldots, w_n\}$, the following system of polynomial equations*

$$x_i^2 - 1 = 0 \;, \quad \textit{for } 1 \le i \le n \;, \qquad \textit{and} \qquad \sum_{i=1}^{n} w_i x_i = 0,$$

has a solution if and only if there exists a partition of W into two sets, $I \subseteq W$ and $W \setminus I$, such that $\sum_{w\in I} w = \sum_{w\in W\setminus I} w$.

Let $[n]$ denote the set of integers $\{1, \ldots, n\}$ and let S_k^n denote the set of k-subsets of $[n]$. For $S \in S_k^n$, let x^S denote the corresponding square-free monomial of degree k in n variables. For example, given $S = \{1, 3, 4\} \subseteq [5]$, the corresponding monomial $x^S = x_1x_3x_4$. Additionally, let $S_k^{n\setminus i}$ denote the k-subsets of $[n] \setminus i$.

THEOREM 1.12. *Given a set of nonpartitionable integers $W = \{w_1, \ldots, w_n\}$ encoded as a system of polynomial equations according to Proposition 1.11, there exists a minimum-degree Nullstellensatz certificate for the* non*existence of a partition of W as follows:*

$$1 = \sum_{i=1}^{n} \Big(\sum_{\substack{k \text{ even} \\ k\le n-1}} \sum_{S\in S_k^{n\setminus i}} c_{i,S}x^S \Big)(x_i^2 - 1) + \Big(\sum_{\substack{k \text{ odd} \\ k\le n}} \sum_{S\in S_k^n} b_S x^S \Big)\Big(\sum_{i=1}^{n} w_i x_i \Big).$$

Moreover, every Nullstellensatz certificate for the system of equations defined by Proposition 1.11 contains one monomial for each of the odd parity subsets of each S_k^n, and one monomial for each of the even parity subsets of each $S_k^{n\setminus i}$.

Theorem 1.12 says that the degree of the certificate is n for n odd, and $n-1$ for n even. By considering the monomials present in the certificate as identifying the integers present on *one* side of a partition, we see that the monomials represent a brute-force iteration over every possible partition of W.

Can one use the Hilbert Nullstellensatz for practical computation? Yes! Theory and practice can actually be far apart. Due to the complexity news above one can be pessimistic, but in practice it is often the case that it is possible to find very low-degree Nullstellensatz certificates. For example, in practice this happens when we certify that a graph is not k-colorable (see [**DLLMM08, DLLMM11**]). To generate the Nullstellensatz certificates fast follows immediately once we guess a degree a Nullstellensatz. Then the recovery of the coefficients is reduced to solving a sequence of linear algebra problems (the knowledge of the degree implies we know the possible monomials, and thus the set up of the linear equations). For the coloring problem the papers [**DLLMM08, DLLMM11**] showed remarkable computational performance (while incorporating mod 2 calculation, branch-and-bound and other heuristics). More can be done to speed up the computation of Nullstellensatz certificates. For example, in [**dRMNR21**] the authors showed that while one can always find a degree d certificate by solving a system of linear equations of size $O(n^d)$ (with n the number of variables), there could exist a Nullstellensatz certificate with far fewer monomials if one is willing to go to higher degree. More deeply, the combinatorics of *designs* (special kind of maps from polynomial rings over K to K) is deeply connected to the constructions of Nullstellensatz certificates. See [**Bus96, BIK$^+$97**].

The foundations of this "linearizations" technique, i.e., setting up a linear system from a Nullstellensatz of fixed degree, can be traced back to [**KK05, KKR05, Mou99, MT08**]. Variants of this technique have been applied to stable sets and vertex coloring [**DLLMO09, Mar08a**], satisfiability (see, e.g.,[**CEI96**]) and cryptography (see, for example, [**CKPS00**]). Using linear algebra (in high-dimension) to calculate with systems of polynomials with finitely many roots is also popular and effective in border basis and Gröbner basis computations (see, e.g., [**Fau99, FM17, KK05, KKR05, MT08, Ste04**]).

1.2. Adding polynomial inequalities: Semi-algebraic discrete math. Much of what we have been discussing so far are representations of combinatorial problems using polynomial equations and we only worry about solutions that are complex or over finite fields. But there is an alternative world when we also use inequalities and/or inequations and we insist on using the *real* solutions to describe a combinatorial problem. In what follows we travel rapidly through *real* nonlinear algebra (see [**BPT, BCR87, Las01, Lau07, Mar08b, Par03**] and the references therein for its foundations).

Instead of varieties and ideals, the key objects are *basic semi-algebraic sets*: Given by a system of real equations and inequalities $S = f_1, \dots, f_s, g_1, \dots, g_k \in \mathbb{R}[x]$, its semi-algebraic set is the set

$$\{x \in \mathbb{R}^n \ s.t. \ f_1(x) = 0, \dots, f_s(x) = 0, \ g_1(x) \geq 0, \dots, g_k(x) \geq 0\}.$$

Computing with semi-algebraic sets is not so easy!

There are so many problems in discrete mathematics that require combinatorial analysis using semi-algebraic sets to solve them that I will only scratch the surface here. Let us start with problems in discrete geometry. Consider first a problem in (real) enumerative geometry. The idea is to enumerate the possible configurations of geometric objects that can be described by polynomial equations. A very good example is, given four unit spheres in R^3, how many lines can be simultaneously tangent to all? In this regard we have [**MPT01**] is the following theorem.

THEOREM 1.13 (MacDonald–Pach–Theobald). *There are at most 12 (real) lines tangent to 4 unit spheres when the centers are not collinear.*

More generally, Sottile and Theobald extended this result for $2n - 2$ spheres in n-dimensional Euclidean space [**ST02**]. A survey of this type of enumerative problems appeared in [**Sot01**].

In many applications bounding the number of combinatorial objects one has a family of polynomials $G = \{f_1, \ldots, f_m\} \subset R[X_1, \ldots X_n]$ (here $m \geq n$) where each polynomial has degree at most d. By a choice of signed for each f_i namely $f_i \geq 0$ or $f_i \leq 0$, the polynomials break R^k into cells (that are defined by points that evaluate to those signs). Basu, Pollack, and Roy [**BPR96**] proved that the number of cells defined by sign choices using P_i is $O(d)^k$. Note that the combinatorial part of the bound depends on the dimension of the variety, rather than on the dimension of the ambient space. This theorem follows earlier work of Milnor and Warren, who in the 1960s used techniques from real algebraic geometry and now has evolved into a useful tool for discrete mathematics [**BPR06**].

This type of analysis can be extended to any field F. Indeed, in general fields one does not have a notion of inequality, but we can define the *zero-pattern* of the polynomial G at $u \in F^n$, this is the set of those index for which $f_i(u) = 0$. Let $ZF(G)$ denote the number of zero-patterns of G as the evaluation point u ranges over F^n. In [**RBG01**] the authors proved almost optimal bound for the size of $ZF(G)$. Sharper results were later found in a rather nice paper by Lauder [**Lau03**]. These zero-pattern bounds can be used in combinatorial applications. One particularly nice application gives nearly tight Ramsey bounds for matrices whose entries are defined by zero-patterns of a sequence of polynomials. This has recently grown into a subfield some reseachers call *semi-algebraic combinatorics*.

One of the central questions in Ramsey theory asks how small the largest clique and independent set in a graph on N vertices can be. By a celebrated result of Erdös, a random graph on N vertices with edge probability 1/2 contains no clique or independent set larger than $2 \log_2 N$, with high probability. Finding explicit constructions of graphs with similar Ramsey-type properties is a famous open problem. A natural approach is to construct such graphs using algebra. This is what we call an *algebraic graph* or hypergraph. We say that an r-uniform hypergraph $\mathcal{H}$ is algebraic of complexity (n, d, m) if the vertices of $\mathcal{H}$ are elements of $\mathbb{F}^n$ for some field $\mathbb{F}$, and there exist m polynomials $f_1, \ldots, f_m : (\mathbb{F}^n)^r \to \mathbb{F}$ of degree at most d such that the edges of $\mathcal{H}$ are determined by the zero-patterns of $f_1, \ldots, f_m$. A number of authors have worked on this topic. See, e.g., [**CFP$^+$14, ST22, Sau21b**].

The aim of the work by some authors has been to show that if an algebraic graph (or hypergraph) of complexity (n, d, m) has good Ramsey properties, then at least one of the parameters n, d, m must be large. In their paper [**RBG01**] Rónyai, Babai, and Ganapathy considered the bipartite variant of the Ramsey problem and proved that if G is an algebraic graph of complexity (n, d, m) on N vertices, then either G or its complement contains a complete balanced bipartite graph of size $\Omega_{n,d,m}(N^{1/(n+1)})$. Sudakov and Tomon [**ST22**] extended this result by showing that such a graph contains either a clique or an independent set of size $N^{\Omega(1/ndm)}$

and proved similar results for algebraic hypergraphs of constant complexity. They obtained a polynomial regularity lemma for r-uniform algebraic hypergraphs that are defined by a single polynomial.

Geometric graphs are particularly nice for analysis. For instance, by Ramsey's theorem, any system of n segments in the plane has roughly $\log n$ members that are either pairwise disjoint or pairwise intersecting. Analogously, any set of n points $p(1), \ldots, p(n)$ in the plane has a subset of roughly $\log \log n$ elements with the property that the orientation of $p(i)p(j)p(k)$ is the same for all triples from this subset with $i < j < k$. For geometric problems the underlying graph can be defined by a small number of polynomial equations and inequalities in terms of the coordinates of the segments and points. There are many fascinating structural properties of "semi-algebraically" defined graphs and hypergraphs; see [**APP$^+$05**, **CFP$^+$14**].

Next, I recount a connection of nonlinear algebra to problems of realizability of combinatorial manifolds, polytopes and matroids. These were the original motivations of Bernd Sturmfels in his graduate student years [**BS86**, **Stu87b**]. In the 1980s, discrete mathematicians realized the importance of polynomials in the context of realization spaces of polyhedra and combinatorial manifolds. Let $P \subset \mathbb{R}^d$ be a d-dimensional convex polytope. The *realization space* of P is the space of all polytopes $P' \subset \mathbb{R}^d$ that are combinatorially equivalent to P, modulo affine transformations. One can see that this set is a semi-algebraic set. For simplicial and for simple polytopes, the realization spaces can be seen as open subspaces of Euclidean space. Researchers, including Bernd, who saw this very early on [**Stu87c**, **Stu87a**], realized that many discrete-geometric decision problems such as "is this combinatorial sphere realizable as a convex polytope?" are equivalent to the solvability of some system of polynomial equations and inequalities (see [**BS89**]). Among significant theory work we have Mnëv's universality theorem (see [**Mn83**, **Mn85**]). It says, given any basic semi-algebraic set W defined by polynomials with integer coefficients there is a polytope $P(W)$ whose realization space is "stably equivalent" to W. This implies that the realization space of a polytope can have the homotopy type of an arbitrary finite simplicial complex (e.g., it can be disconnected), and that all algebraic numbers must appear as coordinates of vertices when we realize all convex polytopes. Later on Richter and Gebert showed (in a lovely constructive proof) that even realization spaces of 4-dimensional polytopes can also be arbitrarily complicated [**RG96**]. In comparison, realization spaces of 3-dimensional polytopes are contractible and all 3-polytopes are realizable with integer coordinates (Steinitz's theorem) (see [**OS94**, **RG99**]).

There is recent work continuing this fascinating topic of realization. In [**AKN20**] the authors prove the realization space of cubes are contractible and a more refined analysis of Mnëv's construction was done in [**AP17**, **AZ15**]. Recently, another way to understand realization spaces was done via the the *slack matrix* of a polytope P. This is a nonnegative real matrix whose rows and columns are indexed by the vertices and facets of P, respectively, whose (i, j) entry is the slack (the value) of the ith vertex in the jth facet inequality. In particular, the (i, j) entry is zero if and only if the ith vertex lies on the jth facet, so the slack matrix determines the vertex-facet incidences, hence the face lattice, of the polytope. Moreover, the convex hull of the rows of the slack matrix is a polytope which is affinely equivalent to P. Therefore a realization space of a d-dimensional polytope

P, modulo affine equivalence, is given by the set of slack matrices of d-dimensional polytopes which have the same support (i.e., the same zero pattern) as the slack matrix of P. Thus the realization space of a polytope (modulo projective transformations) can be defined by the determinantal conditions determining the rank of the slack matrices, together with the zero/positivity pattern of entries of the matrix (modulo scaling of rows and columns). See [**GMTW19**, **GMTW20**, **GMW23a**] for this rich new way to study realizations spaces, especially via computation. Matroid realizations and the associated nonlinear algebra is also quite interesting (see, e.g., [**BLVS**$^+$**99**, **FMM13**, **STW21**] and the references therein) and we must stress that computation with polynomial systems (sometimes combined with optimization or SAT software) allows one to generate *automated proofs* of many remarkable geometric theorems; see, e.g., [**BRS90**, **Bok00**, **SNR11**, **GMW23b**, **Pfe22**].

Next we briefly outline how real algebraic and semi-algebraic geometry can help in the understanding of the solution of combinatorial optimization problems. A central preoccupation of combinatorial optimizers has always been the understanding of the facets that describe the integer hull of a combinatorial problem when the possible outcomes are represented as binary vectors [**Sch03a**, **Sch03b**, **Sch03c**]. It turns out, one can recover quite a bit of information about the integer hull of combinatorial problems from a sequence of combinatorially controlled linear or semi-definite programming problems that give better and better approximations. This type of approach was pioneered in the lift-and-project method of Balas, Ceria, and Cornuéjols [**BCC93**], the matrix-cut method of Lovász and Schrijver [**LS91**], the linearization technique of Sherali and Adams [**SA90**] and the *sum of squares method* of Parrilo [**Par00**] and Lasserre [**Las01**]. It can be seen that the nonlinear algebra point of view leads to all these methods. Here I present a few pointers to that vast literature.

Such an approach has been used in many problems, for example, the study of *stable sets polytopes* [**Lov94**]: A graph G is *perfect* if for every induced subgraph H, the chromatic number of H equals the size of the largest complete subgraph of H. [**eRAeR01**, **Tro04**, **CRST06**]. Perfect graphs are important, among other reasons, because they are graphs for which hard combinatorial optimization problems like coloring and the stable set problem are polynomial-time solvable and they have small extended formulations. In [**LS91**] a general method was developed to construct higher-dimensional polyhedra (or, in some cases, convex sets) whose projection approximates the convex hull of 0–1 valued solutions of a system of linear inequalities. This connects polynomial identities to efficient extended formulations of stable set polytopes of t-perfect graphs (for general graphs, this is quite complex). We say a graph is **t-perfect** if for a linear form $f(z) \geq 0$ over all incidence vectors of stable sets (vertices of the stable set polytope) if and only if there exist polynomials g_i, of degree $\leq t$, such that

$$f = g_1^2 + \dots g_k^2 + \sum a_{ij} x_i x_j + \sum b_i (x_i^2 - x_i).$$

When $t = 1$, these are precisely the perfect graphs.

THEOREM 1.14 (Lovász–Schrijver). *If there exists an integer constant t for which the graph G is t-perfect, then the stable set polytope of G is the projection of a polytope whose number of facets is $O(mn)$ (number of edges m and nodes n of G).*

For more on this topic see recent work [**AD23**] where many statements about perfect graphs are rephrased in terms of polynomials identities. In [**GPT10, GLPT12**] the authors considered a sequence of semidefinite relaxations of the convex hull of the real solutions of a polynomial system encoding a combinatorial problem. They called these approximations *theta bodies* because, for stable sets of graphs, the first theta body in this hierarchy is exactly Lovász's theta body of a graph [**Lov94**] (see [**Par03, Lau07, GPT10**] and references therein for a very extensive survey and the book [**BPT**]).

A multivariate polynomial $p(x)$ is a *sum of squares* (SOS) if it can be written as a sum of squares of other polynomials, that is, $p(x) = \sum_i q_i^2(x), \quad q_i(x) \in \mathbb{R}[x]$. If $p(x)$ is SOS, then clearly $p(x) \geq 0$ for all $x \in \mathbb{R}^n$. The condition that a polynomial is a sum of squares is a natural sufficient (although not necessary) test for nonnegativity of a polynomial. The existence of a sum of squares decomposition can be decided via semidefinite programming. More strongly we can use SDPs to decide the feasibility of a system of polynomial inequalities. Exactly as it happened for the Nullstellensatz, a semi-algebraic set is empty if there is a *Positivstellensatz* certificate (see [**BCR87**]). For fixed degree, one can reformulate the problem as a semidefinite programming problem. Software implementations are available at SOSTOOLS [**PPSP04**], GloptiPoly [**HL03**], and YALMIP [**Löf04**].

We will outline an algorithmic approximation of the convex hull of a finite set of points S, denoted $\text{conv}(S)$, by a sequence of convex bodies recovered from "degree truncations" of the defining polynomial systems. In what follows I will be a radical polynomial ideal. A polynomial f is *nonnegative* modulo I, written as $f \geq 0 \bmod I$, if $f(s) \geq 0$ for all $s \in \mathcal{V}_{\mathbb{R}}(I)$. More strongly, the polynomial f is a *sum of squares (sos)* mod I if there exists $h_j \in \mathbb{R}[x]$ such that $f \equiv \sum_{j=1}^t h_j^2 \bmod I$ for some t, or equivalently, $f - \sum_{j=1}^t h_j^2 \in I$. If, in addition, each h_j has degree at most k, then we say that f is *k-sos* mod I. The ideal I is *k-sos* if *every* polynomial that is nonnegative mod I is k-sos mod I. If every polynomial of degree at most d that is nonnegative mod I is k-sos mod I, we say that I is *(d, k)-sos*.

We say that a polynomial ideal $I \subseteq \mathbb{R}[x]$ is TH_k-exact if every linear polynomial that is nonnegative over $\mathcal{V}_{\mathbb{R}}(I)$, the real variety of I, is a sum of squares of polynomials of degree at most k modulo I. Note that $\text{conv}(\mathcal{V}_{\mathbb{R}}(I))$, the convex hull of $\mathcal{V}_{\mathbb{R}}(I)$, is described by the linear polynomials f such that $f \geq 0 \bmod I$. A certificate for the nonnegativity of f mod I is the existence of a sos-polynomial $\sum_{j=1}^t h_j^2$ that is congruent to f mod I. One can now investigate the convex hull of S through the hierarchy of nested closed convex sets defined by the semidefinite programming relaxations of the set of $(1, k)$-sos polynomials.

DEFINITION 1.15. Let $I \subseteq \mathbb{R}[x]$ be an ideal, and let k be a positive integer. Let $\Sigma_k \subset \mathbb{R}[x]$ be the set of all polynomials that are k-**sos** mod I.

(1) The kth *theta body* of I is

$$\text{TH}_k(I) := \{x \in \mathbb{R}^n \ : \ f(x) \geq 0 \text{ for every linear } f \in \Sigma_k\}.$$

(2) The ideal I is TH_k*-exact* if the kth theta body $\text{TH}_k(I)$ coincides with the closure of $\text{conv}(\mathcal{V}_{\mathbb{R}}(I))$.

(3) The *theta-rank* of I is the smallest k such that $\text{TH}_k(I)$ coincides with the closure of $\text{conv}(\mathcal{V}_{\mathbb{R}}(I))$.

Take, for example, the *max-cut problem*, the defining vanishing ideal is $I(SG) = \langle x_e^2 - x_e \ \forall \ e \in E, \ x^T \forall \ T$ an odd cycle in $G\rangle$. In this case one can prove that the ideal $I(SG)$ is TH_1-exact if and only if G is a bipartite graph. In general the theta-rank of $I(SG)$ is bounded above by the size of the max-cut in G. There is no constant k such that $\mathrm{TH}_k(I(SG)) = \mathrm{conv}(SG)$, for all graphs G. Other formulations of max-cut are studied in [**GLPT12**]. Recall that when $S \subset \mathbb{R}^n$ is a finite set, its vanishing ideal $I(S)$ is zero-dimensional and real radical (see [**Mar08b**, Section 12.5] for a definition of the real radical). One says that a finite set $S \subset \mathbb{R}^n$ is *exact* if its vanishing ideal $I(S) \subseteq \mathbb{R}[x]$ is TH_1-exact. Here are some results on exactness.

THEOREM 1.16 ([**GPT10**]). *For a finite set $S \subset \mathbb{R}^n$, the following are equivalent.*

(1) *S is exact.*
(2) *There is a finite linear inequality description of* $\mathrm{conv}(S)$ *in which for every inequality $g(x) \geq 0$, g is 1-sos mod $I(S)$.*
(3) *There is a finite linear inequality description of* $\mathrm{conv}(S)$ *such that for every inequality $g(x) \geq 0$, every point in S lies either on the hyperplane $g(x) = 0$ or on a unique parallel translate of it.*
(4) *The polytope* $\mathrm{conv}(S)$ *is affinely equivalent to a compressed lattice polytope (every reverse lexicographic triangulation of the polytope is unimodular with respect to the defining lattice).*

More strongly, there are results about the geometry of TH_t-exact sets.

PROPOSITION 1.17. *Suppose $S \subseteq \mathbb{R}^n$ is a finite point set such that for each facet F of* $\mathrm{conv}(S)$ *there is a hyperplane H_F such that $H_F \cap \mathrm{conv}(S) = F$ and S is contained in at most $t+1$ parallel translates of H_F. Then $I(S)$ is TH_t-exact.*

Gouveia, Parrilo, and Thomas [**GPT10**] further showed that theta bodies can be computed as projections to the feasible set of a semidefinite program. These SDPs are constructed using the *combinatorial moment matrices* introduced by [**Lau07**]. I must mention that an important related problem to computing theta bodies is the challenge of representing a "complicated" polytope as the projection of a higher-dimensional, but simpler, polytope or convex set. This is called finding an *extended formulation* and it is a powerful tool in combinatorial optimization. Today we know that many important combinatorial polytopes, such as stable set polytopes, traveling salesperson polytope, matching polytopes, cut and correlation polytopes do not admit small extended formulations (for an overview of this large topic see [**FMP$^+$15, Tho18**] and the breakthroughs [**FMP$^+$15, Rot17**]).

Polynomial inequalities and semi-algebraic sets are also useful to classify combinatorial types or structure. Polytope theory has produced a great number of remarkably simple and complete semi-algebraic characterization results for face-number sets or f-vector sets of classes of polytopes, for example describing the possible f-vectors of simplicial polytopes (see [**Zie95**]). We observe that in most cases these sets can be described as the intersection of a semi-algebraic set with an integer lattice. Such semi-algebraic sets of lattice points have not received enough attention. In [**SZ19**] Sjöberg and Ziegler showed some f-vector sets are not semi-algebraic sets of lattice points. They proved it for the set of all pairs (f_1, f_2) of 4-dimensional polytopes, the set of all f-vectors of simplicial d-polytopes for $d \geq 6$, and the set of all f-vectors of general d-polytopes for $d \geq 6$.

Another very interesting semi-algebraic set comes from *graph homomorphisms.* For two graphs H and G, the homomorphism density $t(H;G)$ is the probability that a random map from the vertex set of H to the vertex set of G maps edges of H to edges of G. For instance, $t(K_2;G) = \frac{2|E|}{n^2}$ where K_2 is the complete graph on two vertices (that is, just an edge), E is the edge set of G and n is the number of vertices of G. Many statements in extremal graph theory can be written as inequalities where the homomorphism densities are variables. For instance, the inequality $t(K_3;G) - t(K_2;G)(2t(K_2;G) - 1) \geq 0$ says that any graph G with n vertices and m edges contains at least $\frac{m}{3n}(4m - n^2)$ triangles. An approach for proving such homomorphism density inequalities has been through deducing the nonnegativity of an inequality involving homomorphism densities by expressing it as a sum of squares. While not all homomorphism density inequalities can be certified this way [**BRST20**], thinking of densities as polynomials has been very fruitful [**BR22**].

Here is a final example of using real-valued nonlinear algebraic methods: The algebra of symmetric functions has several familiar homogeneous bases, for example, the monomial basis, the elementary symmetric polynomial basis, and the Schur polynomial basis. The bases are commonly indexed by partitions λ. Fixing a positive integer n, each basis gives homogeneous polynomials in $\mathbb{Q}[x_1, \ldots, x_n]$ that span the algebra of symmetric polynomials. Let $\{g_\lambda\}$ are the nonzero polynomials obtained from any of the bases, normalized so that $g_\lambda(1, \ldots, 1) = 1$. It is known that if λ and μ are comparable in the dominance order (dominance order on partitions is induced by comparing partial sums), then $g_\lambda(x) \leq g_\mu(x)$ for all positive x (a polynomial inequality) and the converse was known for the most important bases (e.g., the monomial basis, the elementary symmetric polynomial basis, and the Schur polynomial basis) but not for the complete homogeneous symmetric functions basis $\{h_\lambda\}$ [**CGS11**]. Recently, for the basis $\{h_\lambda\}$, Heaton and Shankar finally provided a counterexample to that conjecture. They used methods of sums of squares [**HS21**].

2. The Nonlinear algebra (hidden?) structure in discrete mathematics

In this second half we will look at four topics where a (sometimes hidden) nonlinear algebra structure appears and provides elegant solutions to hard combinatorial problems. Two of the stories will be applied discrete optimization, the other two show the opportunities in algebraic combinatorics.

2.1. Nonlinear discrete optimization. Nonlinear algebra is useful for *integer optimization* is an important part of topic dedicated to optimizing a function over the lattice points of a region in space delimited by some constraints (see [**BW05, DLHK12, Onn10, Sch86**] for introductions). Most commonly, the region over which we optimize is a convex polyhedron. More precisely, the *integer linear programming problem* (ILP(A,c,b)) asks to find the solution of the following query: Given $A \in \mathbb{Z}^{m\times n}$, $\mathbf{b} \in \mathbb{Z}^m$, and $\mathbf{c} \in \mathbb{Z}^n$, find $\mathbf{x} \in \mathbb{Z}^n$ that solves

$$\text{(ILP(A,c,b))} \qquad \min\{\,\mathbf{c}^T\mathbf{x} : A\mathbf{x} = \mathbf{b},\ \mathbf{x} \in \mathbb{Z}^n\ \mathbf{x} \geq 0\,\}$$

or report INFEASIBLE or UNBOUNDED.

It is well known that this problem is, in general, NP-hard and most of the famous theoretical results in the field find useful structure of the polyhedron that may yield speed ups or, in nice cases, polynomial-time algorithms. For instance, when the matrix A is totally unimodular there are well-known results that guarantee an efficient solution [**FT87**]. Not surprisingly, the nonlinear algebra way to look at problems reveals new structure to aid on designing algorithms. We present some examples here.

2.1.1. *Integer feasibility and polynomials.* First, let us talk about a variation of the Nullstellensatz method we saw before, this time applied to solve the *integer feasibility problem*, i.e., the problem of deciding the existence of *integer solutions* of the system of linear equations (ILP(A,c,b)). In [**Las04**] Lasserre considered this problem and reduced it to a sequence of (easier, continuous) linear programs via ideal membership identities.

THEOREM 2.1 (Lasserre). *Let $A \in \mathbb{N}^{m\times n}, b \in \mathbb{N}^m$. Then the following two statements* (i) *and* (ii) *are equivalent:*

(i) *The linear system $Ax = b$ has a solution $x \in \mathbb{N}^n$.*
(ii) *The real-valued polynomial $z \mapsto z^b - 1 := z_1^{b_1} \cdots z_m^{b_m} - 1$ can be written*

$$z^b - 1 = \sum_{j=1}^{n} Q_j(z)(z^{A_j} - 1) \tag{2.1}$$

for some real-valued polynomials $Q_j \in \mathbb{R}[z_1, \ldots, z_m]$, $j = 1, \ldots, n$, all of them with nonnegative coefficients.

In addition, the degree of the Q_j's in (2.1) *is bounded by*

$$b^* := \sum_{j=1}^{m} b_j - \min_k \sum_{j=1}^{m} A_{jk}. \tag{2.2}$$

Let $s(u) := \binom{m+u}{u}$ the dimension of the vector space of polynomials of degree u in m variables. In view of Theorem 2.1, and with b^* as in (2.2), checking the existence of a solution $x \in \mathbb{N}^n$ to $Ax = b$ reduces to checking whether or not there exists a nonnegative solution y to a system of (real nonnegative) linear equations with:

- $n \times s(b^*)$ variables, the nonnegative coefficients of the Q_j's.
- $s(b^* + \max_k \sum_{j=1}^{n} A_{jk})$ equations to identify the terms of same power in both sides of (2.1).

In this reformulation Equation (ILP(A,c,b)) reduces to solving a *linear programming problem* with $ns(b^*)$ variables and $s(b^* + \max_k \sum_j A_{jk})$ equality constraints. Observe that in view of (2.1), this LP has a matrix of constraints with only 0 and ± 1 coefficients. The strength of Theorem 2.1 is precisely to allow $Q_j \in \mathbb{R}[z_1, \ldots, z_m]$ as it permits to check feasibility by solving a (continuous) linear programs, which are efficient to solve both theoretically and practically [**Sch86**].

Note also how Theorem 2.1 reduces the issue of existence of an integer solution $x \in \mathbb{N}^n$ to a particular type of *polynomial ideal membership problem*, that is, $Ax = b$ has a solution $x \in \mathbb{N}^n$ if and only if the polynomial $z^b - 1$ belongs to the *ideal* generated by binomials $I = \langle z^{A_j} - 1\rangle_{j=1,\ldots,n} \subset \mathbb{R}[z_1, \ldots, z_m]$ *and* for some weights Q_j all with *nonnegative coefficients*. In other words, for the ideal $J \subset \mathbb{R}[z_1, \ldots, z_m, y_1, \ldots, y_n]$ generated by the binomials $z^{A_j} - y_j$, $j = 1, \ldots, n$ we wish to decide whether $z^{A_j} - 1$ belongs to it.

Nonlinear algebra provides algorithms for deciding ideal membership and for solving systems of equations. There are many excellent textbooks explaining these methods [**CLO92, CLO05, DE05, MlS21, Stu02, Ste04**]. Applications to discrete mathematics are abundant. For example, to list all the colorings of a graph or all the Ramsey graphs can be done in principle by solving the systems of equations we saw earlier. The importance of solving polynomial equations in discrete mathematics is also evident in coding theory and cryptography (see, e.g., [**CKPS00, GR18**]). A widely popular technique is computing *Gröbner bases* for an ideal I of the ring of polynomials. This is a rewriting of generators, just like Gaussian elimination is a way to rewrite a linear system into a triangular form, giving a new set of generators with additional computational properties (see [**AL94, CLO92**]). Gröbner bases have multiple applications (see, e.g., [**CLO05**] and references therein). We continue with their use in optimization.

2.1.2. *Gröbner bases of toric ideals and integer programming.* Another exciting approach to solve integer optimization problems, using again ideals in polynomial rings, was started in the 1990s [**CT91, HS95a, HT98, Tho95**]. Integer feasibility becomes the challenge of deciding the residue of a monomial modulo an ideal. This is can be approached with methods of Gröbner bases through the Gröbner bases of *toric ideals* (see [**Stu95a, MT07**]). We outline this now.

Let $\mathbb{N}$ denote the set of natural numbers, including 0, and let $A = (\mathbf{a}_1|\cdots|\mathbf{a}_n) \in \mathbb{N}^{d\times n}$ be a matrix of rank d given by its columns. The cone of A in $\mathbb{R}^d$ is $C(A) = \{\sum_{i=1}^n \lambda_i \mathbf{a}_i : \lambda_i \geq 0, i = 1, \ldots, n\}$. For the polynomial ring in n variables over a field K, the toric ideal associated to A, is given by

$$I_A = \langle \mathbf{x^u} - \mathbf{x^v} : \sum_{i=1}^n u_i\mathbf{a}_i = \sum_{i=1}^n v_i\mathbf{a}_i \rangle \subseteq k[x_1, \ldots, x_n],$$

where $\mathbf{x^u} = x_1^{u_1}\cdots x_n^{u_n}$ and $\mathbf{x^v} = x_1^{v_1}\cdots x_n^{v_n}$, as usual. The toric ideal of A provides intimate information about the integer kernel of A, the lattice $\mathbb{Z}A$, and the semigroup $\mathbb{N}A$, generated by the columns of A, as well as ways to test for the solvability of integer programs. For instance, $Ax = b$ has a solution $x \in \mathbb{N}^n$ if and only if the monomial z^b is reduced, with respect to a Gröbner basis of the toric ideal of A, to some nonzero monomial z^α, in which case $x = \alpha \in \mathbb{N}^n$ is a feasible solution.

Geometrically, monomials are naturally identified to the lattice points of the polytope $P_A(b) = \{x : Ax = b, x \geq 0\}$ and one can think of the generators of the *toric ideal* associated to A as moves or arcs that connect all the lattice point solutions inside the polytope $P_A(b)$. A term order necessary for a Gröbner basis is given by the (linear) cost function c^T to be optimized that when we have a Gröbner basis assures we can reach the optimal solution that minimizes c^Tx via a directed path (see [**Stu95a**]). The practical limitations of Gröbner bases for computation are their huge size and the fact that Buchberger's algorithm is not easy to improve.

It is important to note toric ideals give also interesting structural information on the integer programs. For example, it is known that the toric variety $V(I_A)$ is normal if $C(A) \cap \mathbb{Z}A = \mathbb{N}A$, which means the columns of A form *Hilbert bases* (see [**Stu95a**]). In [**DHS09**], the authors proved that if $n - d$ is fixed, then there is an algorithm to decide whether I_A is normal whose running time is polynomial in n and the bit size of A. That is to say, if $n - d$ is fixed one can decide in polynomial time whether $\{\mathbf{a}_1, \ldots, \mathbf{a}_n\}$ is a Hilbert basis of the cone they generate. In [**ST97**] Thomas and Sturmfels discussed *parametric integer programming*. Recall the problem (ILP(A,c,b)), for a choice of A, b, c, this is one problem but let IP_A denote the family of all such programs with fixed matrix A. Two cost functions c and c' are equivalent if $ILP(A, c, b)$ and $ILP(A, c', b)$ have the same set of optimal solutions for each b. These equivalence classes are relatively open polyhedral cones in $\mathbb{R}^n$. They fit together to form a polyhedral fan, called the *Gröbner fan of A*. It is known there is an $(n - d)$-dimensional polytope $\text{St}(A)$ in $\mathbb{R}^n$, called the *state polytope* of A, which has that Gröbner fan as its normal fan. Explicit inequality presentations of these cones are given by the reduced Gröbner basis associated with A. The union of all the reduced Gröbner bases as c varies consists precisely of all edge directions of $\text{St}(A)$ (it is called the *universal Gröbner basis*).

Thirty years later some variations of the 1990s Gröbner techniques provided substantial progress on algorithms for solving integer programs. I will next sketch the updated version and how they yield efficient polynomial-time algorithms when the structure of the matrix A is "nice".

2.1.3. *Graver bases an advances in computational complexity.* Around 2005 a new breed of Gröbner-style techniques emerged and provided a breakthrough in computational complexity. The trade-offs are that we need to make assumptions about the block structure of the matrix A and we need to use a stronger type of (universal) Gröbner bases called *Graver bases.* Next we outline these ideas. See [**DLHK12**, Chapter 11] and [**Onn10**] for a more detailed presentation.

Here is the definition of a Graver basis. The lattice $L(A) = \{\mathbf{x} \in \mathbb{Z}^n : A\mathbf{x} = 0\}$ contains all solutions and it has a natural partial order as follows: For $\mathbf{u}, \mathbf{v} \in \mathbb{Z}^n$ the vector $\mathbf{u}$ is *conformally smaller* than $\mathbf{v}$, denoted $\mathbf{u} \sqsubset \mathbf{v}$, if $|u_i| \leq |v_i|$ and $u_i v_i \geq 0$ for $i = 1, \ldots, n$. For instance, $(3, -2, -8, 0, 8) \sqsubset (4, -3, -9, 0, 9)$, but it is incomparable to $(-4, -3, 9, 1, -8)$. Equivalently, for $\mathbf{u}$ to be conformally smaller than $\mathbf{v}$, they must lie in the same orthant of $\mathbb{R}^n$ and each component of $\mathbf{u}$ must be bounded by the corresponding component of $\mathbf{v}$ in absolute value. The *Graver basis* of an integer matrix A is the set of all conformally minimal nonzero integer linear dependences on A. For example, if the matrix $A = (1\ 2\ 1)$, then its Graver basis is given by the vectors $(2, -1, 0), (0, -1, 2), (1, 0, -1), (1, -1, 1)$ and their negatives. For a fixed cost vector $\mathbf{c}$, the Graver basis of an integer program creates a graph over the lattice points $\{\mathbf{x} \in \mathbb{Z}^n : A\mathbf{x} = \mathbf{b},\ \mathbf{x} \geq 0\}$. The nodes are vectors in $L(\mathbf{b}) := \{\mathbf{x} :\ A\mathbf{x} = \mathbf{b},\ \mathbf{x} \geq 0,\ \mathbf{x} \in \mathbb{Z}^n\}$ and we use the elements of the Graver basis as directed edges (direction changes guided by $\mathbf{c}$) departing from each lattice point $\mathbf{u} \in L(\mathbf{b})$ when they connect to another member of $L(\mathbf{b})$.

Graver bases are indeed special cases of Gröbner bases [**Stu95a**] of the same toric ideal. Thus Graver bases can be used as *test sets*, just like the original idea of Gröbner bases, namely a finite collection of integral vectors with the property that every nonoptimal feasible solution of an integer program can be improved by adding some suitable vector from the Graver basis. The first person to look at this larger set was Graver [**Gra75**], ergo their name. For an nice example of Graver bases and the usefulness of block-structure, consider the famous *transportation problem* where we send products from factories to cities (a bipartite graph) and there is a cost at each edge, thus we optimize a linear function summing the cost of edges used. For the transportation problem A is the node edge incidence matrix of the complete bipartite graph. In that example, the Graver basis consists of all cycles in the graph. That set grows exponentially for general graphs, but if we assume the graph has a constant number k of factories, a nice n-fold structure emerges in

A and we see the set of cycles is polynomial-size because all cycles have length at most k. Block structure is what we need to exploit to save the day.

The main obstacle for using Graver bases is that they, same as Gröbner bases, can be exponentially large, even in fixed dimension, and very hard to compute. In fact, for general matrices it is an NP-complete problem to decide whether a list of vectors is a complete Graver basis. We do not seem to have won anything so far by going these special Gröbner bases. The good news is that Graver bases do become very manageable and efficient to generate for highly block-structured matrices.

A *Graver decomposition theory* has developed over the last twenty years by Onn and collaborators in a series of papers (see [**Onn10, EHK**$^+$**19**] and the references therein). The first example of such good structure came from *n-fold matrices.* Fix any pair of integer matrices A and D with the same number of columns, of dimensions $r \times q$ and $s \times q$, respectively. The n-fold matrix of the ordered pair A, D is the $(s + nr) \times nq$ matrix constructed by stacking copies of A, D as shown below,

$$[A, D]^{(n)} \quad := \quad (\mathbf{1}_n \otimes D) \oplus (I_n \otimes A) \quad = \quad \begin{pmatrix} D & D & \cdots & D \\ A & 0 & \cdots & 0 \\ 0 & A & \cdots & 0 \\ \vdots & \vdots & \ddots & \vdots \\ 0 & 0 & \cdots & A \end{pmatrix}.$$

Naturally, n-fold matrices do appear in applications, our main example being the transportation problems with fixed number of factories on suppliers! For these kind of matrices there is a polynomial time algorithm that, given n, computes the Graver basis $G([A, D]^{(n)})$ of the n-fold matrix $[A, D]^{(n)}$. In particular, the cardinality and the bit size of $G([A, D]^{(n)})$ are bounded by a polynomial function of n (implying that separable convex integer minimization problems over n-fold matrices can be solved in polynomial time). The key idea to prove this goes back to work in combinatorial commutative algebra. A result by Santos and Sturmfels [**SS03**] and Hoşten and Sullivant [**HS07**] says that for every pair of fixed integer matrices $A \in \mathbb{Z}^{r \times q}$ and $D \in \mathbb{Z}^{s \times q}$, there exists a constant $g(A, D)$ such that for all n, the Graver basis of $[A, D]^{(n)}$ consists of vectors with at most $g(A, D)$ nonzero components. Since those first papers remarkable work has happened and n-fold matrices were further generalized and now include other types of decomposable matrices. A full description is included in [**EHK**$^+$**19, KLO18, KO21**]. Although, Graver bases increase exponentially in size (with respect to n), the sparse block structure of the matrices implies structure of the Graver bases, which then can be exploited to find a best augmenting Graver basis vector direction for a nonoptimal feasible solution efficiently. See [**EHK**$^+$**19, GKK22, HOR13, KO21, CEH**$^+$**21**] and references therein.

There is an ever growing demand to solve *nonlinear* discrete optimization problems, not just those with linearly objectives. There are many natural situations when one may want to optimize a nonlinear function. E.g., due to congestion or heavy traffic or heavy communication load the transportation cost on an edge may be a nonlinear function of the flow at each edge. For example, say the cost at each edge is $f_{ij}(x_{ij}) = c_{ij}|x_{ij}|^{a_{ij}}$ for suitable constant a_{ij}. This results in a nonlinear function $\sum f_{ij}(x_{ij})$ which is much harder to minimize. Surprisingly, using Graver bases, even nonlinear separable convex integer minimization problems over block-structured matrices can be solved efficiently (see [**HKW10, HOW11, GKK22,**

KO21] and references therein; see also [**DLHK12**, Chapters 3 and 4]). Graver bases have been a successful method, but convex separable nonlinear functions are still too special. We need to understand other nonlinear objective functions. We see some success again by using algebra.

2.1.4. *Nonlinear algebra for nonlinear discrete optimization.* We switch to discuss a different nonlinear algebraic structure and its consequences for optimization: It is well known that when the dimension is fixed, the integer linear optimization problem (ILP(A,c,b)) can be solved efficiently using Lenstra's famous algorithm (see [**Len83**] and the excellent recent survey by Eisenbrand in [**JLN$^+$10**]). Algebra and geometry were necessary for this historic result. For, instance, approximating the shortest vector problem in a lattice $\mathcal{L}$ was done using the LLL-reduced basis of $\mathcal{L}$ [**LLL82**] which was also crucial in polynomial factorization algorithms. Lenstra's algorithm also relied on the possibility of decomposing the problem into a "small" family of hyperplanes that pass through the feasible polyhedral region P. Khinchin's flatness theorem [**Khi48**] guarantees there is a primitive lattice vector direction $\mathbf{d} \in \mathbb{Z}^n$ such that there are "very few" lattice hyperplanes of the form $\mathbf{d}^\top \mathbf{x} = \gamma$ (with $\gamma \in \mathbb{Z}$) that have a nonempty intersection with a convex body (or polyhedron) P. Lenstra also uses transformation of a feasible polyhedral region P to a more "rounded" shape (this requires finding concentric, inscribed and circumscribed ellipsoids to P that differ by some dilation factor β that only depends on the dimension n). More recent improvements in complexity for (linear objective) Lenstra-type algorithms appear in [**DPV11**]. In the context of this survey it is natural to ask *what happens if one aims to optimize a* nonlinear, nonconvex *objective function in a fixed number of variables?*

Lenstra's result has been extended in several *nonlinear algebra* ways, allowing the objective function to be an arbitrary polynomial. Techniques in nonlinear algebra played a central role in the proofs of these new algorithms. Consider first the optimization problem

$$\begin{aligned} \max/\min \quad & f(x_1, \dots, x_n) \\ \text{subject to} \quad & A\mathbf{x} \leq \mathbf{b} \\ & \mathbf{x} = (x_1, \dots, x_n) \in \mathbb{Z}^n. \end{aligned} \tag{2.3}$$

Here A is a rational matrix, $\mathbf{b}$ is a rational vector. The set of feasible integer solutions is a convex polyhedron we denote by P. The function f is a polynomial function of maximum total degree D with rational coefficients. We make no convexity assumptions about f. Already optimizing polynomials of degree four over problems with two integer variables is an NP-hard problem (see [**DLHKW06**, **AOPT13**]). Thus, even when we fix the dimension, we cannot expect to get a polynomial-time algorithm for solving this type of optimization problem. The best we can hope for a generalization of Lenstra's theorem is an *approximation algorithm.* Nonlinear algebra provides a solution.

We say an algorithm $\mathcal{A}$ is an *ϵ-approximation algorithm* for a maximization problem with optimal cost $f_{\max}$, if $\mathcal{A}$ runs in polynomial time in the encoding length and returns a feasible solution with cost $f_{\mathcal{A}}$, such that $f_{\mathcal{A}} \geq (1-\epsilon) f_{\max}$. A family $\{\mathcal{A}_\epsilon\}_\epsilon$ of ϵ-approximation algorithms is a *fully polynomial time approximation scheme (or for short an FPTAS)* if the running time of $\mathcal{A}_\epsilon$ is polynomial in the encoding length and $1/\epsilon$. The polynomial dependence of the running time in $1/\epsilon$ is a strong requirement, and for many NP-hard problems, efficient approximation algorithms of this type do not exist, unless P = NP. Surprisingly, it was proved in [**DLHKW06**] that there is an FPTAS for the maximization problem (2.3) for all polynomial functions $f(x_1, \ldots, x_n)$ with rational coefficients that are nonnegative on the feasible polyhedral region P, more precisely.

THEOREM 2.2. *Let the dimension d be fixed. There exists an algorithm with input*

- *a polytope $P \subset \mathbb{R}^d$, given by rational linear inequalities, and*
- *a polynomial $f \in \mathbb{Z}[x_1, \ldots, x_d]$ with integer coefficients and maximum total degree D that is nonnegative on $P \cap \mathbb{Z}^d$*

with the following properties.

(1) *For a given k, it computes in running time polynomial in k, the encoding size of P and f, and D lower and upper bounds $L_k \leq f(\mathbf{x}^{\max}) \leq U_k$ satisfying*

$$U_k - L_k \leq \left(\sqrt[k]{|P \cap \mathbb{Z}^d|} - 1 \right) \cdot f(\mathbf{x}^{\max}).$$

(2) *For $k = (1 + 1/\epsilon)\log(|P \cap \mathbb{Z}^d|)$, the bounds satisfy*

$$U_k - L_k \leq \epsilon\, f(\mathbf{x}^{\max}),$$

and they can be computed in time polynomial in the input size, the total degree D, and $1/\epsilon$.

(3) *By iterated bisection of $P \cap \mathbb{Z}^d$, it constructs a feasible solution $\mathbf{x}_\epsilon \in P \cap \mathbb{Z}^d$ with*

$$\left| f(\mathbf{x}_\epsilon) - f(\mathbf{x}^{\max}) \right| \leq \epsilon f(\mathbf{x}^{\max}).$$

We will next state yet another algorithmic generalization of Lenstra's results. *Multi-objective linear optimization problems* are those that have more than one linear objective function to be optimized. This is a natural model when many agents compete for the same resources. When more than one linear objective function is optimized, what does it mean to reach an optimal solution? What if we have three or four objectives to optimize? We will explain the basics now (for a better introduction to multi-objective discrete optimization problems, see [**EG00**]).

Let $A = (a_{ij})$ be an integral $m \times n$-matrix and $\mathbf{b} \in \mathbb{Z}^m$ such that the convex polyhedron $P = \{\mathbf{u} \in \mathbb{R}^n : A\mathbf{u} \leq \mathbf{b}\}$, is bounded. Given k linear functionals $f_1, f_2, \ldots, f_k \in \mathbb{Z}^n$, we consider the *multi-objective integer linear programming problem*

$$\begin{array}{rl} \text{Pareto-min} & \left(f_1(\mathbf{u}), f_2(\mathbf{u}), \ldots, f_k(\mathbf{u})\right) \\ \text{subject to} & A\mathbf{u} \leq \mathbf{b} \\ & \mathbf{u} \in \mathbb{Z}^n \end{array} \tag{2.4}$$

where Pareto-min is defined as the problem of finding all Pareto optima and a corresponding Pareto strategy: For a lattice point $\mathbf{u}$ the vector $\mathbf{f}(\mathbf{u}) = \left(f_1(\mathbf{u}), \ldots, f_k(\mathbf{u})\right)$ is called an *outcome vector*. Such an outcome vector is a *Pareto optimum* if and only if there is no other $\tilde{\mathbf{u}}$ in the feasible set such that $f_i(\tilde{\mathbf{u}}) \leq f_i(\mathbf{u})$ for all i and $f_j(\tilde{\mathbf{u}}) < f_j(\mathbf{u})$ for at least one index j. The corresponding feasible point $\mathbf{u}$ is called a *Pareto strategy*. Thus a feasible vector is a Pareto strategy if no feasible vector can decrease some objective function in the list without causing a simultaneous increase in at least one other objective function.

In general, the number of Pareto optimal solutions may be infinite, but in our discrete situation the number of Pareto optima and strategies is finite. There are several well-known techniques to generate Pareto optima. Some popular methods used to solve such problems include weighting the objectives or using a so-called global criterion approach (see [**EG00**]). In some nice situations, such as multi-objective *linear* programs [**Ise74**], one knows a way to generate all Pareto optima, but most techniques reach only a few of the Pareto optima. The great news are, in fixed dimension, nonlinear algebra can encode and compute *all* the Pareto optima and strategies of a multi-criterion integer linear program in polynomial time! How is this possible? This seems contradictory as we just mention the number of pareto optima can be exponentially large.

The secret is to encode all Pareto optima and Pareto strategies as monomials in a large multivariate polynomials

$$\sum\big\{\, \mathbf{z}^{\mathbf{v}} : \ \mathbf{u} \in P \cap \mathbb{Z}^n \text{ and } \mathbf{v} = \mathbf{f}(\mathbf{u}) \in \mathbb{Z}^k \text{ is a Pareto optimum} \,\big\}. \tag{2.5}$$

$$\sum\big\{\, \mathbf{x}^{\mathbf{u}} : \ \mathbf{u} \in P \cap \mathbb{Z}^n \text{ and } \mathbf{f}(\mathbf{u}) \text{ is a Pareto optimum} \,\big\}. \tag{2.6}$$

Under the assumption that the number of variables is fixed, we can compute in polynomial time a short expression for these huge polynomials. The expression is short and surprisingly efficient. Here is the precise statement of the theorem.

THEOREM 2.3. *Let* $A \in \mathbb{Z}^{m\times n}$, *an* m*-vector* $\mathbf{b}$, *and linear functions* $f_1, \dots, f_k \in \mathbb{Z}^n$ *be given. There are algorithms to perform the following tasks:*

(1) *Compute the generating function* (2.5) *of all the Pareto optima as a sum of rational functions. In particular, we can count how many Pareto optima are there. If we assume* k *and* n *are fixed, the algorithm runs in time polynomial in the size of the input data.*
(2) *Compute the generating function* (2.6) *of all the Pareto strategies as a sum of rational functions. In particular, we can count how many Pareto strategies are there in* P. *If we assume* k *and* n *are fixed, the algorithm runs in time that is bounded polynomially by the size of the input data.*
(3) *Generate the full sequence of Pareto optima ordered lexicographically. If we assume* k *and* n *are fixed, the algorithm runs in polynomial time on the input size and the number of Pareto optima.*

Fixed dimension is crucial for the results. It is known that for nonfixed dimension it is $\#P$-hard to count Pareto optima and NP-hard to find them [**EP92, SP91**]. Recently Nguyen and Pak completed a comprehensive theory of the complexity of lattice problems over polyhedra with many quantifiers [**NP22**].

To conclude this portion we point to other complexity results in fixed dimension that generalize Lenstra's result: An efficient convex integer minimization algorithm was announced by Khachiyan in [**Kha84**]. Later the paper by Khachiyan and Porkolab [**KP00**] includes the case of convex domains and arbitrary quantifier sequences. The case of optimization of quasi-convex polynomial functions appeared in [**Hei05**] and was improved in [**HK13**]. New related results appeared in [**OWW12**]. All these results require some tools from nonlinear algebra. Next we give more detail about the tools used to prove Theorems 2.2,2.3.

2.1.5. *Lattice points of polytopes as polynomials.* How does one prove algorithmic results like Theorems 2.2,2.3? The key ingredient is to be able to compute, from any given input polytope P, a multivariate polynomial that encodes all the lattice points $P \cap \mathbb{Z}^d$, we call it the *lattice point generating function* of P

$$\text{(LattPtGF)} \qquad f(P;\mathbf{z}) = \sum_{\alpha \in P \cap \mathbb{Z}^d} \mathbf{z}^\alpha, \text{ where } \mathbf{z}^\alpha = z_1^{\alpha_1} z_2^{\alpha_2} \ldots z_d^{\alpha_d}.$$

The name comes because $f(P, \mathbf{z})$ could be an infinite formal power series if P is unbounded, but if P is a polytope, then it is a multivariate (Laurent) polynomial with one monomial per lattice point. Here is an example: consider the rectangle P with vertices $V_1 = (0,0)$, $V_2 = (5000, 0)$, $V_3 = (0, 5000)$, and $V_4 = (5000, 5000)$, the generating function $f(P, \mathbf{z})$ is a polynomial in two variables with over 25,000,000 monomials.

$$f(P; z_1, z_2) = 1 + z_1 + z_2 + z_1 z_2^2 + z_1^2 z_2 + \cdots + z_1^{5000} z_2^{5000}.$$

Note how the Equations (2.5) and (2.6) are examples of (LattPtGF). The computation of the generating function of all lattice points of a convex polyhedron P assume we have a formal multivariate generating function but it is not explicitly monomial by monomial. The number of monomials in Equation (LattPtGF) is huge and possibly infinite, thus monomials are an impractical representation. Instead we describe how $f(P;\mathbf{z})$ can be algorithmically rewritten as a sum of multivariate rational functions. These rational functions are recovered from the rays and vertices of cones associated to the polyhedral region P. In the 1990's Barvinok proposed a method [**Bar94, BP99**] that in fixed dimension yields a polynomial-size sum of rational functions. As example, only four rational functions suffice to represent the over 25 million monomials of the example rectangle. Indeed $f(P, z_1, z_2)$ equals

$$\frac{1}{(1-z_1)(1-z_2)} + \frac{z_1{}^{5000}}{(1-z_1{}^{-1})(1-z_2)} + \frac{z_2{}^{5000}}{(1-z_2{}^{-1})(1-z_1)} + \frac{z_1{}^{5000} z_2{}^{5000}}{(1-z_1{}^{-1})(1-z_2{}^{-1})}.$$

It is important to note that if we wish to know the number of lattice points in P, and we knew $f(P;\mathbf{z})$, we could compute it as the limit when the vector $(z_1, \ldots, z_n)$ goes to $(1, 1, \ldots, 1)$. Similarly the maximum of a linear functional over the lattice points equals the highest degree of the univariate polynomial one gets after doing a *monomial substitution* $z_i \to t^{c_i}$ (See [**BW03**]). These two calculations can be difficult because of the poles of the rational functions. One uses complex analysis (residue calculations) to find the answer.

The first remarkable combinatorial application of this representation was done by A. Barvinok who crafted an algorithm to *count* integer points inside polyhedra that runs in polynomial-time for fixed dimension (see [**Bar94, BP99**]. Shortly after Barvinok's breakthrough, Dyer and Kannan [**DK97**] modified the original algorithm of Barvinok, which originally relied on Lenstra's result, giving a new proof that integer programming problems with a fixed number of variables can be solved in polynomial time. This gives a brand new proof of Lenstra's theorem (see [**DLHK12**, Chapter 7]).

A theorem of Brion [**Bri88**] says that to compute the rational function representation of $f(P;\mathbf{z})$ it is enough to do it for tangent cones at each vertex of P. Let P be a convex polytope and let $V(P)$ be the vertex set of P. Let $K_\mathbf{v}$ be the tangent cone at $\mathbf{v} \in V(P)$. This is the (possibly translated) cone defined by the facets touching vertex $\mathbf{v}$. Then the following nice formula holds:

$$f(P;\mathbf{z}) = \sum_{\mathbf{v} \in V(P)} f(K_\mathbf{v};\mathbf{z}).$$

Since it is enough to compute everything for cones, we first explain how to compute the rational function for the "smallest" cones, simple cones. A cone is said to be *simple* if its rays are linearly independent vectors. For instance all the tangent cones of a of a cube are simple. Obtaining the rational function representation of the lattice points of a simple cone $K \subset \mathbb{R}^d$, is easy (see, for example, [**Sta97**, Chapter IV]. Another exposition is given in [**DLHK12**, Chapters 5, 6, and 7] where also recently developed speedups are discussed). Here is the formula, that one can write directly from the coordinates of the rays of the cone and the lattice points in its *fundamental parallelepiped* Π:

$$f(K;\mathbf{z}) = \frac{\sum_{\mathbf{u}\in\Pi\cap\mathbb{Z}^d} \mathbf{z}^{\mathbf{u}}}{(1-\mathbf{z}^{\mathbf{c}_1})(1-\mathbf{z}^{\mathbf{c}_2})\dots(1-\mathbf{z}^{\mathbf{c}_d})}.$$

Here Π is the half open parallelepiped $\{\mathbf{x} : \mathbf{x} = \alpha_1\mathbf{c}_1 + \dots + \alpha_d\mathbf{c}_d,\ 0 \le \alpha_i < 1\}$.

But what to do if the cone K is not simple? Break it into simple cones! We could simply triangulate by adding inner rays, but The efficiency comes from an idea of A. Barvinok: Although triangulating the cone K may be an inefficient way to subdivide the cone (i.e., exponentially many pieces), if one adds *and* subtracts cones, for fixed dimension d, there exists a polynomial time algorithm which decomposes a rational polyhedral cone $K \subset \mathbb{R}^d$ into simple unimodular cones K_i. A simple cone is *unimodular* if its fundamental parallelepiped there is a single lattice point. In fact, via the decomposition, with numbers $\epsilon_i \in \{-1, 1\}$, we can write an expression

$$f(K;\mathbf{z}) = \sum_{i\in I} \epsilon_i f(K_i;\mathbf{z}), \quad |I| < \infty.$$

the index set I is of size polynomial in the input data as long as we work in fixed dimension. For more on how to compute and manipulate generating functions $f(P, \mathbf{z})$ see [**DLHK12, Onn10**].

2.2. Nonlinear methods in algebraic and enumerative combinatorics. In the previous subsection we reviewed a number of ways polynomial ideals, and polynomials over polyhedra are used in discrete optimization. Now we review a few methods from nonlinear algebra that arise in algebraic and enumerative combinatorics.

Polynomials have been used in combinatorial problems for some time, e.g., at least going back to the use of chromatic polynomials to count colorings of graphs. Algebraic geometry has a long history of connections with combinatorics, such as the Schubert calculus and determinantal varieties [**BJS93, Sta99, Ful97**]. Similarly, the theory of toric varieties connects algebraic geometry with combinatorial geometry [**CLS11, Stu08**]. Applications of algebraic geometry to combinatorics were explored starting with the influential work of Richard Stanley [**Sta96, Stu08**]. This important methodology is, given a combinatorial structure, one can associate a monomial ideal and its quotient ring and then obtain a graded standard K-algebra of a quotient polynomial ring over a field K. In that way the Hilbert function of these rings offers rich combinatorial information, and it is well known that for large enough values the Hilbert function of a standard graded K-algebra is, in fact, a polynomial function, the Hilbert polynomial. A famous example is the use of the Stanley–Reisner polynomial rings in bounding f-vectors of simplicial spheres ([**Sta96, MS05**]). Many other sequences arising in combinatorics are Hilbert functions [**Bre98**], including the chromatic polynomial of a graph, the order polynomial of a partially ordered set, the characteristic polynomial of a hyperplane arrangement, and, last but not least, the Ehrhart polynomial of a lattice polytope which we will explore next.

2.2.1. *Ehrhart quasipolynomials in enumerative combinatorics.* We saw the generating function (LattPtGF) of all lattice points of a polyhedron is useful for optimization, but we will see it is also relevant in enumerative combinatorics. Recall that polyhedra are, by definition, given by a finite system of equations and inequalities and they can be rewritten in the popular standard form $P_A(b) = \{x | Ax = b, x \geq 0\}$, for a suitable integral matrix A, and integer vector b where we may add slack variables. It is important to stress the generating function of Equation (LattPtGF) can be used to study the family of parametric polyhedra $P_A(b)$ one obtains from fixing the matrix A and moving the entries of b. The resulting polynomial $f(P_A(b), z)$ will change with b. It turns out that for many algebraic combinatorics applications A is fixed, but the vector b is dynamic. A famous example in algebraic combinatorics of parametric polyhedra are the *flow polytopes*. For a loopless graph G with vertex set $[n+1]$ and with edges directed from smaller to larger vertices, its flow polytope $\mathcal{F}_G(\mathbf{b})$ for the (parameter) *netflow vector* $\mathbf{b} = (a_1, \ldots, a_{n+1}) \in \mathbb{Z}^{n+1}$ is the polyhedron

$$\mathcal{F}_G(\mathbf{b}) = \{f \in \mathbb{R}_{\geq 0}^{E(G)} : M_G f = \mathbf{b}\},$$

where M_G denotes the incidence matrix of G. Flow polytopes appear, for example, in the study of the Kostant partition functions (see, e.g., [**BGDH⁺19**, **MM19**] and the many references therein).

The first functions that concern us here are the *Ehrhart counting functions*

$$E_A(b) = \#\{x : Ax = b, x \geq 0, \quad x \quad \text{integral}\}$$

that count the number of lattice points inside convex polyhedra with fixed matrix A but varying right-hand vector b. The simplest example is when we change the entries of b by the same amount n. In that case, the $E_A(nb)$ is a function of one variable n. If you do this for the unit square $[0,1]^2$ this is a simply a polynomial $E_P(n) = (n+1)^2$. In the case when P is a d-dimensional *lattice polytope* (i.e., all vertices have integer coordinates) its Ehrhart counting function is an univariate polynomial of degree equal to the dimension of P and its leading coefficient is the (normalized) volume of P.

Ehrhart functions can be recovered from the generating functions (LattPtGF) and in fact they can be computed explicitly using software (see, e.g., [**DLHH⁺05**]). The most general theorem describing Ehrhart counting functions says that they are described by finitely many multivariate polynomials that are valid inside polyhedral pieces.

THEOREM 2.4 (Sturmfels [**Stu95b**]). *Given an integral $d \times n$ matrix A and an integral n-vector b.*

$$E_A(b) = |\{x : Ax = b, x \geq 0 \quad x \quad \textit{integral}\}|$$

- *The function $E_A(b)$ is a piecewise quasi-polynomial function. It is only nonzero inside $b \in$* **cone**(A).

  This means each piece of $E_A(b)$ is a polynomial function of degree $n-d$ in the variables $b_1, \ldots, b_d$ plus a a periodic polynomial of smaller degree (its coefficients are periodic in the coordinates of b).

- *The domain or region of validity of each of these quasi-polynomial is a convex polyhedral subcone of cone(A). These subcones are called* chambers.
- *The chambers are the open conical subsets of cone(A), for any pair of vectors b, b' the polytopes $P_A(b)$, $P_A(b')$ are normally (and thus combinatorially) equivalent. The chambers all together form a polyhedral complex that subdivides the interior of cone(A) and their union is cone(A).*

Here is an example to illustrate the above mega-theorem describing all Ehrhart functions.

$$A = \begin{bmatrix} 2 & 1 & 1 & 0 & 0 & 0 \\ 0 & 1 & 0 & 2 & 1 & 0 \\ 0 & 0 & 1 & 0 & 1 & 2 \end{bmatrix}$$

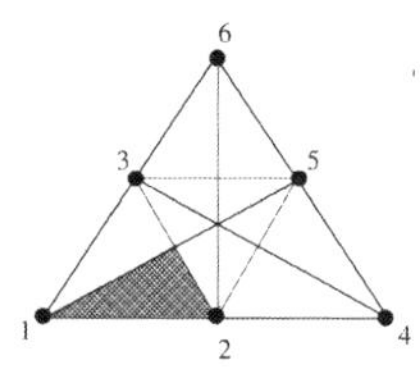

FIGURE 1. The 12 chambers associated to $E_A(b)$, for example.

Here is the formula for the chamber marked in Figure 2.2.1.

$$E_A(b_1, b_2, b_3) = \frac{b_2 b_3}{2} + \frac{b_2 b_3^2}{8} - \frac{b_3^2}{24} + correction,$$

$$correction = \begin{cases} 1 + \frac{b_2}{2} + \frac{2b_3}{3} & \text{if } b_1 = 0 \text{ and } b_2 = 0 \ mod 2, \\ \frac{1}{2} + \frac{b_2}{2} + \frac{5b_3}{12} & \text{if } b_1 = 1 \text{ and } b_2 = 1 \ mod 2, \\ \frac{1}{2} + \frac{3b_2}{8} + \frac{13b_3}{24} & otherwise. \end{cases}$$

2.2.2. *Ehrhart quasi-polynomials in representation theory.* Let us see Ehrhart quasipolynomials in action through a lovely connection to combinatorial representation theory: Every *Lie group* (i.e., a group with a compatible complex manifold structure) has always an associated *Lie algebra*. For example, the *special linear group* $\mathbf{SL}_n(\mathbb{C})$ of $n \times n$ matrices with determinant 1 has an associated Lie algebra $\mathfrak{sl}_n(\mathbb{C})$ the vector space of traceless $n \times n$ matrices. Mathematicians understand that *simple Lie algebras* are building blocks of all other Lie algebras. The *semisimple Lie algebras* are direct sums of simple Lie algebras.

Recall a representation of a Lie algebra $\mathfrak{g}$ on a vector space V is a linear transformation $\rho : \mathfrak{g} \to \mathfrak{gl}(V)$, such that $\rho([x, y]) = [\rho(x), \rho(y)]$. The representations of semisimple Lie algebras decompose into direct sums of *irreducible representations*. The irreducible representations are indexed by *dominant weight vectors* or *highest weight vectors*.They belong to the *weight lattice*. Denoting by λ, μ, and ν the highest weights for an irreducible representation of a finite dimensional complex semisimple Lie algebra it is known,

$$V_\lambda \otimes V_\mu = \bigoplus_\nu C_{\lambda\mu}^\nu V_\nu. \tag{2.7}$$

The nonnegative integers $C^{\nu}_{\lambda\mu}$ are the *Clebsch–Gordan coefficients*. For Lie algebras of type A_r, these are the *Littlewood–Richardson coefficients*. These numbers are quite important. These numbers appear in quantum mechanics and high-energy physics (e.g., Angular moment problems). Some computer scientists are interested because of the connection to the conjecture $P \neq NP$ (see [**BIP19**] and its references). The integers $C^{\nu}_{\lambda\mu}$ are important and one can ask about how to compute them or their properties, such as what is the largest value they can take as ν, λ, μ change.

A key idea to understand and compute the Clebsch–Gordan coefficients is to represent them the number of lattice points of some special convex polytopes. This idea goes back at least to the 1950s in the work of Gelf'and–Tsetlin [**GfC50**] but has since generated a lot of work from the combinatorial representation theory community. See the classic papers of Knutson and Tao [**KT99**] and Berenstein and Zelevinsky [**BZ01**].

A special important example we briefly describe the *Hive polytopes* of Knutson and Tao (1999). A hive pattern of rank r is a triangular array of real numbers

$$\begin{array}{ccccccccc} & & & & h_{0,0,r} & & & & \\ & & & h_{1,0,r-1} & & h_{0,1,r-1} & & & \\ & & h_{2,0,r-2} & & h_{1,1,r-2} & & h_{0,2,r-2} & & \\ & \iddots & & & \vdots & & & \ddots & \\ h_{r,0,0} & & h_{r-1,1,0} & & \cdots & & h_{1,r-1,0} & & h_{0,r,0} \end{array}$$

such that, in every "little rhombus" of entries

$$\begin{array}{ccc} c & & b \\ & a & & d \end{array} \qquad \begin{array}{ccc} & c & \\ a & & b \\ & d & \end{array} \qquad \begin{array}{cccc} & a & & c \\ d & & b & \end{array}$$

we have $a + b \geq c + d$. Here is an example of a hive pattern with $r = 4$:

$$\begin{array}{ccccccccc} & & & & 0 & & & & \\ & & & 5 & & 8 & & & \\ & & 8 & & 12 & & 13 & & \\ & 11 & & 15 & & 17 & & 18 & \\ 12 & & 16 & & 18 & & 20 & & 20. \end{array}$$

When the Lie algebra $\mathfrak{g}$ is of type A_r, i.e., that $\mathfrak{g}$ is isomorphic to $\mathfrak{sl}_{r+1}(\mathbb{C})$, the highest weights are (with respect to the standard basis) partitions of length $r+1$, i.e., sequences λ of integers $\lambda_1 \geq \cdots \geq \lambda_{r+1} \geq 0$. Given partitions $\lambda, \mu, \nu \in \mathbb{Z}^n_{\geq 0}$, put $|\lambda| = \sum_i \lambda_i$, $|\mu| = \sum_i \mu_i$, and $|\nu| = \sum_i \nu_i$. The *hive polytope* $H^{\nu}_{\lambda\mu}$ is the set of those hive patterns with boundary entries fixed as below.

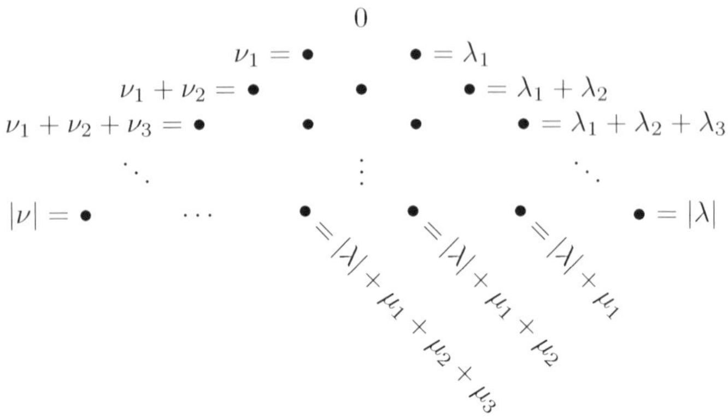

The fundamental theorem that connects the Clebsch–Gordan numbers with polynomials, via polyhedra, is the following (here in increasing generality) [**KT99**, **BZ01**].

THEOREM 2.5 ((Knutson and Tao 1999), (Berenstein and Zelevinsky 2001)). *For type A_r, the Clebsch–Gordan coefficient (also called the Littlewood–Richardson coefficient) $C^{\nu}_{\lambda\mu}$ equals the number of integer lattice points in the* hive polytope $H^{\nu}_{\lambda\mu}$.

More generally, for a finite dimensional complex semisimple Lie algebra $\mathfrak{g}$ and a triple of highest weights (λ, μ, ν) for $\mathfrak{g}$. Then the Clebsch–Gordan coefficient $C^{\nu}_{\lambda\mu}$ equals the number of integer lattice points in the so called BZ-polytope.

Of course, this has two consequences. First, counting lattice points in polyhedra is a way to compute Clebsch–Gordan coefficients. Second, if we imagine that the parameters that define polytopes above are dynamic, the type of functions we obtain are built by polynomials and we can try to answer questions about Clebsch–Gordan coefficients using properties of polynomials. A famous example is the (one parameter) *stretching function of Clebsch–Gordan coefficients*, which is the function $e\colon \mathbb{Z}_{>0} \to Z_{\geq 0}$ given by $e(n) = C^{n\nu}_{n\lambda,n\mu}$. In other words, for any complex semisimple Lie algebra associated to root systems A_r, B_r, C_r, D_r, the stretching function of Clebsch–Gordan coefficients

$$e(n) = C^{n\nu}_{n\lambda,n\mu}.$$

These univariate functions are closely related to the question of saturation (see [**KT99**, **BZ01**, **Sam12**] and references therein). We want to understand these functions. From Theorem 2.5 these are univariate quasi-polynomials. They turn out to be special examples of Ehrhart counting functions. For instance, Derksen and Weyman showed that for the root system A_r, $e(n)$ is a **polynomial** [**DW00**]. What about general root systems? In Table 1 we show three examples of stretched Clebsch-Gordan coefficients for fixed choices of λ, ν, μ and the dilation parameter n. Remember, Ehrhart counting functions are always described by finitely many polynomials as well as modular options. But in this case it turns out the quasi-polynomials are always described by exactly two polynomials, depending only on the parity of n for some root systems. See [**DLM06**] for details and more conjectures.

Ehrhart quasi-polynomials can be quite powerful for counting, but there is more! They can be used for weighted counting and integration too. For $P_A(b) = \{x : Ax = b, \geq 0\}$ a parametric polytope where $b \in \mathbb{R}^N$ is a real multi-parameter, we

TABLE 1. Stretched Clebsch–Gordan coefficients for B_3.

λ, μ, ν	$C^{n\mu}_{n\lambda,n\nu}$
(0, 15, 5) (12, 15, 3) (6, 15, 6)	$\begin{cases} \frac{68339}{64}n^5 + \frac{407513}{384}n^4 + \frac{13405}{32}n^3 + \frac{9499}{96}n^2 + \frac{107}{8}n + 1, & n \text{ even} \\ \frac{68339}{64}n^5 + \frac{407513}{384}n^4 + \frac{13405}{32}n^3 + \frac{16355}{192}n^2 + \frac{659}{64}n + \frac{75}{128}, & n \text{ odd} \end{cases}$
(8, 1, 3) (11, 13, 3) (8, 6, 14)	$\begin{cases} \frac{121}{576}n^6 + \frac{1129}{640}n^5 + \frac{6809}{1152}n^4 + \frac{163}{16}n^3 + \frac{2771}{288}n^2 + \frac{191}{40}n + 1, & n \text{ even} \\ \frac{121}{576}n^6 + \frac{1129}{640}n^5 + \frac{6809}{1152}n^4 + \frac{1933}{192}n^3 + \frac{659}{72}n^2 + \frac{8003}{1920}n + \frac{93}{128}, & n \text{ odd} \end{cases}$
(8, 9, 14) (8, 4, 5) (1, 5, 15)	$\begin{cases} \frac{4117}{192}n^6 + \frac{50369}{640}n^5 + \frac{14829}{128}n^4 + \frac{703}{8}n^3 + \frac{3541}{96}n^2 + \frac{341}{40}n + 1, & n \text{ even} \\ \frac{4117}{192}n^6 + \frac{50369}{640}n^5 + \frac{14829}{128}n^4 + \frac{5599}{64}n^3 + \frac{3451}{96}n^2 + \frac{5001}{640}n + \frac{97}{128}, & n \text{ odd} \end{cases}$

can consider *weighted* sums or integrals over $P_A(b)$. Here the weight is a polynomial or even a quasi-polynomial function $h(x)$ on $\mathbb{R}^d$. One can study a weighted Ehrhart function:

$$E(P_A(b), h) = \sum_{x \in P_A(b) \cap \mathbb{Z}^d} h(x).$$

The behavior of $E(P_A(b), h)$ in the parameter vector b have been the central theme of several works [**Bar06, Bar08, BBDL$^+$12, BBDL$^+$19**]. For an application, we note the leading coefficient is the integral of h over the polytope. In particular, the volume of the polytope is the leading coefficient for the constant function. This fact has been used to compute volumes and integrals over polyhedral domains in record time [**BBDL$^+$11, DLDK$^+$13**]. In combinatorial representation theory, this technology allows us to compute integrals and sums of Clebsch–Gordan coefficients over polyhedral domains (see [**DLEKW23**] and references therein). Finally, reaching greater generality, several authors [**Bar06, BBDL$^+$19**] have computed "interpolating" expressions that combine sums and integrals as follows

$$S^L(P_A(b), h) = \sum_{y \in (b) \cap \mathbb{Z}^d} \int_{P_A(b) \cap (y+L)} h(x)\, dx.$$

Here the integration is over the intersections of $P_A(b)$ with subspaces parallel to a fixed rational subspace L passing through all lattice points in $P_A(b)$, and the sum of integrals is taken over the lattice points. Note that when $L = 0$ and the weight h is 1, $S^0(P_A(b), 1)$ is just the number of integral points in $E(P_A(b))$. One can greatly extend Theorem 2.4 and in [**BBDL$^+$19**] the authors showed on every chamber, $S^L(P_A(b), h)$ is given by a quasi-polynomial function of $b \in \mathbb{R}^N$. In particular, this implies that the weighted $E(P_A(b), h)$ and the integral functions $\int_{P_A(b)} h(x)dx$ are piece-wise quasi-polynomials too.

2.2.3. *Patterns on the coefficients and zeros of combinatorial polynomials.* We say a sequence of numbers $a_0, a_1, \ldots, a_n$ is *unimodal* if there is some k, $0 \le k \le n$ such that

$$a_0 \le a_1 \le \cdots \le a_k \ge \cdots \ge a_{n-1} \ge a_n$$

and it is *log-concave* if

$$a_i^2 \ge a_{i-1}a_{i+1}$$

for all $1 \le i < n$. If all a_i are positive we say the sequence is *positive*. Note that log-concavity implies unimodality. A polynomial $a_0 + a_1 q + \cdots + a_n q^n$ has the property if its sequence $a_0, a_1, \cdots, a_n$ of coefficients has that property. A popular and fruitful activity in algebraic combinatorics is to determine whether a given sequence is unimodal, log-concave, or positive. See [**Sta89, Br15, Bra16**] among many others for a review of that literature. Of particular interest here is the case when the sequence comes from the coefficients of a polynomial, especially those arising from various areas of algebra, combinatorics, and geometry.

It is known that understanding coefficients of polynomials is closely tied to its roots and the evaluation of the polynomial over the complex numbers. A result going back to Newton states that the coefficients of a real-rooted polynomial form a log-concave sequence. Additionally, if the coefficients are symmetric (i.e., $a_i = a_{n-i}$ for all $i \leq n$), then this independently establishes the unimodality of the coefficients. There are many remarkable results about coefficients and roots of combinatorial polynomials. For example, Huh proved log-concavity of the sequence of coefficients of the chromatic polynomial and in absolute value the coefficients are unimodal [**Huh12**]. This exciting work has been extended to all matroids in [**AHK18**].

In general, we are given a family F of subsets of a finite set $S = \{1, \ldots, n\}$, we define the combinatorial polynomial $p_F(x)$ of F or *generating partition function* of F is the polynomial in n variables $x_1, \ldots, x_n$,

$$p_F(x_1, x_2, \ldots, x_n) = \sum_{S \in F} \prod_{i \in S} x_i,$$

we saw a version before where each S in F is a multiset, i.e., elements i appears with multiplicity α_i:

$$P_F(x_1, x_2, \ldots, x_n) = \sum_{S \in F} \prod_{i \in S} x_i^{\alpha_i}.$$

Examples include the matroid bases polynomials where F is the set of bases of a matroid (spanning tree polynomials are special case of matroid bases polynomials), perfect matching polynomials (which have nice representations in terms of Kasteleyn determinants), independent set polynomials of graphs, and we saw the important Equation (LattPtGF) is a type of partition function of lattice points inside a polyhedron that fits the case with multiplicity.

Note that via substitution of variables one can reduced the multivariate case to the univariate case. Setting $x_i = t$ reduces the expression to a univariate polynomial

$$p_F(t) = \sum_{S \in F} t^{|S|} = \sum_{k=1}^{R} a_k t^k, \quad R = max_{S \in F}|S|,$$

where a_k counts the subsets S of cardinality k inside the family F.

The polynomial p_F contains information about the sets in F. Again, note that if we are interested in counting, it means evaluating the corresponding polynomials at some value, say $x_i = 1$, but in many cases we do not know p_F explicitly as the set F maybe too large and hard to enumerate. For instance, the number of independent sets of a graph can be exponentially large in the size of the graph, but one is still interested on at least approximating the count.

Researchers have studied the following problems about p_F: (1) describing special properties about the coefficients of combinatorial polynomials; (2) finding nice short descriptions; and (3) approximating p_F or its evaluation. The evaluation of combinatorial polynomials at nonnatural numbers may sometimes count interesting quantities too, despite there being no reason for it to do so. Prime examples of the so-called *combinatorial reciprocity theorems* (see [**BS18**]).

The main message here is that understanding p_F, or its univariate version, requires understanding its zeros and properties. For instance, the problem of approximately evaluating combinatorial polynomials in a region of the complex plane is reduced to the problem of establishing that the polynomial has no zeros in that region. Combinatorial polynomials have also been studied in the statistical physics community For some models/polynomials such as the matching polynomial we know that absence/presence of zeros on exactly corresponds to easiness/hardness of approximation. Many polynomials are known to have zero sets confined to a prescribed region in the complex plane. My favorite examples of this type of work regards the roots of Ehrhart polynomials of lattice polytopes; see [**BDLD**$^+$**05**, **Bra08**, **BD08**].

Borcea and Brändén [**BB09a**, **BB09b**, **BB10**] developed the theory of real-rootedness in multivariate polynomials. These ideas are captured in the notions of hyperbolic/stable polynomials which have had remarkable applications in combinatorics (and other fields). For a subset S of $\mathbb{C}^n$, a polynomial $p(z_1, \ldots, z_n)$ is called S-stable if $p(s) \neq 0$ for all $s \in S$. Let $H := \{z \in \mathbb{C} : Im(z) > 0\}$ denote the open upper complex half-plane. H^n-stable polynomials are called *stable*. If p is a stable polynomial with only real coefficients, then p is a real stable polynomial. Note the real stable polynomials in one variable are precisely the real-rooted polynomials. Real stability is a multivariate generalization of the notion of real-rootedness.

There are several combinatorial applications of real stability in polynomials. For instance, the *van der Waerden conjecture* on the permanent of a negative and doubly stochastic matrix A (namely, one where the sum of entries in each row and column is 1), stated that

$$per(A) \geq \frac{n!}{n^n},$$

the minimum being reached when all elements are $1/n$. This was solved in the 1980s by Falikman and Egorychev, but later, Gurvits [**Gur08**] gave a remarkable proof using properties of real stability of polynomials. Similarly, in [**MSS15a**], the authors used the analysis of polynomials to show the existence of infinite families of regular bipartite Ramanujan graphs of every degree greater than two. Ramanujan graphs are examples of expander graphs, important in various algorithmic and modeling contexts. The power of analysis of roots of polynomials was also demonstrated in the solution of the *Kadison–Singer conjecture* [**MSS15b**].

Related to zeros of multivariate polynomials there are two other classes, that in a way extend stable polynomials, which have also seen great activity. A *hyperbolic polynomial* with respect to $e \in \mathbb{R}^n$ is a form $f \in \mathbb{R}[x_1, \ldots, x_n]$, such that $f(e) \neq 0$ and all roots of the univariate polynomial $f(te + a)$ are real $\forall a \in \mathbb{R}^n$. The hyperbolicity cone of f with respect to e is defined as

$$C(f, e) = \{a \in \mathbb{R}^n : f(te - a) \neq 0 \text{ when } t \leq 0\}.$$

Note a real homogeneous polynomial is stable if and only if it is hyperbolic in every direction in the positive orthant. Hyperbolicity is a tool to show real stability of polynomials and also establishes interesting connections to optimization and probability [**Pem12**].

In groundbreaking work Brändén and Huh introduced *Lorentzian polynomials* [**BH20**], which generalize the notion of log-concavity of sequences to multivariate homogeneous polynomials. Examples of Lorentzian polynomials include homogeneous stable polynomials as well as volume polynomials of convex bodies. Their theory has many remarkable consequences including showing that the

Alexandrov–Fenchel inequalities on mixed volumes of Minkowski sums of polytopes follow from the Lorentzian property of the volume polynomial. Lorentzian polynomials have been used in randomized algorithms to count bases of a matroid [**AGV21**, **ALGV19**] and to give a generalization of Postnikov's formula for the volume of a generalized permutahedron (the normalization of the integer point generating function of a generalized permutahedron is Lorentzian) [**BES20**]. Similarly, normalizations of certain projections of the generating function of integer points of generalized permutahedra with zero-one vertices are also Lorentzian. In [**HMMSD22**], Huh et al. showed that the normalization of certain projections of the generating function of the lattice points of Gel′fand–Tsetlin polytopes is Lorentzian too. The paper [**MS21**] extended this to flow polytopes and they proved the Kostant partition function (another Ehrhart quasipolynomial) is log-concave along root directions.

From the algorithmic point of view zeros of polynomials can be used to design approximation algorithms to compute p_F (see [**Bar16**,**BR19**,**PR22**] and references therein). One very exciting special application of understanding the geometry of zeros of matroid polynomials appears in [**AGV21**]. The authors used properties of matroid bases polynomials to give a deterministic $2^{O(r)}$-approximation algorithm for the number of bases of a given matroid of rank r and the number of common bases of any two matroids of rank r. A key tool in this algorithmic result was the fact, shown by Adiprasito, Huh, and Katz in [**AHK18**], that the matroid basis polynomial of any matroid is a (completely) log-concave polynomial. Formally, they prove that the multivariate generating polynomial of the bases of any matroid is (and all of its directional derivatives along the positive orthant are) log-concave as functions over the positive orthant. Another powerful view of many combinatorial inequalities was recently crafted by Chan and Pak in [**CP22**].

To conclude the survey, I must mention another connection of zeros of polynomials to enumerative combinatorics. For years generating functions have been used to analyze combinatorial sequences and in the univariate case have a long rich history [**Wil06**]. But when the combinatorial sequences are multivariate, the asymptotics for the Taylor coefficients $a_{\mathbf{r}}$ of functions of several variables with multi-index r are sometimes connected to the properties of singularities define by a certain algebraic variety. Yet another point of contact between nonlinear algebra and combinatorics. See [**PW13**] for a survey of multivariate generating functions and its applications.

3. Epilogue

To conclude, let us say this was just a taste of the very strong interaction between nonlinear algebra and computational discrete mathematics. Trying to keep the survey under control, I had to leave out several "discrete" topics including nonlinear algebra and coding theory cryptography (e.g., [**AFF$^+$16**, **CFP21**, **GR18**, **LSV21**, **TVN07**]), nonlinear algebra and data science/machine learning [**ASS17**, **AKRS21**, **BKSW18**, **CPU20**, **CJM$^+$21**, **HL13**], nonlinear algebra and game theory [**Dat03**, **Dat10**, **TY19**], etc. Fortunately, some topics will be treated in this volume. I hope that you are now curious enough to learn more through the references listed here and contribute new results yourself!

Acknowledgments

I thank NSF for the support I received through DMS grant 1818969. I am grateful to CINVESTAV-IPN and IMATE-UNAM Juriquilla in México, where the first version of this survey was crafted (Fall 2022) and to ICERM, where the final versions of this survey was worked out (Winter/Spring 2023). These institutions and their members/staff provided support and a wonderful creative environment.

I thank Edinah Gnang, Adriana Hansberger, Chiara Meroni, Amanda Montejano, Alejandro Morales, Jakob Nördstrom, Deborah Oliveros, Igor Pak, Annie Raymond, Rafael H. Villarreal, and Eduardo S. Zeron for comments and useful conversations. I am also grateful to the two anonymous reviewers and the editors of this volume for their work and effort.

References

[AD23] A. A. Ahmadi and C. Dibek, *A sum of squares characterization of perfect graphs*, SIAM J. Appl. Algebra Geom. **7** (2023), no. 4, 685–715, DOI 10.1137/22M1530410. MR4654605

[AES20] B. Aronov, E. Ezra, and M. Sharir, *Testing polynomials for vanishing on Cartesian products of planar point sets*, 36th International Symposium on Computational Geometry, LIPIcs. Leibniz Int. Proc. Inform., vol. 164, Schloss Dagstuhl. Leibniz-Zent. Inform., Wadern, 2020, pp. Art. No. 8, 14. MR4117721

[AFF$^+$16] M. R. Albrecht, J.-C. Faugère, P. Farshim, G. Herold, and L. Perret, *Polly Cracker, revisited*, Des. Codes Cryptogr. **79** (2016), no. 2, 261–302, DOI 10.1007/s10623-015-0048-8. MR3483945

[AGV21] N. Anari, S. O. Gharan, and C. Vinzant, *Log-concave polynomials, I: Entropy and a deterministic approximation algorithm for counting bases of matroids*, Duke Math. J. **170** (2021), no. 16, 3459–3504, DOI 10.1215/00127094-2020-0091. MR4332671

[AHK18] K. Adiprasito, J. Huh, and E. Katz, *Hodge theory for combinatorial geometries*, Ann. of Math. (2) **188** (2018), no. 2, 381–452, DOI 10.4007/annals.2018.188.2.1. MR3862944

[AKN20] K. Adiprasito, D. Kalmanovich, and E. Nevo, *On the realization space of the cube* (English, with English and French summaries), Sém. Lothar. Combin. **84B** (2020), Art. 80, 12. MR4138707

[AKRS21] C. Améndola, K. Kohn, P. Reichenbach, and A. Seigal, *Invariant theory and scaling algorithms for maximum likelihood estimation*, SIAM J. Appl. Algebra Geom. **5** (2021), no. 2, 304–337, DOI 10.1137/20M1328932. MR4274830

[AL94] W. W. Adams and P. Loustaunau, *An introduction to Gröbner bases*, Graduate Studies in Mathematics, vol. 3, American Mathematical Society, Providence, RI, 1994, DOI 10.1090/gsm/003. MR1287608

[ALGV19] N. Anari, K. Liu, S. O. Gharan, and C. Vinzant, *Log-concave polynomials II: High-dimensional walks and an FPRAS for counting bases of a matroid*, STOC'19—Proceedings of the 51st Annual ACM SIGACT Symposium on Theory of Computing, ACM, New York, 2019, pp. 1–12, DOI 10.1145/3313276.3316385. MR4003314

[Alo95] N. Alon, *Tools from higher algebra*, Handbook of combinatorics, Vol. 1, 2, Elsevier Sci. B. V., Amsterdam, 1995, pp. 1749–1783. MR1373688

[Alo99] N. Alon, *Combinatorial Nullstellensatz*, Combin. Probab. Comput. **8** (1999), no. 1-2, 7–29, DOI 10.1017/S0963548398003411. Recent trends in combinatorics (Mátraháza, 1995). MR1684621

[Alo00] N. Alon, *Algebraic and probabilistic methods in discrete mathematics*, Geom. Funct. Anal. **Special Volume** (2000), 455–470, DOI 10.1007/978-3-0346-0425-3_1. GAFA 2000 (Tel Aviv, 1999). MR1826264

[AOPT13] A. A. Ahmadi, A. Olshevsky, P. A. Parrilo, and J. N. Tsitsiklis, *NP-hardness of deciding convexity of quartic polynomials and related problems*, Math. Program. **137** (2013), no. 1-2, Ser. A, 453–476, DOI 10.1007/s10107-011-0499-2. MR3010433

[AP17] K. A. Adiprasito and A. Padrol, *The universality theorem for neighborly polytopes*, Combinatorica **37** (2017), no. 2, 129–136, DOI 10.1007/s00493-016-3253-9. MR3638337

[APP+05] N. Alon, J. Pach, R. Pinchasi, R. Radoičić, and M. Sharir, *Crossing patterns of semi-algebraic sets*, J. Combin. Theory Ser. A **111** (2005), no. 2, 310–326, DOI 10.1016/j.jcta.2004.12.008. MR2156215

[Arr06] E. Arrondo, *Another elementary proof of the Nullstellensatz*, Amer. Math. Monthly **113** (2006), no. 2, 169–171, DOI 10.2307/27641869. MR2203239

[AS16] N. Alon and J. H. Spencer, *The probabilistic method*, 4th ed., Wiley Series in Discrete Mathematics and Optimization, John Wiley & Sons, Inc., Hoboken, NJ, 2016. MR3524748

[ASS17] H. Abo, A. Seigal, and B. Sturmfels, *Eigenconfigurations of tensors*, Algebraic and geometric methods in discrete mathematics, Contemp. Math., vol. 685, Amer. Math. Soc., Providence, RI, 2017, pp. 1–25. MR3625569

[AT92] N. Alon and M. Tarsi, *Colorings and orientations of graphs*, Combinatorica **12** (1992), no. 2, 125–134, DOI 10.1007/BF01204715. MR1179249

[AZ15] K. A. Adiprasito and G. M. Ziegler, *Many projectively unique polytopes*, Invent. Math. **199** (2015), no. 3, 581–652, DOI 10.1007/s00222-014-0519-y. MR3314513

[Bar94] A. I. Barvinok, *A polynomial time algorithm for counting integral points in polyhedra when the dimension is fixed*, Math. Oper. Res. **19** (1994), no. 4, 769–779, DOI 10.1287/moor.19.4.769. MR1304623

[Bar06] A. Barvinok, *Computing the Ehrhart quasi-polynomial of a rational simplex*, Math. Comp. **75** (2006), no. 255, 1449–1466, DOI 10.1090/S0025-5718-06-01836-9. MR2219037

[Bar08] A. Barvinok, *Integer points in polyhedra*, Zurich Lectures in Advanced Mathematics, European Mathematical Society (EMS), Zürich, 2008, DOI 10.4171/052. MR2455889

[Bar16] A. Barvinok, *Combinatorics and complexity of partition functions*, Algorithms and Combinatorics, vol. 30, Springer, Cham, 2016, DOI 10.1007/978-3-319-51829-9. MR3558532

[Bay82] D. A. Bayer, *The division algorithm and the Hilbert scheme*, ProQuest LLC, Ann Arbor, MI, 1982. Thesis (Ph.D.)–Harvard University. MR2632095

[BB09a] J. Borcea and P. Brändén, *The Lee-Yang and Pólya-Schur programs. I. Linear operators preserving stability*, Invent. Math. **177** (2009), no. 3, 541–569, DOI 10.1007/s00222-009-0189-3. MR2534100

[BB09b] J. Borcea and P. Brändén, *The Lee-Yang and Pólya-Schur programs. II. Theory of stable polynomials and applications*, Comm. Pure Appl. Math. **62** (2009), no. 12, 1595–1631, DOI 10.1002/cpa.20295. MR2569072

[BB10] J. Borcea and P. Brändén, *Hyperbolicity preservers and majorization* (English, with English and French summaries), C. R. Math. Acad. Sci. Paris **348** (2010), no. 15-16, 843–846, DOI 10.1016/j.crma.2010.07.006. MR2677977

[BBDL+11] V. Baldoni, N. Berline, J. A. De Loera, M. Köppe, and M. Vergne, *How to integrate a polynomial over a simplex*, Math. Comp. **80** (2011), no. 273, 297–325, DOI 10.1090/S0025-5718-2010-02378-6. MR2728981

[BBDL+12] V. Baldoni, N. Berline, J. A. De Loera, M. Köppe, and M. Vergne, *Computation of the highest coefficients of weighted Ehrhart quasi-polynomials of rational polyhedra*, Found. Comput. Math. **12** (2012), no. 4, 435–469, DOI 10.1007/s10208-011-9106-4. MR2946460

[BBDL+19] V. Baldoni, N. Berline, J. A. De Loera, M. Köppe, and M. Vergne, *Three Ehrhart quasi-polynomials*, Algebr. Comb. **2** (2019), no. 3, 379–416, DOI 10.5802/alco.46. MR3968744

[BCC93] E. Balas, S. Ceria, and G. Cornuéjols, *A lift-and-project cutting plane algorithm for mixed* 0-1 *programs*, Math. Programming **58** (1993), no. 3, Ser. A, 295–324, DOI 10.1007/BF01581273. MR1216789

[BCC+17] J. Blasiak, T. Church, H. Cohn, J. A. Grochow, E. Naslund, W. F. Sawin, and C. Umans, *On cap sets and the group-theoretic approach to matrix multiplication*, Discrete Anal., posted on 2017, Paper No. 3, 27, DOI 10.19086/da.1245. MR3631613

[BCR87] J. Bochnak, M. Coste, and M.-F. Roy, *Géométrie algébrique réelle* (French), Ergebnisse der Mathematik und ihrer Grenzgebiete (3) [Results in Mathematics and Related Areas (3)], vol. 12, Springer-Verlag, Berlin, 1987. MR949442

[BD08] B. Braun and M. Develin, *Ehrhart polynomial roots and Stanley's non-negativity theorem*, Integer points in polyhedra—geometry, number theory, representation theory, algebra, optimization, statistics, Contemp. Math., vol. 452, Amer. Math. Soc., Providence, RI, 2008, pp. 67–78, DOI 10.1090/conm/452/08773. MR2405764

[BDLD+05] M. Beck, J. A. De Loera, M. Develin, J. Pfeifle, and R. P. Stanley, *Coefficients and roots of Ehrhart polynomials*, Integer points in polyhedra—geometry, number theory, algebra, optimization, Contemp. Math., vol. 374, Amer. Math. Soc., Providence, RI, 2005, pp. 15–36, DOI 10.1090/conm/374/06897. MR2134759

[BES20] S. Backman, C. Eur, and C. Simpson, *Simplicial generation of Chow rings of matroids*, Sém. Lothar. Combin. **84B** (2020), Art. 52, 11. MR4138680

[BGDH+19] C. Benedetti, R. S. González D'León, C. R. H. Hanusa, P. E. Harris, A. Khare, A. H. Morales, and M. Yip, *A combinatorial model for computing volumes of flow polytopes*, Trans. Amer. Math. Soc. **372** (2019), no. 5, 3369–3404, DOI 10.1090/tran/7743. MR3988614

[BH20] P. Brändén and J. Huh, *Lorentzian polynomials*, Ann. of Math. (2) **192** (2020), no. 3, 821–891, DOI 10.4007/annals.2020.192.3.4. MR4172622

[BIK+96] P. Beame, R. Impagliazzo, J. Krajíček, T. Pitassi, and P. Pudlák, *Lower bounds on Hilbert's Nullstellensatz and propositional proofs*, Proc. London Math. Soc. (3) **73** (1996), no. 1, 1–26, DOI 10.1112/plms/s3-73.1.1. MR1387081

[BIK+97] S. Buss, R. Impagliazzo, J. Krajíček, P. Pudlák, A. A. Razborov, and J. Sgall, *Proof complexity in algebraic systems and bounded depth Frege systems with modular counting*, Comput. Complexity **6** (1996/97), no. 3, 256–298, DOI 10.1007/BF01294258. MR1486929

[BIP19] P. Bürgisser, C. Ikenmeyer, and G. Panova, *No occurrence obstructions in geometric complexity theory*, J. Amer. Math. Soc. **32** (2019), no. 1, 163–193, DOI 10.1090/jams/908. MR3868002

[Bj95] A. Björner, *Topological methods*, Handbook of combinatorics, Vol. 1, 2, Elsevier Sci. B. V., Amsterdam, 1995, pp. 1819–1872. MR1373690

[BJMS15] S. Brodsky, M. Joswig, R. Morrison, and B. Sturmfels, *Moduli of tropical plane curves*, Res. Math. Sci. **2** (2015), Art. 4, 31, DOI 10.1186/s40687-014-0018-1. MR3333700

[BJS93] S. C. Billey, W. Jockusch, and R. P. Stanley, *Some combinatorial properties of Schubert polynomials*, J. Algebraic Combin. **2** (1993), no. 4, 345–374, DOI 10.1023/A:1022419800503. MR1241505

[BKSW18] P. Breiding, S. Kališnik, B. Sturmfels, and M. Weinstein, *Learning algebraic varieties from samples*, Rev. Mat. Complut. **31** (2018), no. 3, 545–593, DOI 10.1007/s13163-018-0273-6. MR3847076

[BLVS+99] A. Björner, M. Las Vergnas, B. Sturmfels, N. White, and G. M. Ziegler, *Oriented matroids*, 2nd ed., Encyclopedia of Mathematics and its Applications, vol. 46, Cambridge University Press, Cambridge, 1999, DOI 10.1017/CBO9780511586507. MR1744046

[BLW86] N. L. Biggs, E. K. Lloyd, and R. J. Wilson, *Graph theory. 1736–1936*, 2nd ed., The Clarendon Press, Oxford University Press, New York, 1986. MR879117

[Bok00] J. Bokowski, *Effective methods in computational synthetic geometry*, Automated deduction in geometry (Zurich, 2000), Lecture Notes in Comput. Sci., vol. 2061, Springer, Berlin, 2001, pp. 175–192, DOI 10.1007/3-540-45410-1_11. MR1908032

[BP96] S. R. Buss and T. Pitassi, *Good degree bounds on Nullstellensatz refutations of the induction principle*, J. Comput. System Sci. **57** (1998), no. 2, 162–171, DOI 10.1006/jcss.1998.1585. Complexity 96—The Eleventh Annual IEEE Conference on Computational Complexity (Philadelphia, PA). MR1661788

[BP99] A. Barvinok and J. E. Pommersheim, *An algorithmic theory of lattice points in polyhedra*, New perspectives in algebraic combinatorics (Berkeley, CA, 1996), Math. Sci. Res. Inst. Publ., vol. 38, Cambridge Univ. Press, Cambridge, 1999, pp. 91–147. MR1731815

[BPR96] S. Basu, R. Pollack, and M.-F. Roy, *On the number of cells defined by a family of polynomials on a variety*, Mathematika **43** (1996), no. 1, 120–126, DOI 10.1112/S0025579300011621. MR1401711

[BPR06] S. Basu, R. Pollack, and M.-F. Roy, *Algorithms in real algebraic geometry*, 2nd ed., Algorithms and Computation in Mathematics, vol. 10, Springer-Verlag, Berlin, 2006. MR2248869

[BPS98] D. Bayer, I. Peeva, and B. Sturmfels, *Monomial resolutions*, Math. Res. Lett. **5** (1998), no. 1-2, 31–46, DOI 10.4310/MRL.1998.v5.n1.a3. MR1618363

[BPS21] G. Balletti, M. Panizzut, and B. Sturmfels, *K3 polytopes and their quartic surfaces*, Adv. Geom. **21** (2021), no. 1, 85–98, DOI 10.1515/advgeom-2020-0016. MR4203287

[BPT] G. Blekherman, P. A. Parrilo, and R. R. Thomas (eds.), *Semidefinite optimization and convex algebraic geometry*, MOS-SIAM Series on Optimization, vol. 13, Society for Industrial and Applied Mathematics (SIAM), Philadelphia, PA; Mathematical Optimization Society, Philadelphia, PA, 2013. MR3075433

[Br15] P. Brändén, *Unimodality, log-concavity, real-rootedness and beyond*, Handbook of enumerative combinatorics, Discrete Math. Appl. (Boca Raton), CRC Press, Boca Raton, FL, 2015, pp. 437–483. MR3409348

[BR19] A. Barvinok and G. Regts, *Weighted counting of solutions to sparse systems of equations*, Combin. Probab. Comput. **28** (2019), no. 5, 696–719, DOI 10.1017/s0963548319000105. MR3991383

[BR22] G. Blekherman and A. Raymond, *A path forward: Tropicalization in extremal combinatorics*, Adv. Math. **407** (2022), Paper No. 108561, 68, DOI 10.1016/j.aim.2022.108561. MR4453032

[Bra08] B. Braun, *Norm bounds for Ehrhart polynomial roots*, Discrete Comput. Geom. **39** (2008), no. 1-3, 191–193, DOI 10.1007/s00454-008-9049-y. MR2383758

[Bra16] B. Braun, *Unimodality problems in Ehrhart theory*, Recent trends in combinatorics, IMA Vol. Math. Appl., vol. 159, Springer, [Cham], 2016, pp. 687–711, DOI 10.1007/978-3-319-24298-9_27. MR3526428

[Bre98] F. Brenti, *Hilbert polynomials in combinatorics*, J. Algebraic Combin. **7** (1998), no. 2, 127–156, DOI 10.1023/A:1008656320759. MR1609885

[Bri88] M. Brion, *Points entiers dans les polyèdres convexes* (French), Ann. Sci. École Norm. Sup. (4) **21** (1988), no. 4, 653–663. MR982338

[Bro87] W. D. Brownawell, *Bounds for the degrees in the Nullstellensatz*, Ann. of Math. (2) **126** (1987), no. 3, 577–591, DOI 10.2307/1971361. MR916719

[BRS90] J. Bokowski, J. Richter, and B. Sturmfels, *Nonrealizability proofs in computational geometry*, Discrete Comput. Geom. **5** (1990), no. 4, 333–350, DOI 10.1007/BF02187794. MR1043715

[BRST20] G. Blekherman, A. Raymond, M. Singh, and R. R. Thomas, *Simple graph density inequalities with no sum of squares proofs*, Combinatorica **40** (2020), no. 4, 455–471, DOI 10.1007/s00493-019-4124-y. MR4150878

[BS86] J. Bokowski and B. Sturmfels, *Reell realisierbare orientierte Matroide* (German), Bayreuth. Math. Schr. **21** (1986), 1–13. Diskrete Strukturen, algebraische Methoden und Anwendungen (Bayreuth, 1985). MR864084

[BS89] J. Bokowski and B. Sturmfels, *Computational synthetic geometry*, Lecture Notes in Mathematics, vol. 1355, Springer-Verlag, Berlin, 1989, DOI 10.1007/BFb0089253. MR1009366

[BS18] M. Beck and R. Sanyal, *Combinatorial reciprocity theorems*, Graduate Studies in Mathematics, vol. 195, American Mathematical Society, Providence, RI, 2018. An invitation to enumerative geometric combinatorics, DOI 10.1090/gsm/195. MR3839322

[Bus96] S. R. Buss, *Lower bounds on Nullstellensatz proofs via designs*, Proof complexity and feasible arithmetics (Rutgers, NJ, 1996), DIMACS Ser. Discrete Math. Theoret. Comput. Sci., vol. 39, Amer. Math. Soc., Providence, RI, 1998, pp. 59–71, DOI 10.1090/dimacs/039/04. MR1486614

[BW03] A. Barvinok and K. Woods, *Short rational generating functions for lattice point problems*, J. Amer. Math. Soc. **16** (2003), no. 4, 957–979, DOI 10.1090/S0894-0347-03-00428-4. MR1992831

[BW05] D. Bertsimas and R. Weismantel, *Optimization over integers*, Dynamic Ideas, Belmont, Massachusetts, USA, 2005.

[BZ01] A. Berenstein and A. Zelevinsky, *Tensor product multiplicities, canonical bases, and totally positive varieties*, Invent. Math. **143** (2001), no. 1, 77–128, DOI 10.1007/s002220000102. MR1802793

[CDRD$^+$14] A. Conca, S. Di Rocco, J. Draisma, J. Huh, B. Sturmfels, and F. Viviani, *Combinatorial algebraic geometry*, Lecture Notes in Mathematics, vol. 2108, Springer, Cham; Centro Internazionale Matematico Estivo (C.I.M.E.), Florence, 2014. Lecture notes from the CIME-CIRM Summer School held in Levico Terme, June 10–15, 2013; Edited by Sandra Di Rocco and Bernd Sturmfels; Fondazione CIME/CIME Foundation Subseries. MR3309087

[CEH$^+$21] J. Cslovjecsek, F. Eisenbrand, C. Hunkenschröder, L. Rohwedder, and R. Weismantel, *Block-structured integer and linear programming in strongly polynomial and near linear time*, Proceedings of the 2021 ACM-SIAM Symposium on Discrete Algorithms (SODA), [Society for Industrial and Applied Mathematics (SIAM)], Philadelphia, PA, 2021, pp. 1666–1681, DOI 10.1137/1.9781611976465.101. MR4262534

[CEI96] M. Clegg, J. Edmonds, and R. Impagliazzo, *Using the Gröbner basis algorithm to find proofs of unsatisfiability*, Proceedings of the Twenty-eighth Annual ACM Symposium on the Theory of Computing (Philadelphia, PA, 1996), ACM, New York, 1996, pp. 174–183, DOI 10.1145/237814.237860. MR1427512

[CFGH19] D. Conlon, J. Fox, A. Grinshpun, and X. He, *Online Ramsey numbers and the subgraph query problem*, Building bridges II—mathematics of László Lovász, Bolyai Soc. Math. Stud., vol. 28, Springer, Berlin, [2019] ©2019, pp. 159–194. MR4297780

[CFP$^+$14] D. Conlon, J. Fox, J. Pach, B. Sudakov, and A. Suk, *Ramsey-type results for semi-algebraic relations*, Trans. Amer. Math. Soc. **366** (2014), no. 9, 5043–5065, DOI 10.1090/S0002-9947-2014-06179-5. MR3217709

[CFP21] O. Chakraborty, J.-C. Faugère, and L. Perret, *Cryptanalysis of the extension field cancellation cryptosystem*, Des. Codes Cryptogr. **89** (2021), no. 6, 1335–1364, DOI 10.1007/s10623-021-00873-9. MR4264980

[CGS11] A. Cuttler, C. Greene, and M. Skandera, *Inequalities for symmetric means*, European J. Combin. **32** (2011), no. 6, 745–761, DOI 10.1016/j.ejc.2011.01.020. MR2821548

[CJM$^+$21] T. Ö. Çelik, A. Jamneshan, G. Montúfar, B. Sturmfels, and L. Venturello, *Wasserstein distance to independence models*, J. Symbolic Comput. **104** (2021), 855–873, DOI 10.1016/j.jsc.2020.10.005. MR4180150

[CKPS00] N. Courtois, A. Klimov, J. Patarin, and A. Shamir, *Efficient algorithms for solving overdefined systems of multivariate polynomial equations*, Advances in cryptology—EUROCRYPT 2000 (Bruges), Lecture Notes in Comput. Sci., vol. 1807, Springer, Berlin, 2000, pp. 392–407, DOI 10.1007/3-540-45539-6_27. MR1772028

[CLO92] D. Cox, J. Little, and D. O'Shea, *Ideals, varieties, and algorithms: An introduction to computational algebraic geometry and commutative algebra*, Undergraduate Texts in Mathematics, Springer-Verlag, New York, 1992, DOI 10.1007/978-1-4757-2181-2. MR1189133

[CLO05] D. A. Cox, J. Little, and D. O'Shea, *Using algebraic geometry*, 2nd ed., Graduate Texts in Mathematics, vol. 185, Springer, New York, 2005. MR2122859

[CLS11] D. A. Cox, J. B. Little, and H. K. Schenck, *Toric varieties*, Graduate Studies in Mathematics, vol. 124, American Mathematical Society, Providence, RI, 2011, DOI 10.1090/gsm/124. MR2810322

[CP22] S. H. Chan and I. Pak, *Introduction to the combinatorial atlas*, Expo. Math. **40** (2022), no. 4, 1014–1048, DOI 10.1016/j.exmath.2022.08.003. MR4518642

[CPU20] M. Casanellas, S. Petrović, and C. Uhler, *Algebraic statistics in practice: Applications to networks*, Annu. Rev. Stat. Appl. **7** (2020), 227–250, DOI 10.1146/annurev-statistics-031017-100053. MR4104192

[CR06] C. W. Curtis and I. Reiner, *Representation theory of finite groups and associative algebras*, AMS Chelsea Publishing, Providence, RI, 2006. Reprint of the 1962 original, DOI 10.1090/chel/356. MR2215618

[CRST06] M. Chudnovsky, N. Robertson, P. Seymour, and R. Thomas, *The strong perfect graph theorem*, Ann. of Math. (2) **164** (2006), no. 1, 51–229, DOI 10.4007/annals.2006.164.51. MR2233847

[CT91] P. Conti and C. Traverso, *Buchberger algorithm and integer programming*, Applied algebra, algebraic algorithms and error-correcting codes (New Orleans, LA, 1991), Lecture Notes in Comput. Sci., vol. 539, Springer, Berlin, 1991, pp. 130–139, DOI 10.1007/3-540-54522-0_102. MR1229314

[Dat03] R. S. Datta, *Universality of Nash equilibria*, Math. Oper. Res. **28** (2003), no. 3, 424–432, DOI 10.1287/moor.28.3.424.16397. MR1997243

[Dat10] R. S. Datta, *Finding all Nash equilibria of a finite game using polynomial algebra*, Econom. Theory **42** (2010), no. 1, 55–96, DOI 10.1007/s00199-009-0447-z. MR2551729

[DE05] A. Dickenstein and I. Z. Emiris (eds.), *Solving polynomial equations*, Algorithms and Computation in Mathematics, vol. 14, Springer-Verlag, Berlin, 2005. Foundations, algorithms, and applications, DOI 10.1007/b138957. MR2161984

[DHS09] P. Dueck, S. Hoşten, and B. Sturmfels, *Normal toric ideals of low codimension*, J. Pure Appl. Algebra **213** (2009), no. 8, 1636–1641, DOI 10.1016/j.jpaa.2008.11.045. MR2517998

[DK97] M. Dyer and R. Kannan, *On Barvinok's algorithm for counting lattice points in fixed dimension*, Math. Oper. Res. **22** (1997), no. 3, 545–549, DOI 10.1287/moor.22.3.545. MR1467384

[DL95] J. A. De Loera, *Gröbner bases and graph colorings*, Beiträge Algebra Geom. **36** (1995), no. 1, 89–96. MR1337123

[DLDK+13] J. A. De Loera, B. Dutra, M. Köppe, S. Moreinis, G. Pinto, and J. Wu, *Software for exact integration of polynomials over polyhedra*, Comput. Geom. **46** (2013), no. 3, 232–252, DOI 10.1016/j.comgeo.2012.09.001. MR2994431

[DLEKW23] J. A. De Loera, L. Escobar, N. Kaplan, and C. Wang, *Sums of weighted lattice points of polytopes*, Sém. Lothar. Combin. **89B** (2023), Art. 17, 12. MR4659525

[DLGMM19] J. A. De Loera, X. Goaoc, F. Meunier, and N. H. Mustafa, *The discrete yet ubiquitous theorems of Carathéodory, Helly, Sperner, Tucker, and Tverberg*, Bull. Amer. Math. Soc. (N.S.) **56** (2019), no. 3, 415–511, DOI 10.1090/bull/1653. MR3974609

[DLHH+05] J. A. De Loera, D. C. Haws, R. Hemmecke, P. Huggins, J. Tauzer, and R. Yoshida, *LattE, version 1.2*, Available at `http://www.math.ucdavis.edu/~latte/`, 2005.

[DLHK12] J. A. De Loera, R. Hemmecke, and M. Köppe, *Algebraic and geometric ideas in the theory of discrete optimization*, MOS-SIAM Series on Optimization, vol. 14, Society for Industrial and Applied Mathematics (SIAM), Philadelphia, PA; Mathematical Optimization Society, Philadelphia, PA, 2013. MR3024570

[DLHKW06] J. A. De Loera, R. Hemmecke, M. Köppe, and R. Weismantel, *Integer polynomial optimization in fixed dimension*, Math. Oper. Res. **31** (2006), no. 1, 147–153, DOI 10.1287/moor.1050.0169. MR2205524

[DLHMO10] J. A. De Loera, C. J. Hillar, P. N. Malkin, and M. Omar, *Recognizing graph theoretic properties with polynomial ideals*, Electron. J. Combin. **17** (2010), no. 1, Research Paper 114, 26, DOI 10.37236/386. MR2679568

[DLLMM08] J. A. De Loera, J. Lee, P. N. Malkin, and S. Margulies, *Hilbert's Nullstellensatz and an algorithm for proving combinatorial infeasibility*, ISSAC 2008, ACM, New York, 2008, pp. 197–206, DOI 10.1145/1390768.1390797. MR2513506

[DLLMM11] J. A. De Loera, J. Lee, P. N. Malkin, and S. Margulies, *Computing infeasibility certificates for combinatorial problems through Hilbert's Nullstellensatz*, J. Symbolic Comput. **46** (2011), no. 11, 1260–1283, DOI 10.1016/j.jsc.2011.08.007. MR2846906

[DLLMO09] J. A. Loera, J. Lee, S. Margulies, and S. Onn, *Expressing combinatorial problems by systems of polynomial equations and Hilbert's Nullstellensatz*, Combin. Probab. Comput. **18** (2009), no. 4, 551–582, DOI 10.1017/S0963548309009894. MR2507737

[DLM06] J. A. De Loera and T. B. McAllister, *On the computation of Clebsch-Gordan coefficients and the dilation effect*, Experiment. Math. **15** (2006), no. 1, 7–19. MR2229381

[DLW22] J. A. De Loera and W. J. Wesley, *Ramsey numbers through the lenses of polynomial ideals and nullstellensätze*.

[DPV11] D. Dadush, C. Peikert, and S. Vempala, *Enumerative lattice algorithms in any norm via M-ellipsoid coverings*, 2011 IEEE 52nd Annual Symposium on Foundations of

Computer Science—FOCS 2011, IEEE Computer Soc., Los Alamitos, CA, 2011, pp. 580–589, DOI 10.1109/FOCS.2011.31. MR2932734

[dRMNR21] S. F. De Rezende, O. Meir, J. Nordström, and R. Robere, *Nullstellensatz size-degree trade-offs from reversible pebbling*, Comput. Complexity **30** (2021), no. 1, Paper No. 4, 45, DOI 10.1007/s00037-020-00201-y. MR4218368

[DW00] H. Derksen and J. Weyman, *Semi-invariants of quivers and saturation for Littlewood-Richardson coefficients*, J. Amer. Math. Soc. **13** (2000), no. 3, 467–479, DOI 10.1090/S0894-0347-00-00331-3. MR1758750

[EG00] M. Ehrgott and X. Gandibleux, *A survey and annotated bibliography of multiobjective combinatorial optimization* (English, with English and German summaries), OR Spektrum **22** (2000), no. 4, 425–460, DOI 10.1007/s002910000046. MR1798676

[EG17] J. S. Ellenberg and D. Gijswijt, *On large subsets of* $\mathbb{F}_q^n$ *with no three-term arithmetic progression*, Ann. of Math. (2) **185** (2017), no. 1, 339–343, DOI 10.4007/annals.2017.185.1.8. MR3583358

[EHK+19] F. Eisenbrand, C. Hunkenschröder, K.-M. Klein, M. Koutecký, A. Levin, and S. Onn, *An algorithmic theory of integer programming*, 2019.

[Eli92] S. Eliahou, *An algebraic criterion for a graph to be four-colourable*, International Seminar on Algebra and its Applications (Spanish) (México City, 1991), Aportaciones Mat. Notas Investigación, vol. 6, Soc. Mat. Mexicana, México, 1992, pp. 3–27. MR1310369

[EP92] V. A. Emelichev and V. A. Perepelitsa, *On the power of sets of alternatives in discrete multicriterial problems* (Russian), Diskret. Mat. **3** (1991), no. 3, 3–12, DOI 10.1515/dma.1992.2.5.461; English transl., Discrete Math. Appl. **2** (1992), no. 5, 461–471. MR1138087

[eRAeR01] J. L. Ramírez Alfonsín and B. A. Reed (eds.), *Perfect graphs*, Wiley-Interscience Series in Discrete Mathematics and Optimization, John Wiley & Sons, Ltd., Chichester, 2001. MR1858793

[ES94] D. Eisenbud and B. Sturmfels, *Finding sparse systems of parameters*, J. Pure Appl. Algebra **94** (1994), no. 2, 143–157, DOI 10.1016/0022-4049(94)90029-9. MR1282836

[Fau99] J.-C. Faugére, *A new efficient algorithm for computing Gröbner bases* (F_4), J. Pure Appl. Algebra **139** (1999), no. 1-3, 61–88, DOI 10.1016/S0022-4049(99)00005-5. Effective methods in algebraic geometry (Saint-Malo, 1998). MR1700538

[Fis88] K. G. Fischer, *Symmetric polynomials and Hall's theorem*, Discrete Math. **69** (1988), no. 3, 225–234, DOI 10.1016/0012-365X(88)90051-9. MR940078

[FM17] J.-C. Faugère and C. Mou, *Sparse FGLM algorithms*. part 3, J. Symbolic Comput. **80** (2017), no. part 3, 538–569, DOI 10.1016/j.jsc.2016.07.025. MR3574527

[FMM13] K. Fukuda, H. Miyata, and S. Moriyama, *Complete enumeration of small realizable oriented matroids*, Discrete Comput. Geom. **49** (2013), no. 2, 359–381, DOI 10.1007/s00454-012-9470-0. MR3017917

[FMP+15] S. Fiorini, S. Massar, S. Pokutta, H. R. Tiwary, and R. de Wolf, *Exponential lower bounds for polytopes in combinatorial optimization*, J. ACM **62** (2015), no. 2, Art. 17, 23, DOI 10.1145/2716307. MR3346156

[FT87] A. Frank and É. Tardos, *An application of simultaneous Diophantine approximation in combinatorial optimization*, Combinatorica **7** (1987), no. 1, 49–65, DOI 10.1007/BF02579200. MR905151

[Ful97] W. Fulton, *Young tableaux*, London Mathematical Society Student Texts, vol. 35, Cambridge University Press, Cambridge, 1997. With applications to representation theory and geometry. MR1464693

[GfC50] I. M. Gel′fand and M. L. Cetlin, *Finite-dimensional representations of the group of unimodular matrices* (Russian), Doklady Akad. Nauk SSSR (N.S.) **71** (1950), 825–828. MR35774

[GGL95] R. L. Graham, M. Grötschel, and L. Lovász (eds.), *Handbook of combinatorics. Vol. 1, 2*, Elsevier Science B.V., Amsterdam; MIT Press, Cambridge, MA, 1995. MR1373655

[GHZ19] D. Gonzalez, X. He, and H. Zheng, *An upper bound for the restricted online Ramsey number*, Discrete Math. **342** (2019), no. 9, 2564–2569, DOI 10.1016/j.disc.2019.05.024. MR3959680

[GJ79] M. R. Garey and D. S. Johnson, *Computers and intractability: A guide to the theory of NP-completeness*, A Series of Books in the Mathematical Sciences, W. H. Freeman and Co., San Francisco, CA, 1979. MR519066

[GK10] L. Guth and N. H. Katz, *Algebraic methods in discrete analogs of the Kakeya problem*, Adv. Math. **225** (2010), no. 5, 2828–2839, DOI 10.1016/j.aim.2010.05.015. MR2680185

[GK15] L. Guth and N. H. Katz, *On the Erdős distinct distances problem in the plane*, Ann. of Math. (2) **181** (2015), no. 1, 155–190, DOI 10.4007/annals.2015.181.1.2. MR3272924

[GKK22] T. Gavenčiak, M. Koutecký, and D. Knop, *Integer programming in parameterized complexity: Five miniatures*. part 1, Discrete Optim. **44** (2022), no. part 1, Paper No. 100596, 33, DOI 10.1016/j.disopt.2020.100596. MR4414790

[GLPT12] J. Gouveia, M. Laurent, P. A. Parrilo, and R. Thomas, *A new semidefinite programming hierarchy for cycles in binary matroids and cuts in graphs*, Math. Program. **133** (2012), no. 1-2, Ser. A, 203–225, DOI 10.1007/s10107-010-0425-z. MR2921097

[GMTW19] J. Gouveia, A. Macchia, R. R. Thomas, and A. Wiebe, *The slack realization space of a polytope*, SIAM J. Discrete Math. **33** (2019), no. 3, 1637–1653, DOI 10.1137/18M1233649. MR4002715

[GMTW20] J. Gouveia, A. Macchia, R. R. Thomas, and A. Wiebe, *Projectively unique polytopes and toric slack ideals*, J. Pure Appl. Algebra **224** (2020), no. 5, 106229, 14, DOI 10.1016/j.jpaa.2019.106229. MR4046235

[GMW23a] J. Gouveia, A. Macchia, and A. Wiebe, *General non-realizability certificates for spheres with linear programming*, J. Symbolic Comput. **114** (2023), 172–192, DOI 10.1016/j.jsc.2022.04.013. MR4416745

[GMW23b] J. Gouveia, A. Macchia, and A. Wiebe, *General non-realizability certificates for spheres with linear programming*, J. Symbolic Comput. **114** (2023), 172–192, DOI 10.1016/j.jsc.2022.04.013. MR4416745

[Gna23] E. K. Gnang, *A proof of the kotzig-ringel-rosa conjecture*, 2023.

[GPT10] J. Gouveia, P. A. Parrilo, and R. R. Thomas, *Theta bodies for polynomial ideals*, SIAM J. Optim. **20** (2010), no. 4, 2097–2118, DOI 10.1137/090746525. MR2630035

[GR18] E. Gorla and A. Ravagnani, *An algebraic framework for end-to-end physical-layer network coding*, IEEE Trans. Inform. Theory **64** (2018), no. 6, 4480–4495, DOI 10.1109/TIT.2017.2778726. MR3809754

[Gra75] J. E. Graver, *On the foundations of linear and integer linear programming. I*, Math. Programming **9** (1975), no. 2, 207–226, DOI 10.1007/BF01681344. MR386673

[GRS90] R. L. Graham, B. L. Rothschild, and J. H. Spencer, *Ramsey theory*, 2nd ed., Wiley-Interscience Series in Discrete Mathematics and Optimization, John Wiley & Sons, Inc., New York, 1990. A Wiley-Interscience Publication. MR1044995

[Gur08] L. Gurvits, *Van der Waerden/Schrijver-Valiant like conjectures and stable (aka hyperbolic) homogeneous polynomials: one theorem for all*, Electron. J. Combin. **15** (2008), no. 1, Research Paper 66, 26, DOI 10.37236/790. With a corrigendum. MR2411443

[Gut16] L. Guth, *Polynomial methods in combinatorics*, University Lecture Series, vol. 64, American Mathematical Society, Providence, RI, 2016, DOI 10.1090/ulect/064. MR3495952

[GV02] D. Grigoriev and N. Vorobjov, *Complexity of Null- and Positivstellensatz proofs*, Ann. Pure Appl. Logic **113** (2002), no. 1-3, 153–160, DOI 10.1016/S0168-0072(01)00055-0. First St. Petersburg Conference on Days of Logic and Computability (1999). MR1875740

[Hei05] S. Heinz, *Complexity of integer quasiconvex polynomial optimization*, J. Complexity **21** (2005), no. 4, 543–556, DOI 10.1016/j.jco.2005.04.004. MR2152721

[HK13] R. Hildebrand and M. Köppe, *A new Lenstra-type algorithm for quasiconvex polynomial integer minimization with complexity* $2^{O(n \log n)}$, Discrete Optim. **10** (2013), no. 1, 69–84, DOI 10.1016/j.disopt.2012.11.003. MR3003413

[HKW10] R. Hemmecke, M. Köppe, and R. Weismantel, *A polynomial-time algorithm for optimizing over N-fold 4-block decomposable integer programs*, Integer programming and combinatorial optimization, Lecture Notes in Comput. Sci., vol. 6080, Springer, Berlin, 2010, pp. 219–229, DOI 10.1007/978-3-642-13036-6_17. MR2661100

[HL03] D. Henrion and J.-B. Lasserre, *GloptiPoly: Global optimization over polynomials with Matlab and SeDuMi*, ACM Trans. Math. Software **29** (2003), no. 2, 165–194, DOI 10.1145/779359.779363. MR2000881

[HL13] C. J. Hillar and L.-H. Lim, *Most tensor problems are NP-hard*, J. ACM **60** (2013), no. 6, Art. 45, 39, DOI 10.1145/2512329. MR3144915

[HMMSD22] J. Huh, J. P. Matherne, K. Mészáros, and A. St. Dizier, *Logarithmic concavity of Schur and related polynomials*, Trans. Amer. Math. Soc. **375** (2022), no. 6, 4411–4427, DOI 10.1090/tran/8606. MR4419063

[HOR13] R. Hemmecke, S. Onn, and L. Romanchuk, *n-fold integer programming in cubic time*, Math. Program. **137** (2013), no. 1-2, Ser. A, 325–341, DOI 10.1007/s10107-011-0490-y. MR3010429

[HOW11] R. Hemmecke, S. Onn, and R. Weismantel, *A polynomial oracle-time algorithm for convex integer minimization*, Math. Program. **126** (2011), no. 1, Ser. A, 97–117, DOI 10.1007/s10107-009-0276-7. MR2764341

[HS95a] S. Hoşten and B. Sturmfels, *GRIN: An implementation of Gröbner bases for integer programming*, Integer programming and combinatorial optimization (Copenhagen, 1995), Lecture Notes in Comput. Sci., vol. 920, Springer, Berlin, 1995, pp. 267–276, DOI 10.1007/3-540-59408-6_57. MR1367987

[HS95b] B. Huber and B. Sturmfels, *A polyhedral method for solving sparse polynomial systems*, Math. Comp. **64** (1995), no. 212, 1541–1555, DOI 10.2307/2153370. MR1297471

[HS97] B. Huber and B. Sturmfels, *Bernstein's theorem in affine space*, Discrete Comput. Geom. **17** (1997), no. 2, 137–141, DOI 10.1007/BF02770870. MR1424821

[HS07] S. Hoşten and S. Sullivant, *A finiteness theorem for Markov bases of hierarchical models*, J. Combin. Theory Ser. A **114** (2007), no. 2, 311–321, DOI 10.1016/j.jcta.2006.06.001. MR2293094

[HS21] A. Heaton and I. Shankar, *An SOS counterexample to an inequality of symmetric functions*, J. Pure Appl. Algebra **225** (2021), no. 8, Paper No. 106656, 10, DOI 10.1016/j.jpaa.2020.106656. MR4190610

[HSYY08] P. Huggins, B. Sturmfels, J. Yu, and D. S. Yuster, *The hyperdeterminant and triangulations of the 4-cube*, Math. Comp. **77** (2008), no. 263, 1653–1679, DOI 10.1090/S0025-5718-08-02073-5. MR2398786

[HT98] S. Hoşten and R. Thomas, *Gröbner bases and integer programming*, Gröbner bases and applications (Linz, 1998), London Math. Soc. Lecture Note Ser., vol. 251, Cambridge Univ. Press, Cambridge, 1998, pp. 144–158. MR1699817

[Huh12] J. Huh, *Milnor numbers of projective hypersurfaces and the chromatic polynomial of graphs*, J. Amer. Math. Soc. **25** (2012), no. 3, 907–927, DOI 10.1090/S0894-0347-2012-00731-0. MR2904577

[HW08] C. J. Hillar and T. Windfeldt, *Algebraic characterization of uniquely vertex colorable graphs*, J. Combin. Theory Ser. B **98** (2008), no. 2, 400–414, DOI 10.1016/j.jctb.2007.08.004. MR2389606

[IPP22] C. Ikenmeyer, I. Pak, and G. Panova, *Positivity of the symmetric group characters is as hard as the polynomial time hierarchy*, Proceedings of the 2023 Annual ACM-SIAM Symposium on Discrete Algorithms (SODA), SIAM, Philadelphia, PA, 2023, pp. 3573–3586, DOI 10.1137/1.9781611977554.ch136. MR4538090

[IPS99] R. Impagliazzo, P. Pudlák, and J. Sgall, *Lower bounds for the polynomial calculus and the Gröbner basis algorithm*, Comput. Complexity **8** (1999), no. 2, 127–144, DOI 10.1007/s000370050024. MR1724604

[IRNR18] A. Iosevich, O. Roshe-N′yuton, and M. Rudnev, *On discrete values of bilinear forms* (Russian, with Russian summary), Mat. Sb. **209** (2018), no. 10, 71–88, DOI 10.4213/sm8966; English transl., Sb. Math. **209** (2018), no. 10, 1482–1497. MR3859410

[Ise74] H. Isermann, *Proper efficiency and the linear vector maximum problem*, Operations Res. **22** (1974), no. 1, 189–191, DOI 10.1287/opre.22.1.189. MR408806

[JLN+10] M. Jünger, T. Liebling, D. Naddef, G. Nemhauser, W. Pulleyblank, G. Reinelt, G. Rinaldi, and L. Wolsey (eds.), *50 years of integer programming 1958–2008*, Springer-Verlag, Berlin, 2010. From the early years to the state-of-the-art; Papers from the

12th Combinatorial Optimization Workshop (AUSSOIS 2008) held in Aussois, January 7–11, 2008, DOI 10.1007/978-3-540-68279-0. MR2640549

[Kha84] L. G. Khachiyan, *Convexity and complexity in polynomial programming*, Proceedings of the International Congress of Mathematicians, Vol. 1, 2 (Warsaw, 1983), PWN, Warsaw, 1984, pp. 1569–1577. MR804797

[Khi48] A. Ya. Hinčin, *A quantitative formulation of the approximation theory of Kronecker* (Russian), Izv. Akad. Nauk SSSR Ser. Mat. **12** (1948), 113–122. MR24925

[KK05] A. Kehrein and M. Kreuzer, *Characterizations of border bases*, J. Pure Appl. Algebra **196** (2005), no. 2-3, 251–270, DOI 10.1016/j.jpaa.2004.08.028. MR2116166

[KKR05] A. Kehrein, M. Kreuzer, and L. Robbiano, *An algebraist's view on border bases*, Solving polynomial equations, Algorithms Comput. Math., vol. 14, Springer, Berlin, 2005, pp. 169–202, DOI 10.1007/3-540-27357-3_4. MR2161988

[KLO18] M. Koutecký, A. Levin, and S. Onn, *A parameterized strongly polynomial algorithm for block structured integer programs*, 45th International Colloquium on Automata, Languages, and Programming, LIPIcs. Leibniz Int. Proc. Inform., vol. 107, Schloss Dagstuhl. Leibniz-Zent. Inform., Wadern, 2018, pp. Art. No. 85, 14. MR3830016

[KO21] M. Koutecký and S. Onn, *Sparse integer programming is FPT*, Bull. Eur. Assoc. Theor. Comput. Sci. EATCS **134** (2021), 69–71. MR4566637

[Kol88] J. Kollár, *Sharp effective Nullstellensatz*, J. Amer. Math. Soc. **1** (1988), no. 4, 963–975, DOI 10.2307/1990996. MR944576

[KP00] L. Khachiyan and L. Porkolab, *Integer optimization on convex semialgebraic sets*, Discrete Comput. Geom. **23** (2000), no. 2, 207–224, DOI 10.1007/PL00009496. MR1739606

[KSS10] H. Kaplan, M. Sharir, and E. Shustin, *On lines and joints*, Discrete Comput. Geom. **44** (2010), no. 4, 838–843, DOI 10.1007/s00454-010-9246-3. MR2728035

[KSZ92] M. M. Kapranov, B. Sturmfels, and A. V. Zelevinsky, *Chow polytopes and general resultants*, Duke Math. J. **67** (1992), no. 1, 189–218, DOI 10.1215/S0012-7094-92-06707-X. MR1174606

[KT99] A. Knutson and T. Tao, *The honeycomb model of* $\mathrm{GL}_n(\mathbf{C})$ *tensor products. I. Proof of the saturation conjecture*, J. Amer. Math. Soc. **12** (1999), no. 4, 1055–1090, DOI 10.1090/S0894-0347-99-00299-4. MR1671451

[Las01] J. B. Lasserre, *Global optimization with polynomials and the problem of moments*, SIAM J. Optim. **11** (2000/01), no. 3, 796–817, DOI 10.1137/S1052623400366802. MR1814045

[Las04] J. B. Lasserre, *A discrete Farkas lemma*, Discrete Optim. **1** (2004), no. 1, 67–75, DOI 10.1016/j.disopt.2004.04.002. MR2098081

[Lau03] A. G. B. Lauder, *Zero-patterns of polynomials and Newton polytopes*, J. Combin. Theory Ser. A **102** (2003), no. 1, 10–15, DOI 10.1016/S0097-3165(03)00007-4. MR1970974

[Lau07] M. Laurent, *Semidefinite representations for finite varieties*, Math. Program. **109** (2007), no. 1, Ser. A, 1–26, DOI 10.1007/s10107-004-0561-4. MR2291590

[Laz77] D. Lazard, *Algèbre linéaire sur* $K[X_1, \cdots, X_n]$, *et élimination* (French), Bull. Soc. Math. France **105** (1977), no. 2, 165–190. MR491702

[Len83] H. W. Lenstra Jr., *Integer programming with a fixed number of variables*, Math. Oper. Res. **8** (1983), no. 4, 538–548, DOI 10.1287/moor.8.4.538. MR727410

[LL81] S.-Y. R. Li and W. C. W. Li, *Independence numbers of graphs and generators of ideals*, Combinatorica **1** (1981), no. 1, 55–61, DOI 10.1007/BF02579177. MR602416

[LLL82] A. K. Lenstra, H. W. Lenstra Jr., and L. Lovász, *Factoring polynomials with rational coefficients*, Math. Ann. **261** (1982), no. 4, 515–534, DOI 10.1007/BF01457454. MR682664

[LLO16] B. Li, B. Lowenstein, and M. Omar, *Low degree Nullstellensatz certificates for 3-colorability*, Electron. J. Combin. **23** (2016), no. 1, Paper 1.6, 12, DOI 10.37236/5103. MR3484711

[LN17] M. Lauria and J. Nordström, *Graph colouring is hard for algorithms based on Hilbert's Nullstellensatz and Gröbner bases*, 32nd Computational Complexity Conference, LIPIcs. Leibniz Int. Proc. Inform., vol. 79, Schloss Dagstuhl. Leibniz-Zent. Inform., Wadern, 2017, pp. Art. No. 2, 20. MR3691127

[Löf04] J. Löfberg, *YALMIP: A toolbox for modeling and optimization in MATLAB*, Proceedings of the CACSD Conference (Taipei, Taiwan), 2004.

[Lov94] L. Lovász, *Stable sets and polynomials*, Discrete Math. **124** (1994), no. 1-3, 137–153, DOI 10.1016/0012-365X(92)00057-X. Graphs and combinatorics (Qawra, 1990). MR1258849

[LS91] L. Lovász and A. Schrijver, *Cones of matrices and set-functions and* 0–1 *optimization*, SIAM J. Optim. **1** (1991), no. 2, 166–190, DOI 10.1137/0801013. MR1098425

[LSV21] H. H. López, I. Soprunov, and R. H. Villarreal, *The dual of an evaluation code*, Des. Codes Cryptogr. **89** (2021), no. 7, 1367–1403, DOI 10.1007/s10623-021-00872-w. MR4272610

[Mar08a] S. Margulies, *Computer algebra, combinatorics, and complexity: Hilbert's Nullstellensatz and NP-complete problems*, ProQuest LLC, Ann Arbor, MI, 2008. Thesis (Ph.D.)–University of California, Davis. MR2712545

[Mar08b] M. Marshall, *Positive polynomials and sums of squares*, Mathematical Surveys and Monographs, vol. 146, American Mathematical Society, Providence, RI, 2008, DOI 10.1090/surv/146. MR2383959

[Mat10] J. Matoušek, *Thirty-three miniatures*, Student Mathematical Library, vol. 53, American Mathematical Society, Providence, RI, 2010. Mathematical and algorithmic applications of linear algebra, DOI 10.1090/stml/053. MR2656313

[Mİ10] M. Michałek, *A short proof of combinatorial Nullstellensatz*, Amer. Math. Monthly **117** (2010), no. 9, 821–823, DOI 10.4169/000298910X521689. MR2760383

[MİS21] M. Michałek and B. Sturmfels, *Invitation to nonlinear algebra*, Graduate Studies in Mathematics, vol. 211, American Mathematical Society, Providence, RI, [2021] ©2021, DOI 10.1090/gsm/211. MR4423369

[MM19] K. Mészáros and A. H. Morales, *Volumes and Ehrhart polynomials of flow polytopes*, Math. Z. **293** (2019), no. 3-4, 1369–1401, DOI 10.1007/s00209-019-02283-z. MR4024590

[Mn83] N. E. Mnëv, *On the realizability over fields of the combinatorial types of convex polytopes* (Russian, with English summary), Zap. Nauchn. Sem. Leningrad. Otdel. Mat. Inst. Steklov. (LOMI) **123** (1983), 203–207. Differential geometry, Lie groups and mechanics, V. MR697253

[Mn85] N. E. Mnëv, *Varieties of combinatorial types of projective configurations and convex polyhedra* (Russian), Dokl. Akad. Nauk SSSR **283** (1985), no. 6, 1312–1314. MR802868

[MO22] B. McLaughlin and M. Omar, *On distinct distances between a variety and a point set*, Electron. J. Combin. **29** (2022), no. 3, Paper No. 3.21, 9, DOI 10.37236/8956. MR4452629

[Mou99] B. Mourrain, *A new criterion for normal form algorithms*, Applied algebra, algebraic algorithms and error-correcting codes (Honolulu, HI, 1999), Lecture Notes in Comput. Sci., vol. 1719, Springer, Berlin, 1999, pp. 430–443, DOI 10.1007/3-540-46796-3_41. MR1846517

[MPT01] I. G. Macdonald, J. Pach, and T. Theobald, *Common tangents to four unit balls in* $\mathbb{R}^3$, Discrete Comput. Geom. **26** (2001), no. 1, 1–17, DOI 10.1007/s004540010090. MR1832726

[MS05] E. Miller and B. Sturmfels, *Combinatorial commutative algebra*, Graduate Texts in Mathematics, vol. 227, Springer-Verlag, New York, 2005. MR2110098

[MS15] D. Maclagan and B. Sturmfels, *Introduction to tropical geometry*, Graduate Studies in Mathematics, vol. 161, American Mathematical Society, Providence, RI, 2015, DOI 10.1090/gsm/161. MR3287221

[MS21] K. Mészáros and L. Setiabrata, *Lorentzian polynomials from polytope projections*, Algebr. Comb. **4** (2021), no. 4, 723–739, DOI 10.5802/alco.179. MR4311086

[MSS15a] A. W. Marcus, D. A. Spielman, and N. Srivastava, *Interlacing families I: Bipartite Ramanujan graphs of all degrees*, Ann. of Math. (2) **182** (2015), no. 1, 307–325, DOI 10.4007/annals.2015.182.1.7. MR3374962

[MSS15b] A. W. Marcus, D. A. Spielman, and N. Srivastava, *Interlacing families II: Mixed characteristic polynomials and the Kadison-Singer problem*, Ann. of Math. (2) **182** (2015), no. 1, 327–350, DOI 10.4007/annals.2015.182.1.8. MR3374963

[MT07] D. Maclagan and R. R. Thomas, *Computational algebra and combinatorics of toric ideals*, Commutative algebra and combinatorics, Ramanujan Math. Soc. Lect. Notes Ser., vol. 4, Ramanujan Math. Soc., Mysore, 2007, pp. Part I: vi+106. With the co-operation of Sara Faridi, Leah Gold, A. V. Jayanthan, Amit Khetan and Tony Puthenpurakal. MR2406243

[MT08] B. Mourrain and P. Trébuchet, *Stable normal forms for polynomial system solving*, Theoret. Comput. Sci. **409** (2008), no. 2, 229–240, DOI 10.1016/j.tcs.2008.09.004. MR2474338

[NP22] D. Nguyen and I. Pak, *Short Presburger arithmetic is hard*, SIAM J. Comput. **51** (2022), no. 2, STOC17-1–STOC17-30, DOI 10.1137/17M1151146. MR4413074

[Oma22] M. Omar, *Partition rank and partition lattices*, 2022.

[Oma23] M. Omar, *Sets avoiding full-rank three-point patterns in* $(\mathbb{F}_q^n)^k$ *are exponentially small*, Australas. J. Combin. **86** (2023), 423–429. MR4630085

[Onn04] S. Onn, *Nowhere-zero flow polynomials*, J. Combin. Theory Ser. A **108** (2004), no. 2, 205–215, DOI 10.1016/j.jcta.2004.06.011. MR2098841

[Onn10] S. Onn, *Nonlinear discrete optimization*, Zurich Lectures in Advanced Mathematics, European Mathematical Society (EMS), Zürich, 2010. An algorithmic theory, DOI 10.4171/093. MR2724387

[OS94] S. Onn and B. Sturmfels, *A quantitative Steinitz' theorem*, Beiträge Algebra Geom. **35** (1994), no. 1, 125–129. Festschrift on the occasion of the 65th birthday of Otto Krötenheerdt. MR1287206

[OWW12] T. Oertel, C. Wagner, and R. Weismantel, *Convex integer minimization in fixed dimension*, Preprint, arXiv:1203.4175, 2012.

[Par00] P. A. Parrilo, *Structured semidefinite programs and semialgebraic geometry methods in robustness and optimization*, Ph.D. thesis, California Institute of Technology, May 2000.

[Par03] P. A. Parrilo, *Semidefinite programming relaxations for semialgebraic problems*, Math. Program. **96** (2003), no. 2, Ser. B, 293–320, DOI 10.1007/s10107-003-0387-5. Algebraic and geometric methods in discrete optimization. MR1993050

[Pem12] R. Pemantle, *Hyperbolicity and stable polynomials in combinatorics and probability*, Current developments in mathematics, 2011, Int. Press, Somerville, MA, 2012, pp. 57–123. MR3098077

[Pet00] J. Petersen, *Die Theorie der regulären graphs* (German), Acta Math. **15** (1891), no. 1, 193–220, DOI 10.1007/BF02392606. MR1554815

[Pfe22] J. Pfeifle, *Large final polynomials from integer programming*, ACM Commun. Comput. Algebra **55** (2021), no. 3, 82–86. MR4363366

[PPSP04] S. Prajna, A. Papachristodoulou, S. Seiler, and P. A. Parrilo, *SOSTOOLS: Sum of squares optimization toolbox for MATLAB*, 2004.

[PR22] V. Patel and G. Regts, *Approximate counting using Taylor's theorem: A survey*, Bull. Eur. Assoc. Theor. Comput. Sci. EATCS **138** (2022), 40–70. MR4608123

[PS93] P. Pedersen and B. Sturmfels, *Product formulas for resultants and Chow forms*, Math. Z. **214** (1993), no. 3, 377–396, DOI 10.1007/BF02572411. MR1245200

[PW13] R. Pemantle and M. C. Wilson, *Analytic combinatorics in several variables*, Cambridge Studies in Advanced Mathematics, vol. 140, Cambridge University Press, Cambridge, 2013, DOI 10.1017/CBO9781139381864. MR3088495

[Qui10] R. Quilodrán, *The joints problem in* $\mathbb{R}^n$, SIAM J. Discrete Math. **23** (2009/10), no. 4, 2211–2213, DOI 10.1137/090763160. MR2594983

[Raz98] A. A. Razborov, *Lower bounds for the polynomial calculus*, Comput. Complexity **7** (1998), no. 4, 291–324, DOI 10.1007/s000370050013. MR1691494

[RBG01] L. Rónyai, L. Babai, and M. K. Ganapathy, *On the number of zero-patterns of a sequence of polynomials*, J. Amer. Math. Soc. **14** (2001), no. 3, 717–735, DOI 10.1090/S0894-0347-01-00367-8. MR1824986

[RG96] J. Richter-Gebert, *Realization spaces of polytopes*, Lecture Notes in Mathematics, vol. 1643, Springer-Verlag, Berlin, 1996, DOI 10.1007/BFb0093761. MR1482230

[RG99] J. Richter-Gebert, *The universality theorems for oriented matroids and polytopes*, Advances in discrete and computational geometry (South Hadley, MA, 1996), Contemp. Math., vol. 223, Amer. Math. Soc., Providence, RI, 1999, pp. 269–292, DOI 10.1090/conm/223/03144. MR1661389

[Rom21] J. A. Romero-Barbosa, *Convex algebraic geometry approaches to graph coloring and stable set problems*, Ph.D. thesis, University of Waterloo, 2021.

[Rot17] T. Rothvoss, *The matching polytope has exponential extension complexity*, J. ACM **64** (2017), no. 6, Art. 41, 19, DOI 10.1145/3127497. MR3713797

[SA90] H. D. Sherali and W. P. Adams, *A hierarchy of relaxations between the continuous and convex hull representations for zero-one programming problems*, SIAM J. Discrete Math. **3** (1990), no. 3, 411–430, DOI 10.1137/0403036. MR1061981

[Sag01] B. E. Sagan, *The symmetric group*, 2nd ed., Graduate Texts in Mathematics, vol. 203, Springer-Verlag, New York, 2001. Representations, combinatorial algorithms, and symmetric functions, DOI 10.1007/978-1-4757-6804-6. MR1824028

[Sam12] S. V. Sam, *Symmetric quivers, invariant theory, and saturation theorems for the classical groups*, Adv. Math. **229** (2012), no. 2, 1104–1135, DOI 10.1016/j.aim.2011.10.009. MR2855088

[Sau21a] L. Sauermann, *On the size of subsets of $\mathbb{F}_p^n$ without p distinct elements summing to zero*, Israel J. Math. **243** (2021), no. 1, 63–79, DOI 10.1007/s11856-021-2145-x. MR4299141

[Sau21b] L. Sauermann, *On the speed of algebraically defined graph classes*, Adv. Math. **380** (2021), Paper No. 107593, 55, DOI 10.1016/j.aim.2021.107593. MR4205109

[Sch86] A. Schrijver, *Theory of linear and integer programming*, Wiley-Interscience Series in Discrete Mathematics, John Wiley & Sons, Ltd., Chichester, 1986. A Wiley-Interscience Publication. MR874114

[Sch95] A. Schrijver, *Polyhedral combinatorics*, Handbook of combinatorics, Vol. 1, 2, Elsevier Sci. B. V., Amsterdam, 1995, pp. 1649–1704. MR1373686

[Sch03a] A. Schrijver, *Combinatorial optimization. Polyhedra and efficiency. Vol. A*, Algorithms and Combinatorics, vol. 24, Springer-Verlag, Berlin, 2003. Paths, flows, matchings; Chapters 1–38. MR1956924

[Sch03b] A. Schrijver, *Combinatorial optimization. Polyhedra and efficiency. Vol. B*, Algorithms and Combinatorics, vol. 24, Springer-Verlag, Berlin, 2003. Matroids, trees, stable sets; Chapters 39–69. MR1956925

[Sch03c] A. Schrijver, *Combinatorial optimization. Polyhedra and efficiency. Vol. C*, Algorithms and Combinatorics, vol. 24, Springer-Verlag, Berlin, 2003. Disjoint paths, hypergraphs; Chapters 70–83. MR1956926

[SdZ18] J. Solymosi and F. de Zeeuw, *Incidence bounds for complex algebraic curves on Cartesian products*, New trends in intuitive geometry, Bolyai Soc. Math. Stud., vol. 27, János Bolyai Math. Soc., Budapest, 2018, pp. 385–405. MR3889269

[SNR11] P. Schreck, J. Narboux, and J. Richter-Gebert (eds.), *Automated deduction in geometry*, Lecture Notes in Computer Science, vol. 6877, Springer, Heidelberg, 2011. Revised selected papers from the 8th International Workshop (ADG 2010) held at the Technische Universität München, Munich, July 22–24, 2010; Lecture Notes in Artificial Intelligence, DOI 10.1007/978-3-642-25070-5. MR2906059

[Sot01] F. Sottile, *Enumerative real algebraic geometry*, Algorithmic and quantitative real algebraic geometry (Piscataway, NJ, 2001), DIMACS Ser. Discrete Math. Theoret. Comput. Sci., vol. 60, Amer. Math. Soc., Providence, RI, 2003, pp. 139–179, DOI 10.1090/dimacs/060/11. MR1995019

[SP91] I. V. Sergienko and V. A. Perepelitsa, *Finding the set of alternatives in discrete multi-criterion problems*, Cybernetics **3** (1991), 673–683.

[SS03] F. Santos and B. Sturmfels, *Higher Lawrence configurations*, J. Combin. Theory Ser. A **103** (2003), no. 1, 151–164, DOI 10.1016/S0097-3165(03)00092-X. MR1986836

[ST97] B. Sturmfels and R. R. Thomas, *Variation of cost functions in integer programming*, Math. Programming **77** (1997), no. 3, Ser. A, 357–387, DOI 10.1016/S0025-5610(96)00068-8. MR1447563

[ST02] F. Sottile and T. Theobald, *Lines tangent to $2n-2$ spheres in $\mathbb{R}^n$*, Trans. Amer. Math. Soc. **354** (2002), no. 12, 4815–4829, DOI 10.1090/S0002-9947-02-03014-3. MR1926838

[ST08] B. Sturmfels and J. Tevelev, *Elimination theory for tropical varieties*, Math. Res. Lett. **15** (2008), no. 3, 543–562, DOI 10.4310/MRL.2008.v15.n3.a14. MR2407231

[ST22] B. Sudakov and I. Tomon, *Ramsey properties of algebraic graphs and hypergraphs*, Forum Math. Sigma **10** (2022), Paper No. e95, 20, DOI 10.1017/fms.2022.85. MR4502021

[Sta89] R. P. Stanley, *Log-concave and unimodal sequences in algebra, combinatorics, and geometry*, Graph theory and its applications: East and West (Jinan, 1986), Ann. New York Acad. Sci., vol. 576, New York Acad. Sci., New York, 1989, pp. 500–535, DOI 10.1111/j.1749-6632.1989.tb16434.x. MR1110850

[Sta96] R. P. Stanley, *Combinatorics and commutative algebra*, 2nd ed., Progress in Mathematics, vol. 41, Birkhäuser Boston, Inc., Boston, MA, 1996. MR1453579

[Sta97] R. P. Stanley, *Enumerative combinatorics. Vol. 1*, Cambridge Studies in Advanced Mathematics, vol. 49, Cambridge University Press, Cambridge, 1997. With a foreword by Gian-Carlo Rota; Corrected reprint of the 1986 original, DOI 10.1017/CBO9780511805967. MR1442260

[Sta99] R. P. Stanley, *Enumerative combinatorics. Vol. 2*, Cambridge Studies in Advanced Mathematics, vol. 62, Cambridge University Press, Cambridge, 1999. With a foreword by Gian-Carlo Rota and appendix 1 by Sergey Fomin, DOI 10.1017/CBO9780511609589. MR1676282

[Ste04] H. J. Stetter, *Numerical polynomial algebra*, Society for Industrial and Applied Mathematics (SIAM), Philadelphia, PA, 2004, DOI 10.1137/1.9780898717976. MR2048781

[Stu87a] B. Sturmfels, *Boundary complexes of convex polytopes cannot be characterized locally*, J. London Math. Soc. (2) **35** (1987), no. 2, 314–326, DOI 10.1112/jlms/s2-35.2.314. MR881520

[Stu87b] B. Sturmfels, *Computational synthetic geometry*, ProQuest LLC, Ann Arbor, MI, 1987, Thesis (Ph.D.)–University of Washington. MR2716247

[Stu87c] B. Sturmfels, *On the decidability of Diophantine problems in combinatorial geometry*, Bull. Amer. Math. Soc. (N.S.) **17** (1987), no. 1, 121–124, DOI 10.1090/S0273-0979-1987-15532-7. MR888886

[Stu94a] B. Sturmfels, *On the Newton polytope of the resultant*, J. Algebraic Combin. **3** (1994), no. 2, 207–236, DOI 10.1023/A:1022497624378. MR1268576

[Stu94b] B. Sturmfels, *On the number of real roots of a sparse polynomial system*, Hamiltonian and gradient flows, algorithms and control, Fields Inst. Commun., vol. 3, Amer. Math. Soc., Providence, RI, 1994, pp. 137–143, DOI 10.1090/fic/003/10. MR1297991

[Stu94c] B. Sturmfels, *Viro's theorem for complete intersections*, Ann. Scuola Norm. Sup. Pisa Cl. Sci. (4) **21** (1994), no. 3, 377–386. MR1310632

[Stu95a] B. Sturmfels, *Gröbner bases and convex polytopes*, University Lecture Series, vol. 8, American Mathematical Society, Providence, RI, 1996, DOI 10.1090/ulect/008. MR1363949

[Stu95b] B. Sturmfels, *On vector partition functions*, J. Combin. Theory Ser. A **72** (1995), no. 2, 302–309, DOI 10.1016/0097-3165(95)90067-5. MR1357776

[Stu98] B. Sturmfels, *Polynomial equations and convex polytopes*, Amer. Math. Monthly **105** (1998), no. 10, 907–922, DOI 10.2307/2589283. MR1656923

[Stu02] B. Sturmfels, *Solving systems of polynomial equations*, CBMS Regional Conference Series in Mathematics, vol. 97, Conference Board of the Mathematical Sciences, Washington, DC; by the American Mathematical Society, Providence, RI, 2002, DOI 10.1090/cbms/097. MR1925796

[Stu08] B. Sturmfels, *Algorithms in invariant theory*, 2nd ed., Texts and Monographs in Symbolic Computation, SpringerWienNewYork, Vienna, 2008. MR2667486

[STW21] J. Sidman, W. Traves, and A. Wheeler, *Geometric equations for matroid varieties*, J. Combin. Theory Ser. A **178** (2021), Paper No. 105360, 15, DOI 10.1016/j.jcta.2020.105360. MR4175891

[STY07] B. Sturmfels, J. Tevelev, and J. Yu, *The Newton polytope of the implicit equation* (English, with English and Russian summaries), Mosc. Math. J. **7** (2007), no. 2, 327–346, 351, DOI 10.17323/1609-4514-2007-7-2-327-346. MR2337885

[SW22] L. Sauermann and Y. Wigderson, *Polynomials that vanish to high order on most of the hypercube*, J. Lond. Math. Soc. (2) **106** (2022), no. 3, 2379–2402, DOI 10.1112/jlms.12637. MR4498556

[SY08] B. Sturmfels and J. Yu, *Tropical implicitization and mixed fiber polytopes*, Software for algebraic geometry, IMA Vol. Math. Appl., vol. 148, Springer, New York, 2008, pp. 111–131, DOI 10.1007/978-0-387-78133-4_7. MR2410718

[SZ93] B. Sturmfels and A. Zelevinsky, *Maximal minors and their leading terms*, Adv. Math. **98** (1993), no. 1, 65–112, DOI 10.1006/aima.1993.1013. MR1212627

[SZ19] H. Sjöberg and G. M. Ziegler, *Semi-algebraic sets of f-vectors*, Israel J. Math. **232** (2019), no. 2, 827–848, DOI 10.1007/s11856-019-1888-0. MR3990960

[Tao14] T. Tao, *Algebraic combinatorial geometry: The polynomial method in arithmetic combinatorics, incidence combinatorics, and number theory*, EMS Surv. Math. Sci. **1** (2014), no. 1, 1–46, DOI 10.4171/EMSS/1. MR3200226

[Tho95] R. R. Thomas, *A geometric Buchberger algorithm for integer programming*, Math. Oper. Res. **20** (1995), no. 4, 864–884, DOI 10.1287/moor.20.4.864. MR1378110

[Tho18] R. R. Thomas, *Spectrahedral lifts of convex sets*, Proceedings of the International Congress of Mathematicians—Rio de Janeiro 2018. Vol. IV. Invited lectures, World Sci. Publ., Hackensack, NJ, 2018, pp. 3819–3842. MR3966553

[Tro04] N. Trotignon, *Graphes parfaits: Structure et algorithmes. (perfect graphs : structure and algorithms)*, Ph.D. thesis, Joseph Fourier University, Grenoble, France, 2004.

[TVN07] M. Tsfasman, S. Vlăduţ, and D. Nogin, *Algebraic geometric codes: Basic notions*, Mathematical Surveys and Monographs, vol. 139, American Mathematical Society, Providence, RI, 2007, DOI 10.1090/surv/139. MR2339649

[TY19] N. M. Tran and J. Yu, *Product-mix auctions and tropical geometry*, Math. Oper. Res. **44** (2019), no. 4, 1396–1411, DOI 10.1287/moor.2018.0975. MR4032448

[Wil06] H. S. Wilf, *generatingfunctionology*, 3rd ed., A K Peters, Ltd., Wellesley, MA, 2006. MR2172781

[Zie95] G. M. Ziegler, *Lectures on polytopes*, Graduate Texts in Mathematics, vol. 152, Springer-Verlag, New York, 1995, DOI 10.1007/978-1-4613-8431-1. MR1311028

Department of Mathematics, University of California Davis, Davis, California 95616

Email address: deloera@math.ucdavis.edu

Proceedings of Symposia in Pure Mathematics
Volume **111**, 2025
https://doi.org/10.1090/pspum/111/02041

Hypergeometric systems, statistics, and algorithms

Nobuki Takayama

ABSTRACT. This article is a tutorial and a survey on a numerical evaluation method for normalizing constants and cumulative distribution functions expressed in terms of hypergeometric integrals of several variables. The method is called the holonomic gradient method.

1. Introduction

Normalizing constants in statistics translate non-negative functions into probability distributions. For example, the normalizing constant of the non-negative function $x^{\beta-1}\exp(-x/2)$ on $\mathbf{R}_{\geq 0}$ is

$$Z(\beta) = \int_0^\infty x^{\beta-1}\exp(-x/2)dx = 2^\beta\Gamma(\beta) \quad \text{for } \mathrm{Re}\,\beta > 0.$$

Then,

$$\frac{x^{\beta-1}\exp(-x/2)}{Z(\beta)}$$

is a probability distribution on $\mathbf{R}_{\geq 0}$. Here, β is a parameter of the distribution. The cumulative distribution function of this distribution is

$$\frac{1}{Z(\beta)}\int_0^y x^{\beta-1}\exp(-x/2)dx.$$

This article is a tutorial and a survey on a numerical evaluation method for normalizing constants and cumulative distribution functions expressed in terms of integrals. The method is called the holonomic gradient method (HGM) introduced by a group including the author [**27**], which will be elementarily illustrated with an example of the χ^2-distribution in Sections 2 and 4. In classical statistics, special functions such as the Γ function, the Beta function, the error function, and the confluent hypergeometric functions are used as central mathematical ingredients. However, modern statistics utilizes multiple integrals with several parameters, which are usually evaluated numerically by Monte Carlo methods. On the other hand, modern mathematics has developed the theory of holonomic systems, the

2020 *Mathematics Subject Classification.* Primary 62-02, 62H10, 33C70; Secondary 32C38, 13P10.

Key words and phrases. Holonomic system, hypergeometric integral, normalizing constant, Weyl algebra.

This work is supported by JST CREST grant number JP19209317 and JSPS KAKENHI grant number 21K03270.

theory of hypergeometric functions of several variables, and algorithms for holonomic systems. See, e.g., [**30**], [**35**], [**36**], [**41**] and their references on these developments. The HGM is an attempt to apply these ideas to integrals in statistics. The framework of the HGM is analogous to the homotopy continuation method to solve systems of algebraic equations. The homotopy method utilizes a system of differential equations to extend a set of solutions of a starting system of binomial equations; (see the celebrated work by Huber and Sturmfels [**12**]). The HGM constructs a holonomic system of linear partial differential equations to extend numerical values of integrals at a starting point. Although the choice of a starting system is systematic in the homotopy continuation method, finding a good starting point for the HGM is an open question. For now, the starting points are chosen by human insights.

Sections 2, 3, and 4 describe the nature of the HGM and explain why numerical evaluations are important in statistics. The chi-squared (χ^2) distribution is used as a running example and as a motivation for the use of HGM. Section 2 is an introduction to chi-squared distribution and the need for numerical evaluations of integrals. Section 3 is an introduction to an algebraic part of the HGM and a description of a general scheme of the HGM. The HGM will be applied to our running example in Section 4. These sections are enlarged English translations of a part of my lecture notes for a summer school class.[1] In Section 5, we present a notable application of the HGM; An efficient evaluation method for the cumulative distribution function (CDF) of the maximal eigenvalue of Wishart random matrices is briefly explained. The CDF is a generalization of the CDF of the chi-squared distribution. The final section 6 provides a list of some known results on statistical distributions addressed by the HGM. It is a revised and enlarged version of a part of my survey in Japanese that appeared in an earlier proceedings.[2]

The article [**36**] by A. L. Sattelberger and B. Sturmfels explains analogous content as this article from a different angle. By reading this article together, readers will be able to get an overview of this field and the issues involved. For readers interested in rigorous proofs on fundamental facts and algorithms; see [**30**], [**35**], and [**41**].

2. The χ^2 distribution

The chi-squared distribution is one of the most widely used probability distributions. For example, the chi-squared test is used to examine whether two categorical variables in contingency tables are independent when the sample sizes are large (see, e.g., [**39**, Chapters 2 and 5]). To perform the chi-squared test, we need to calculate the following integral numerically:

$$\int_0^x t^{\frac{n}{2}-1} e^{-\frac{t}{2}} dt. \tag{2.1}$$

This is a function of x. We will use this integral to illustrate the holonomic gradient method (HGM), which is a new approach to evaluating integrals in statistics.

[1] Hypergeometric school 2013/2014 (in Japanese), Rokko lecture note series 23, 3–25, `http://www.math.kobe-u.ac.jp/HOME/home-j/rokko.html`.

[2] Proceedings of the Twenty-Seventh RAMP Symposium, Shizuoka University, Japan, October 15-16, 2015.

We define a function $T_n(x)$ on $\mathbf{R}$ as

$$T_n(x) = \begin{cases} x^{\frac{n}{2}-1}e^{-\frac{x}{2}}/Z_T(n) & x > 0 \\ 0 & x \le 0. \end{cases}$$

Here, $Z_T(n)$ is the normalizing constant

$$Z_T(n) = \int_0^\infty x^{\frac{n}{2}-1}e^{-\frac{x}{2}} = 2^{\frac{n}{2}}\Gamma\left(\frac{n}{2}\right)$$

so that $\int_{\mathbf{R}} T_n(x)dx = 1$. The function $T_n(x)$ is a probability density function; this distribution is the χ^2-distribution. The cumulative distribution function (CDF) of the χ^2-distribution is the integral (2.1) divided by the normalizing constant $Z_T(n)$. Experts in hypergeometric functions see at a glance that the integral (2.1) is the incomplete Gamma function.[3]

By way of introduction to those who have not yet taken a course in statistical tests, let me explain why the χ^2 distribution is important. The density function of the normal distribution $N(m, \sigma^2)$ of mean m and variance σ^2 is

$$e^{-\frac{(x-m)^2}{2\sigma^2}}/N, \quad N = \sqrt{2\pi\sigma^2}.$$

The normal distribution is the most fundamental distribution in statistics. F. Gauss firstly showed that any random experimental error follows a normal distribution (see, e.g., [**13**, 4.8]). It is essentially the central limit theorem (see, e.g., [**6**, VIII], [**39**, Chapter 2]), which states that the distribution of the average of many random samples converges to a normal distribution under some conditions when the number of samples increases. The χ^2 distribution is the probability distribution of the sum of the square of independent random variables having a standard normal distribution.

PROPOSITION 2.1 ([**6**, §2.3 (p. 48)]). *Let $X_1, \ldots, X_n$ be independent random variables whose distribution is $N(0,1)$. Then, the distribution of the random variable $Y = \sum_{i=1}^n X_i^2$ is χ^2 with n degrees of freedom.*

Before proving this proposition, we will first provide a program to simulate this idea. The notion of a random variable is fundamental in modern statistics. A measurable map X from a probability space $(\Omega, \mathcal{F}, P)$ to $(\mathbf{R}, \mathcal{B})$ is called a random variable. Here Ω is a sample space, $\mathcal{F}$ is an event space (σ-algebra), P is a probability measure, and $\mathcal{B}$ is the Borel set of $\mathbf{R}$ (see, e.g., [**6**, IV], [**39**, Chapter 2]). The cumulative distribution $F_X(x)$ of the random variable is

$$F_X(x) := P(X \le x) = P(\Omega_x), \ \Omega_x := \{\omega \in \Omega \mid X(w) \le x\}. \tag{2.2}$$

The function $p(x) := \frac{dF_x}{dx}$ is called the distribution of X and we have $P(X \le x) = \int_{-\infty}^x p(x)dx$.

Intuitively speaking, we may regard the random variable X as a random number generator. The statement "the distribution of the random number generator is a function $p(x)$" means that the histogram of the random number generator fits the graph of the function $p(x)$. Then, this proposition is interpreted as a behavior of random number generators. The program Listing 1 is an R program [**32**] for simulating this proposition where `rnorm(n)` generates n random numbers of the normal distribution with mean$=0$ and variance$=1$. The degrees n of freedom

[3] An expression in terms of a hypergeometric function is given in (4.3).

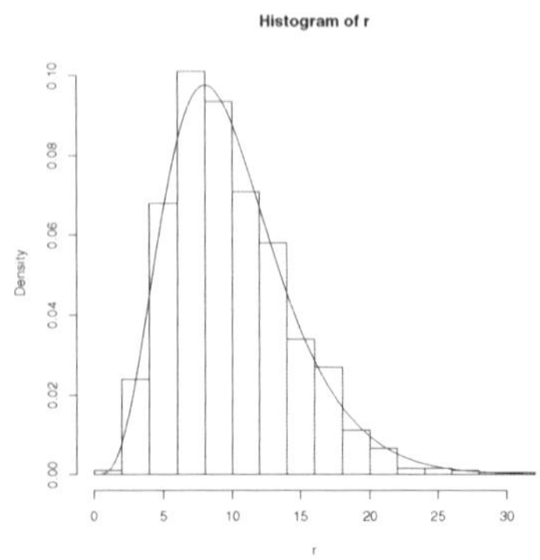

FIGURE 1. Output of the listing "χ^2 simulation"

are set to 10; dchisq(x,10) is the density function of the χ^2 distribution with 10 degrees of freedom.

LISTING 1. χ^2 simulation

```
mychi2<-function(n) {
 a<-rnorm(n);
 return(sum(a^2));
}
r<-NULL;
for (i in seq(1,1000,1)) { r<-cbind(r,mychi2(10)); }
hist(r,prob=TRUE);
curve(dchisq(x,10),add=TRUE)
```

The output of the program is shown in Figure 1.

We can prove Proposition 2.1 using induction with respect to n with a help of the following Theorem.

THEOREM 2.2 ([**6**, §1.2 (p. 6)]). *Let X, Y be independent random variables. The distributions of X and Y are $f(x)$ and $g(y)$ respectively. Then the distribution of the random variable $Z = X + Y$ is the convolution $(f * g)(z) = \int_{\mathbf{R}} f(z-t)g(t)dt$.*

PROOF. Since the joint probability distribution of X and Y is $f(x)g(y)$, the probability $P(X + Y \leq z)$ is $\int_{x+y\leq z} f(x)g(y)dxdy$. Change the variables as $u = x + y, v = y$. Then, the integral is equal to

$$\int_{u\leq z} f(u-v)g(v)dudv = \int_{u\leq z} \left(\int_{\mathbf{R}} f(u-v)g(v)dv \right) du. \qquad \square$$

PROOF OF PROPOSITION 2.1. Suppose $n = 1$. The probability that the random variable $Y = X^2$ is smaller than $c \geq 0$ is

$$P(Y < c) = P(-\sqrt{c} < X < \sqrt{c}).$$

From the definition that the distribution of the random variable X is $N(0,1)$, we have

$$P(a \leq X \leq b) = \int_a^b \frac{1}{\sqrt{2\pi}} e^{-\frac{1}{2}x^2} dx \quad (a \leq b).$$

Applying the change of variables $y = x^2$, we have

$$\begin{aligned} P(Y < c) &= \int_{-\sqrt{c}}^{\sqrt{c}} \frac{1}{\sqrt{2\pi}} e^{-\frac{1}{2}x^2} dx \\ &= 2\int_0^{\sqrt{c}} \frac{1}{\sqrt{2\pi}} e^{-\frac{1}{2}x^2} dx \\ &= \int_0^c \frac{1}{\sqrt{2\pi}} y^{-\frac{1}{2}} e^{-\frac{1}{2}y} dy \\ &= \int_0^c T_1(y) dy \end{aligned}$$

which means that the distribution of Y is the χ^2 distribution with 1 degree of freedom. Suppose the statement holds up to n. By applying Theorem 2.2, our proof reduces to prove that $T_n * T_1 = T_{n+1}$. It is a calculation of the beta function. Noting that the supports of T_n and T_1 is $x \geq 0$, we have

$$(T_n * T_1)(x) = \frac{1}{Z_T(n)Z_T(1)} e^{-\frac{x}{2}} \int_0^x (x-t)^{\frac{n}{2}-1} t^{-\frac{1}{2}} dt.$$

Applying the change of variables $t = xs$, we conclude that the integral of the right-hand side is

$$x^{\frac{n-1}{2}} \int_0^1 (1-s)^{\frac{n}{2}-1} s^{-\frac{1}{2}} dt = x^{\frac{n-1}{2}} \frac{\Gamma(\frac{n}{2})\Gamma(\frac{1}{2})}{\Gamma\left(\frac{n+1}{2}\right)}.$$

Thus, we have completed the proof of Proposition 2.1. □

Numerical evaluation of the cumulative distribution function is important in statistical tests. We start with a simple example.

Suppose we toss a coin 4 times and obtain a head on all 4 tosses. Is the coin fair? In order to conduct our test, we will first assume that the coin is fair. In other words, we will assume that the probability of producing a head is $1/2$ and that of producing a tail is $1/2$. Statisticians call such an assumption the null hypothesis. Under the null hypothesis, the probability of getting the head on all 4 tosses is $(1/2)^4 = 1/16 = 0.0625$. If producing a head all 4 times is considered very rare, we should reject the null hypothesis. Statisticians usually set the threshold of rareness at 0.05 or 0.01. Since we computed the probability of producing a head all 4 times to be $0.0625 > 0.05$, we cannot say that the result is rare (under the null hypothesis). Accordingly, we would conclude that we do not have sufficient evidence that the coin is not fair and thus cannot reject the null hypothesis. The probability $1/16$ is called *the p-value* under the null hypothesis. Note that this test does not claim that the coin is fair. It only claims that there is no strong reason to say that it is not fair coin.

Let us present a second example of a statistical test. Assume that software system F has a random number generator **rk**() defined as

```
rk<-function(n) { return(runif(n)-0.5); }
```

It is advertised that the distribution of the generator `rk` of Software F is the normal distribution $N(0, 0.25)$. In this exercise, our task is whether this statement is true or not by using the χ^2 distribution. Since variance of a random variable X is the expectation $E[(X - E[X])^2]$, the distribution of the generator `rk()/0.25` is $N(0, 1)$. We call `rk()/0.25` m times and calculate the sum of the squared values. Let $\bar{y}$ be the sum. If the p-value $\int_{\bar{y}}^{\infty} T_m(x)dx$, which is the probability of producing a value as rare or rarer than $\bar{y}$, is smaller than 0.05, the advertisement will be rejected. Below is a program to test when $m = 500$.

```
m<-500
a<- rk(m); hist(a); #We see from the histogram
 #that the ad is fake.
b<- sum((rk(m)/0.25)^2);
p<- 1-pchisq(b,df=m); #1-(the CDF value at b)
# In fact, the p-value p is 1.038966e-05
```

The output $1.03\cdots \times 10^{-5}$ is smaller than 0.05; thus the advertisement's claim should be rejected. Testing random number generators is an important research topic and we note that more clever and advanced methods are available.

The χ^2 test is often used to test the assumption that the distribution of observed values minus expected values is approximately a normal distribution (see, e.g., [**39**, 5.4]).

3. Introduction to the holonomic gradient method

We have seen that the numerical evaluation of the cumulative distribution (CDF) function is important in statistics. The holonomic gradient method (HGM) is a new algorithmic method to evaluate CDFs. The key notion for the HGM is a holonomic ideal.[4] Let $D_n = \mathbf{C}\langle x_1, \dots, x_n, \partial_1, \dots, \partial_n \rangle$ be the ring of differential operators of n variables.[5] It is the quotient of a free algebra generated by $x_1, \dots, x_n, \partial_1, \dots, \partial_n$ divided by the relations $x_i x_j = x_j x_i$, $\partial_i \partial_j = \partial_j \partial_i$, and $\partial_i x_j = x_j \partial_i + \delta_{ij}$ where δ_{ij} is Kronecker's delta.

We use the multi-index notations $x^\alpha = \prod_{i=1}^n x_i^{\alpha_i}$, $\partial^\beta = \prod_{i=1}^n \partial_i^{\beta_i}$, and $|\alpha| = \alpha_1 + \cdots + \alpha_n$ in the sequel. Let $\ell = \sum c_{\alpha,\beta} x^\alpha \partial^\beta$, $c_{\alpha,\beta} \in \mathbf{C}$ be an element of the Weyl algebra D_n. It acts on a distribution (or an analytic function) f by

$$\ell \bullet f = \sum c_{\alpha,\beta} x^\alpha \frac{\partial^{|\beta|} f}{\partial x_1^{\beta_1} \cdots \partial x_n^{\beta_n}}. \tag{3.1}$$

The action is denoted by $\bullet$ and if there is no risk of confusion with the multiplication in D_n, we omit it. Let I be a left ideal of D_n. The distribution f is called "annihilated by I" or "a solution of I" when $\ell \bullet f = 0$ for all $\ell \in I$.

The family of $\mathbf{C}$-vector subspaces of D_n

$$\mathcal{F}_k = \bigoplus_{|\alpha|+|\beta| \le k} \mathbf{C} x^\alpha \partial^\beta \tag{3.2}$$

is called the Bernstein filtration.

DEFINITION 3.1. A left ideal I of D_n is called *a holonomic ideal* when

$$\dim_{\mathbf{C}} \mathcal{F}_k / (\mathcal{F}_k \cap I) = \text{a degree } n \text{ polynomial in } k \tag{3.3}$$

for sufficiently large k.

[4] It is also called holonomic system in this article.

[5] Do not confuse the notation $\langle\ \rangle$ here with the notation of the ideal $\langle f_1, \dots, f_m \rangle$ generated by $f_1, \dots, f_m$. The symbol $[\ ]$ is used for the (commutative) polynomial ring and $\langle\ \rangle$ is used to say that it is a noncommutative ring.

For example, the left ideal generated by $\partial_1, \ldots, \partial_n$ is a holonomic ideal because the complete representatives of the quotient vector space in (3.3) is $\oplus_{|\alpha|\leq k} \mathbf{C} x^\alpha$ and then the dimension is $\binom{n+k}{n}$. Note that there are several equivalent definitions for holonomic ideals and these equivalences can be proved, e.g., by a Gröbner fan [**35**, 2.2]. Although the definition is a little complicated, holonomic ideals satisfy the following nice properties.

THEOREM 3.2 ([**41**, Chapter 6]). *Let I be a holonomic ideal.*

(1) *The intersection*

$$I \cap \mathbf{C}\langle x_1, \ldots, x_n, \partial_j \rangle \tag{3.4}$$

contains nonzero elements for any j. In other words, the holonomic ideal I contains an ordinary differential operator with respect to the variable x_j by regarding other x variables as parameters.

(2) *The left ideal of D_{n-1}*

$$(I + \partial_n D_n) \cap D_{n-1} \tag{3.5}$$

is a holonomic ideal in D_{n-1}. Note that $(I + \partial_n D_n)$ is the sum of the left ideal I and the right ideal $\partial_n D_n$.

PROOF. Let J be intersection (3.4). Consider the $\mathbf{C}$-linear map

$$\mathcal{F}_k \cap \mathbf{C}\langle x_1, \ldots, x_n, \partial_j \rangle \ni \ell \mapsto [\ell] \in \mathcal{F}_k / (\mathcal{F}_k \cap I).$$

Since the dimension of the left hand side is $\binom{n+1+k}{n+1}$, there exists a non-zero kernel.

See, e.g., [**30**, Theorem 5.1] or [**41**, 6.10] as to a proof of the second property. □

The second property is crucial in the HGM and algorithms to find elements in (3.5) are called *integration algorithms* (see, e.g., [**30**, 5.2], [**35**, Algorithm 5.5.4, p. 230], [**36**], [**41**, 6.10]) or *creative telescoping* (see, e.g., [**3**], [**15**]).

The left ideal of D_m

$$J := (I + \partial_{m+1} D_n + \cdots + \partial_n D_n) \cap D_m \tag{3.6}$$

is called *the integration ideal* of I with respect to $x_{m+1}, \ldots, x_n$. If I is an holonomic ideal in D_n, it is an holonomic ideal in D_m by Theorem 3.2.

PROPOSITION 3.3. *Assume that a rapidly decreasing function $f(x_1, \ldots, x_n)$ is annihilated by an holonomic ideal I in D_n. Then, the integral*

$$g(x_1, \ldots, x_m) := \int_{\mathbf{R}^{n-m}} f(x_1, \ldots, x_n) dx_{m+1} \cdots dx_n \tag{3.7}$$

is annihilated by the integration ideal J of I.

PROOF. Let ℓ be in J. Then ℓ can be expressed as $\ell' + \partial_{m+1} r_{m+1} + \partial_n r_n$ where $\ell' \in I$ and $r_j \in D_n$. Then,

$$\ell \bullet g = \int_{\mathbf{R}^{n-m}} \sum_{j=m+1}^{n} \frac{\partial(r_j \bullet f)}{\partial x_j} dx_{m+1} \cdots dx_n.$$

The last integral is 0 because f is a rapidly decreasing function. □

As to more general form of this Proposition, see, e.g., [**30**, §6].

We denote by $\mathbf{C}(x) := \mathbf{C}(x_1, \ldots, x_n)$ the rational function field. Let

$$R_n = \mathbf{C}(x)\langle \partial_1, \ldots, \partial_n \rangle = \bigoplus_{\beta \in \mathbf{N}_0^n} \mathbf{C}(x)\partial^\beta \tag{3.8}$$

be the rational Weyl algebra. We regard the rational functions as coefficients and the multiplication in R_n is defined by $\partial_i c(x) = c(x)\partial_i + \frac{\partial c(x)}{\partial x_i}$ where $c(x)$ is a rational function. The Weyl algebra D_n is a subalgebra of R_n. For a holonomic ideal I, the quotient space $R_n/(R_n I)$ is finite dimensional as a vector space over $\mathbf{C}(x)$ by the first property of holonomic ideals in Theorem 3.2. Let $s_1, \ldots, s_r$ be a basis of $R_n/(R_n I)$ as the vector space. Then, there exist rational functions $p_{ij}^k(x)$ such that

$$\partial_i s_j = \sum_{k=1}^r p_{ij}^k s_k \tag{3.9}$$

in $R_n/(R_n I)$.

Definition 3.4. The system of vector valued differential equations

$$\frac{\partial Y}{\partial x_i} = P_i Y, \quad P_i = (p_{ij}^k \mid j, k = 1, \ldots, r) \tag{3.10}$$

is called *a system of Pfaffian equations* associated to I and $s_1, \ldots, s_r$. Here Y is a column vector function of r entries. The $r \times r$ matrix P_i is called *the i-th Pfaffian matrix.*

Example 3.5. Consider the left ideal I generated by

$$(x^2 - y^2)\partial_y^2 - y\partial_y, x\partial_x + y\partial_y - 1 \tag{3.11}$$

in the Weyl algebra $\mathbf{C}\langle x, y, \partial_x, \partial_y \rangle$. Note that $f = x \arcsin(y/x)$ satisfies

$$(x^2 - y^2)\frac{\partial^2 f}{\partial y^2} - y\frac{\partial f}{\partial y} = 0, x\frac{\partial f}{\partial x} + y\frac{\partial f}{\partial y} - f = 0$$

The theory of Gröbner basis works in the rational Weyl algebra $\mathbf{C}(x,y)\langle \partial_x, \partial_y \rangle$ analogously with the ring of polynomials in two variables (see, e.g., [**36**], [**41**, Chapter 6]). The set (3.11) is already a Gröbner basis for the lexicographic order $\partial_x \succ \partial_y$. Note that Buchberger's criterion 1 does not work and readers are encouraged to calculate the S-polynomial and reduce it. The set of the standard monomials is $\{\partial_y, 1\}$. The system of Pfaffian equations associated to the set is

$$\frac{\partial Y}{\partial x} = \begin{pmatrix} \frac{-y^2}{x^3 - y^2 x} & 0 \\ \frac{-y}{x} & \frac{1}{x} \end{pmatrix} Y, \quad \frac{\partial Y}{\partial y} = \begin{pmatrix} \frac{y}{x^2 - y^2} & 0 \\ 1 & 0 \end{pmatrix} Y.$$

Note that $Y = (\frac{\partial f}{\partial y}, f)^T$ is a solution. There are several software systems to perform Gröbner basis computations in the Weyl algebra and the rational Weyl algebra [**36**]. The system developed by us is [**33**].

We are ready to explain the HGM. A distribution is called *a holonomic function* when it is annihilated by a holonomic ideal. The HGM is for a definite integral with parameters of a holonomic function. The HGM consists of 3 steps.

STEP 1. Find a holonomic ideal I which annihilates the kernel holonomic function of a given definite integral. Apply integration algorithms to I or find elements of the integration ideal of I through theoretical consideration and find a holonomic ideal J whose elements annihilate a given definite integral with parameters (see, e.g., [**15**], [**35**, Chapter 5], [**36**, 2.11]). The ideal J is regarded as a system of linear partial differential equations on the parameter space.

STEP 2. Translate J into a system of Pfaffian equations. See [**1**] on an efficient method.

STEP 3a. Evaluate numerically the integral and its derivatives at a special point of the parameters. We can utilize series expansions or numerical integrations.

STEP 3b. Perform a numerical integration of the system of Pfaffian equations with initial values at a special point or with boundary values at some points and evaluate the integral on a broader area of the parameters.

We will give examples of the HGM in Sections 4 and 5. See also expositions of the HGM given in [**27**, Appendix] and [**41**, Chapter 6].

The system of Pfaffian equations for a principal ideal generated by an ordinary differential operator is nothing but the system of first order ordinary differential equations for it learned in elementary courses of ODEs. For the n variable case, the system of Pfaffian equations is expressed as

$$dY = (P_1 dx_1 + \cdots + P_n dx_n)Y. \tag{3.12}$$

Let $c(t) \in \mathbf{R}^n$ be a smooth curve parametrized by $t \in \mathbf{R}$. The system of Pfaffian equations can be restricted to the curve; the restricted system is a system of first order ODEs.[6] The column vector function Y satisfies the following ODE on the curve.

$$\frac{dY(c(t))}{dt} = \left(\sum_{i=1}^{n} P_i(c(t)) \frac{dc_i}{dt} \right) Y(c(t))$$

See, e.g., [**41**, §6.2, §6.3] for details. Although we can utilize several ODE solvers in the last step, there are several difficulties of numerical analysis in step 3b. These are discussed in [**43**] with a sparse interpolation/extrapolation method and a defusing method.

NOTE FOR ADVANCED TOPICS 3.6. Our HGM utilizes a system linear partial differential equations. The HGM using difference equations is also possible. The article [**24**] discusses an interesting method for finding a system of linear difference equations that satisfies a given hypergeometric integral from the solutions of a likelihood equation.

4. Revisiting the χ^2 distribution by using HGM

In the previous section, we have introduced the HGM. Here, we evaluate the CDF of the χ^2 distribution by using the HGM to provide a demonstration of a numerical evaluation.

We want to evaluate numerically the integral

$$\gamma(x) = \int_0^x t^{\frac{n}{2}-1} e^{-\frac{t}{2}} dt.$$

[6]We will call both a system of first order ODEs and an ODE (ordinary differential equation) simply an ODE in the sequel if no confusion arises.

This can be expressed as a definite integral with parameter x with the Heaviside function:

$$\gamma(x) = \int_{-\infty}^{\infty} H(x-t)H(t)t^{\frac{n}{2}-1}e^{-\frac{t}{2}}dt. \tag{4.1}$$

Here, the Heaviside function is defined as

$$H(t) = \begin{cases} 1 & t \geq 0, \\ 0 & t < 0. \end{cases}$$

NOTE FOR ADVANCED TOPICS 4.1. An algorithm for finding an annihilating holonomic ideal of the Heaviside function for a semi-algebraic set is given in [**29**]. See also, e.g., [**30**, §6].

STEP 1. We will now derive a linear ordinary differential equation (ODE) satisfied by the integral $\gamma(x)$. The existence of the linear ODE can be proved by the theory of holonomic systems. See Theorem 3.2. Several algorithms have been proposed to derive an ODE annihilating a given definite integral with a parameter. See, e.g., [**15**], [**35**, Chapter 5], [**36**]. We will apply one of the algorithms to $\gamma(x)$. First, we need a holonomic ideal annihilating the kernel function.[7] Note

$$\partial_t \bullet H(t) = \delta(t)$$

$$t\delta(t) = 0$$

where $\delta(t)$ is the delta function, $\partial_t = \frac{\partial}{\partial t}$, and $\partial_t \bullet H(t)$ is defined as $\frac{\partial H}{\partial t}$. Then, we have $t\partial_t H(t) = 0$. Put $f(t,x) = H(x-t)H(t)g(t)$, $g(t) = t^{\frac{n}{2}-1}e^{-\frac{t}{2}}$. It follows from the chain rule that

$$\partial_t f(t,x) = H'(t)H(x-t)g(t) - H(t)H'(x-t)g(t) + H(t)H(x-t)g'(t).$$

Multiplying both sides by $(x-t)t$, we obtain

$$(x-t)t\partial_t f(t,x) = H(t)H(x-t)g'(t)$$

since $xH'(x) = x\delta(x) = 0$. The function $g'(t)$ is equal to a rational function multiplied by $g(t)$. We can then conclude that the following differential operators annihilate the function $f(t,x)$.

$$\ell_1 := (x-t)\left(t\partial_t - \left(\frac{n}{2} - 1\right) + \frac{1}{2}t\right)$$

$$\ell_2 := (x-t)\partial_x.$$

The left ideal $I = D\ell_1 + D\ell_2$ of the ring of differential operators $D = \mathbf{C}\langle x, t, \partial_x, \partial_t \rangle$ is a holonomic ideal. We can check it by an algorithm (see, e.g., [**35**, Alg 2.2.5, p.66]). Its elements annihilate the function f. Since I is a holonomic left ideal, $J = (I + \partial_t D) \cap K\langle x, \partial_x \rangle$ is not 0 ideal by Theorem 3.2. Let us take a non-zero element ℓ of it. Since we have $\ell = P_1 + \partial_t P_2 (P_1 \in I, P_2 \in D)$, the operator ℓ annihilates the integral $\gamma(x)$:

$$\ell \bullet \int_{-\infty}^{\infty} f(x,t)dt = \int_{-\infty}^{\infty} P_1 f(x,t)dt + \int_{-\infty}^{\infty} \partial_t P_2 f(x,t)dt = [P_2 f]_{-\infty}^{\infty} = 0.$$

[7] See [**30**, 4.4] as to an algorithmic method.

Thus, $\ell\gamma = 0$ is the linear ODE we want. With help from the implementation of an elimination algorithm.[8] of the sum of the left ideal I and the right ideal $\partial_t D$, we conclude $J = \langle 2x\partial_x^2 + (x - n + 2)\partial_x \rangle$. Having completed step 1, we can now apply steps 2 and 3.

STEP 2. From $\left(2x\partial_x^2 + (x - n + 2)\partial_x\right)\gamma = 0$, we have

$$\partial_x \begin{pmatrix} \gamma \\ \partial_x \gamma \end{pmatrix} = \begin{pmatrix} 0 & 1 \\ 0 & -\frac{1}{2x}(x - n + 2) \end{pmatrix} \begin{pmatrix} \gamma \\ \partial_x \gamma \end{pmatrix} \tag{4.2}$$

STEP 3a. We will compute a series expansion of the integral at point $x = 0$ as follows;

$$\begin{aligned} \int_0^x t^{\frac{n}{2}-1} e^{-\frac{t}{2}} dt &= \sum_{k=0}^{\infty} \int_0^x t^{\frac{n}{2}-1} \frac{1}{k!} \left(-\frac{t}{2}\right)^k dt \\ &= \sum_{k=0}^{\infty} \frac{1}{k!} \left(-\frac{1}{2}\right)^k \int_0^x t^{\frac{n}{2}-1+k} dt \\ &= \sum_{k=0}^{\infty} \frac{1}{k!} \left(-\frac{1}{2}\right)^k \frac{x^{\frac{n}{2}+k}}{\frac{n}{2}+k} \\ &= x^{\frac{n}{2}} {}_1F_1\left(\frac{n}{2}; \frac{n}{2}+1; -\frac{x}{2}\right) / \frac{n}{2}. \end{aligned} \tag{4.3}$$

Here, the hypergeometric function ${}_1F_1$ is defined as

$${}_1F_1(a, c; z) = \sum_{i=0}^{\infty} \frac{(a)_i}{(c)_n n!} z^i \tag{4.4}$$

where $(a)_i = a(a+1)\cdots(a+i-1), (a)_0 = 1$. $(a)_i$ is called the Pochhammer symbol. When we make an analogous construction of the χ^2 distribution for multivariate normal distributions, we obtain a matrix hypergeometric function ${}_1F_1$ as discussed later in Section 5.

Numerical evaluation of the series gives values of the integral for small x efficiently.

STEP 3b. We will apply numerical solvers to the system of the ODE (4.2).

Note that when degrees of freedom n become larger, we need ingenuity to produce reasonably accurate values efficiently. In other words, we should modify the ODE so that numerical solvers work accurately. We introduce a new dependent variable $h(x)$ as $\gamma(x) = e^{-\frac{x}{2}} x^{\frac{n}{2}} h(x)$. $h(x)$ satisfies the ODE

$$(\theta(\theta + n/2) - (x/2)(\theta + 1))\, h = 0, \theta = x\partial_x.$$

The vector-valued function $H(x) = (h, \partial_x h)^T$ satisfies

$$\partial_x H = \begin{pmatrix} 0 & 1 \\ \frac{1}{2x} & -\frac{1}{x}(n/2 + 1 - x/2) \end{pmatrix} H.$$

[8]See, e.g.,[**41**, §7.4.9] for commands for this on Macaulay2, Risa/Asir, and Singular. As to the creative telescoping method, there are `SumTools`, `Ore_algebra` packages in Maple and `HolonomicFunctions` package in Mathematica.

The function $h(x)$ has the following series expansion

$$
\begin{aligned}
e^{x/2}x^{-n/2}\int_0^x t^{\frac{n}{2}-1}e^{-\frac{t}{2}}dt &= x^{-n/2}\int_0^x t^{\frac{n}{2}-1}e^{-(t-x)/2}dt \\
&= x^{-n/2}\sum_{k=0}^{\infty}\int_0^x t^{\frac{n}{2}-1}(x-t)^k\frac{1}{k!}\left(\frac{1}{2}\right)^k dt \\
&= \sum_{k=0}^{\infty}\frac{1}{k!}\left(\frac{1}{2}\right)^k x^k\int_0^1 s^{\frac{n}{2}-1}(1-s)^k ds, \quad \text{by } t = xs \\
&= \sum_{k=0}^{\infty}\left(\frac{x}{2}\right)^k\frac{1}{k!}\frac{\Gamma(n/2)\Gamma(k+1)}{\Gamma(n/2+k+1)} \\
&= \frac{1}{n/2}\sum_{k=0}^{\infty}\left(\frac{x}{2}\right)^k/(n/2+1)_k \\
&= \frac{1}{n/2}{}_1F_1\left(1;\frac{n}{2}+1;\frac{x}{2}\right) \qquad (4.5)
\end{aligned}
$$

which is not the sum of terms $(-x/2)^i$, $i = 0,1,2,\ldots$ with alternating sign and converges faster. By solving the ODE for H, we obtain a code that works well even when n is relatively large. An implementation `evalig3.py` by Python of this evaluation method is available from [**7**]. Below is an example input for the program.

```
x,g=ig(70,50)
print([x[-1],g[-1]])
# Output is [70.0, 2.0144809414397258e+31]
```

Of course, there are several better special methods to evaluate the function $\gamma(x)$. However, the HGM is a general framework to evaluate CDF's and normalizing constants in statistics. Moreover, if we add ideas to each step of the HGM for a specific problem to obtain better performance and accuracy, we can produce special methods for the problem.

5. Wishart distribution

Let X_i be a random column vector whose distribution is the m-dimensional multivariate normal (or Gaussian) distribution with mean vector 0 and covariance matrix Σ. We take n such independent random vectors $X_1,\ldots,X_n$. Put $X = (X_1,\ldots,X_n)$, which is an $m\times n$ matrix, and define an $m\times m$ random matrix $W = XX^T$. When $m=1$, we have $W = X_1^2+\cdots+X_n^2$ and we have seen that the CDF of W can be expressed in terms of the hypergeometric function ${}_1F_1$ in (4.3). The distribution of W is called the Wishart distribution. It is known that the CDF of the maximal eigenvalue of W can be expressed in terms of the matrix hypergeometric function ${}_1F_1$.

THEOREM 5.1 ([**4**]). *Let ℓ_1 be the maximal eigenvalue of W. Then the probability that ℓ_1 is smaller than x is*

$$
P(\ell_1 < x) = C\exp\left(-\frac{x}{2}\operatorname{Tr}\Sigma^{-1}\right)x^{\frac{1}{2}nm}{}_1F_1\left(\frac{m+1}{2};\frac{n+m+1}{2};\frac{x}{2}\Sigma^{-1}\right) \qquad (5.1)
$$

where ${}_1F_1$ is the hypergeometric function of a matrix argument and

$$
C = \frac{\Gamma_m\left(\frac{m+1}{2}\right)}{2^{\frac{1}{2}nm}(\det\Sigma)^{\frac{1}{2}n}\Gamma_m\left(\frac{n+m+1}{2}\right)}, \quad \Gamma_m(t) = \pi^{m(m-1)/4}\prod_{i=1}^m\Gamma(t-(i-1)/2).
$$

The function ${}_1F_1(a; c; Y)$ is

$$\sum_{k=0}^{\infty} \sum_{\kappa, |\kappa|=k} \frac{(a)_\kappa}{(c)_\kappa} \frac{C_\kappa(Y)}{k!}, \quad (a)_\kappa = \frac{\Gamma_m(a,\kappa)}{\Gamma_m(a)}$$

where $\kappa = (\kappa_1, \ldots, \kappa_m)$ is a partition of k, $\Gamma_m(a,\kappa) = \pi^{m(m-1)/4} \prod_{i=1}^m \Gamma(a + \kappa_i - (i-1)/2)$, and $C_\kappa(Y)$ is the (C-normalized) zonal polynomial for the $m \times m$ real symmetric matrix Y. It is an invariant function of an action of the orthogonal group and satisfies

$$C_\kappa(H^T Y H) = C_\kappa(Y), \quad H \in O(m).$$

Then, it depends only on the eigenvalues of Y; the hypergeometric function in (5.1) also depends only on the eigenvalues of $\frac{x}{2}\Sigma^{-1}$. For the definition and properties of zonal polynomials, see, e.g., [**26**, §7.2]. One of the important properties of the zonal polynomial is that it behaves like a monomial with respect to the Laplace transformation on positive definite symmetric matrices. An analogous formula of $\int_0^\infty \exp(-rs)s^{t-1}ds = \Gamma(t)r^{-t}$ holds for $|S|^{t-(m+1)/2}C_\kappa(T^{1/2}ST^{1/2})$ where S, T are positive definite symmetric matrices.[9]

The hypergeometric function ${}_1F_1$ of the matrix argument has the following integral representation:

$$(5.2) \quad {}_1F_1(a; c; Y) = \frac{\Gamma_m(c)}{\Gamma_m(a)\Gamma_m(c-a)} \cdot \int_{0<X<I_m} \exp(\operatorname{Tr} XY)|X|^{a-(m+1)/2}|I_m - X|^{c-a-(m+1)/2}dX$$

where $0 < X < I_m$ means that X and $I_m - X$ are positive definite and $dX = \prod_{i \le j} dx_{ij}$

We want to apply the HGM to evaluate $P(\ell_1 < x)$. This is a generalization of our method for the χ^2 distribution given in Section 4. We will use the following theorem for step 1 of the HGM.

THEOREM 5.2 ([**25**]). *The function ${}_1F_1(a; c; \mathrm{diag}(y_1, \ldots, y_m))$ is annihilated by*

$$(5.3) \qquad g_i = y_i\partial_i^2 + (c - y_i)\partial_i + \frac{1}{2}\sum_{j=1, j\neq i}^{m} \frac{y_j}{y_i - y_j}(\partial_i - \partial_j) - a$$

For a proof of this fact, see the book by Muirhead [**26**, §7.5], a comprehensive reference on statistics and hypergeometric functions with matrix arguments.

Step 2 of the HGM is described in [**9**, §4]. This is a technical step constructing a sparse Pfaffian equation of the form (3.10) and an ODE obtained by a restriction of it to $\frac{x}{2}\Sigma^{-1}$. For details, see also our source code [**10**, `code-n.c`].

Let us proceed to step 3. We need to evalute ${}_1F_1$, which is a sum of zonal polynomials, and its derivatives at a point. Let us briefly explain what kind of polynomial the zonal polynomial is. Let c_κ be the C-normalizing constant

$$c_\kappa = \frac{2^{|\kappa|}(|\kappa|)!}{j_\kappa}, \quad j_\kappa = \prod_{(i,j)\in\kappa} (\kappa'_j - i + 2(\kappa_i - j + 1))(\kappa'_j - i + 1 + 2(\kappa_i - j))$$

---

[9] $C_\kappa(T^{1/2}ST^{1/2})$ is denoted by $C_\kappa(TS)$ or $C_\kappa(ST)$ in the literature.

where κ' is the conjugate partition and $(i,j) \in \kappa$ means that the product is taken over all coordinates (i,j) of boxes in the Young diagram of the partitions κ and κ' respectively. For example, when

$$\kappa = (3,1) =$$ 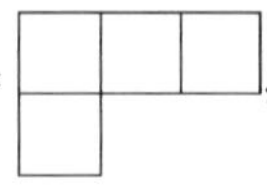 ,

we have

$$\kappa' =$$ 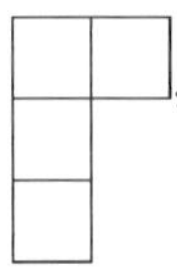 $$, (i,j) = (1,1),(1,2),(1,3),(2,1),(2,2),(2,3),$$

and $\kappa_1 = 3, \kappa_2 = 1$, $\kappa'_1 = 2, \kappa'_2 = 1, \kappa'_3 = 1$. Below is a table of $\alpha = 2$ Jack polynomials $J_\kappa = \frac{1}{c_\kappa} C_\kappa$ when $m = 2$:

$$\begin{aligned}
J_{(1,0)} &= x_1 + x_2 \\
J_{(2,0)} &= 3x_1^2 + 2x_1x_2 + 3x_2^2 \\
J_{(1,1)} &= 2x_1x_2 \\
J_{(3,0)} &= 15x_1^3 + 9x_1^2x_2 + 9x_1x_2^2 + 15x_2^3 \\
J_{(2,1)} &= 4x_1^2x_2 + 4x_1x_2^2 \\
&= \cdots \\
J_{(6,0)} &= 10395x_1^6 + 5670x_1^5x_2 + 4725x_1^4x_2^2 + 4500x_1^3x_2^3 + 4725x_1^2x_2^4 + 5670x_1x_2^5 \\
&\quad + 10395x_2^6
\end{aligned}$$

Here, x_i are eigenvalues of the 2×2 symmetric matrix Y. The reader should be convinced that evaluation of ${}_1F_1$ by series expansion is not easy from the table.

The hypergeometric series in terms of zonal polynomials converges relatively faster when the eigenvalues of Y are small. We will use the series for step 3a following the method of [**14**]. The key of the method is an efficient evaluation of zonal polynomials. In [**14**], the authors utilize the Pieri type formula

$$J_\kappa(x_1, \ldots, x_m) = \sum_\mu J_\mu(x_1, \ldots, x_{m-1}) x_m^{|\kappa/\mu|} \beta_{\kappa\mu} \tag{5.4}$$

where μ runs over the set $\{\mu \mid \kappa_1 \geq \mu_1 \geq \kappa_2 \geq \mu_2 \geq \cdots \geq \mu_{m-1} \geq \kappa_m\}$ and $\beta_{\kappa\mu}$ is a constant [**14**]. We extend values of the series expansion at small x to larger x by the ODE derived in step 2.

Listing 3 evaluates $P(\ell_1 < q)$ by our HGM for $m = 3$, $n = 5$, and $\Sigma = \beta^{-1}/2$ where $\beta = \mathrm{diag}(1,2,3)$. Since it numerically solves an ODE given in [**9**], it evaluates all values of the probability in $[0, q]$. Listing 2 is a simulation code using a random number generator for comparison with the HGM.

Although the evaluation by `hgm.twishart()` works well for several cases, there are limitations. Let y_i be eigenvalues of the matrix $\frac{x}{2}\Sigma^{-1}$. They are $m \leq 10$ (a larger m requires more computational resources), $y_i \neq y_j$, $n \neq$ (large number) and the y_i must not contain both tiny and huge numbers. An evaluation method for the case $y_i = y_j$ is given in [**28**], which utilizes a system of linear partial differential equations on $y_i = y_j$ satisfied by ${}_1F_1$.

LISTING 2. Simulation by R

```
# simulation by random matrices.
rpwishart<-function(lambda=3,sigma=diag(c(1/2,1/4,1/6)),
 n=5,try=10000)
{ count<-0;r<-rWishart(try,df=n,Sigma=sigma);
 for (k in seq(1,try)) {
 ell1 <- max(eigen(r[,,k])$values);
 if (ell1 < lambda) count<-count+1; }
 return(count/try) }
rpwishart()
 [1] 0.5202
```

LISTING 3. HGM on R

```
#install.packages("hgm")
library("hgm")
hgm.pwishart(m=3,n=5,beta=c(1,2,3),q=3)
 [1] 3.0242949 0.5247871 ...#it means P(ell<3.024)=0.524
plot(hgm.pwishart(m=3,n=5,beta=c(1,2,3),q=10,autoplot=1))
```

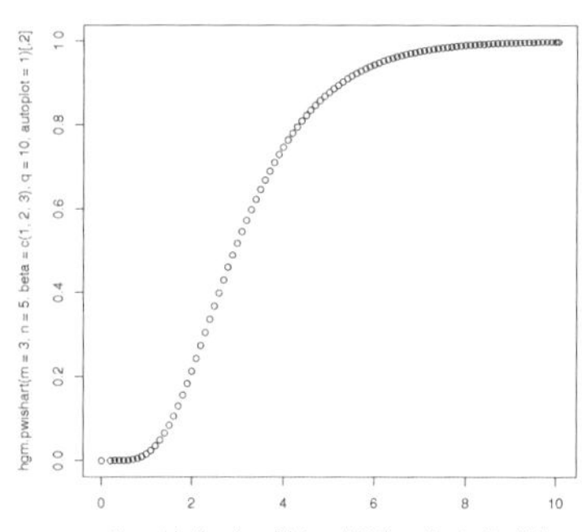

FIGURE 2. A graph of $P(\ell_1 < x)$ by Listing 3

6. HGM dictionary

To this point, we have provided an overview of the HGM using the χ^2 distribution and the Wishart distribution. In this section, we offer an overview of some of the probability distributions studied by the HGM. The theory of holonomic systems certifies the existence of a holonomic system and a Pfaffian equation satisfied by integrals with parameters as in the case of the χ^2 distribution and the Wishart distribution. Although there are algorithms to find the holonomic system and the Pfaffian equation, an interesting point is that it is sometimes possible to find holonomic systems and Pfaffian equations *by hand* for certain classes of probability distributions. We present such distributions in this section. Additional topics related to the HGM can be found from the list in [**11**].

6.1. Fisher–Bingham distribution, Bingham distribution.

6.1.1. *Distribution and normalizing constant.* The Fisher–Bingham distribution is the restriction of the multivariate normal distribution on the d-dimensional sphere $S^d(r) = \{(t_1, \ldots, t_{d+1}) \mid \sum_{i=1}^{d+1} t_i^2 = r^2, r > 0\}$. The density function is $\frac{1}{Z(x,y,r)} \exp\left(\sum_{1 \le i \le j \le d+1} x_{ij} t_i t_j + \sum_{i=1}^{d+1} y_i t_i\right) |dt|$ where Z is the normalizing constant

$$Z(x,y,r) = \int_{S^d(r)} \exp\left(\sum_{1\le i\le j\le d+1} x_{ij} t_i t_j + \sum_{i=1}^{d+1} y_i t_i\right) |dt|. \tag{6.1}$$

We denote by $|dt|$ the invariant measure on a sphere of radius r satisfying

$$\int_{S^d(r)} |dt| = r^d \frac{2\pi^{(d+1)/2}}{\Gamma((d+1)/2)}.$$

The Bingham distribution is the restriction of the Fisher-Bingham distribution to $r = 1$, $y_i = 0$, $x_{ij} = 0$, $i \neq j$. The variable x_{ii} is denoted by θ_i in [**37**].

6.1.2. *Holonomic system.* Let $\partial_{ij} = \partial/\partial x_{ij}$, $\partial_k = \partial/\partial y_k$, $\partial_r = \partial/\partial r$. The normalizing constant Z satisfies the following holonomic system [**27**]

$$\partial_{ij} - \partial_i \partial_j, \quad \sum_{i=1}^{d+1} \partial_i^2 - r^2, \quad r\partial_r - 2\sum_{i\le j} x_{ij}\partial_i\partial_j - \sum_i y_i \partial_i - d,$$

$$x_{ij}\partial_i^2 + 2(x_{jj} - x_{ii})\partial_i\partial_j - x_{ij}\partial_j^2 + \sum_{s\neq i,j} (x_{sj}\partial_i\partial_s - x_{is}\partial_j\partial_s) + y_j\partial_i - y_i\partial_j.$$

The normalizing constant of the Bingham distribution satisfies the following holonomic system

$$\sum_{i=1}^{d} \partial_{ii} - 1, \quad 2(x_{ii} - x_{jj})\partial_{ii}\partial_{jj} - (\partial_{ii} - \partial_{jj}), \ (1 \le i < j \le d).$$

6.1.3. *Pfaffian equation.* When we restrict the equation to $x_{ij} = 0$, $i \neq j$, the Pfaffian equation for the direction r is $\partial_r \bullet \tilde{F} = P^{(r)} \tilde{F}$ [**18**] where we put $\tilde{F} = (\partial_1, \ldots, \partial_{d+1}, \partial_1^2, \ldots, \partial_{d+1}^2)^T \bullet Z$. The ij element of the $(2d+2) \times (2d+2)$ matrix $P^{(r)} = (p_{ij}^{(r)})$ is

$$rp_{ij}^{(r)} = (2x_{ii}r^2 + 1)\delta_{ij} + \sum_{k=1}^{d+1} y_i \delta_{j(k+d+1)} \quad (1 \le i \le d+1),$$

$$rp_{(i+d+1)j}^{(r)} = y_i r^2 \delta_{ij} + (2x_i r^2 + 2)\delta_{j(i+d+1)} + \sum_{k\neq i} \delta_{j(k+d+1)} \quad (1 \le i \le d+1).$$

Pfaffian equations for other directions are given in [**21**]. For the normalizing constant of the Bingham distribution, see [**37**, Theorem 1].

6.1.4. *Implementation.* The normalizing constant of the Bingham distribution can be evaluated by the function `hgm.ncBingham` in the R package hgm [**10**]. The normalizing constant of the Fisher-Bingham distribution can be evaluated by the program provided in [**21**].

6.2. Orthant probability.

6.2.1. *Distribution and normalizing constant.* Consider the restriction of the multivariate normal distribution to the first orthant $t_i \geq 0$. The normalizing constant is

$$Z(x,y) = \int_{\mathbf{R}^d} H(t) \exp\left(\sum_{1\leq i\leq j\leq d} x_{ij} t_i t_j + \sum_{i=1}^{d} y_i t_i \right) dt_1 \cdots dt_d \tag{6.2}$$

where $H(t)$ is the Heaviside function that takes the value 1 on the first orthant and takes the value 0 out of it.

6.2.2. *Holonomic system.* [**19**, Theorems 2 and 5] gives a holonomic system when $H(t)$ is a holonomic distribution annihilated by $P_1, \ldots, P_s$. Then, the holonomic system for Z is

$$\begin{aligned} \varphi(P_k) &\quad (1 \leq k \leq s), \quad \varphi \text{ is defined in } [\mathbf{19}], \\ \partial_{x_{ij}} - 2\partial_{y_i}\partial_{y_j} &\quad (1 \leq i < j \leq d), \\ \partial_{x_{ii}} - \partial_{y_i}^2 &\quad (1 \leq i \leq d). \end{aligned}$$

6.2.3. *Pfaffian equation.* For $J \subset [d] = \{1, \ldots, d\}$, we put

$$\begin{aligned} h_J(x,y,t) &= \sum_{i\in J}\sum_{j\in J} x_{ij} t_i t_j + \sum_{k\in J} y_k t_k, \\ g_J(x,y) &= \int_0^\infty \cdots \int_0^\infty \exp\left(h_J(x,y,t)\right) dt_J, \end{aligned}$$

where $dt_J = \prod_{j\in J} dt_j$ and put $g_\emptyset = 1$ for the empty J. The Pfaffian equation is

$$\partial_{y_i} g_J = \begin{cases} \mu_i^J g_J + \sum_{j\in J} \sigma_{ij}^J g_{J\setminus\{j\}} & i \in J \\ 0 & i \notin J \end{cases} \tag{6.3}$$

$$\partial_{x_{ij}} g_J = \begin{cases} 2\partial_{y_i}\partial_{y_j} g_J & \{i,j\} \subset J,\ i<j \\ \partial_{y_i}^2 g_J & \{i\} \subset J,\ i=j \\ 0 & else. \end{cases} \tag{6.4}$$

6.2.4. *Implementation.* The function `hgm.northant` [**10**] can be used to evaluate the normalizing constant. Below is an example.

```
x <- matrix(c(15,26,23,19, 26,47,46,35,
 23,46,65,38, 19,35,38,33), nrow =4)
y <- c(1,2,3,4); hgm.ncorthant(x,y)
```

Note that there is a limitation to the current algorithm and its implementation: When y is apart from the origin, `hgm.ncorthant` becomes slower to maintain accuracy.

6.3. Fisher distribution.

6.3.1. *Distribution and normalizing constant.* The Fisher distribution is defined on $SO(p) = \{X \in \mathbf{R}^{p\times p} \mid X^T X = I_p, \det(X) = 1\}$ as

$$\exp(\mathrm{Tr}\, \Theta^\top X)\delta_{SO(n)}(X)/Z(\Theta).$$

Here, $\delta_{SO(p)}(X)$ is the delta function whose support is $SO(p)$, $\Theta = (\theta_{ij})$, and Z is the normalizing constant defined by

$$Z(\Theta) = \int_{X=(X_{ij})\in \mathbf{R}^{p\times p}} \exp(\mathrm{Tr}\, \Theta^\top X)\delta_{SO(p)}(X)dX.$$

6.3.2. *Holonomic system.* The holonomic system satisfied by $Z(\Theta)$ is

$$A_{ij}^{(1)} = \sum_{k=1}^{p} \partial_{ik}\partial_{jk} - \delta_{ij}, \quad \tilde{A}_{ij}^{(1)} = \sum_{k=1}^{p} \partial_{ki}\partial_{kj} - \delta_{ij} \quad (i \le j), \quad A^{(2)} = \det(\partial_{ij}) - 1,$$

$$A_{ij}^{(3)} = \sum_{k=1}^{p} (-\theta_{jk}\partial_{ik} + \theta_{ik}\partial_{jk}), \quad \tilde{A}_{ij}^{(3)} = \sum_{k=1}^{p} (-\theta_{kj}\partial_{ki} + \theta_{ki}\partial_{kj}) \quad (i < j),$$

where $\partial_{ij} = \partial/\partial\theta_{ij}$ [**17**].

The restriction of this system to the diagonal Θ is interesting and is obtained in [**38**] when $p = 3$ as

$$\ell_{ij} = (\theta_{ii}^2 - \theta_{jj}^2)\partial_{ii}\partial_{jj} - (\theta_{jj}\partial_{ii} - \theta_{ii}\partial_{jj}) - (\theta_{ii}^2 - \theta_{jj}^2)\partial_{kk}, \quad i \neq j$$

Here, k is chosen as $\{i, j, k\} = \{1, 2, 3\}$.

6.3.3. *Pfaffian equation.* The Pfaffian equation is unknown for $p > 3$. When $p = 3$ and Θ is diagonal, the Pfaffian equation is given in [**38**, Theorem 2]. In particular, when we parametrize θ_{ii} by constants a, b, c as $\theta_{11} = at$, $\theta_{22} = bt$, $\theta_{33} = ct$, the ODE with respect to t is

$$\frac{dC}{dt} = \left(A - \frac{2}{t}\,\mathrm{diag}(0, 1, 1, 1)\right) C \tag{6.5}$$

where A is a 4×4 matrix $[[0, a, b, c], [a, 0, c, b], [b, c, 0, a], [c, b, a, 0]]$ and $C = ((1, \partial_{11}, \partial_{22}, \partial_{33})^T \bullet Z)(at, bt, ct)$.

6.3.4. *Implementation.* A series expansion of the normalizing constant is given in [**38**]. The function `hgm.ncso3` [**10**] utilizes the series to evaluate the normalizing constant and its derivatives when the θ_{ij} are small and extends these values by the ODE (6.5).

```
hgm.ncso3(a=1,b=2,c=3,t0=0,q=3);
[1] 11.8731
```

The output is the normalizing constant for $(a, b, c) = (1, 2, 3)$, $t = \mathsf{q} = 3$ in (6.5).

6.4. Discrete A-hypergeometric distribution.

6.4.1. *Distribution and normalizing constant.* Let $A = (a_{ij})$ be a $d \times n$ matrix with integer entries. We denote by a_i the i-th column vector of A. We suppose that the column vectors generate $\mathbf{Z}^d$ and there exists an i-th row such that $a_{ij} > 0$, $\forall j$. Let $\mathbf{N}_0$ be the set of non-negative integers. For $\beta \in \mathbf{N}_0 A = \mathbf{N}_0 a_1 + \cdots + \mathbf{N}_0 a_n$, we define a polynomial

$$Z_A(\beta; x) = \sum_{Au=\beta, u\in \mathbf{N}_0^d} \frac{x^u}{u!} \tag{6.6}$$

called an A-hypergeometric polynomial [**34**]. We can regard x as a parameter vector and let $\{U \in \mathbf{N}_0^n \mid AU = \beta\}$ be a sample space. The discrete A-hypergeometric distribution is $P(u) = \frac{x^u}{u!}/Z_A(\beta; x)$. See [**42**] on the relation to the toric model and the asymptotic properties of the distribution.

6.4.2. *Holonomic system.* Let $c_1, \ldots, c_d$ be indeterminate and S_i be a difference operator satisfying $S_i \bullet f(c_i) = f(c_i - 1)$. Define the ring of differential-difference operator $D = \mathbf{C}\langle c_1, \ldots, c_d, S_1, \ldots, S_d\rangle\langle x_1, \ldots, x_n, \partial_1, \ldots, \partial_n\rangle$. The normalizing constant $Z_A(c; x)$ is annihilated by

$$\sum_{j=1}^{n} a_{ij} x_j \partial_j - c_i =: E_i - c_i, \quad \partial_i - S^{a_i} \qquad (i = 1, \ldots, d). \tag{6.7}$$

The GKZ system or A-hypergeometric system by Gel'fand-Kapranov-Zelevinsky (see, e.g., [**35**]) is obtained by eliminating the difference operators.

6.4.3. *Pfaffian equation.* When A is a product of a p-simplex and a q-simplex, the discrete A-hypergeometric distribution is the hypergeometric distribution on the $p \times q$ contingency table with fixed marginal sums. The Pfaffian equation for this case is given in [**8**] by utilizing the intersection numbers of twisted cohomology groups. A general algorithm to derive the Pfaffian equation and the difference equation by utilizing a Macaulay matrix is given in [**1**] and [**31**].

A Markov chain Monte-Carlo (MCMC) algorithm for the discrete A-hypergeometric distribution is given in [**5**], where the authors report the breakthrough of finding that generators of affine toric ideals give Markov bases. Their study leads us to study the discrete A-hypergeometric distribution by utilizing A-hypergeometric polynomials.

A new application of Pfaffian equations for hypergeometric polynomials is a sampler. An efficient exact sampler for partitions is given in [**22**] by utilizing a Pfaffian equation for

$$A = \begin{pmatrix} 1 & 1 & \cdots 1 \\ 0 & m_2 & \cdots m_n \end{pmatrix}, 0 < m_2 < \cdots < m_n.$$

An exact sampler algorithm for the A-hypergeometric distribution by utilizing a Pfaffian equation is given in [**23**]. It works in parallel and outputs high quality samples, but is, for now, slow for general A.

6.4.4. *Implementation.* The Risa/Asir [**33**] package `gtt_ekn3.rr` gives an evaluator of A-hypergeometric polynomials for two-by-two contingency tables and several utility functions [**40**]. Below are some examples.

```
import("gtt_ekn3.rr");
//Data of 2 by 4 contingency table
U=[[1,1,2,3],[1,3,1,1]];
//conditional maximal likelihood estimation
gtt_ekn3.cmle(U);
// Data, row sum, column sum
 [[ 1 1 2 3 ]
 [ 1 3 1 1 ],[[7,6],[2,4,3,4]],
// Estimated probability.
 [ 1 67147/183792 120403/64148 48801/17869 ]
 [ 1 1 1 1 ]]
```

```
load("ot_hgm_ahg.rr");
A= [[1,1,1,1,1,0,0,0,0,0],
  [1,0,0,0,0,1,0,0,0,1],
  [0,1,0,0,0,1,1,0,0,0],
  [0,0,1,0,0,0,1,1,0,0],
  [0,0,0,1,0,0,0,1,1,0],
  [0,0,0,0,1,0,0,0,1,1]];
U= [23,21,20,34,30,42,11,10,20,29];
B=vtol(matrix_list_to_matrix(A)*ltov(U));
SumU = oh_base.sum(U);
X = vtol((1/SumU)*ltov(U)); R = [x1,x2];
//Expectation w.r.t. u1,u2
EV=hgm_ahg_expected_values_contiguity(A,B,X,R|msize=3);
```

The Macaulay matrix method is implemented in the Risa/Asir package `mt_mm.rr`. Algorithms and implementations to restrict GKZ hypergeometric systems to subvarieties are given in [**2**].

6.5. Polyhedron probability.

6.5.1. *Distribution and normalizing constant.* We mean by the polyhedron probability the restriction of a multivariate normal distribution to a polyhedron. It is a generalization of the orthant probability. It is studied in terms of HGM in [**16**]. When the polyhedron is defined as

$$P = \{x \in \mathbf{R}^d : \sum_{i=1}^{d} a_{ij}x_i + b_j \geq 0,\ 1 \leq j \leq n\}, \tag{6.8}$$

by parameters a_{ij}, b_j, the normalizing constant is

$$\frac{1}{(2\pi)^{d/2}} \int_{x\in P} \exp(-\frac{1}{2}\sum_{i=1}^{d} x_i^2)dx_1 \cdots dx_d. \tag{6.9}$$

6.5.2. *Pfaffian equation.* Let F_j be the facet of P standing for $\sum_{i=1}^d a_{ij}x_i+b_j = 0$. The set of all non-empty intersections of facets is denoted by $\mathcal{F}$. In other words, $\mathcal{F}$ is defined as $\mathcal{F} = \{J \subset \{1,\dots,n\} \mid F_J \neq \emptyset\}$, $F_J = \cap_{j\in J}F_j$. We add the empty index set to $\mathcal{F}$ for convenience. We can define a basis of a Pfaffian equation indexed as $\{g^J,\ J \in \mathcal{F}\}$ by $\mathcal{F}$. Our Pfaffian equation is

$$\partial_{a_{ij}} g^J = \sum_{k=1}^{n} a_{ik}\partial_{b_k}\partial_{b_j} g^J \quad (1 \leq i \leq d,\ 1 \leq j \leq n,\ J \in \mathcal{F}),$$

$$\partial_{b_j} g^J = g^{J\cup\{j\}} \quad (j \in J^c,\ J \in \mathcal{F}),$$

$$\partial_{b_j} g^J = -\sum_{k\in J} \alpha_J^{jk}(a) \left(b_k g^J + \sum_{\ell\in J^c} \alpha_{k\ell}(a) g^{J\cup\{\ell\}} \right) \quad (j \in J,\ J \in \mathcal{F}).$$

Here, $(\alpha_F^{ij}(a))_{i,j\in F}$ is the ij element of the inverse matrix of the matrix $\alpha_F(a)$ and the ij element of $\alpha_F(a)$ is $\sum_{k=1}^d a_{ki}a_{kj}$. Note that they give denominators of Pfaffian matrices.

6.5.3. *Implementation.* When P is a simplex, an implementation is obtainable from [**20**].

Let us present an example. When $a = [[1,0],[0,1],[-1,-1]], b = [0,0,2]$, we have $\mathcal{F} = \{\emptyset, 1, 2, 3, 12, 13, 23\}$ where, e.g., 12 denotes the set of indexes $\{1,2\}$. We evaluate the cumulative value of the multivariate normal distribution with mean 0 and covariance diag(1, 1) on $x > 0, y > 0, -x - y + 2 > 0$.

```
./a.out 1 2 1 0 0 0 1 0 -1 -1 2
result:
 0.1775361552 0.4772498666 0.4772498666
 1.0000000000 0.3100122972 0.1353352805 0.1353352805
Probability = 0.177536
```

The arguments of the example above are $S = 1, d = 2, a_{11} = 1, a_{21} = 0, b_1 = 0, a_{21} = 1, a_{22} = 1, b_2 = 0, a_{31} = -1, a_{32} = -1, b_3 = 2$.

References

[1] V. Chestnov, F. Gasparotto, M. K. Mandal, P. Mastrolia, S. J. Matsubara-Heo, H. J. Munch, and N. Takayama, *Macaulay matrix for Feynman integrals: Linear relations and intersection numbers*, J. High Energy Phys. **9** (2022), Paper No. 187, 55, DOI 10.1007/jhep09(2022)187. MR4490523

[2] V. Chestnov, S. J. Matsubara-Heo, H. J. Munch, and N. Takayama, *Restrictions of Pfaffian systems for Feynman integrals*, J. High Energy Phys. **11** (2023), Paper No. 202, 80, DOI 10.1007/jhep11(2023)202. MR4672770

[3] F. Chyzak, *The ABC of creative telescoping—Algorithms, bounds, complexity*, 2014. HAL Id: tel-0106931.

[4] A. G. Constantine, *Some non-central distribution problems in multivariate analysis*, Ann. Math. Statist. **34** (1963), 1270–1285, DOI 10.1214/aoms/1177703863. MR181056

[5] P. Diaconis and B. Sturmfels, *Algebraic algorithms for sampling from conditional distributions*, Ann. Statist. **26** (1998), no. 1, 363–397, DOI 10.1214/aos/1030563990. MR1608156

[6] W. Feller, *An introduction to probability theory and its applications. Vol. II*, 2nd ed., John Wiley & Sons, Inc., New York-London-Sydney, 1971. MR270403

[7] Sample codes in "Hypergeometric systems, statistics, and algorithms", Available at `https://github.com/nobuki-takayama/h-s-alg`.

[8] Y. Goto and K. Matsumoto, *Pfaffian equations and contiguity relations of the hypergeometric function of type $(k+1, k+n+2)$ and their applications*, Funkcial. Ekvac. **61** (2018), no. 3, 315–347, DOI 10.1619/fesi.61.315. MR3889669

[9] H. Hashiguchi, Y. Numata, N. Takayama, and A. Takemura, *The holonomic gradient method for the distribution function of the largest root of a Wishart matrix*, J. Multivariate Anal. **117** (2013), 296–312, DOI 10.1016/j.jmva.2013.03.011. MR3053549

[10] HGM: Holonomic gradient method and gradient descent—A package in cran. Available at `https://cran.r-project.org/web/packages/hgm/index.html`.

[11] References for the holonomic gradient method (HGM) and the holonomic gradient descent method (HGD). Available at `http://www.math.kobe-u.ac.jp/OpenXM/Math/ref-hgm`.

[12] B. Huber and B. Sturmfels, *A polyhedral method for solving sparse polynomial systems*, Math. Comp. **64** (1995), no. 212, 1541–1555, DOI 10.2307/2153370. MR1297471

[13] K. Itō, *Introduction to probability theory*, Cambridge University Press, Cambridge, 1984. Translated from the Japanese by the author. MR777504

[14] P. Koev and A. Edelman, *The efficient evaluation of the hypergeometric function of a matrix argument*, Math. Comp. **75** (2006), no. 254, 833–846, DOI 10.1090/S0025-5718-06-01824-2. MR2196994

[15] C. Koutschan, *A fast approach to creative telescoping*, Math. Comput. Sci. **4** (2010), no. 2-3, 259–266, DOI 10.1007/s11786-010-0055-0. MR2775992

[16] T. Koyama, *Holonomic modules associated with multivariate normal probabilities of polyhedra*, Funkcial. Ekvac. **59** (2016), no. 2, 217–242, DOI 10.1619/fesi.59.217. MR3560500

[17] T. Koyama, *The annihilating ideal of the Fisher integral*, Kyushu J. Math. **74** (2020), no. 2, 415–427. MR4235577

[18] T. Koyama, H. Nakayama, K. Nishiyama, and N. Takayama, *Holonomic gradient descent for the Fisher-Bingham distribution on the d-dimensional sphere*, Comput. Statist. **29** (2014), no. 3-4, 661–683, DOI 10.1007/s00180-013-0456-z. MR3261833

[19] T. Koyama and A. Takemura, *Calculation of orthant probabilities by the holonomic gradient method*, Jpn. J. Ind. Appl. Math. **32** (2015), no. 1, 187–204, DOI 10.1007/s13160-015-0166-8. MR3318908

[20] T. Koyama, *Simplex probability*, Available at `https://github.com/tkoyama-may10/simplex`.

[21] A. Kume and T. Sei, *On the exact maximum likelihood inference of Fisher-Bingham distributions using an adjusted holonomic gradient method*, Stat. Comput. **28** (2018), no. 4, 835–847, DOI 10.1007/s11222-017-9765-3. MR3766046

[22] S. Mano, *Partition structure and the A-hypergeometric distribution associated with the rational normal curve*, Electron. J. Stat. **11** (2017), no. 2, 4452–4487, DOI 10.1214/17-EJS1361. MR3724486

[23] S. Mano and N. Takayama, *Algebraic algorithm for direct sampling from toric models*, Preprint, `arXiv:2110.14992`, 2021.

[24] S.-J. Matsubara-Heo and S. Telen, *Twisted cohomology and likelihood ideals*, Adv. in Appl. Math. **165** (2025), Paper No. 102832, 32, DOI 10.1016/j.aam.2024.102832. MR4851756

[25] R. J. Muirhead, *Systems of partial differential equations for hypergeometric functions of matrix argument*, Ann. Math. Statist. **41** (1970), 991–1001, DOI 10.1214/aoms/1177696975. MR264799

[26] R. J. Muirhead, *Aspects of multivariate statistical theory*, Wiley Series in Probability and Mathematical Statistics, John Wiley & Sons, Inc., New York, 1982. MR652932

[27] H. Nakayama, K. Nishiyama, M. Noro, K. Ohara, T. Sei, N. Takayama, and A. Takemura, *Holonomic gradient descent and its application to the Fisher-Bingham integral*, Adv. in Appl. Math. **47** (2011), no. 3, 639–658, DOI 10.1016/j.aam.2011.03.001. MR2822207

[28] M. Noro, *System of partial differential equations for the hypergeometric function ${}_1F_1$ of a matrix argument on diagonal regions*, Proceedings of the 2016 ACM International Symposium on Symbolic and Algebraic Computation, ACM, New York, 2016, pp. 381–388, DOI 10.1145/2930889.2930905. MR3565738

[29] T. Oaku, *Algorithms for integrals of holonomic functions over domains defined by polynomial inequalities*, J. Symbolic Comput. **50** (2013), 1–27, DOI 10.1016/j.jsc.2012.05.004. MR2996866

[30] T. Oaku, *Algorithms for D-modules, integration, and generalized functions with applications to statistics*, The 50th anniversary of Gröbner bases, Adv. Stud. Pure Math., vol. 77, Math. Soc. Japan, Tokyo, 2018, pp. 253–352, DOI 10.2969/aspm/07710253. MR3839713

[31] T. Hibi, K. Nishiyama, and N. Takayama, *Pfaffian systems of A-hypergeometric equations I: Bases of twisted cohomology groups*, Adv. Math. **306** (2017), 303–327, DOI 10.1016/j.aim.2016.10.021. MR3581304

[32] "The R project for statistical computing", Available at `https://www.r-project.org/`.

[33] R. Asir, *a computer algebra system*, Available at `https://github.com/openxm-org/OpenXM`.

[34] M. Saito, B. Sturmfels, and N. Takayama, *Hypergeometric polynomials and integer programming*, Compositio Math. **115** (1999), no. 2, 185–204, DOI 10.1023/A:1000645909064. MR1668990

[35] M. Saito, B. Sturmfels, and N. Takayama, *Gröbner deformations of hypergeometric differential equations*, Algorithms and Computation in Mathematics, vol. 6, Springer-Verlag, Berlin, 2000, DOI 10.1007/978-3-662-04112-3. MR1734566

[36] A. Sattelberger and B. Sturmfels, *D-modules and holonomic functions*, Preprint, `arXiv:1910.01395`, 2019.

[37] T. Sei and A. Kume, *Calculating the normalising constant of the Bingham distribution on the sphere using the holonomic gradient method*, Stat. Comput. **25** (2015), no. 2, 321–332, DOI 10.1007/s11222-013-9434-0. MR3306709

[38] T. Sei, H. Shibata, A. Takemura, K. Ohara, and N. Takayama, *Properties and applications of Fisher distribution on the rotation group*, J. Multivariate Anal. **116** (2013), 440–455, DOI 10.1016/j.jmva.2013.01.010. MR3049915

[39] S. Sullivant, *Algebraic statistics*, Graduate Studies in Mathematics, vol. 194, American Mathematical Society, Providence, RI, 2018, DOI 10.1090/gsm/194. MR3838364

[40] Y. Tachibana, Y. Goto, T. Koyama, and N. Takayama, *Holonomic gradient method for two-way contingency tables*, Algebr. Stat. **11** (2020), no. 2, 125–153, DOI 10.2140/astat.2020.11.125. MR4229207

[41] N. Takayama, *Gröbner basis for rings of differential operators and applications*, Gröbner bases, Springer, Tokyo, 2013, pp. 279–344, DOI 10.1007/978-4-431-54574-3_6. MR3204385

[42] N. Takayama, S. Kuriki, and A. Takemura, *A-hypergeometric distributions and Newton polytopes*, Adv. in Appl. Math. **99** (2018), 109–133, DOI 10.1016/j.aam.2018.05.001. MR3806759

[43] N. Takayama, T. Yaguchi, and Y. Zhang, *Comparison of numerical solvers for differential equations for holonomic gradient method in statistics*, Preprint, `arXiv:2111.10947`, 2021.

Department of Mathematics, Kobe University, Rokko, Nada, Kobe, 657-8501, Japan
Email address: `takayama@math.kobe-u.ac.jp`

Proceedings of Symposia in Pure Mathematics
Volume **111**, 2025
https://doi.org/10.1090/pspum/111/02042

Bernd's contributions to algebraic statistics

Seth Sullivant

ABSTRACT. Bernd has had an enormous influence on the development of the field of algebraic statistics. This paper highlights some of the key results in his two seminal papers *Algebraic algorithms for sampling from conditional distributions* (with Persi Diaconis) and *On the toric algebra of graphical models* (with Dan Geiger and Chris Meek) which have had an outsize influence on the field. We also discuss related papers which were inspired by or developed from those seeds.

1. Introduction

Bernd has contributed to many different aspects of algebraic statistics and it would be difficult to summarize everything in a short paper. Still, there are some key contributions that have had an outsize role in the development of the field and also have developed my own personal interests in the field. I will also try to highlight how some of those ideas fit into the larger framework of some of Bernd's interests. My discussion will revolve around a couple of key papers.

The first paper, "Algebraic algorithms for sampling from conditional distributions", was written jointly with Persi Diaconis [**12**]. While there have certainly been papers before this one that used algebraic tools for statistics in some way, this is widely considered the paper that started the field of algebraic statistics (in the context of using algebraic geometry to study statistical models and procedures). At its heart, this paper addresses a rather simple and specialized problem in statistics, which is phrased in a combinatorial language and is just the question of generating (uniform) random samples of the lattice points in a polytope, using a simple Markov chain. The novel part of this paper is that it connected the problem to toric ideals, which opened up a range of computational and theoretical tools to address problems in statistics. Furthermore, the idea that computational algebra and Gröbner bases could be used to address problems in statistics seems to originate in this paper.

The second paper, "On the toric algebra of graphical models", was written jointly with Dan Geiger and Chris Meek [**23**]. I think of this paper as one of the first to begin studying the algebraic geometry of statistical models. The paper studies the class of (undirected, discrete) graphical models, and the equations that vanish on the probability distributions in those models. Another great insight

2020 *Mathematics Subject Classification.* Primary 62R01; Secondary 13P10, 14M25, 62H22.
Key words and phrases. Algebraic statistics, Markov basis, toric ideal, binomial ideal, maximum likelihood degree.

is that the conditional independence structures that are commonly studied along with graphical models can be interpreted as algebraic conditions, specifically as determinantal ideals of certain associated matrices. And finally, this paper is the one that introduced the algebraic study of maximum likelihood estimation. All of these starting points have led to numerous follow-up papers by many authors.

Another key similarity between [**12**] and [**23**] is that, because they were so novel, it took a long time for them to finally get accepted and published in the *Annals of Statistics*. The delay to publication in both cases was so long that in many cases papers that are clearly followups to the main ideas in [**12**] or [**23**] often ended up in print before them!

2. Algebraic algorithms for sampling from conditional distributions

My understanding is that the genesis for the paper "Algebraic algorithms for sampling from conditional distributions" started with a very simple sounding combinatorial question. The project started when Bernd and Persi were in residence at the Mittag-Leffler Institute in Stockholm, Sweden for the 1991–1992 year-long program on combinatorics. Persi had been interested in analyzing ranked choice voting, which leads to a particular combinatorial problem on permutations, and was asking around the combinatorics community visiting the institute if anyone had any ideas on this problem.

Imagine there are k candidates, call the set of them $[k] = \{1, 2, \ldots, k\}$. A ranking of the candidates from best to worst is just a permutation $\sigma \in S_k$. If there are n voters, and each voter gives a ranking, we get a summary of the votes, which is the function $u : S_k \to \mathbb{N}$, where $u(\sigma)$ is the number of voters that chose ranking σ. One might like to know if information from this voter population profile carries some deeper information about the feelings of the population on the particular candidates. So one could try to come up with different models for this permutation data, and analyze how well those models fit the data.

Among the different models for analyzing the permutation data is one that is solely based on the ranking of individual candidates. Specifically, for each candidate $i \in [k]$ and each possible ranking position $j \in [k]$ we introduce a positive real parameter $\theta_{ij} \in \mathbb{R}_{>0}$. Think of θ_{ij} as a sort of propensity measure of candidate i to be ranked in position j. A large value of θ_{ij} means i is very likely to be ranked in position j, whereas as small value of θ_{ij} means that i is unlikely to be ranked in position j. Then, for each $\sigma \in S_n$ we define the probability of observing the ranking σ in this model to be

$$P(\sigma) = \frac{1}{Z} \prod_{i \in [k]} \theta_{i,\sigma(i)}.$$

The quantity Z is a normalizing factor (also called the partition function) that guarantees that $P : S_k \to \mathbb{R}$ is actually a probability distribution on S_k. That is,

$$Z = \sum_{\sigma \in S_k} \prod_{i \in [k]} \theta_{i,\sigma(i)}.$$

This model is a first order model, in that the probability of a ranking only depends on whether candidate i is ranked in position j or not. There is no interaction between the candidates. For example, a second order model might give probability weights to pairs of candidates. Maybe a voter that always rates candidate 1 highly, also rates candidate 2 poorly. So correlation and anti-correlation between those candidates could be modeled in a second order model.

Now, if we want to test whether the first order model fits our data well, we have different statistical approaches. We will now highlight the approach which was explored in [**12**].

Each permutation $\sigma \in S_k$ has a different probability in terms of the parameters in the model. But if we look at a voting summary $u : S_k \to \mathbb{N}$, we see that many different voting summaries have the same probability under this model.

EXAMPLE 2.1. Suppose there are three candidates, and consider the voting summary u that has

$$u(123) = 1, u(132) = 0, u(213) = 0, u(231) = 1, u(312) = 1, u(321) = 0.$$

The probability of this voting summary under the model is

$$\frac{6}{Z^3}\theta_{11}\theta_{22}\theta_{33}\theta_{12}\theta_{23}\theta_{33}\theta_{13}\theta_{21}\theta_{32}.$$

Here we assume that the individual votes are independent of each other, so the probability of a summary is the product of probabilities for each individual voter. The factor of 6 is a multinomial coefficient which allows us to ignore the order of the voters. Note that this is the same probability as for the voting summary v with

$$v(123) = 0, v(132) = 1, v(213) = 1, v(231) = 0, v(312) = 0, v(321) = 1.$$

A basic combinatorial/probabilistic question that arises in this statistical analysis of this data is: How do we generate (all/some/random) voting summaries that have the same probability under the first order model? Since the probability of each individual permutation is just determined by which i is in which position j, we can summarize each individual σ by its permutation matrix $M(\sigma) \in \mathbb{N}^{k\times k}$, where

$$M(\sigma)_{ij} = \begin{cases} 1 & \text{if } \sigma(i) = j \\ 0 & \text{otherwise.} \end{cases}$$

Then the probability of observing the data u is just determined by the *first order summary* matrix

$$M(u) = \sum_{\sigma\in S_k} u(\sigma)M(\sigma).$$

So, in other words, our basic problem is to determine how to generate voting summaries v such that $M(u) = M(v)$. The generation of random voting summaries with fixed first order summary is an important step in Monte Carlo algorithms for performing Fisher's exact test. This allows us to test the hypothesis that the data was generated from the model by seeing how the given data table compares to other random tables that have the same probability under the model. One of the key advantages of this method is that it does not depend on knowing or estimating the parameters θ that may have produced the data.

The map $M(\cdot)$ is a linear transformation that takes $u \in \mathbb{N}^{k!}$ and produces a first order summary matrix $M(u) \in \mathbb{N}^{k\times k}$. Thinking about the permutation matrices $M(\sigma)$ as vectors of length k^2, the matrix representing the linear transformation M is a $k^2 \times k!$ matrix, whose columns are just all permutation matrices, considered as vectors. For instance, for $k = 3$, we have

$$M = \begin{pmatrix} 1 & 1 & 0 & 0 & 0 & 0 \\ 0 & 0 & 1 & 1 & 0 & 0 \\ 0 & 0 & 0 & 0 & 1 & 1 \\ 0 & 0 & 1 & 0 & 1 & 0 \\ 1 & 0 & 0 & 0 & 0 & 1 \\ 0 & 1 & 0 & 1 & 0 & 0 \\ 0 & 0 & 0 & 1 & 0 & 1 \\ 0 & 1 & 0 & 0 & 1 & 0 \\ 1 & 0 & 1 & 0 & 0 & 0 \end{pmatrix} \tag{2.1}$$

where the columns correspond to the permutations $\{123, 132, 213, 231, 312, 321\}$, in that order. The set of all voting summaries with a given first order summary is the set of points

$$\mathcal{F}(u) = \{v \in \mathbb{N}^{k!} : M(v) = M(u)\}.$$

The strategy that Persi was interested in employing to generate a random v with the same first order summary as u was based on finding a random walk over the set $\mathcal{F}(u)$. Specifically, suppose that $\mathcal{B} \subseteq \ker_{\mathbb{Z}} M$ is a finite set of integral elements in the kernel of M. If we take any element $b \in \mathcal{B}$, and add it to $v \in \mathcal{F}$, the new summary $v + b \in \mathcal{F}(u)$ since $M(v + b) = M(v) + M(b) = M(v)$ where $b \in \ker_{\mathbb{Z}} M$. Well, $v + b \in \mathcal{F}(u)$ *provided* that $v + b \in \mathbb{N}^{k!}$, that is, we need all entries to be nonnegative. This observation leads to a Markov chain on $\mathcal{F}(u)$ as follows:

Starting from $v \in \mathcal{F}(u)$, we choose a random $b \in \mathcal{B}$. If $v + b \in \mathbb{N}^{k!}$, then we move to $v + b$, otherwise stay at v. Keep doing this process for a long enough time, and this random walk will converge to the uniform distribution on $\mathcal{F}(u)$: *provided* that the walk is actually guaranteed to reach everywhere in $\mathcal{F}(u)$. That leads to the main definition from [**12**].

DEFINITION 2.2. Let $A \in \mathbb{N}^{d \times r}$. A *Markov basis* for A is a finite subset $\mathcal{B} \subseteq \ker_{\mathbb{Z}}(A)$ such that for all $u, v \in \mathbb{N}^r$, with $Au = Av$, there is a sequence of elements $b_1, b_2, \ldots, b_m \in \mathcal{B}$ such

(1) $v = u + \sum_{i=1}^{m} b_i$, and
(2) for all $j = 1, \ldots, m$, $u + \sum_{i=1}^{j} b_i \in \mathbb{N}^r$.

Once one has a Markov basis, then each fiber

$$\mathcal{F}(u) = \{v \in \mathbb{N}^r : Au = Av\}$$

is connected by the elements of the Markov basis $\mathcal{B}$. So with a Markov basis one is guaranteed that the Markov chain on each fiber is connected and will converge to the uniform distribution over the whole fiber.[1] So Persi's original question specifically wanted to know the Markov basis for the $k^2 \times k!$ matrices corresponding to the voter problem above, so that statistical tests could be performed for some specific voter data sets.

I think we are probably lucky that Bernd did not just come up with a nice simple description of the Markov basis for the voting problem. In fact, it was not until 2013 when Yamaguchi, Ogawa, and Takemura gave a complete description of the Markov basis for these matrices [**44**], resolving a refined conjecture of Diaconis

[1] Modifications to the Markov chain can produce arbitrary distributions over the fiber $\mathcal{F}(u)$ using Metropolis weights.

and Eriksson [**11**]. Instead, what emerged from the collaboration between Bernd and Persi that started this project was a connection between the idea of a Markov basis and objects of commutative algebra called *toric ideals* which we explain now.

A toric ideal is a certain ideal in a polynomial ring, that is also associated to the matrix $A \in \mathbb{N}^{d\times r}$. In particular, consider the polynomial ring $\mathbb{C}[p] := \mathbb{C}[p_1, p_2, \dots, p_r]$. For a nonnegative integer vector u let $p^u = p_1^{u_1} \cdots p_r^{u_r}$.

DEFINITION 2.3. Let $A \in \mathbb{N}^{d\times r}$. The *toric ideal* associated to A is the ideal

$$I_A = \langle p^u - p^v : u, v \in \mathbb{N}^r \text{ and } Au = Av\rangle.$$

While Definition 2.3 is certainly compact, it does not tell the whole story about where these ideals come from, and why they are called toric, nor how exactly one would think that the toric ideal is related to Markov bases. To get a better sense of the interconnection of ideas, consider the following polynomial map associated to the matrix A:

$$\phi^A : \mathbb{C}^d \to \mathbb{C}^r, \quad (\theta_1, \dots, \theta_d) \mapsto \left(\prod_{i=1}^d \theta_i^{a_{i1}}, \cdots, \prod_{i=1}^d \theta_i^{a_{ir}}\right).$$

The Zariski closure of the image of ϕ^A is called the *toric variety* V_A. This variety is "toric" because there is a natural action of the algebraic torus $(\mathbb{C}^*)^d$ on it. In particular, if $\theta \in (\mathbb{C}^*)^d$, and $y \in V_A$, then $\phi^A(\theta) \cdot y \in V_A$ as well (where $\phi^A(\theta) \cdot y$ denotes the entry-wise product of the vector $\phi^A(\theta)$ with the vector y). Furthermore, the image of the algebraic torus is a Zariski dense subset of V_A.

The toric ideal I_A is the vanishing ideal of V_A, that is

$$I_A = \{f \in \mathbb{C}[p] : f(y) = 0 \text{ for all } y \in V_A\}.$$

We can get a glimpse of why this must be the case from the definition of I_A and the fact that ϕ^A has coordinate functions that are monomials in θ. Any relation on those monomials must be of the form that two monomials p^u and p^v evaluate to the same expression in terms of the θ, when we make the substitution $p = \phi_A(\theta)$. But when we make that substitution on a single monomial p^u, we get the monomial θ^{Au}.

Toric varieties are related to statistics because if we take the toric variety V_A, and intersect with the probability simplex

$$\Delta_{r-1} := \{(p_1, \dots, p_r) \in \mathbb{R}^r : \sum_i p_i = 1, p_i \geq 0 \text{ for all } i\}$$

we get a statistical model

$$\mathcal{M}^A = V_A \cap \Delta_{r-1}$$

called a *log-linear model* or a *discrete exponential family.* They are among the most important basic models in statistics.

So the first order model for voting data is just a discrete exponential family: we associate a $k^2 \times k!$ matrix, who columns are just the $k!$ permutation matrices of size $k \times k$. For $k = 3$, the toric ideal for this model is in a polynomial ring with $3! = 6$ indeterminates, indexed by the permutations:

$$\mathbb{C}[p] = \mathbb{C}[p_{123}, p_{132}, p_{213}, p_{231}, p_{312}, p_{321}].$$

It is a principal ideal

$$I_A = \langle p_{123}p_{231}p_{312} - p_{132}p_{213}p_{321}\rangle.$$

That makes sense because the kernel of the matrix M in (2.1) is one dimensional, equal to the span of the vector $(1, -1, -1, 1, 1, -1)$, which naturally corresponds to that cubic binomial.

From this one example, it might be tempting to think that a minimal generating set of a toric ideal I_A always corresponds to a basis for the lattice $\ker_{\mathbb{Z}} A$, but this is usually not the case. For example, for the first order model with $k = 4$ candidates, the toric ideal has minimal generating set consisting of 18 binomials of degree two and 160 binomials of degree 3. Whereas $\ker_{\mathbb{Z}} A$ in this case is a lattice of dimension 14. So many, many more polynomials are needed to generate the toric ideal than just those corresponding to a generating set of the lattice.

What do these toric ideals have to do with the statistical problem and Markov bases? That was provided by the main theorem of [**12**]. To state this, we need one more piece of notation. If $b \in \mathbb{Z}^r$ is a vector, we can write $b = b^+ - b^-$ where b^+ and b^- are nonnegative vectors and so that for all i, $b_i^+ b_i^- = 0$. That is, the expression $b = b^+ - b^-$ involves no cancellation.

THEOREM 2.4 (Fundamental theorem of Markov bases [**12**]). *Let $A \in \mathbb{N}^{d\times r}$, and $\mathcal{B} \subseteq \ker_{\mathbb{Z}} A$ be a finite set. Then $\mathcal{B}$ is a Markov basis for A if and only if $\{p^{b^+} - p^{b^-} : b \in \mathcal{B}\}$ is a generating set of the toric ideal I_A.*

So the fundamental theorem of Markov bases allows us to find these special sets of vectors for running Markov chains on fibers just by computing a generating set of a toric ideal. That allows for tools from computational algebra to be employed to solve statistical problems. I think the first use of the expression "fundamental theorem of Markov bases" comes in [**16**], when Mathias Drton, Bernd, and I were writing up our Oberwolfach lecture notes. Maybe it should be called the Diaconis–Sturmfels theorem instead, but I just like the word "fundamental", I guess. I even heard at least one person call it the "fundamental theorem of algebraic statistics" before.

Of course, Bernd's discovery of the connection of Persi's Markov basis problem to the generating sets of toric ideals didn't come out of the blue, and fit in with other problems that Bernd had been working on in relation to toric ideals, the study of toric varieties, and their connections to problems in optimization. While the fundamental theorem of Markov bases turns each fiber into a connected graph using the generating set of the toric ideal, Gröbner bases take that a step further. Using the Groöbner basis produces a directed graph on the fiber, whose unique sink is the optimal solution of an associated integer program (originally discovered by Conti and Traverso [**7**]).

Persi and Bernd's paper opened up a wealth of applications of Markov bases, and the paper itself studies a number of interesting examples which have been studied by many authors. Key examples include Markov bases for contingency tables and the related undirected graphical models (mentioned briefly in Section 4), logistic regressions, and hierarchical models. The most comprehensive summary is the book [**3**], which gives a survey of Markov basis results up to that time. A more recent survey is [**1**].

3. Topics around Markov bases

The quest to study Markov bases led to projects in a number of other areas that are of interest and value in their own right both for algebraic statistics and as areas of pure mathematics. In this section, I highlight a couple of these topics which emerged from the algebraic study of Markov bases.

As we have seen in the fundamental theorem of Markov bases, a set of moves $\mathcal{B} \subseteq \ker_{\mathbb{Z}} A$ yields connected fibers for all u, if and only if this set of moves corresponds to a generating set of the toric ideal. However, there might be situations where it is too difficult to determine the entire generating set of I_A, or there might be a natural subset of $\ker_{\mathbb{Z}} A$ that we wish to use to run a Markov chain. In these situations, even though our set of moves does not connect every fiber, one hopes that it connects enough of the fiber to be useful in practice for applications. This leads to the question of how to determine how well connected those fibers are by a given set of moves. A simple generalization of the fundamental theorem of Markov bases is the following.

PROPOSITION 3.1. *Let $A \in \mathbb{N}^{d\times r}$, $\mathcal{B} \subseteq \ker_{\mathbb{Z}} A$ be a finite set,*

$$J_{\mathcal{B}} = \langle p^{b^+} - p^{b^-} : b \in \mathcal{B}\rangle,$$

and $u, v \in \mathbb{N}^r$ such that $Au = Av$. Then u and v can be connected using the set of moves $\mathcal{B}$ if and only if

$$p^u - p^v \in J_{\mathcal{B}}.$$

While Proposition 3.1 is a straightforward generalization of Theorem 2.4, what is not at all clear is how to check when $p^u - p^v \in J_{\mathcal{B}}$. What Bernd, Persi, and David Eisenbud brought to this problem [**10**] was to connect it to the theory of binomial ideals that Bernd and David had been developing in the seminal paper [**18**]. A binomial is a polynomial of the form $p^u - cp^v$, where $c \in \mathbb{K}$. A binomial ideal is an ideal with a generating set of binomials. One of the most significant results about binomial ideals is the following.

THEOREM 3.2 ([**18**]). *Let $\mathbb{K}$ be an algebraically closed field and $I \subseteq \mathbb{K}[p]$ be a binomial ideal. Then I has a primary decomposition*

$$I = I_1 \cap I_2 \cap \cdots \cap I_k$$

where all the ideals $I_1, \ldots, I_k$ are binomial ideals, and all the associated primes of I are also binomial ideals.

The hope of [**10**] was that from the primary decomposition of $J_{\mathcal{B}}$, it might be more easily possible to tell whether or not $p^u - p^v \in J_{\mathcal{B}}$. Indeed, the primary ideals that arise in the decomposition are often easier to understand the containment therein, and if

$$J_{\mathcal{B}} = I_1 \cap I_2 \cap \cdots \cap I_k$$

then $p^u - p^v \in J_{\mathcal{B}}$ if and only if $p^u - p^v \in I_i$ for all $i \in [k]$.

Binomial primary decomposition ended up receiving a lot of attention from different authors including former students and many collaborators of Bernd. Some key papers include [**13, 19, 26, 30**].

Another direction that arose from the study of Markov bases is the notion of ideals in infinite dimensional polynomial rings that are finitely generated up to symmetry. While this area, and associated works of representation stability are now quite far from algebraic statistics, a number of the original motivating problems in the area started with some problems there, specifically working with Markov bases. To put this in context, and see Bernd's contributions to the area, we must highlight a surprising result that first materialized in the study of Markov bases of contingency table problems. To explain this, consider the following linear map

$$\pi^{(r_1,r_2,r_3)} : \mathbb{C}^{r_1 \times r_2 \times r_3} \to \mathbb{C}^{r_1 \times r_2} \oplus \mathbb{C}^{r_1 \times r_3} \oplus \mathbb{C}^{r_2 \times r_3},$$

$$u \mapsto (u|_{12}, u|_{13}, u|_{23}).$$

In this representation $u|_{12} \in \mathbb{C}^{r_1 \times r_2}$ denotes the marginal of u, obtained by summing out over the third index. If we let $u(i,j,k)$ denote the (i,j,k) entry of u, and $u|_{12}(i,j)$ the (i,j) entry of $u|_{12}$, then we can calculate

$$u|_{12}(i,j) = \sum_{k=1}^{r_3} u(i,j,k)$$

with similar formulas for $u|_{13}$ and $u|_{23}$. Formally, as an operation on tensors we could write the map on basis vectors as

$$a_i \otimes b_j \otimes c_k \mapsto (a_i \otimes b_j) \oplus (a_i \otimes c_k) \oplus (b_j \otimes c_k).$$

The map $\pi^{(r_1,r_2,r_3)}$ that computes 2-way marginals of a 3-way table u is a linear map, and so it should have an associated Markov basis. But this resulting Markov basis is one that varies with the dimensions r_1, r_2, r_3. Bernd and Persi had calculated a Markov basis for this problem in [**12**], for $3 \times 3 \times 3$ tables, and found that minimal generating set consisted of binomials of degrees 4 and 6. Satoshi Aoki and Akimichi Takemura began a project to understand what these Markov bases were like for larger values of r_1, r_2, and r_3. Their first big breakthrough was the paper [**4**] which proved the following striking result.

THEOREM 3.3. *For any r_3, the minimal Markov basis for $\pi^{(3,3,r_3)}$ consists of binomials of degree ≤ 10.*

If we only record where there are nonzero entries in the elements of our Markov bases of $\pi^{(r_1,r_2,r_3)}$, we can speak of the "format" of a move, which would be the smallest $s_1 \times s_2 \times s_3$ box that all the nonzero entries could fit. The result of Aoki and Takemura also found that the largest format that appears for any r_3 is $3 \times 3 \times 5$.

Of course, when that result was first discovered it was so striking, there was kind of a rush to figure out if this was just a special case, or if it would generalize to other 3-way tables, or to other settings beside this. This was all in the spring of 2002, I was just finishing my master's degree with Serkan Hoşten and had committed to working with Bernd at UC Berkeley to start in Fall 2002. That summer, Bernd went to Europe to attend conferences, and I was attending a conference at the University of Western Ontario.[2] There I was at my first conference, and Lalo González Vega introduces himself to me and hands me a photocopy of a handwritten sheet in Bernd's beautiful handwriting that has the outline of a theory that explains Theorem 3.3. That handwritten document had the beginnings of the development of the theory of higher Lawrence configurations of Bernd and Paco Santos [**40**]. Higher Lawrence configurations end up being a special case of the matrix structures

[2]Does anyone else remember that the sign outside the university said "Strategic Computational Algebra" instead of "Symbolic Computational Algebra"?

that appear in n-fold integer programming [**9**]. In the context of Markov bases of 3-way contingency tables, their result implies the following.

THEOREM 3.4. *For any fixed r_1, r_2, there is a constant $C(r_1, r_2)$ such that for all r_3, the degree of the binomials in the Markov basis of for $\pi^{(r_1,r_2,r_3)}$ consists of binomials of degree $\leq C(r_1, r_2)$.*

For general multiway tables, Bernd and Paco's main theorem implies an analogous finiteness result for logit complexes. That result was subsequently generalized (using the same proof ideas) by Serkan Hoşten and myself to cover the analogous result for all contingency table Markov basis problems in the situation where all but one table dimension is fixed and the last dimension is allowed to vary [**28**].

Theorem 3.4 was striking especially in comparison to the result of Jesús De Loera and Shmuel Onn, who had recently shown that Markov basis problems for 3-way contingency tables can be arbitrarily complicated, and that that complexity holds even for $3 \times r_2 \times r_3$ tables [**8**]. In other words, we have a dichotomy between letting one table dimension tend to infinity, versus letting two table dimensions tend to infinity. In the first case, we get finite complexity, in the second case we get arbitrarily bad complexity.

One key observation in these problems is that the Markov basis problems for tables are invariant under permuting the levels. That is, swapping slices in a table does not affect the Markov basis problem at all. Converting the Markov basis to a problem about generating sets of ideals via the fundamental theorem of Markov bases, we have for each (r_1, r_2, r_3), a polynomial ring

$$R_{(r_1,r_2,r_3)} = \mathbb{C}[p_{i_1,i_2,i_3} : i_1 \in [r_1], i_2 \in [r_2], i_3 \in [r_3]]$$

and an ideal $I_{(r_1,r_2,r_3)} \subseteq R_{(r_1,r_2,r_3)}$. Each of these ideals is invariant under the action of the group $S_{r_1} \times S_{r_2} \times S_{r_3}$. Furthermore, the whole family of ideals are nested as

$$I_{(r_1,r_2,r_3)} \subseteq I_{(s_1,s_2,s_3)} \text{ if } r_1 \leq s_1, r_2 \leq s_2, r_3 \leq s_3.$$

One can of course ask: "How special are the results of Theorem 3.4? Do they only apply for Markov bases, or is there a general phenomenon at play?" Thus the study of ascending chains of ideals (or varieties) invariant of symmetry groups was born. This work has been greatly extended by Jan Draisma and collaborators to problems in algebraic statistics [**14, 15, 24**]. But, in my mind, this first paper of Bernd and Paco really set the stage for this whole area of research. In addition, it is one of the only papers in the area that actually has bounds on stabilization. While many of the papers in the area prove a theoretical result (based on Noetherianity) that stabilization must occur *at some point* in a chain of symmetric ideals, Bernd and Paco's paper is one of the only ones that actually gives any sort of bound in terms of the input data of the problem for when stabilization is guaranteed to occur. The argument is a model of simplicity, which I hope that other projects in that area could strive to emulate.

4. On the toric algebra of graphical models

The first paper of Bernd and Persi is the one that most people credit with starting the field of algebraic statistics, and many other research projects were started that originated from ideas in that paper. A notable project was the eventual book of Giovanni Pistone, Eva Riccomagno, and Henry Wynn that used tools from computational algebra to study design of experiments and coined the term *algebraic statistics* [**38**].

But, I would say that the field of algebraic statistics really took off with the paper "On the toric algebra of graphical models" by Dan Geiger, Chris Meek, and Bernd. It finally appeared in 2006, but the work for this project was laid by Bernd's visit to Microsoft Research, I believe sometime in 2001. I first heard about this project in 2001, at any rate. I was just starting research with Serkan Hoşten on my master's degree at San Francisco State University at that time, and we were looking for different projects. I remember that in Fall 2001 and Bernd, Serkan, and I went to UC Davis for a talk that Bernd gave on this paper. Serkan and I started working on related projects, which lead to our first paper together [**27**], based on hearing this talk. This was one of these classic Bernd events where, after giving the talk, interested parties went to another room where people could take turns presenting different problems to work on. Eventually, a larger group returned to Davis in Spring 2002 to talk about these ideas further in the early growth of the field of algebraic statistics.

I think that "On the toric algebra of graphical models" was so important in the development of algebraic statistics because it really started the idea of studying the algebraic geometry of the statistical model itself, in addition to pursuing applications of the specific equations. This paper concerns the class of undirected graphical models, but there are graphical models for all kinds of other graphs, including those with directed edges, bidirected edges, and mixtures of these [**32**].

Let $G = (V, E)$ be an undirected graph. For each $v \in V$, we have a random variable X_v. For [**23**], those random variables are assumed to be discrete random variables with finitely many states $X_v \in [r_v]$, though the theory of graphical models makes sense without those assumptions. We will assume finitely many states in this paper.

Recall that a clique in a graph G is a subgraph of G that is complete. Let $\mathcal{C}(G)$ denote the set of maximal cliques in G. To each clique C in the graph, we associate a potential function $\psi_C(x_C)$, which is a nonnegative function on $\prod_{c \in C}[r_c]$. Note that $x_C = (x_c : c \in C)$ denotes the vector which is indexed just by the variables in C. We define a joint probability density function of the random vector $X = (X_v : v \in V)$ by the product formula:

$$f(x) = \frac{1}{Z} \prod_{C \in \mathcal{C}(G)} \psi_C(x_C). \tag{4.1}$$

As in Section 2, Z denotes a normalizing constant to make this a probability distribution. Specifically,

$$Z = \sum_{x \in \prod_{v \in V}[r_v]} \prod_{C \in \mathcal{C}(G)} \psi_C(x_C).$$

For more general random variables, Z is obtained by integrating out over the state space. The discrete undirected graphical model consists of all probability distributions that arise with a fixed graph G, and all choices of possible potential functions.

While (4.1) looks very probabilistic in nature, we can see that it is actually giving a monomial parametrization, and hence is parametrizing a toric variety.

EXAMPLE 4.1. Let G be the 4-cycle graph with $V = [4]$ and $E = \{12, 23, 34, 14\}$. The parametrization of the undirected graphical model is

$$f(x_1, x_2, x_3, x_4) = \frac{1}{Z}\psi_{12}(x_1, x_2)\psi_{23}(x_2, x_3)\psi_{34}(x_3, x_4)\psi_{14}(x_1, x_4).$$

This does not look very algebraic, but as algebraists we would write this in an equivalent representation as a parametrization of a projective toric variety:

$$p_{i_1,i_2,i_3,i_4} = a_{i_1,i_2}b_{i_2,i_3}c_{i_3,i_4}d_{i_1,i_4}$$

for $i_1 \in [r_1], i_2 \in [r_2], i_3 \in [r_3]$, and $i_4 \in [r_4]$. The vanishing ideal of this model, I_G, is a toric ideal in the polynomial ring

$$\mathbb{C}[p] := \mathbb{C}[p_{i_1,i_2,i_3,i_4} : i_1 \in [r_1], i_2 \in [r_2], i_3 \in [r_3], i_4 \in [r_4]].$$

For example, if $r_1 = r_2 = r_3 = r_4 = 2$, then the polynomial ring $\mathbb{C}[p]$ has 16 indeterminates. The vanishing ideal I_G has 8 minimal generators of degree two, for instance:

$$p_{1111}p_{1212} - p_{1112}p_{1211}$$

and 8 minimal generators of degree 4, for instance:

$$p_{1111}p_{1222}p_{2112}p_{2221} - p_{1112}p_{1221}p_{2111}p_{2222}.$$

So, on the one hand, [**23**] is another paper about toric ideals, and this all fits in the framework started in [**12**]. In fact, sampling problems for contingency tables do make an appearance in [**12**], and hence the corresponding toric ideals were already interesting there. But the real contribution of [**23**] is the study of the model structure of these varieties.

In fact, experts on graphical models had already been studying the equations that vanish on the graphical models through the concept of conditional independence. One goal of the paper [**23**] was to understand these conditional independence structures in an algebraic context.[3] Let X, Y, Z be random variables with a joint density function $f(x, y, z)$. We say that X and Y are independent given Z if

$$f_{X,Y|Z}(x, y|z) = f_{X|Z}(x|z)f_{Y|Z}(y|z)$$

for all x, y, z. An equivalent representation is that:

$$f(x_1, y_1, z)f(x_2, y_2, z) = f(x_1, y_2, z)f(x_2, y_1, z)$$

for all x_1, x_2, y_1, y_2, z. For discrete random variables, this translates into quadratic constraints on the joint probability distribution. Indeed, this turns out to be equivalent to the degree 2 (determinantal) equations:

$$p_{i_1j_1k}p_{i_2j_2k} - p_{i_1j_2k}p_{i_2j_1k} = 0$$

for all i_1, i_2, j_1, j_2, k.

These constraints arise in the study of graphical models because to every undirected graph G, we get a collection of conditional independence statements that every probability distribution associated to the graph must satisfy. Specifically, we have the *global Markov statements* associated to the graph G, which are described as follows.

For every triple of disjoint subsets A, B, C of $[n]$, such that C separates A and B in G, we get a conditional independence statement, $X_A \perp\!\!\!\perp X_B|X_C$ (or just

[3] A completely different algebraic perspective on conditional independence comes in the work of Studený [**41**].

$A \perp\!\!\!\perp B|C$ for short). Each conditional independence statement gives a collection of quadratic polynomials in the probabilities p_i. A probability distribution which satisfies those conditional independence statements must satisfy the corresponding quadratic polynomials. We can collect these all together, for all the separating triples A, B, C on the graph G, and this gives an ideal which we denote $I_{\mathrm{global}(G)}$, the *ideal of global Markov statements of* G. Since every probability distribution coming from the graphical model satisfies those global Markov statements, we always have that

$$I_{\mathrm{global}(G)} \subseteq I_G$$

where I_G denotes the toric ideal that is the vanishing ideal of the undirected graphical model. One of the big theorems from [**23**] concerns when equality occurs.

THEOREM 4.2. *For an undirected graph* G*, the toric ideal* I_G *equals the global Markov ideal* $I_{\mathrm{global}(G)}$ *if and only if* G *is a chordal graph.*

For experts on graphical models, this perhaps does not come as a surprise because chordal graphs are well known to be the nicest family of graphs as far as graphical models are concerned. But this was one of the first characterization theorems in algebraic statistics, classifying when a certain family of models satisfies some nice property. There are many different variations on graphical models both in types of graphs that can arise and the types of random variables that are studied, and, perhaps surprisingly, only in this first studied case of discrete random variables with undirected graphs is there a complete characterization. Bernd, Luis Garcia-Puente, and Mike Stillman studied the problem for directed graphical models with discrete random variables but not much progress had been made since then [**22**]. The case of Gaussian graphical models is also very interesting, for both directed and undirected graphs. Bernd and Caroline Uhler studied this problem for undirected graphs [**42**] and I had initiated the study for directed graphical models [**43**] but the problem of classifying graphs such that the global Markov ideal equals the vanishing ideal of the model remains open in all of these cases. See [**34**] for recent progress.

Of course, Theorem 4.2 is not the end of the story when it comes to the conditional independence statements. In fact, graphical modelers already knew an important theorem: The Hammersley–Clifford theorem shows that conditional independence statements implied by an undirected graph give exactly the positive probability distributions that come from the parametrization. A natural question to ask about in that context then is: what other probability distributions satisfy the global Markov statements but do not come from the model? This question was again naturally addressed by the tool of binomial primary decomposition, since $I_{\mathrm{global}(G)}$ is a binomial ideal. Indeed, among the primary components of $I_{\mathrm{global}(G)}$ is the ideal I_G, but many other ideals that also reveal useful structural information about the distributions that satisfy the global Markov statements but don't come from the model. See, e.g., [**31**] for further developments. Other exploration of the primary decomposition of conditional independence ideals appears in [**20, 21**].

A second question addressed in "On the toric algebra of graphical models" concerns the algebraic complexity of maximum likelihood estimation. Graphical modelers had known that if the underlying graph is a chordal graph, then there are simple closed formulas for the maximum likelihood estimate in terms of the data. Dan and Chris were interested in knowing if those were the only graphs where there were such nice formulas. That question lead to one of the most innovative ideas that came out of this paper. In fact, it was "On the toric algebra of graphical models" that introduced the notion of maximum likelihood degree, which turns out to have many deep and interesting connections to other problems in algebraic geometry.

To state how the maximum likelihood degree works we first consider a general discrete statistical model. Let $\mathcal{M} \subseteq \Delta_{r-1}$ be a statistical model. So a probability distribution $p \in \mathcal{M}$ has $p_i = \operatorname{Prob}(X = i)$ for a random variable X which has r states. Independent identically distributed data consists of a sequence of trials from the underlying random variable X, so we get a string $i_1, i_2, \ldots, i_n$ of different outcomes where $i_k \in [r]$ for all k. Maximum likelihood estimation tries to find an estimate for an underlying distribution p in the model $\mathcal{M}$. The probability of observing the data string is

$$\prod_{k=1}^{n} p_{i_k}.$$

We can summarize this probability by using the vector of counts $u \in \mathbb{N}^r$ defined by $u_i = \#\{j : i_j = i\}$. Often u is just called the data vector in algebraic papers on maximum likelihood. Then, in terms of the vector of counts u, the likelihood of observing the sequence $i_1, i_2, \ldots, i_n$ is

$$p^u := \prod_{i=1}^{r} p_i^{u_i}.$$

In maximum likelihood estimation, we think of this monomial with u fixed and the $p \in \mathcal{M}$ as a variable and we want to solve the optimization problem

$$\max_{p \in \mathcal{M}} p^u.$$

The value of $p \in \mathcal{M}$ that maximizes the likelihood function is called the *maximum likelihood estimate* and is denoted $\hat{p}$.

Maximum likelihood estimation is an algebraic optimization problem. At first glance, it looks like the degree of the function being optimized changes with the data vector u, in fact, upon taking the logarithm of the likelihood function, we end up with the log-likelihood function

$$\ell_u(p) = \sum_{i=1}^{r} u_i \log p_i.$$

If the model $\mathcal{M}$ is an algebraic statistical model, then the critical equations of the optimization problem

$$\max_{p \in \mathcal{M}} \ell_u(p)$$

are rational functions of p, and depend linearly on u. It is not hard to see that for generic data u, the number of complex critical points to this system is a fixed number. This number is the *maximum likelihood degree* of the model and denoted $\operatorname{mld}(\mathcal{M})$. General principles from algebraic geometry say that this ml-degree will also be the algebraic degree of the function that maps the data u to the maximum likelihood estimate $\hat{p}$.

In the context of maximum likelihood estimation, it was known that chordal graphical models have maximum likelihood estimates that can be written down as a rational formula in terms of the data. And so the question was: Are these the only graphs with that property? Said in terms of this new concept of maximum likelihood degree, chordal graphical models have ml-degree 1. The second major theorem from [**23**] is the following.

THEOREM 4.3. *The ml-degree of an undirected discrete graphical model is equal to one if and only if the graph is a chordal graph.*

The concept of the ml-degree was officially named in the papers [**6, 25**]. The study of models with ml-degree 1 is quite interesting, and Huh [**29**] showed that the algebraic varieties with ml-degree equal to 1 are precisely the reduced A-discriminants, making a surprising connection between problems of algebraic statistics and interesting areas of algebraic geometry. See [**17**] for recent work. Of course, it was Bernd who played "matchmaker" and helped connect Huh's work in tropical geometry and intersection theory to this problem in algebraic statistics.

5. Conclusions

These two papers [**12, 23**], while playing an outsize role in the development of algebraic statistics, and in my own personal research development, represent only a small portion of Bernd's contribution to the field. Bernd wrote two books on the subject [**16, 37**], and ideas from algebraic statistics play into other other areas of his research as well. Among the contributions worth further note are the connections between tropical geometry and MAP estimation in sequence alignment problems [**36**], applications of discrete geometry to log-concave density estimation [**39**], tropical geometry and its connection to phylogenetics [**33**], algebraic geometry for the method of moments [**2**], the connection between rank testing and conditional independence structures [**35**], and the combinatorial structure of epistasis [**5**], among many other projects. What is clear from all these projects is how Bernd has been able to make unexpected connections between applied research areas and tools from algebra and combinatorics, and the large number of research collaborators who he draws into the field.

References

[1] F. Almendra-Hernández, J. A. De Loera, and S. Petrović, *Markov bases: A 25 year update*, J. Amer. Statist. Assoc. **119** (2024), no. 546, 1671–1686, DOI 10.1080/01621459.2024.2310181. MR4766018

[2] C. Améndola, J.-C. Faugère, and B. Sturmfels, *Moment varieties of Gaussian mixtures*, J. Algebr. Stat. **7** (2016), no. 1, 14–28, DOI 10.18409/jas.v7i1.42. MR3529332

[3] S. Aoki, H. Hara, and A. Takemura, *Markov bases in algebraic statistics*, Springer Series in Statistics, Springer, New York, 2012, DOI 10.1007/978-1-4614-3719-2. MR2961912

[4] S. Aoki and A. Takemura, *Minimal basis for a connected Markov chain over* $3 \times 3 \times K$ *contingency tables with fixed two-dimensional marginals*, Aust. N. Z. J. Stat. **45** (2003), no. 2, 229–249, DOI 10.1111/1467-842X.00278. MR1983834

[5] N. Beerenwinkel, L. Pachter, and B. Sturmfels, *Epistasis and shapes of fitness landscapes*, Statist. Sinica **17** (2007), no. 4, 1317–1342. MR2398598

[6] F. Catanese, S. Hoşten, A. Khetan, and B. Sturmfels, *The maximum likelihood degree*, Amer. J. Math. **128** (2006), no. 3, 671–697. MR2230921

[7] P. Conti and C. Traverso, *Buchberger algorithm and integer programming*, Applied algebra, algebraic algorithms and error-correcting codes (New Orleans, LA, 1991), Lecture Notes in Comput. Sci., vol. 539, Springer, Berlin, 1991, pp. 130–139, DOI 10.1007/3-540-54522-0_102. MR1229314

[8] J. De Loera and S. Onn, *The complexity of three-way statistical tables*, SIAM J. Comput. **33** (2004), no. 4, 819–836, DOI 10.1137/S0097539702403803. MR2065334

[9] J. A. De Loera, R. Hemmecke, S. Onn, and R. Weismantel, *n-fold integer programming*, Discrete Optim. **5** (2008), no. 2, 231–241, DOI 10.1016/j.disopt.2006.06.006. MR2408419

[10] P. Diaconis, D. Eisenbud, and B. Sturmfels, *Lattice walks and primary decomposition*, Mathematical essays in honor of Gian-Carlo Rota (Cambridge, MA, 1996), Progr. Math., vol. 161, Birkhäuser Boston, Boston, MA, 1998, pp. 173–193. MR1627343

[11] P. Diaconis and N. Eriksson, *Markov bases for noncommutative Fourier analysis of ranked data*, J. Symbolic Comput. **41** (2006), no. 2, 182–195, DOI 10.1016/j.jsc.2005.04.009. MR2197154

[12] P. Diaconis and B. Sturmfels, *Algebraic algorithms for sampling from conditional distributions*, Ann. Statist. **26** (1998), no. 1, 363–397, DOI 10.1214/aos/1030563990. MR1608156

[13] A. Dickenstein, L. F. Matusevich, and E. Miller, *Combinatorics of binomial primary decomposition*, Math. Z. **264** (2010), no. 4, 745–763, DOI 10.1007/s00209-009-0487-x. MR2593293

[14] J. Draisma, *Noetherianity up to symmetry*, Combinatorial algebraic geometry, Lecture Notes in Math., vol. 2108, Springer, Cham, 2014, pp. 33–61, DOI 10.1007/978-3-319-04870-3_2. MR3329086

[15] J. Draisma and J. Kuttler, *Bounded-rank tensors are defined in bounded degree*, Duke Math. J. **163** (2014), no. 1, 35–63, DOI 10.1215/00127094-2405170. MR3161311

[16] M. Drton, B. Sturmfels, and S. Sullivant, *Lectures on algebraic statistics*, Oberwolfach Seminars, vol. 39, Birkhäuser Verlag, Basel, 2009, DOI 10.1007/978-3-7643-8905-5. MR2723140

[17] E. Duarte, O. Marigliano, and B. Sturmfels, *Discrete statistical models with rational maximum likelihood estimator*, Bernoulli **27** (2021), no. 1, 135–154, DOI 10.3150/20-BEJ1231. MR4177364

[18] D. Eisenbud and B. Sturmfels, *Binomial ideals*, Duke Math. J. **84** (1996), no. 1, 1–45, DOI 10.1215/S0012-7094-96-08401-X. MR1394747

[19] Z. S. Eser and L. F. Matusevich, *Decompositions of cellular binomial ideals*, J. Lond. Math. Soc. (2) **94** (2016), no. 2, 409–426, DOI 10.1112/jlms/jdw012. MR3556446

[20] A. Fink, *The binomial ideal of the intersection axiom for conditional probabilities*, J. Algebraic Combin. **33** (2011), no. 3, 455–463, DOI 10.1007/s10801-010-0253-5. MR2772542

[21] A. Fink, J. Rajchgot, and S. Sullivant, *Matrix Schubert varieties and Gaussian conditional independence models*, J. Algebraic Combin. **44** (2016), no. 4, 1009–1046, DOI 10.1007/s10801-016-0698-2. MR3566228

[22] L. D. Garcia, M. Stillman, and B. Sturmfels, *Algebraic geometry of Bayesian networks*, J. Symbolic Comput. **39** (2005), no. 3-4, 331–355, DOI 10.1016/j.jsc.2004.11.007. MR2168286

[23] D. Geiger, C. Meek, and B. Sturmfels, *On the toric algebra of graphical models*, Ann. Statist. **34** (2006), no. 3, 1463–1492, DOI 10.1214/009053606000000263. MR2278364

[24] C. J. Hillar and S. Sullivant, *Finite Gröbner bases in infinite dimensional polynomial rings and applications*, Adv. Math. **229** (2012), no. 1, 1–25, DOI 10.1016/j.aim.2011.08.009. MR2854168

[25] S. Hoşten, A. Khetan, and B. Sturmfels, *Solving the likelihood equations*, Found. Comput. Math. **5** (2005), no. 4, 389–407, DOI 10.1007/s10208-004-0156-8. MR2189544

[26] S. Hoşten and J. Shapiro, *Primary decomposition of lattice basis ideals*, J. Symbolic Comput. **29** (2000), no. 4-5, 625–639, DOI 10.1006/jsco.1999.0397. Symbolic computation in algebra, analysis, and geometry (Berkeley, CA, 1998). MR1769658

[27] S. Hoşten and S. Sullivant, *Gröbner bases and polyhedral geometry of reducible and cyclic models*, J. Combin. Theory Ser. A **100** (2002), no. 2, 277–301, DOI 10.1006/jcta.2002.3301. MR1940337

[28] S. Hoşten and S. Sullivant, *A finiteness theorem for Markov bases of hierarchical models*, J. Combin. Theory Ser. A **114** (2007), no. 2, 311–321, DOI 10.1016/j.jcta.2006.06.001. MR2293094

[29] J. Huh, *Varieties with maximum likelihood degree one*, J. Algebr. Stat. **5** (2014), no. 1, 1–17, DOI 10.18409/jas.v5i1.22. MR3279951

[30] T. Kahle, E. Miller, and C. O'Neill, *Irreducible decomposition of binomial ideals*, Compos. Math. **152** (2016), no. 6, 1319–1332, DOI 10.1112/S0010437X16007272. MR3518313

[31] T. Kahle, J. Rauh, and S. Sullivant, *Positive margins and primary decomposition*, J. Commut. Algebra **6** (2014), no. 2, 173–208, DOI 10.1216/JCA-2014-6-2-173. MR3249835

[32] S. L. Lauritzen, *Graphical models*, Oxford Statistical Science Series, vol. 17, The Clarendon Press, Oxford University Press, New York, 1996. Oxford Science Publications. MR1419991

[33] B. Lin, B. Sturmfels, X. Tang, and R. Yoshida, *Convexity in tree spaces*, SIAM J. Discrete Math. **31** (2017), no. 3, 2015–2038, DOI 10.1137/16M1079841. MR3693600

[34] P. Misra and S. Sullivant, *Gaussian graphical models with toric vanishing ideals*, Ann. Inst. Statist. Math. **73** (2021), no. 4, 757–785, DOI 10.1007/s10463-020-00765-0. MR4274248

[35] J. Morton, L. Pachter, A. Shiu, B. Sturmfels, and O. Wienand, *Convex rank tests and semigraphoids*, SIAM J. Discrete Math. **23** (2009), no. 3, 1117–1134, DOI 10.1137/080715822. MR2538642

[36] L. Pachter and B. Sturmfels, *Parametric inference for biological sequence analysis*, Proc. Natl. Acad. Sci. USA **101** (2004), no. 46, 16138–16143, DOI 10.1073/pnas.0406011101. MR2114587

[37] L. Pachter and B. Sturmfels (eds.), *Algebraic statistics for computational biology*, Cambridge University Press, New York, 2005. Edited by Lior Pachter and Bernd Sturmfels, DOI 10.1017/CBO9780511610684. MR2205865

[38] G. Pistone, E. Riccomagno, and H. P. Wynn, *Algebraic statistics*, Monographs on Statistics and Applied Probability, vol. 89, Chapman & Hall/CRC, Boca Raton, FL, 2001. Computational commutative algebra in statistics. MR2332740

[39] E. Robeva, B. Sturmfels, and C. Uhler, *Geometry of log-concave density estimation*, Discrete Comput. Geom. **61** (2019), no. 1, 136–160, DOI 10.1007/s00454-018-0024-y. MR3925548

[40] F. Santos and B. Sturmfels, *Higher Lawrence configurations*, J. Combin. Theory Ser. A **103** (2003), no. 1, 151–164, DOI 10.1016/S0097-3165(03)00092-X. MR1986836

[41] M. Studený, *Probabilistic conditional independence structures*, Information Science and Statistics, Springer, London, 2005. MR3183760

[42] B. Sturmfels and C. Uhler, *Multivariate Gaussian, semidefinite matrix completion, and convex algebraic geometry*, Ann. Inst. Statist. Math. **62** (2010), no. 4, 603–638, DOI 10.1007/s10463-010-0295-4. MR2652308

[43] S. Sullivant, *Algebraic geometry of Gaussian Bayesian networks*, Adv. in Appl. Math. **40** (2008), no. 4, 482–513, DOI 10.1016/j.aam.2007.04.004. MR2412156

[44] T. Yamaguchi, M. Ogawa, and A. Takemura, *Markov degree of the Birkhoff model*, J. Algebraic Combin. **40** (2014), no. 1, 293–311, DOI 10.1007/s10801-013-0488-z. MR3226827

Department of Mathematics, North Carolina State University, Raleigh, North Carolina 27695-8205

Email address: `smsulli2@ncsu.edu`

Proceedings of Symposia in Pure Mathematics
Volume 111, 2025
https://doi.org/10.1090/pspum/111/02043

A primer in likelihood geometry

Carlos Améndola and Jose Israel Rodriguez

Dedicated to Bernd Sturmfels
on the occasion of his sixtieth birthday

ABSTRACT. One of the first points of contact in algebraic statistics championed by Bernd Sturmfels is that of likelihood geometry. It studies the algebraic and geometric aspects of the method for statistical inference called maximum likelihood estimation. The use of algebraic methods in the context of likelihood geometry appears in a wide range of models. These include discrete models with their toric geometry, Gaussian models involving linear spaces of symmetric matrices, and nonparametric inference for log-concave densities. We survey some of the main results and Bernd's contributions in this area. Throughout this chapter we provide snippets of code for the newcomer to adapt for their own examples.

1. Introduction

At the heart of likelihood geometry is maximum likelihood estimation. This statistical problem asks how likely a particular given data is to have occurred according to a model. It is natural to assume that a good estimate for the true distribution is one where the likelihood of the data is maximized given the model constraints. This methodology was championed by Ronald Fisher in the beginning of the twentieth century [**2**].

To find such estimate, one considers the likelihood function on a model with respect to some given data. Maximizing this function becomes an optimization problem which one aims to solve.

We focus on two kinds of models, *log-linear models* which are discrete and live inside a finite dimensional probability simplex, and *Gaussian models* which are continuous and live inside the cone of positive definite matrices. In the first case, the models correspond geometrically to *toric varieties*, a fundamental object studied in algebraic geometry. The second case includes *linear concentration models*, whose concentration matrices lie on a linear space. The maximum likelihood estimation problem for such models invokes methods from the rich field of convex algebraic geometry [**13**].

2020 *Mathematics Subject Classification.* 14J17, 14M25, 14Q15, 62F10, 62H22, 62R01.

Key words and phrases. maximum likelihood estimation, log-linear models, linear concentration models, multivariate Gaussian distribution, maximum likelihood degree, algebraic optimization, toric models, likelihood equations.

This research is funded in part by the Alfred P. Sloan Foundation and the Wisconsin Alumni Research Foundation.

The maximum likelihood (ML) approach for the two types of models discussed here amounts to solving a multivariate polynomial system known as the likelihood equations. The number of solutions to these equations for generic data is called the *maximum likelihood degree* (ML degree) of the model. It is an invariant that measures the algebraic complexity of expressing the maximum likelihood estimator (MLE) for a given statistical model. The maximum likelihood degree was introduced in [**16**].

To solve the likelihood equations one can use *Gröbner bases* [**17**, Chapter 2] to symbolically compute the ML degree. Alternatively, one can use *numerical nonlinear algebra* to obtain accurate numerical estimates for the solutions of the optimization problem.

Bernd has also done work on maximum likelihood estimation outside of the realm of parametric statistics. In the context of nonparametric statistics he has considered models for log-concave densities. In principle, this problem is nonparametric, but using tent functions one can reduce it to an algebraic and combinatorial problem [**29, 45, 46**].

2. Discrete models with their toric geometry

2.1. Model invariants. In this section we consider parametric statistical models in the n-dimensional *probability simplex*

$$\Delta_n := \{(p_0, \ldots, p_n) \in \mathbb{R}^{n+1}_{\geq 0} : p_0 + \cdots + p_n = 1\}.$$

These are parametrized subsets $\phi(\Theta) = \mathcal{M} \subseteq \Delta_n$ with $\Theta \subseteq \mathbb{R}^d$. When ϕ is a polynomial map, the implicit description of $\mathcal{M}$ is given by the ideal of the image of the map, and possibly inequalities, such as those defining the probability simplex. Generators of the image's ideal are known as the *model invariants* and when setting them to zero we have *implicit equations*.

BIG PROBLEM 2.1. Given a parametric model, determine all model invariants.

The standard approach to this problem is to compute a lexicographic Gröbner basis in a polynomial ring in the variables θ and p under an elimination term order, that is, where $\theta > p$. For an introduction to elimination theory see [**17**, Chapter 3].

EXAMPLE 2.2 (Hardy–Weinberg). If $\theta \in \Theta = [0, 1]$ denotes the probability that a biased coin lands on heads, then the probabilities of observing 0, 1, or 2 heads when tossing the coin twice are:

$$\phi(\theta) = ((1 - \theta)^2, 2\theta(1 - \theta), \theta^2) = (p_0, p_1, p_2).$$

As θ varies, one obtains the *statistical model* $\mathcal{M} \subset \Delta_2$. The implicit equation describing this model (apart from $p_0 + p_1 + p_2 = 1$) is the *Hardy-Weinberg* quadric:

$$4p_0p_2 - p_1^2 = 0.$$

The generators of the ideal of implicit equations can be computed using Gröbner basis computations based on elimination theory. For instance, this `Macaulay2` [**28**] code computes a Gröbner basis with respect to an elimination term order of the ideal of the parameterization's graph. The polynomials in the Gröbner basis involving only the variables p_0, p_1, p_2 are a Gröbner basis for the model.

```
i1 : R = QQ[theta, p0,p1,p2, MonomialOrder => Eliminate 1];
i2 : parameterization = ideal(
         p0-(1-theta)^2,
         p1-2*theta*(1-theta),
         p2-theta^2
         );
i3 : G = gb parameterization
o3 = GroebnerBasis[status: done; S-pairs encountered up to degree 2]
o3 : GroebnerBasis

i4 : gens G
o4 = | p0+p1+p2-1 p1^2+4p1p2+4p2^2-4p2 2theta-p1-2p2 |
             1       3
o4 : Matrix R  <--- R
```

The output o4 consists of three polynomials g_1, g_2 and g_3, none of which are the Hardy–Weinberg quadric. To recover this quadric, one homogenizes g_2 with respect to $p_0 + p_1 + p_2 = 1$. Alternatively, one can work projectively as seen in this snippet of code where output o8 is the Gröbner basis. The unique polynomial involving only variables p_0, p_1, p_2 is the Hardy–Weinberg quadric.

```
i5 : R = QQ[t0,t1, p0,p1,p2, MonomialOrder=>Eliminate 2];
i6 : parameterization = ideal( p0-(t0)^2, p1-2*t1*(t0), p2-t1^2 );

i7 : G = gb parameterization
o7 = GroebnerBasis[status: done; S-pairs encountered up to degree 4]
o7 : GroebnerBasis

i8 : gens G
o8 = | p1^2-4p0p2 t1p1-2t0p2 2t1p0-t0p1 t1^2-p2 2t0t1-p1 t0^2-p0 |
             1       6
o8 : Matrix R  <--- R
```

◇

The degree of the Hardy–Weinberg model is two. In general, the degree of the model is defined as the degree of the ideal of model invariants.

By working projectively, the output of model invariants is given by a binomial ideal, which Bernd has studied extensively in [**25**]. If a parametrization induces a monomial map between projective spaces, then the ideal of model invariants is generated by homogeneous binomials and the linear constraint of the variables summing to one. These invariants are understood through the use of toric geometry [**44**, Chapter 8].

2.2. Likelihood geometry of discrete models. The maximum likelihood estimation problem is as follows. We observe i.i.d. data (a sequence of outcomes) $X^{(1)}, \ldots, X^{(N)}$, where $X^{(j)} \in \{0, 1, \ldots, n\}$ and N is the sample size. We summarize this data as the *sufficient statistic* $u = (u_0, \ldots, u_n)$ where u_i counts the number of times outcome i is observed. Note that $u_0 + \cdots + u_n = N$. Each summarized data vector u gives a *likelihood function*

$$L_u : \Delta_n \to \mathbb{R}, \quad L_u(p) := p_0^{u_0} p_1^{u_1} \cdots p_n^{u_n}$$

on the probability simplex Δ_n. Since this function is nonnegative on Δ_n and zero on the boundary, its global maximum over the simplex occurs in the interior (in fact, the maximum is attained at $\frac{1}{N}(u_0, \ldots, u_n)$). Since the logarithm is concave on the positive orthant, the maximum of the likelihood function L_u and the log-likelihood function $\ell_u := \log L_u$ coincide.

We are interested in finding the global optimum of ℓ_u when constrained to the model $\mathcal{M}$. Since our model has a parametrization, we can rewrite the log-likelihood function in the form

$$\ell(\theta \mid u) = (\ell_u \circ \phi)(\theta) = \sum_{j=0}^{n} u_j \log p_j(\theta). \tag{2.1}$$

DEFINITION 2.3. The maximum likelihood estimate (MLE) of the model $\mathcal{M} \subseteq \Delta_n$ given by data u is

$$\hat{\theta} = \arg\max_{\theta \in \Theta} L(\theta|u).$$

We may also call the actual point $\hat{p} \in \mathcal{M}$ corresponding to $\hat{\theta}$, the MLE of the model.

To find this optimum, we set the partial derivatives of ℓ_u to zero, yielding these likelihood equations:

$$\sum_{j=0}^{n} \frac{u_j}{p_j} \frac{\partial p_j}{\partial \theta_i}(\theta) = 0 \qquad i = 1, \ldots, d \tag{2.2}$$

where $\theta = (\theta_1, \ldots, \theta_d)$. In the literature, (2.2) are also known as the *score equations*.

Over $\mathbb{R}$, every solution to the score equations corresponds to either a saddle point or another local extremum, and the number of solutions depends on the choice of u. On the other hand, if we complexify the problem and solve the equations over $\mathbb{C}$, it was proven in [**33**] that the number of solutions is (generically) independent of u and they correspond to complex critical points of the function $L(\theta|u) : \mathbb{C}^d \to \mathbb{C}$.

DEFINITION 2.4. If $\phi : \Theta \to \mathcal{M}$ in (2.1) is generically one to one, then the *maximum likelihood degree (ML degree)* is the number of complex solutions to the score equations (2.2) for generic data vectors u.

The ML degree tells us the algebraic complexity of the maximum likelihood estimation problem. In particular, if we want to give the MLE from u in terms of radicals, it is sufficient that the ML degree is at most four. The most extreme case, when the ML degree is one, means that the likelihood equations have a unique solution and is discussed further in Section 2.4.

EXAMPLE 2.5 (Hardy–Weinberg, revisited). In the Hardy–Weinberg model, suppose one observes a vector of counts $u = (u_0, u_1, u_2)$ and one wishes to estimate the maximum likelihood estimate $\hat{\theta}$. We want to maximize the likelihood function

$$\ell_u(p) = p_0^{u_0} p_1^{u_1} p_2^{u_2} = 2^{u_1} \theta^{u_1 + 2u_2} (1 - \theta)^{2u_0 + u_1}.$$

In this case, the maximizer is given explicitly as a rational function

$$\hat{\theta} = \frac{u_1 + 2u_2}{2(u_0 + u_1 + u_2)}.$$

This means that the *ML degree* of this model is one. ◇

One approach to solve the likelihood equations is to use Gröbner basis computations. This was addressed in [**33**] where algorithms to solve the likelihood equations for parametric models and models described implicitly; see [**36**, Definition 1.5]. These algorithms have been implemented using the algebraic optimization degree `Macaulay2` package [**31**].

EXAMPLE 2.6. The next snippet of code computes the ML degree of the implicit model defined by the vanishing of the determinant of the 3×3 matrix $[p_{i,j}]$.

```
i1 :  R = QQ[p_(1,1)..p_(3,3)];
i2 :  I = ideal det genericMatrix(R,3,3);
i3 :  MLequationsDegree I
o3 =  10
```

This shows that the ML degree of this model is 10. This type of model is called a *mixture model.* More precisely, this example is the 2-mixture model of the *independence model* associated to two discrete random variables with 3 states each. We will generalize this example in the next section. ◇

2.3. ML estimation of toric models and their mixtures. Toric models have the nice property that their ideal of model invariants are generated by binomials, as we saw in Example 2.2. These are examples of *log-linear models.* Let A be a full rank $d \times (n+1)$ integer matrix, then

$$\mathcal{M}_A = \{p \in \Delta_n \mid \log p \in \operatorname{rowspan}(A)\}.$$

Log-linear models have a nice property for maximum likelihood estimation coming from a key result known as *Birch's theorem.*

THEOREM 2.7 ([**22**, Proposition 2.1.5]). *The MLE for a log-linear model $\mathcal{M}_A$ given data u is the unique solution $\hat{p}$ to*

$$A\hat{p} = \frac{1}{N}Au, \qquad \hat{p} \in \mathcal{M}_A.$$

On the algebra side, since Birch's theorem intersects the toric variety $\mathcal{M}_A$ with a linear space of the same codimension, we get the following relation between the *degree* of a toric model and its maximum likelihood degree.

COROLLARY 2.8. *For a log-linear model $\mathcal{M}_A$, the algebraic degree of the model is an upper bound to the maximum likelihood degree.*

Recently, there has been an exciting connection to invariant theory, first suggested by Bernd and studied by Améndola, Kohn, Reichenbach, and Seigal in [**11**]. This exploits the fact that in a log-linear model there is a natural action by the group GT_d of diagonal invertible $d \times d$ matrices. More precisely, given the $d \times (n+1)$ integer matrix A, a torus element $\lambda = (\lambda_1, \ldots, \lambda_d) \in GT_d$, and a *linearization* vector $b \in \mathbb{Z}^d$, the action on a point $x \in \mathbb{C}^{n+1}$ is given by

$$(\lambda \cdot x)_j = \lambda_1^{a_{1j}-b_1} \cdots \lambda_d^{a_{dj}-b_d} x_j$$

for $j = 0, \ldots, n$.

We say that a vector $x \in \mathbb{C}^{n+1}$ is *polystable* with respect to the action of the group G, if $x \neq 0$ and its orbit $G \cdot x$ is closed in the Euclidean topology.

THEOREM 2.9 ([**11**, Theorem 4.3]). *Let $u \in \mathbb{N}^{n+1}$ be a vector of counts and let $N = u_+$. The MLE of a log-linear model $\mathcal{M}_A$ given u exists if and only if the vector of ones $\mathbb{1}$ is polystable under the action of the torus GT_d given by the matrix NA with linearization $b = Au \in \mathbb{Z}^d$.*

The theorem above can be seen as an algebraic invariant theoretical formulation of another polyhedral condition for MLE existence.

THEOREM 2.10 ([**26**]). *The MLE given u exists for $\mathcal{M}_A$ if and only if $\frac{1}{N}Au$ lies in the relative interior of the polytope defined by the convex hull of the columns of A.*

Theorem 2.9 connects maximum likelihood estimation in a log-linear model to the *capacity problem* in invariant theory. Algorithms exist for both problems, notably *iterative proportional scaling* (IPS) and norm minimization, respectively. It is thus interesting to compare them and to study how to pass from one algorithm to the other; see [**11**, Figure 4].

On the topology side, June Huh relates the ML degree of a toric model to an Euler characteristic [**34**]. In fact, his result can be used to compute the ML degree of a *smooth* model. Informally, consider a statistical model $\mathcal{M}$ in Δ_n and its closure $\overline{\mathcal{M}}$ in $\mathbb{P}^n$. If $\mathcal{H}$ denotes the union of coordinate hyperplanes in $\mathbb{P}^n$ and the statistical hyperplane $\{[p_0 : \cdots : p_n] \in \mathbb{P}^n : \sum p_i = 0\}$, then the topological Euler characteristic of $\mathcal{M} \setminus \mathcal{H}$ is the ML degree up to sign.

EXAMPLE 2.11 (Independence model [**32**]). Consider the Segre variety $\mathbb{P}^{m-1} \times \mathbb{P}^{n-1}$ embedded as the smooth variety $\mathcal{V}$ in $\mathbb{P}^{mn-1}$. Fix coordinates $(x_1 : \cdots : x_m)$ on $\mathbb{P}^{m-1}$ and coordinates $(y_1 : \cdots : y_n)$ on $\mathbb{P}^{n-1}$. The open subset $\mathcal{V} \setminus \mathcal{H}$ consists of all points in $\mathbb{P}^{m-1} \times \mathbb{P}^{n-1}$ with $x_1 x_2 \cdots x_m y_1 y_2 \cdots y_n (x_1 + \cdots + x_m)(y_1 + \cdots + y_n) \neq 0$. Hence

$$\mathcal{V}\backslash\mathcal{H} \quad = \quad (\mathbb{P}^{m-1} \text{ minus } m+1 \text{ hyperplanes}) \times (\mathbb{P}^{n-1} \text{ minus } n+1 \text{ hyperplanes}).$$

Each factor has signed Euler characteristic 1, and hence so does their product. $\diamond$

Huh's theorem does not apply to singular models. One important example of a singular model is a mixture model. In statistical terms, mixture models represent a non-homogeneous population consisting of different groups. The probability density function of a mixture is a convex combination of the individual group densities, with weights equal to the proportion of each group. In geometric terms, the Zariski closure of such a model is a *secant variety* [**50**, Chapter 14]. Maximum likelihood estimation for these models can be attacked through the use of the EM algorithm. However, the EM algorithm is well known to get stuck at saddle points or local optima of the likelihood function. Rather than dealing with the fixed points of the EM algorithm [**38**], one can find all critical points of the likelihood function to obtain the global optimum of the likelihood function. The maximum likelihood degree gives an upper bound and an algebraic measure of complexity for this problem. Thus, giving formulas for the ML degree is highly applicable for statistics and algebraic optimization.

This motivates a topological formula for singular models using the *local Euler obstruction function* [**18**, Section 4.1].

THEOREM 2.12 ([**43**, Corollary 1.7]). *Let Y be an irreducible subvariety of $(\mathbb{C}^*)^n$. Assume that $Y^\circ = Y \setminus \mathcal{H}$ is nonempty. Then*

$$\mathrm{MLdeg}(Y) = (-1)^{\dim Y} \chi(Eu_{Y^\circ})$$

where on the right-hand side is the Euler characteristic of the local Euler obstruction function of Y°.

A version of this result first appears in [**15**] to disprove a conjecture [**36**, Conjecture 1.8], and another version [**47**] is used to determine the ML degree of a class of mixture models. The latter was also used to resolve a conjecture of Bernd, Hauenstein, and Rodriguez [**32**, Conjecture 4.1] on the ML degree of rank at most two matrices of size $m \times n$, which we denote by X_{mn}. Furthermore, using the results from [**47**, Section 4], one finds the closed form expressions for the ML degree of X_{mn} with $m \leq n$ and $m = 2, \ldots, 5$:

$$\begin{array}{rcl} \text{MLdeg}(X_{2n}) & = & 1, \\ \text{MLdeg}(X_{3n}) & = & 4 \cdot 2^{n-1} - 6, \\ \text{MLdeg}(X_{4n}) & = & 18 \cdot 3^{n-1} - 40 \cdot 2^{n-1} + 25, \\ \text{MLdeg}(X_{5n}) & = & 96 \cdot 4^{n-1} - 270 \cdot 3^{n-1} + 260 \cdot 2^{n-1} - 90. \end{array}$$

The appendix of [**47**] prescribes a recursion to compute the ML degree of X_{mn} for any m and n, while here we list out specific values of the ML degrees of X_{mn}.

	$m=2$	$m=3$	$m=4$	$m=5$	$m=6$	$m=7$
$n=m$	1	10	191	6776	378477	30305766
$n=m+1$	1	26	843	40924	2865245	274740990
$n=m+2$	1	58	3119	212936	19177197	2244706374
$n=m+3$	1	122	10587	1015564	118430045	17048729886
$n=m+4$	1	250	34271	4586456	692277357	122818757286.

These singularity theory tools [**41**] have found applications in other areas of algebraic statistics and algebraic optimization like Euclidean distance degrees [**20**, **40**] and linear optimization on a variety [**42**] motivated by [**48**]. It is still open how to use these tools or other methods to compute ML degrees of low rank tensors, symmetric matrices with rank constraints, and Grassmannians. Although, exciting recent work motivated by physics [**1**], educational testing [**4**, Section 2], and graphical models [**5**] show exciting new developments and connections to likelihood geometry.

2.4. ML degree one. As in Example 2.2, a natural question is to classify all models with ML degree one. This was done for very affine varieties by Huh [**35**].

This classification is especially important in the context of *maximum likelihood estimators*. A maximum likelihood estimator for a model is a map from $\mathbb{R}^{n+1}_{\geq 0}$ to the model $\mathcal{M}$ taking the data to the maximum likelihood estimate for ℓ_u on $\mathcal{M}$. Here we let $\mathcal{M}$ be a discrete statistical model in the open part of the probability simplex Δ_n. If the $n+1$ coordinates of the maximum likelihood estimator are rational functions in u, then one says that $\mathcal{M}$ has *rational maximum likelihood estimator*. While Huh's results imply that a statistical model with maximum likelihood degree one has a rational maximum likelihood estimator, it is the work of Bernd, Duarte, and Marigliano [**23**] that gives a complete characterization of statistical models (i.e., living in the probability simplex) with rational MLE in terms of discriminantal triples [**23**, Definition 6] and Horn pairs. A *Horn matrix* H is a $m \times (n+1)$ integer matrix whose column sums are zero. Let

$$(Hu)^{h_j} := \prod_{i=1}^{m} (h_{i0}u_0 + h_{i1}u_1 + \cdots + h_{in}u_n)^{h_{ij}}.$$

A *Horn pair* (H, λ) is a Horn matrix H and a vector $\lambda = (\lambda_0, \dots, \lambda_n) \in \mathbb{R}^{n+1}$ with $\lambda_i \neq 0$ such that

$$\lambda_0(Hu)^{h_0} + \lambda_1(Hu)^{h_1} + \cdots + \lambda_n(Hu)^{h_n} = 1$$

and such that the rational function

$$\varphi_{(H,\lambda)}(u) = (\lambda_0(Hu)^{h_0}, \lambda_1(Hu)^{h_1}, \dots, \lambda_n(Hu)^{h_n})$$

has domain and codomain $\mathbb{R}^{n+1}_{>0}$.

The following theorem is extracted from [**23**, Theorem 1].

THEOREM 2.13. Assume that the model $\mathcal{M}$ has a well-defined maximum likelihood estimator $\Phi : \mathbb{R}^{n+1}_{>0} \to \mathcal{M}$. Then Φ is a rational function if and only if there exists a Horn pair (H, λ) such that $\mathcal{M}$ is the image of the Horn map

$$\varphi_{(H,\lambda)} : \mathbb{R}^{n+1}_{>0} \to \mathbb{R}^{n+1}_{>0}.$$

In this case, $\Phi = \varphi_{(H,\lambda)}$.

The equality of the two maps highlights an important aspect of models with rational maximum likelihood estimator. It says that when a model has an MLE of the simplest possible closed form, i.e., a rational function, the polynomials appearing in the numerators and denominators of the rational formulas must factor into linear forms with positive coefficients.

EXAMPLE 2.14 (Hardy–Weinberg, once again). According to Example 2.5, the ML estimator Φ for the Hardy-Weinberg model given data $u = (u_0, u_1, u_2)$ is

$$\Phi(u) = \left(\frac{(2u_0 + u_1)^2}{4(u_0 + u_1 + u_2)^2}, \frac{(u_1 + 2u_2)(2u_0 + u_1)}{2(u_0 + u_1 + u_2)^2}, \frac{(u_1 + 2u_2)^2}{4(u_0 + u_1 + u_2)^2} \right).$$

This is a rational maximum likelihood estimator, and this map is indeed $\varphi_{(H,\lambda)}$, where

$$H = \begin{pmatrix} 2 & 1 & 0 \\ 0 & 1 & 2 \\ -2 & -2 & -2 \end{pmatrix} \qquad \text{and} \qquad \lambda = (1, 2, 1).$$

We mention here that there is a third characterization in [**23**, Theorem 1] involving discriminantal triples that also offers a way for constructing such models using toric geometry.

Before closing this section, we mention an interesting dual problem in the context of maximum likelihood estimation. Given a point $p \in \mathcal{M}$, what is the set of vectors u such that p is the MLE for $\mathcal{M}$ given u? Such set is called the *Log-Voronoi cell* of p and has been studied in [**3**, **6**]. One can show that these cells are always convex, and for several classes of models, such as ML degree one and linear models, they are always polytopes.

3. Gaussian models and symmetric matrices

In this section the ambient space for our statistical models is the cone of positive definite matrices PD. This is natural because Gaussians are determined by their covariance matrix Σ (which is positive definite) and mean, which we will assume to be 0.

In fact, the Gaussian density is then given by

$$f_\Sigma(x) = \frac{1}{\sqrt{\det(2\pi\Sigma)}} \exp\left(-\frac{1}{2}x^\top \Sigma^{-1} x\right), \quad x \in \mathbb{R}^m.$$

There are some similarities and differences in this case to the discrete model case, which will become apparent in the next subsections.

3.1. Likelihood geometry of Gaussian models. The maximum likelihood estimation problem now takes the following form. We again observe i.i.d. data $X^{(1)}, \ldots, X^{(N)}$, where $X^{(j)} \sim \mathcal{N}(0, \Sigma)$ and N is the sample size. The sufficient statistic is now the *sample covariance matrix*

$$S = \sum_{j=1}^{N} \frac{1}{N} X^{(j)} X^{(j)^\top}, \tag{3.1}$$

and the *log-likelihood function* takes the form (see, e.g., [**50**, Section 7.1])

$$\ell(\Sigma) = -\frac{N}{2}(\log\det(\Sigma) + \operatorname{tr}(S\Sigma^{-1})). \tag{3.2}$$

The gradient of ℓ is $\nabla(\ell)(\Sigma) \propto \Sigma - S$, so that $\hat{\Sigma} = S$ is the global optimum whenever $S \in \mathcal{M}$. One subtlety is that S is only guaranteed to be positive semidefinite, and it need not be positive definite. This means that the MLE may not always exist.

3.2. Gaussian graphical models: Directed vs. undirected. We consider two types of graphical models.

3.2.1. *Directed graphical models.* Here we give a brief introduction to this kind of Gaussian models. A textbook reference is [**50**, Chapter 13].

Let G be a directed acyclic graph (DAG) with m vertices. Without loss of generality one can label the vertices with $\{1, \ldots, m\}$ so that if there is an edge $i \to j$ then $i < j$. In this case we say i is a parent of j and j is a child of i.

The directed graphical model given by G can be defined parametrically via the recursive equation:

$$X_j = \sum_{i \in \operatorname{pa}(j)} \lambda_{ij} X_i + \epsilon_j, \tag{3.3}$$

where $\operatorname{pa}(j)$ denotes the parents of j, λ_{ij} are edge weights, and $\epsilon_j \sim \mathcal{N}(0, \omega_j)$, $j = 1, \ldots, m$ are independent errors.

Example 3.1. Consider the directed graph shown in Figure 1. Then the structural equations are given by

$$\begin{aligned} X_1 &= \epsilon_1, \\ X_2 &= \lambda_{12} X_1 + \epsilon_2, \\ X_3 &= \lambda_{13} X_1 + \epsilon_3, \\ X_4 &= \lambda_{24} X_2 + \lambda_{34} X_3 + \epsilon_4. \end{aligned}$$

Note that we may rewrite this in matrix-vector form

$$\begin{pmatrix} X_1 \\ X_2 \\ X_3 \\ X_4 \end{pmatrix} = \begin{pmatrix} 0 & 0 & 0 & 0 \\ \lambda_{12} & 0 & 0 & 0 \\ \lambda_{13} & 0 & 0 & 0 \\ 0 & \lambda_{24} & \lambda_{34} & 0 \end{pmatrix} \begin{pmatrix} X_1 \\ X_2 \\ X_3 \\ X_4 \end{pmatrix} + \begin{pmatrix} \epsilon_1 \\ \epsilon_2 \\ \epsilon_3 \\ \epsilon_4 \end{pmatrix}.$$

◇

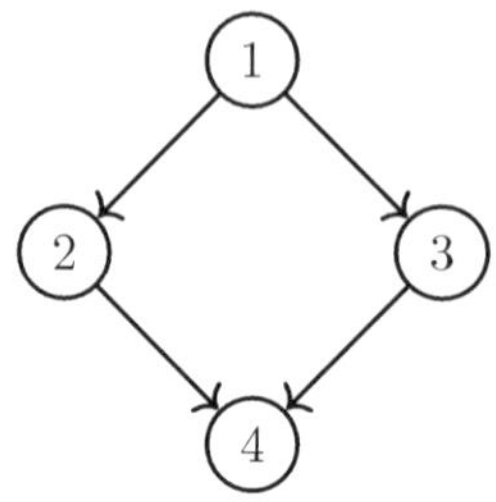

FIGURE 1. The diamond graph, a directed graph on four nodes.

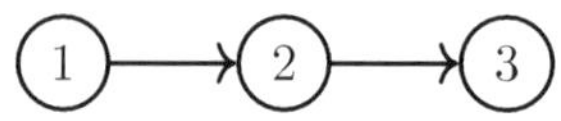

FIGURE 2. The 3-path is a directed acyclic graph.

As in the example, we can always write

$$X = \Lambda^\top X + \epsilon. \qquad (I - \Lambda)^\top X = \epsilon \tag{3.4}$$

This implies that the covariance matrices Σ in the model are parametrized by the lower triangular matrix $\Lambda = (\lambda_{ij})$ and diagonal matrix $\Omega = \operatorname{diag}(\omega_1, \ldots, \omega_m)$ via

$$\Sigma = (I - \Lambda)^{-\top}\Omega(I - \Lambda)^{-1} \tag{3.5}$$

(note that $\det(I - \Lambda) = 1 \neq 0$). That is, we have for the concentration matrix $K = \Sigma^{-1}$:

$$K = (I - \Lambda)\Omega^{-1}(I - \Lambda)^\top. \tag{3.6}$$

Let $Y = [X^{(1)}|X^{(2)}|\ldots|X^{(N)}] \in \mathbb{R}^{m\times N}$ the matrix with the sample as columns. Then we can rewrite (3.1) as $S = \frac{1}{N}YY^\top$. Considering the maximum likelihood estimation problem, when we plug in expression (3.6) for K in (3.2), we find that

$$\begin{aligned}\ell_Y(\Sigma) &= \log\det(\Omega^{-1}) - \operatorname{tr}(Y(I - \Lambda)\Omega^{-1}(I - \Lambda)^\top Y^\top) \\ &= -\sum_{j=1}^m \log\omega_j - \sum_{j=1}^m \frac{1}{n\omega_j}[(Y(I - \Lambda))(Y(I - \Lambda))^\top]_{jj}.\end{aligned}$$

From here, one can see that solving the likelihood equations is equivalent to minimizing $\|X^{(j)} - \sum_{i\in\mathrm{pa}(j)} \lambda_{ij}X^{(i)}\|^2$ for each $j = 1, \ldots, m$, which is exactly the classical least squares problem m times.

This implies the ML degree of any DAG model is 1. For a general formula for the MLE see [**39**, p. 154].

EXAMPLE 3.2. Consider the directed path shown in Figure 2. The MLE for such model is given by

$$\hat{\Sigma} = \begin{pmatrix} s_{11} & s_{12} & \frac{s_{12}s_{23}}{s_{22}} \\ s_{12} & s_{22} & s_{23} \\ \frac{s_{12}s_{23}}{s_{22}} & s_{23} & s_{33} \end{pmatrix},$$

whose determinant is given as a rational function of principal minors of S:

$$\det(\hat{\Sigma}) = \frac{(s_{12}^2 - s_{11}s_{22})(s_{23}^2 - s_{22}s_{33})}{s_{22}}.$$

We make two observations here. One is that many entries of $\hat{\Sigma}$ coincide with the ones of the sample covariance matrix S. This is no coincidence, and will be explained in the next section in the undirected case. The second observation from the determinant is that while S will be generically invertible as soon as the sample size N is at least 3, the MLE exists here already for generic rank 2 matrices S. $\diamond$

DEFINITION 3.3. The *maximum likelihood threshold* of a graph G, denoted $\mathrm{mlt}(G)$ is the minimum sample size N needed so that the MLE exists with probability one.

For instance, in Example 3.2 the maximum likelihood threshold of the path is 2. For DAGs one has the following result, which follows from [**39**, Section 5.4.1] (see also the generalization in [**21**, Theorem 1]).

PROPOSITION 3.4. If G is a DAG with $V = [m]$, then

$$\mathrm{mlt}(G) = \max_{1 \le i \le m} |\mathrm{pa}(i)| + 1.$$

The maximum that appears in the right-hand side is also known as the *in-degree* of G.

If the graph G is a *transitive* DAG (TDAG), which means $i \to j$ and $j \to k$ implies $i \to k$, we can define the group

$$\mathcal{G} = \{g \in \mathrm{GL}_m \mid g_{ij} = 0 \text{ for } i \neq j \text{ and } i \not\to j\} \tag{3.7}$$

which turns the DAG model into a *Gaussian group model*. These were introduced in [**10**] and reformulate the maximum likelihood estimation problem for such Gaussian models in terms of the null cone problem in invariant theory, in direct analogy to Theorem 2.9. There we talked about polystability of a point x respect to an action of a group G by looking at the orbit $G \cdot x$ and its Euclidean closure. If the vector 0 belongs to such orbit closure, then x is called *unstable*, otherwise it is *semistable*. The term *stable* is reserved for a polystable point x that also has finite stabilizer under the action of G.

The main result for Gaussian group models is of the following form, where we use the notation $Y \in \mathbb{R}^{m \times N}$ for the sample matrix and ℓ_Y for the respective log-likelihood function.

THEOREM 3.5. If a group $\mathcal{G} \subseteq \mathrm{GL}_m(\mathbb{C})$ is a Zariski closed self-adjoint group, closed under nonzero scalar multiples, then maximum likelihood estimation for $\mathcal{M}_{\mathcal{G}}$ relates to the action of $\mathcal{G} \cap \mathrm{SL}_m$ as follows:

(1) Y is unstable if and only if ℓ_Y is not bounded from above.
(2) Y is semistable if and only if ℓ_Y is bounded from above.
(3) Y polystable if and only if an MLE exists.
(4) Y stable if and only if there is a unique MLE.

In the case of a TDAG, we have in particular this invariant theory interpretation, matching the intuition from the linear regression at the start of the section.

PROPOSITION 3.6 ([**10**, Theorem 5.3]). Let G be a TDAG and let Y be a matrix of samples in $\mathbb{R}^{m \times N}$. If some ith row of Y is a linear combination of the rows of the parents of i, then Y is unstable and the likelihood function is unbounded from above. Otherwise, Y is polystable and an MLE exists.

In particular, in this TDAG case, semistable implies polystable.

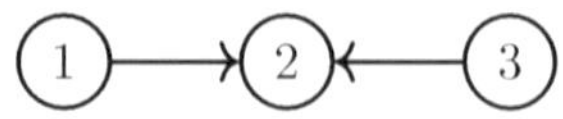

FIGURE 3. A collider triple is an example of a TDAG.

EXAMPLE 3.7. Consider the collider graph G in Figure 3. By Proposition 3.4, the maximum likelihood threshold of G is 3. Indeed, when $n \leq 2$ we expect that the second row of a sample matrix $Y \in \mathbb{R}^{3\times n}$ will be a linear combination of the first and third rows, so that by Proposition 3.6, the MLE will generically not exist. However, the MLE can still exist for special choices of Y. Consider $n = 2$ and

$$Y = \begin{pmatrix} 1 & 0 \\ 1 & 0 \\ 0 & 1 \end{pmatrix}.$$

By Proposition 3.6 the MLE exists and it is computed from the invariant theory perspective in [**10**, Example 5.8]. Alternatively, a direct search of the optimizers of the likelihood function gives that there is no unique MLE and they are all given by

$$2I_3 + 2t \begin{pmatrix} t & -t & 1 \\ -t & t & -1 \\ 1 & -1 & 0 \end{pmatrix}, \quad t \in \mathbb{R}.$$

◇

3.2.2. *Undirected graphical models.* Now let G be an undirected graph. In the context of probabilistic graphical models, whenever there is a missing edge $i - j$, this implies the *pairwise Markov property*:

$$X_i \perp\!\!\!\perp X_j | X_{[m]\setminus\{i,j\}} \tag{3.8}$$

Remarkably, this condition translates precisely into the vanishing $K_{ij} = 0$ of the (i, j) entry of the inverse of the covariance matrix.

DEFINITION 3.8. Let G be an undirected graph with $V = [m]$. The *Gaussian graphical model* given by G consists of all m-dimensional Gaussian distributions with covariance matrix Σ such that $(\Sigma^{-1})_{ij} = 0$ for every missing edge $i - j$.

The main reference for understanding the likelihood geometry behind these models is the paper by Bernd and Uhler [**49**], and the follow-up work by Uhler [**51**].

EXAMPLE 3.9. The undirected version of the graph in Figure 1 is known as the 4-cycle and is shown in Figure 4. The (undirected) Gaussian graphical model consists of $\Sigma \in \mathrm{PD}_4$ such that $K = \Sigma^{-1}$ satisfies $K_{14} = K_{23} = 0$, corresponding to the missing edges $1 - 4$ and $2 - 3$. Note that this is an 8-dimensional model in the 10-dimensional cone of 4×4 symmetric positive definite matrices PD_4.

The 4-cycle is of special significance because it is the smallest example of a graph with ML degree larger than 1. As we will see in Example 3.11, its ML degree turns out to be 5. Furthermore, by Galois theory, the maximum likelihood estimator is in general not rational nor expressible by radicals in the entries of the sample covariance S. Nevertheless, one can use computational algebraic techniques to compute the score equations, followed by numerical algebraic geometry computations to find numerical approximations to the solutions. ◇

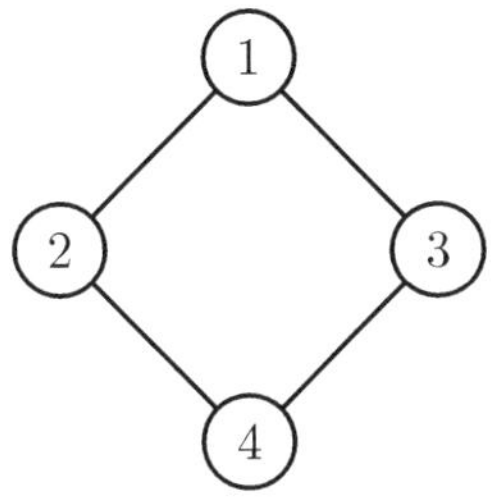

FIGURE 4. The 4-cycle, an undirected graph on four nodes.

We now consider the problem of existence of the MLE for Gaussian graphical models. We denote by S_G the matrix S but with 0 entries whenever the entry (i, j) is not an edge of G. In other words, we are projecting onto the linear space of the nonzero entries.

THEOREM 3.10. The MLE for a Gaussian graphical model exists if and only if the partial matrix S_G can be completed to a positive definite matrix. The MLE will be then the unique completion $\hat{\Sigma}$ such that $\hat{K}_{ij} = 0$ for all missing edges $i - j$.

Next, we illustrate both a computation of the ML degree as well as the MLE with the 4 cycle from Example 3.9.

EXAMPLE 3.11. This code features `Macaulay2` package `GraphicalModelsMLE` [**8**].

```
i1 : loadPackage "GraphicalModelsMLE"
i2 : G = graph{{1,2},{1,3},{2,4},{3,4}}
o2 = Graph{1 => {2, 3}}
           2 => {1, 4}
           3 => {1, 4}
           4 => {2, 3}
o2 : Graph

i3 : MLdegree(gaussianRing G)
o3 = 5

i4 : S = matrix {{2, 3, 5, 7},
                 {3, 11, 13, 17},
                 {5, 13, 19, 23},
                 {7, 17, 23, 29}};
i4 : solverMLE(G,S,SampleData=>false)
o4 = (-6.45848,  | 2       3       5       5.78639 |, 5)
                 | 3       11      13.2923 17      |
                 | 5       13.2923 19      23      |
                 | 5.78639 17      23      29      |
```

The last output is the optimum value of the log-likelihood function, the MLE of the covariance matrix, and the ML degree of the model. Note that, as announced in Theorem 3.10, the MLE $\hat{\Sigma}$ in the output coincides with S in all entries except the ones corresponding to the missing edges. $\diamond$

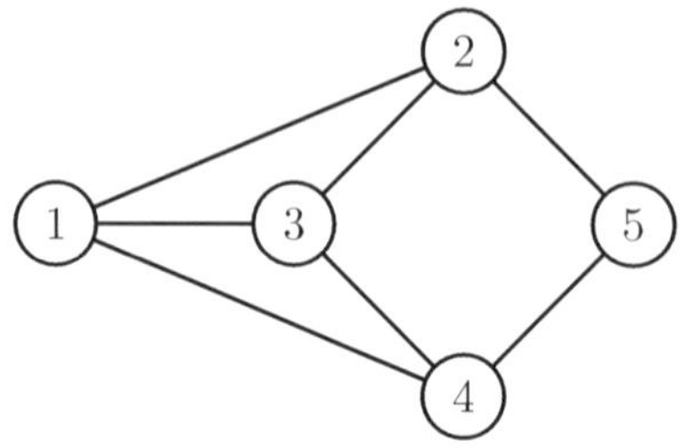

FIGURE 5. Undirected graph on five nodes in Example 3.14.

A conjectured formula by Bernd, Drton, and Sullivant for the ML degree of a cycle has been open since it appeared first in [**22**, Section 7.4].

CONJECTURE 3.12. The ML degree of the m-cycle is given by $(m-3)\cdot 2^{m-2}+1$.

Recently, however, a proof for the conjecture of the *degree* of the cycle was given in [**19**].

The maximum likelihood threshold of undirected graphs is less understood. Some special cases and bounds are given by Gross and Sullivant in [**30**]. One particularly nice application of computational algebra in this direction is the following result from [**51**].

THEOREM 3.13. If the elimination ideal $I_{G,n}$ obtained from the ideal of $(n+1)\times(n+1)$ minors of S of unknowns by eliminating all unknowns corresponding to non-edges of the graph G is the zero ideal then the MLE exists with probability one for n observations.

We illustrate the above theorem with the following example.

EXAMPLE 3.14. Consider the graph G in Figure 5. The following `Macaulay2` code computes the elimination ideal $I_{G,n}$ for $n=1,2,3$. The unknowns to eliminate are s_{15}, s_{24}, s_{35} corresponding to the three missing edges in G.

```
i1 : R=QQ[flatten flatten (for i from 1 to 5 list
                                for j from i to 5 list s_(i,j))]
i2 : S = genericSymmetricMatrix(R,5)

o2 = | s_(1,1) s_(1,2) s_(1,3) s_(1,4) s_(1,5) |
     | s_(1,2) s_(2,2) s_(2,3) s_(2,4) s_(2,5) |
     | s_(1,3) s_(2,3) s_(3,3) s_(3,4) s_(3,5) |
     | s_(1,4) s_(2,4) s_(3,4) s_(4,4) s_(4,5) |
     | s_(1,5) s_(2,5) s_(3,5) s_(4,5) s_(5,5) |

i3 : E = {s_(1,5),s_(2,4),s_(3,5)}

i4 : I_1 = eliminate(E,minors(2,S));
i5 : codim I_1, degree I_1
o5 = (7, 16)

i6 : I_2 = eliminate(E,minors(3,S));
i7 : codim I_2, degree I_2
```

```
o7 = (3, 29)

i9 : codim I_3, degree I_3
o9 = (0, 1)
i10 : I_3
o10 = ideal ()
```

The last line reveals that $I_{G,3}$ is the zero ideal. Therefore we can conclude that the MLE exists with probability one for 3 observations. One can show that the MLE does not exist for $n = 1$ observation, and exists with probability strictly between 0 and 1 for $n = 2$ observations, so that the maximum likelihood threshold for this graph is 3. ◇

Just as in the discrete case, one can also talk about Gaussian *Log-Voronoi cells* of a particular $\Sigma \in \mathcal{M}$. These consist of all S such that Σ is the MLE for $\mathcal{M}$ given S. They have been recently studied in [**7**].

3.3. Other Gaussian models. During the dark times of the lockdown because of the COVID-19 pandemic, Bernd proposed a list of questions circling around *LSSM* models (linear spaces of symmetric matrices). A linear space of symmetric matrices $\mathcal{L}$ defines a Gaussian model by requiring that the inverses $K = \Sigma^{-1}$ of the covariance matrices lie on $\mathcal{L}$ defined by linear constraints on the concentration matrices. LSSMs include undirected Gaussian graphical models as a special case.

The goal of Bernd's list of questions on LSSMs was to bring the nonlinear algebra community together. This was very successful and the papers coming out of this effort were published in the special issue of Le Matematiche. A significant number of the projects [**9, 12, 14, 24, 27, 37**] focused on aspects of likelihood geometry for LSSMs. With many thanks to Bernd, his collaborators and the scientific community he has built, the future is bright.

Acknowledgments

We thank Max Hill for comments on a first draft. We are very thankful to Serkan Hosten, Diane Maclagan, and Frank Sottile for organizing this volume and to the anonymous reviewers for numerous helpful suggestions, corrections, and comments for this contribution.

References

[1] D. Agostini, T. Brysiewicz, C. Fevola, L. Kühne, B. Sturmfels, and S. Telen, *Likelihood degenerations*, Adv. Math. **414** (2023), Paper No. 108863, 39, DOI 10.1016/j.aim.2023.108863. With an appendix by Thomas Lam. MR4539061

[2] J. Aldrich and J. Aldrich, *R. A. Fisher and the making of maximum likelihood 1912–1922*, Statist. Sci. **12** (1997), no. 3, 162–176, DOI 10.1214/ss/1030037906. MR1617519

[3] Y. Alexandr, *Logarithmic Voronoi polytopes for discrete linear models*, Algebr. Stat. **15** (2024), no. 1, 1–13, DOI 10.2140/astat.2024.15.1. MR4687261

[4] Y. Alexandr, M. Bakenhus, M. Curiel, S. K. Deshpande, E. Gross, Y. Gu, M. Hill, J. Johnson, B. Kagy, V. Karwa, J. Li, H. Lyu, S. Petrović, and J. I. Rodriguez, *New directions in algebraic statistics: Three challenges from 2023*, Algebr. Stat. **15** (2024), no. 2, 357–382, DOI 10.2140/astat.2024.15.357. MR4836470

[5] Y. Alexandr, J. I. Coons, and N. Sturma, *Mixtures of discrete decomposable graphical models*, Algebr. Stat. **15** (2024), no. 2, 269–293, DOI 10.2140/astat.2024.15.269. MR4836467

[6] Y. Alexandr and A. Heaton, *Logarithmic Voronoi cells*, Algebr. Stat. **12** (2021), no. 1, 75–95, DOI 10.2140/astat.2021.12.75. MR4251255

[7] Y. Alexandr and S. Hoşten, *Logarithmic Voronoi cells for Gaussian models*, J. Symbolic Comput. **122** (2024), Paper No. 102256, 18, DOI 10.1016/j.jsc.2023.102256. MR4637627

[8] C. Améndola, L. D. García Puente, R. Homs, O. Kuznetsova, and H. J. Motwani, *Computing maximum likelihood estimates for Gaussian graphical models with Macaulay2*, J. Softw. Algebra Geom. **12** (2022), no. 1, 1–10, DOI 10.2140/jsag.2022.12.1. MR4512257

[9] C. Améndola, L. Gustafsson, K. Kohn, O. Marigliano, and A. Seigal, *The maximum likelihood degree of linear spaces of symmetric matrices*, Matematiche (Catania) **76** (2021), no. 2, 535–557, DOI 10.4418/2021.76.2.15. MR4334903

[10] C. Améndola, K. Kohn, P. Reichenbach, and A. Seigal, *Invariant theory and scaling algorithms for maximum likelihood estimation*, SIAM J. Appl. Algebra Geom. **5** (2021), no. 2, 304–337, DOI 10.1137/20M1328932. MR4274830

[11] C. Améndola, K. Kohn, P. Reichenbach, and A. Seigal, *Toric invariant theory for maximum likelihood estimation in log-linear models*, Algebr. Stat. **12** (2021), no. 2, 187–211, DOI 10.2140/astat.2021.12.187. MR4350876

[12] C. Améndola and P. Zwiernik, *Likelihood geometry of correlation models*, Matematiche (Catania) **76** (2021), no. 2, 559–583, DOI 10.4418/2021.76.2.16. MR4334904

[13] G. Blekherman, P. A. Parrilo, and R. R. Thomas (eds.), *Semidefinite optimization and convex algebraic geometry*, MOS-SIAM Series on Optimization, vol. 13, Society for Industrial and Applied Mathematics (SIAM), Philadelphia, PA; Mathematical Optimization Society, Philadelphia, PA, 2013. MR3075433

[14] T. Boege, J. I. Coons, C. Eur, A. Maraj, and F. Röttger, *Reciprocal maximum likelihood degrees of Brownian motion tree models*, Matematiche (Catania) **76** (2021), no. 2, 383–398, DOI 10.4418/2021.76.2.6. MR4334894

[15] N. Budur and B. Wang, *Bounding the maximum likelihood degree*, Math. Res. Lett. **22** (2015), no. 6, 1613–1620, DOI 10.4310/MRL.2015.v22.n6.a4. MR3507253

[16] F. Catanese, S. Hoşten, A. Khetan, and B. Sturmfels, *The maximum likelihood degree*, Amer. J. Math. **128** (2006), no. 3, 671–697. MR2230921

[17] D. A. Cox, J. Little, and D. O'Shea, *Ideals, varieties, and algorithms*, 4th ed., Undergraduate Texts in Mathematics, Springer, Cham, 2015. An introduction to computational algebraic geometry and commutative algebra, DOI 10.1007/978-3-319-16721-3. MR3330490

[18] A. Dimca, *Sheaves in topology*, Universitext, Springer-Verlag, Berlin, 2004, DOI 10.1007/978-3-642-18868-8. MR2050072

[19] R. A. Dinu, M. Michałek, and M. Vodička, *Geometry of the Gaussian graphical model of the cycle*, Preprint, arXiv:2111.02937, 2021.

[20] J. Draisma, E. Horobeţ, G. Ottaviani, B. Sturmfels, and R. R. Thomas, *The Euclidean distance degree of an algebraic variety*, Found. Comput. Math. **16** (2016), no. 1, 99–149, DOI 10.1007/s10208-014-9240-x. MR3451425

[21] M. Drton, C. Fox, A. Käufl, and G. Pouliot, *The maximum likelihood threshold of a path diagram*, Ann. Statist. **47** (2019), no. 3, 1536–1553, DOI 10.1214/18-AOS1724. MR3911121

[22] M. Drton, B. Sturmfels, and S. Sullivant, *Lectures on algebraic statistics*, Oberwolfach Seminars, vol. 39, Birkhäuser Verlag, Basel, 2009, DOI 10.1007/978-3-7643-8905-5. MR2723140

[23] E. Duarte, O. Marigliano, and B. Sturmfels, *Discrete statistical models with rational maximum likelihood estimator*, Bernoulli **27** (2021), no. 1, 135–154, DOI 10.3150/20-BEJ1231. MR4177364

[24] S. Dye, K. Kohn, F. Rydell, and R. Sinn, *Maximum likelihood estimation for nets of conics*, Matematiche (Catania) **76** (2021), no. 2, 399–414, DOI 10.4418/2021.76.2.7. MR4334895

[25] D. Eisenbud and B. Sturmfels, *Binomial ideals*, Duke Math. J. **84** (1996), no. 1, 1–45, DOI 10.1215/S0012-7094-96-08401-X. MR1394747

[26] N. Eriksson, S. E. Fienberg, A. Rinaldo, and S. Sullivant, *Polyhedral conditions for the nonexistence of the MLE for hierarchical log-linear models*, J. Symbolic Comput. **41** (2006), no. 2, 222–233, DOI 10.1016/j.jsc.2005.04.003. MR2197157

[27] C. Eur, T. Fife, J. A. Samper, and T. Seynnaeve, *Reciprocal maximum likelihood degrees of diagonal linear concentration models*, Matematiche (Catania) **76** (2021), no. 2, 447–459, DOI 10.4418/2021.76.2.10. MR4334898

[28] D. R. Grayson and M. E. Stillman, *Macaulay2, a software system for research in algebraic geometry*, Available at http://www2.macaulay2.com.

[29] A. Grosdos, A. Heaton, K. Kubjas, O. Kuznetsova, G. Scholten, and M.-Ş. Sorea, *Exact solutions in log-concave maximum likelihood estimation*, Adv. in Appl. Math. **143** (2023), Paper No. 102448, 32, DOI 10.1016/j.aam.2022.102448. MR4500184

[30] E. Gross and S. Sullivant, *The maximum likelihood threshold of a graph*, Bernoulli **24** (2018), no. 1, 386–407, DOI 10.3150/16-BEJ881. MR3706762

[31] M. Härkönen, B. Hollering, F. T. Kashani, and J. I. Rodriguez, *Algebraic optimization degree*, ACM Commun. Comput. Algebra **54** (2020), no. 2, 44–48. MR4161231

[32] J. Hauenstein, J. I. Rodriguez, and B. Sturmfels, *Maximum likelihood for matrices with rank constraints*, J. Algebr. Stat. **5** (2014), no. 1, 18–38, DOI 10.18409/jas.v5i1.23. MR3279952

[33] S. Hoşten, A. Khetan, and B. Sturmfels, *Solving the likelihood equations*, Found. Comput. Math. **5** (2005), no. 4, 389–407, DOI 10.1007/s10208-004-0156-8. MR2189544

[34] J. Huh, *The maximum likelihood degree of a very affine variety*, Compos. Math. **149** (2013), no. 8, 1245–1266, DOI 10.1112/S0010437X13007057. MR3103064

[35] J. Huh, *Varieties with maximum likelihood degree one*, J. Algebr. Stat. **5** (2014), no. 1, 1–17, DOI 10.18409/jas.v5i1.22. MR3279951

[36] J. Huh and B. Sturmfels, *Likelihood geometry*, Combinatorial algebraic geometry, Lecture Notes in Math., vol. 2108, Springer, Cham, 2014, pp. 63–117, DOI 10.1007/978-3-319-04870-3_3. MR3329087

[37] Y. Jiang, K. Kohn, and R. Winter, *Linear spaces of symmetric matrices with non-maximal maximum likelihood degree*, Matematiche (Catania) **76** (2021), no. 2, 461–481, DOI 10.4418/2021.76.2.11. MR4334899

[38] K. Kubjas, E. Robeva, and B. Sturmfels, *Fixed points EM algorithm and nonnegative rank boundaries*, Ann. Statist. **43** (2015), no. 1, 422–461, DOI 10.1214/14-AOS1282. MR3311865

[39] S. L. Lauritzen, *Graphical models*, Oxford Statistical Science Series, vol. 17, The Clarendon Press, Oxford University Press, New York, 1996. Oxford Science Publications. MR1419991

[40] L. G. Maxim, J. I. Rodriguez, and B. Wang, *Euclidean distance degree of the multiview variety*, SIAM J. Appl. Algebra Geom. **4** (2020), no. 1, 28–48, DOI 10.1137/18M1233406. MR4048615

[41] L. G. Maxim, J. I. Rodriguez, and B. Wang, *Applications of singularity theory in applied algebraic geometry and algebraic statistics*, Handbook of geometry and topology of singularities VII, Springer, Cham, 2025, pp. 767–818, DOI 10.1007/978-3-031-68711-2_14. MR4890439

[42] L. G. Maxim, J. I. Rodriguez, B. Wang, and L. Wu, *Linear optimization on varieties and Chern-Mather classes*, Adv. Math. **437** (2024), Paper No. 109443, 22, DOI 10.1016/j.aim.2023.109443. MR4673253

[43] L. G. Maxim, J. I. Rodriguez, B. Wang, and L. Wu, *Logarithmic cotangent bundles, Chern-Mather classes, and the Huh-Sturmfels involution conjecture*, Comm. Pure Appl. Math. **77** (2024), no. 2, 1486–1508, DOI 10.1002/cpa.22156. MR4673886

[44] M. Michałek and B. Sturmfels, *Invitation to nonlinear algebra*, Graduate Studies in Mathematics, vol. 211, American Mathematical Society, Providence, RI, 2021, DOI 10.1090/gsm/211. MR4423369

[45] E. Robeva, B. Sturmfels, N. Tran, and C. Uhler, *Maximum likelihood estimation for totally positive log-concave densities*, Scand. J. Stat. **48** (2021), no. 3, 817–844, DOI 10.1111/sjos.12462. MR4303559

[46] E. Robeva, B. Sturmfels, and C. Uhler, *Geometry of log-concave density estimation*, Discrete Comput. Geom. **61** (2019), no. 1, 136–160, DOI 10.1007/s00454-018-0024-y. MR3925548

[47] J. I. Rodriguez and B. Wang, *The maximum likelihood degree of mixtures of independence models*, SIAM J. Appl. Algebra Geom. **1** (2017), no. 1, 484–506, DOI 10.1137/16M1088843. MR3686780

[48] B. Sturmfels, *Beyond linear algebra*, ICM—International Congress of Mathematicians. Vol. 6. Sections 12–14, EMS Press, Berlin, 2023, pp. 4820–4840. MR4680425

[49] B. Sturmfels and C. Uhler, *Multivariate Gaussian, semidefinite matrix completion, and convex algebraic geometry*, Ann. Inst. Statist. Math. **62** (2010), no. 4, 603–638, DOI 10.1007/s10463-010-0295-4. MR2652308

[50] S. Sullivant, *Algebraic statistics*, Graduate Studies in Mathematics, vol. 194, American Mathematical Society, Providence, RI, 2018, DOI 10.1090/gsm/194. MR3838364

[51] C. Uhler, *Geometry of maximum likelihood estimation in Gaussian graphical models*, Ann. Statist. **40** (2012), no. 1, 238–261, DOI 10.1214/11-AOS957. MR3014306

Institute of Mathematics, Technical University of Berlin, Berlin, Germany
Email address: amendola@math.tu-berlin.de

Department of Mathematics and Department of Electrical and Computer Engineering, University of Wisconsin–Madison, Madison, Wisconsin 53706
Email address: Jose@math.wisc.edu

Proceedings of Symposia in Pure Mathematics
Volume 111, 2025
https://doi.org/10.1090/pspum/111/02044

Developments in tropical convexity

Michael Joswig

Dedicated to Bernd Sturmfels
on the occasion of his sixtieth birthday.

ABSTRACT. The term "tropical convexity" was coined by Develin and Sturmfels who published a landmark paper with that title in 2004. However, the topic has much older roots and is deeply connected to linear and combinatorial optimization and other areas of mathematics. The purpose of this survey is to sketch how that article contributed to shaping the field of tropical geometry as we know it today.

1. Introduction

In their article, Speyer and Sturmfels wrote [**75**] "... the tropical approach in mathematics [is] now an integral part of geometric combinatorics and algebraic geometry. It has also expanded into mathematical physics, number theory, symplectic geometry, computational biology, and beyond." Why is this the case, and how did that happen? I think the reason is that the principles underlying tropical geometry are at the same time elementary and nonetheless deep. Since basic tropical arithmetic is so simple, it occurs everywhere and thus has been rediscovered independently many times, long before tropical mathematics was recognized as a subject of its own. However, it is the link to geometry, in its many flavors, which reveals the strength of these ideas. The article on *tropical convexity* by Develin and Sturmfels [**32**] provides ample evidence to support these claims.

Tropical convexity may be seen as the linear algebra part of tropical geometry. It has a considerable pre-tropical history under the name of "$(\max,+)$-linear algebra"; see, e.g., the monographs of Cuninghame-Green [**29**]; Baccelli, Cohen, Olsder, and Quadrat [**13**]; and Butkovič [**22**]. The DNA of $(\max,+)$-linear algebra comprises quite a diverse range of topics including combinatorial optimization, functional analysis, discrete event systems and game theory. One way of seeing

2020 *Mathematics Subject Classification.* 14T15, 14T90, 90C24.

Key words and phrases. Tropical geometry, polyhedral geometry, linear programming, combinatorial optimization.

Supported by Deutsche Forschungsgemeinschaft (EXC 2046: "MATH$^+$" and SFB-TRR 195: "Symbolic Tools in Mathematics and their Application"). The hospitality of the Hausdorff Institute for Mathematics in Bonn, during the Trimester Program "Discrete Optimization" is gratefully acknowledged.

I am indebted to George Balla, Anton Dochtermann, and two anonymous reviewers who helped me in improving the exposition.

tropical geometry is as a method to employ techniques from polyhedral geometry to solve problems in algebra and algebraic geometry. As its fundamental contribution the article [**32**] fuses (max, +)-linear algebra with that train of ideas. The benefit is mutual: Principles from algebraic geometry become accessible to the above named fields of application. Conversely, these applications help to identify interesting geometric questions.

The structure of this survey is as follows. In Section 2 we start out with first principles concerning *tropical cones* and *tropical polytopes*, and we finish with the main result of [**32**]. Throughout we occasionally deviate from the original notation and terminology of [**32**]. While some of the original results allow for natural generalizations (which were found later), here we choose a path which is meant to ease the reading. Section 3 is devoted to *tropical linear spaces*, which can be defined and motivated in many ways; here we describe them as a special class of tropical polytopes. Section 4 comprises examples showing the impact of tropical convexity outside tropical geometry.

For more details on tropical geometry in general we refer the reader to the monograph by Maclagan and Sturmfels [**64**] and to the book by Joswig [**48**].

2. Tropically convex sets and their structure

For tropical arithmetic, as in [**32**], we use the notation $\alpha \oplus \beta = \min(\alpha, \beta)$ and $\alpha \odot \beta = \alpha + \beta$ if $\alpha, \beta \in \mathbb{T} := \mathbb{R} \cup \{\infty\}$. The triplet $(\mathbb{T}, \oplus, \odot)$ is the *tropical semiring* (with respect to min). Tropical addition and scalar multiplication can be extended to vectors componentwise by letting $x \oplus y = (x_i \oplus y_i)_{i=1}^d$ and $\lambda \odot x = (\lambda \odot x_i)_{i=1}^d$, where $x, y \in \mathbb{T}^d$ and $\lambda \in \mathbb{T}$.

2.1. Tropical cones. The definition which marks the beginning of our topic is the following. A set $S \subset \mathbb{R}^d$ is called a *tropical cone* if

$$\lambda \odot x \oplus \mu \odot y \in S \quad \text{for all } x, y \in S \text{ and } \lambda, \mu \in \mathbb{R}. \tag{2.1}$$

Tropical addition is idempotent, and so it immediately follows that if a tropical cone S contains a vector $x \in \mathbb{R}^d$, then it also contains $\{\lambda \odot x \mid \lambda \in \mathbb{R}\} = x + \mathbb{R}\mathbb{1}$, where $\mathbb{1}$ is the all-ones-vector (here of length d). The following is [**32**, Proposition 4]; see also [**48**, Lemma 5.7].

PROPOSITION 1. *A subset S of $\mathbb{R}^d$ is a tropical cone if and only if*

$$S = \left\{\lambda_1 \odot x^{(1)} \oplus \ldots \oplus \lambda_k \odot x^{(k)} \mid \lambda_i \in \mathbb{R},\ x^{(i)} \in S\right\}.$$

Tropical cones are precisely the "idempotent semimodules", with respect to the tropical semiring, in the terminology of Cohen, Gaubert, and Quadrat [**25**]; see also [**38**].

A subset of the *tropical projective torus* $\mathbb{R}^d/\mathbb{R}\mathbb{1}$ is *tropically convex* if it arises as the canonical projection $S/\mathbb{R}\mathbb{1}$ of a tropical cone $S \subset \mathbb{R}^d$. For $M \subset \mathbb{R}^d/\mathbb{R}\mathbb{1}$ the *tropical convex hull*, written $\operatorname{tconv}(M)$, is the smallest tropically convex set containing M. The most influential concept from [**32**] until today is the following: A *tropical polytope* is the tropical convex hull of finitely many points. Figure 1 shows six tropically convex sets in the plane, and the three examples in the top rows are finitely generated, i.e., tropical polytopes. Yet the bottom three are not finitely generated. For instance, the unique minimal set of generators of the triangle (e) is the uncountable set of points on the diagonal edge. A first observation reveals a

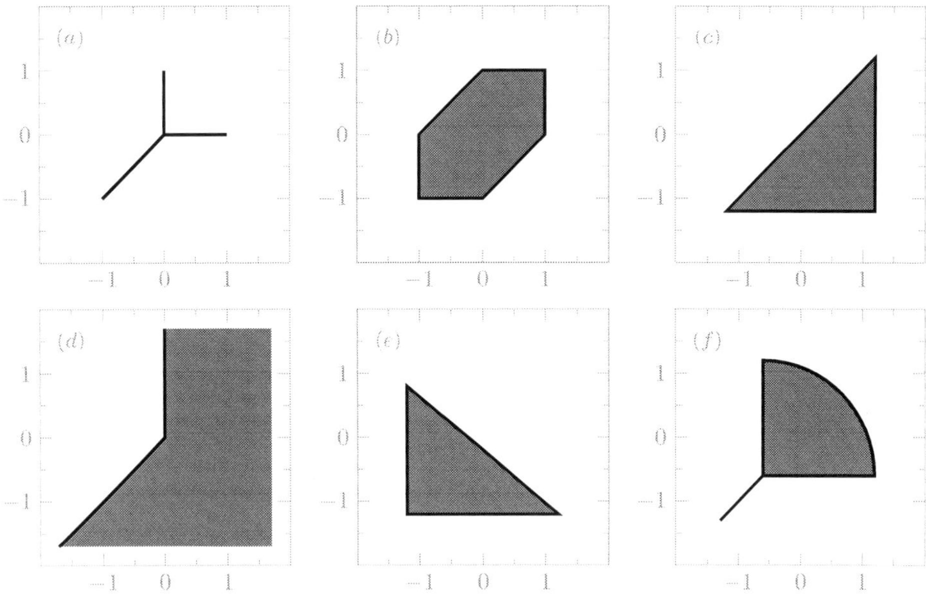

FIGURE 1. Six tropically convex sets in the plane $\mathbb{R}^3/\mathbb{R}\mathbb{1}$. Here and below a point $x \in \mathbb{R}^3$ is drawn in the Euclidean plane $\mathbb{R}^2$ as $(x_2 - x_1, x_3 - x_1)$; i.e., we consistently pick $(0, x_2 - x_1, x_3 - x_1)$ as a representative of $x + \mathbb{R}\mathbb{1}$.

similarity between tropical and ordinary polytopes; see [**32**, Proposition 21]. The statement can be generalized to arbitrary compact tropically convex sets [**48**, Theorem 5.43].

PROPOSITION 2. *For each tropical polytope P in $\mathbb{R}^d/\mathbb{R}\mathbb{1}$ there exists a unique minimal set V such that $P = \operatorname{tconv}(V)$.*

The points in this minimal generating set are the *tropical vertices* of P. It is interesting that tropical cones and tropical polytopes occurred in the literature many years before, e.g., in Gaubert's thesis [**38**]. Yet it is the decidedly combinatorial point of view of [**32**] which reveals crucial new structural information.

Before we explore these structural results, it is worth spending a moment to confirm that the notions "(tropical) cone" and "(tropical) convexity" are adequate. This is best seen by switching to the max-tropical semiring $(\mathbb{R} \cup \{-\infty\}, \max, +)$, which is isomorphic to the min-tropical semiring via passing to negatives. Then any real number $\alpha \in \mathbb{R}$ satisfies $\alpha > -\infty$, and $-\infty$ is the neutral element of the max-tropical addition. In this sense a tropical cone is closed with respect to taking tropical linear combinations with coefficients which are automatically positive. This "built-in positivity" is responsible for the hybrid character of tropically convex sets: as we will see below, to some extent they behave like tropical analogs of linear spaces, and to some extent they behave like tropical analogs of convex sets. Loho and Végh developed a generalization of tropical convexity which takes arbitrary (suitably defined) tropical signs into account [**62**].

In the sequel we want to use the min- and max-tropical semirings together. Whenever we need to distinguish these two tropical additions, we write $\underline{\oplus} = \min$

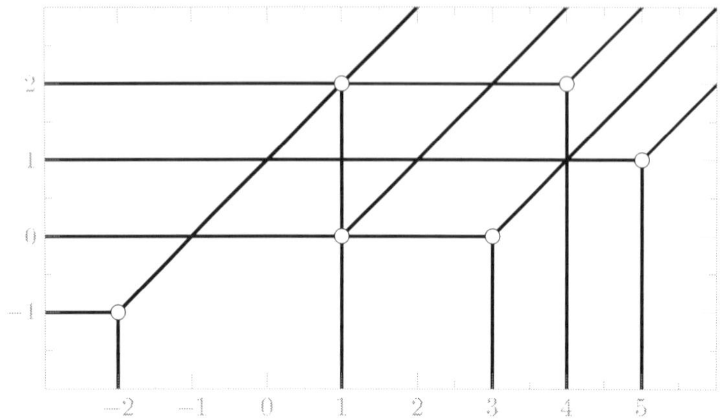

FIGURE 2. Max-tropical hyperplane arrangement in $\mathbb{R}^3/\mathbb{R}\mathbb{1}$.

and $\overline{\oplus}$ = max. Roughly speaking, the min-convention sides well with standard conventions concerning Puiseux series, while tropicalization with respect to max preserves the order relation; so the latter is the usual choice in optimization. Conceptually, it is clearly possible to express all tropical geometry in terms of either min or max. However, as in [**48**], we prefer to explain min-tropical convexity in terms of max-tropical hyperplanes. This is the topic of the subsequent section.

2.2. Arrangements of tropical hyperplanes. For $d = 1$ the tropical projective torus $\mathbb{R}^1/\mathbb{R}\mathbb{1}$ is a single point, which is not very interesting. So we assume that $d \geq 2$. In the sequel we will fix finitely many vectors $v^{(1)}, \ldots, v^{(n)} \in \mathbb{R}^d$, which we may identify with the columns of a matrix $V \in \mathbb{R}^{d \times n}$. We want to study the tropical polytope $P = \operatorname{tconv}(v^{(1)}, \ldots, v^{(n)})$, which we also write as $\operatorname{tconv}(V)$. To this end it turns out to be instrumental that the matrix $V = (v_{ij})$ can be employed in more than one way. For instance, the vector $-v^{(j)}$ gives rise to the homogeneous max-tropical linear form

$$-v_{1j} \odot X_1 \overline{\oplus} \cdots \overline{\oplus} -v_{dj} \odot X_d \tag{2.2}$$

in the d unknowns $X_1, \ldots, X_d$. The *(*max-*)tropical hyperplane* $\mathcal{T}(-v^{(j)})$ is the locus in $\mathbb{R}^d/\mathbb{R}\mathbb{1}$, where this tropical polynomial vanishes, i.e., the maximum in (2.2) is attained at least twice. It is immediate that the point $v^{(j)}$ is contained in $\mathcal{T}(-v^{(j)})$, since in that case the evaluation at each term gives zero; as $d \geq 2$ that zero (which is also the maximum) is attained at least twice. The point $v^{(j)}$ is the *apex* of $\mathcal{T}(-v^{(j)})$. Recall that we defined tropical convexity in terms of min as tropical addition, whereas (2.2) is a max-tropical polynomial.

Let $A = A_V$ be the product of the n linear forms $-v^{(1)}, \ldots, -v^{(n)}$ in the semiring of homogeneous max-tropical polynomials. Then the max-tropical hypersurface

$$\mathcal{T}(A) = \mathcal{T}(-v^{(1)}) \cup \cdots \cup \mathcal{T}(-v^{(n)})$$

is the union of n tropical hyperplanes in $\mathbb{R}^d/\mathbb{R}\mathbb{1}$. The topological closure of a connected component of the complement $\mathbb{R}^d/\mathbb{R}\mathbb{1} \setminus \mathcal{T}(A)$ is called a *region* of A. Each region is an ordinary convex polyhedron in $\mathbb{R}^d/\mathbb{R}\mathbb{1} \simeq \mathbb{R}^{d-1}$, and the regions form a polyhedral subdivision of the tropical projective torus; we denote the latter as $\mathcal{S}_V$.

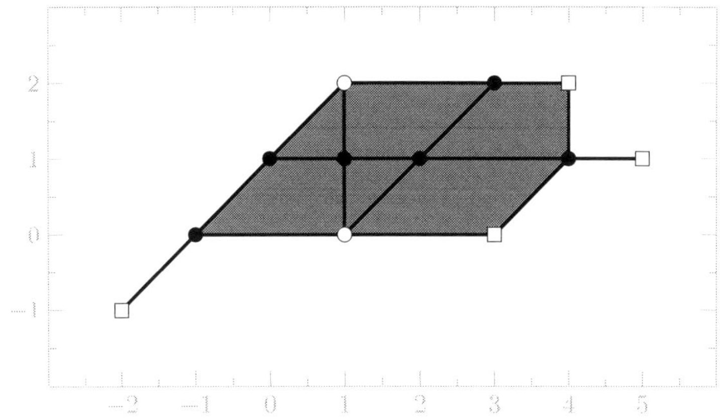

FIGURE 3. Min-tropical polygon in $\mathbb{R}^3/\mathbb{R}\mathbb{1}$. The six generators are white; among these the four tropical vertices are marked by squares. The two circled generators are redundant for the tropical convex hull.

A version of the *structure theorem of tropical convexity* is the following. Its essence is already expressed in [**32**, Theorems 1 and 15, Proposition 16]. However, the connection to tropical hyperplane arrangements was made explicit by Ardila and Develin [**12**].

THEOREM 3. *For* $V \in \mathbb{R}^{d\times n}$ *we have:*

(1) *The polyhedral decomposition* $\mathcal{S}_V$ *of* $\mathbb{R}^d/\mathbb{R}\mathbb{1}$*, which is formed by the regions of the* max*-tropical hyperplane arrangement* A_V*, is dual to the (lower) regular subdivision induced by* V*, considered as a height function on the vertices of the ordinary polytope* $\Delta_{d-1} \times \Delta_{n-1}$*.*
(2) *The* min*-tropical polytope* tconv(V) *agrees with the union of the bounded cells of the polyhedral complex* $\mathcal{S}_V$*.*

In [**32**] the decomposition of tconv(V) as an ordinary polyhedral complex from Theorem 3(2) is called the "tropical complex" induced by V. A more modern term for the same is *covector decomposition*; this refers to Ardila and Develin [**12**], who called a "covector" what was called a "type" in [**32**]. The covector decomposition is linearly isomorphic with the subcomplex of bounded faces of the unbounded ordinary polyhedron dual to $\mathcal{S}_V$.

EXAMPLE 4. We consider the 3×6-matrix

$$V = \begin{pmatrix} 0 & 0 & 0 & 0 & 0 & 0 \\ -2 & 1 & 3 & 4 & 5 & 1 \\ -1 & 0 & 0 & 2 & 1 & 2 \end{pmatrix},$$

whose columns form the apices of six max-tropical lines in $\mathbb{R}^3/\mathbb{R}\mathbb{1}$. We obtain an arrangement A_V with 20 regions; see Figure 2. Six of these regions are bounded, and together with two bounded edges, they form the min-tropical polytope tconv(V); see Figure 3.

Various extensions of Theorem 3 are known. Notably, Ardila and Develin [**12**] introduced "tropical oriented matroids" to generalize tropical hyperplane arrangements. Horn [**43**] showed that tropical oriented matroids correspond bijectively to the subdivisions of products of simplices which are not necessarily regular. In Section 2.4 we will see how infinite coordinates lead to *subpolytopes* of $\Delta_{d-1} \times \Delta_{n-1}$ (i.e., convex hulls of subsets of the vertices) and their subdivisions.

2.3. Products of simplices and their triangulations. Let $V \in \mathbb{R}^{d\times n}$. The structure theorem (Theorem 3) ties the tropical convex hull $\operatorname{tconv}(V)$ to the regular subdivision $\mathcal{S}_V$ of $\Delta_{d-1} \times \Delta_{n-1}$. If $\mathcal{S}_V$ is a triangulation, the matrix V and the tropical polytopes $\operatorname{tconv}(V)$ are called *generic*. The product $S_d \times S_n$ of symmetric groups operates naturally on the $d \cdot n$ vertices of $\Delta_{d-1} \times \Delta_{n-1}$. For generic $V, W \in \mathbb{R}^{d\times n}$ the two triangulations $\mathcal{S}_V$ and $\mathcal{S}_W$, are *equivalent* if they lie in the same orbit with respect to the induced action. It is an interesting task, but difficult, to enumerate these equivalence classes, also called *combinatorial types*, for given d and n.

EXAMPLE 5. For $d = 2$ the product $\Delta_{d-1} \times \Delta_{n-1}$ is a prism over Δ_{n-1}. This polytope has exactly $n!$ labeled triangulations, which form a single orbit; see [**31**, §6.2.1] and [**48**, §4.6]. So the first row of Table 1 contains ones.

Standard software for enumerating (regular) triangulations up to symmetry includes TOPCOM [**69, 70**] and mptopcom [**45, 46**]. The known values for products of simplices are given in Table 1. The corresponding table in [**32**, p. 20] contains two typos, which have been corrected in [**64**, p. 241]. The first nontrivial values were probably computed by a Maple program called PUNTOS, which was written by Jesús De Loera. The table [**64**, p. 241] was obtained with TOPCOM, and the additional values, for $(d, n) \in \{(3, 7), (4, 5)\}$, have been found with mptopcom [**45**]. With TOPCOM v1.1.0, Rambau recently computed the total number of symmetry classes of all triangulations, including the nonregular ones, of $\Delta_3 \times \Delta_5$ as 25,606,173,722; yet the corresponding regular count, which would be the entry (4,6) in Table 1, seems to be currently unknown.

TABLE 1. The known numbers of combinatorial types of regular triangulations of $\Delta_{d-1} \times \Delta_{n-1}$, up to symmetry.

$d \setminus n$	2	3	4	5	6	7
2	1	1	1	1	1	1
3		5	35	530	13 621	531 862
4			7 869	7 051 957		

REMARK 6. The product of two simplices $\Delta_{d-1} \times \Delta_{n-1}$ is known to admit nonregular triangulations if and only if $(d-2)(n-2) \geq 4$; see [**31**, Theorem 6.2.19]. While such nonregular triangulations are not directly relevant for tropical geometry, their mere existence (and abundance, in higher dimensions) is one of the obstacles to overcome in obtaining full enumerations of the regular triangulations.

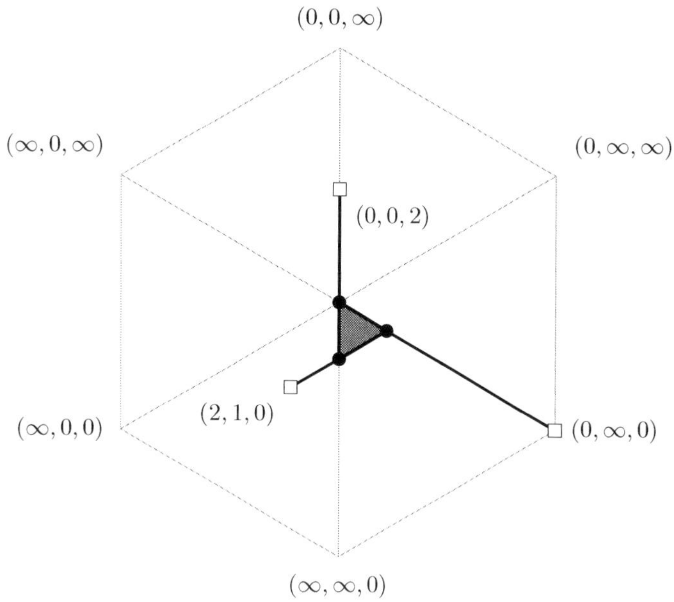

FIGURE 4. Tropical triangle in $\mathbb{TP}^2$ (isometric sketch). Here $\mathbb{TP}^2$ takes the shape of a regular hexagon, and $\mathbb{R}^3/\mathbb{R}\mathbb{1}$ is its interior.

2.4. Infinite coordinates. In the matrices V that we considered so far the neutral element, ∞, of the tropical addition $\oplus = \min$ was explicitly forbidden. Let the *support* of $V \in \mathbb{T}^{d\times n}$ be the subset of $[d] \times [n]$ chosen by the finite coefficients. Fink and Rincón generalized part (1) of the structure theorem (Theorem 3) to arbitrary V, where $\Delta_{d-1} \times \Delta_{n-1}$ is replaced by the subpolytope defined by the support of V; see [**37**, §4] and [**48**, Corollary 4.13]. In order to understand how part (2) can be generalized, we first need to re-examine our basic definitions.

Clearly, we can replace $\mathbb{R}^d$ by $\mathbb{T}^d$ to obtain tropical cones in $\mathbb{T}^d$, with the analogous definition to (2.1). Let

$$\mathbb{TP}^{d-1} := \left(\mathbb{T}^d \setminus \{\infty\mathbb{1}\}\right)/\mathbb{R}\mathbb{1}$$

be the *tropical projective space* of rank $d-1$, where $\infty\mathbb{1}$ is the vector of length d where each coefficient is ∞. Now passing from $\mathbb{T}^d \setminus \{\infty\mathbb{1}\}$ to $\mathbb{TP}^{d-1}$ by taking the quotient provides notions of tropical convexity and tropical polytopes in tropical projective spaces.

The construction above is beneficial from a topological point of view as $\mathbb{TP}^{d-1}$ compactifies $\mathbb{R}^d/\mathbb{R}\mathbb{1}$. In fact, the pair $(\mathbb{TP}^{d-1}, \mathbb{R}^d/\mathbb{R}\mathbb{1})$ is homeomorphic with the pair formed by a $(d-1)$-dimensional closed ball and its interior, see [**48**, Proposition 5.3]; the planar case $d = 3$ is sketched in Figure 4. The boundary equals the set difference $\partial\mathbb{TP}^{d-1} = \mathbb{TP}^{d-1} \setminus (\mathbb{R}^d/\mathbb{R}\mathbb{1})$, and this is naturally stratified by the supports of vectors in $\mathbb{T}^d$. Here the *support* of $v \in \mathbb{T}^d$ is the set of indices i such that v_i is finite. For $Z \subsetneq [d]$ let $\mathbb{TP}(Z)$ be the *stratum* of points whose support equals the complement $[d] \setminus Z$. This set is a copy of $\mathbb{R}^{d-k}/\mathbb{R}\mathbb{1}$, where k is the cardinality of Z. Since the strata in $\mathbb{TP}^{d-1}$ correspond bijectively to the proper subsets of $[d]$, the tropical projective space $\mathbb{TP}^{d-1}$ inherits the combinatorial structure of a simplex.

OBSERVATION 7. Let $P \subseteq \mathbb{TP}^{d-1}$ be a tropical polytope. For any $Z \subsetneq [d]$ the intersection $P \cap \mathbb{TP}(Z)$ is tropically convex.

We call $P \cap \mathbb{TP}(Z)$ a *stratum* of P. In this way, by applying Theorem 3(1) to the strata and taking suitable unbounded hyperplane regions into account, we arrive at a generalization of Theorem 3(2) to tropical polytopes in tropical projective space. The following example shows that the strata of a tropical polytope are not necessarily finitely generated.

EXAMPLE 8. We consider the matrix

$$V = \begin{pmatrix} 0 & 0 & 2 \\ \infty & 0 & 1 \\ 0 & 2 & 0 \end{pmatrix} \in \mathbb{T}^{3\times 3},$$

and $T = \operatorname{tconv}(V)$ is the tropical triangle in $\mathbb{TP}^2$ shown in Figure 4. For instance, the stratum $\mathbb{TP}(\{2\})$ is the set of points with homogeneous coordinates (v_1, ∞, v_3) where $v_1, v_3 \in \mathbb{R}$; the intersection $P \cap \mathbb{TP}(\{2\})$ comprises the single point $(0, \infty, 0)$. All other boundary strata of P are empty. The generic stratum $P \cap \mathbb{TP}(\emptyset)$ is tropically convex, but not a tropical polytope in the tropical projective torus $\mathbb{R}^3/\mathbb{R}\mathbb{1} = \mathbb{TP}(\emptyset)$.

2.5. Eigenvalues and eigenvectors. We start out with studying another tropical analog of ordinary linear algebra. Let $A = (a_{ij})$ be an $n\times n$-matrix with coefficients in the tropical semiring $\mathbb{T}$. A real number λ is a *tropical eigenvalue* of A if there is a vector $x \in \mathbb{T}^n \setminus \{\infty\mathbb{1}\}$ such that

$$A \odot x = \lambda \odot x, \tag{2.3}$$

in which case x is a *tropical eigenvector* of A with respect to the eigenvalue λ. One can read the matrix A as the weighted adjacency matrix of a directed graph $\Gamma = \Gamma(A)$ on n nodes. Here $a_{ij} = \infty$ signals the absence of the arc (i, j). However, when a_{ij} is finite, this value is the weight of that arc. The *mean weight* of a directed path π with k arcs $(i_0, i_1), (i_1, i_2), \ldots, (i_{k-1}, i_k)$ is the value $c(\pi) = \frac{1}{k}(a_{i_0,i_1} + \cdots + a_{i_{k-1},i_k})$. For $i_k = i_0$ the path π is a cycle, and the mean weight $c(\pi)$ is called the *cycle mean* of π. Now the *minimum cycle mean* of A is defined as

$$\lambda(A) = \min\{c(\zeta) \mid \zeta \text{ cycle in } \Gamma(A)\}.$$

THEOREM 9. *Let $\Gamma(A)$ be strongly connected. Then the minimum cycle mean $\lambda(A)$ is the only tropical eigenvalue of A.*

Karp [**55**] gave a polynomial time algorithm for computing $\lambda(A)$; see also [**48**, §3.6]. This tropical eigenvalue is also the smallest root of its characteristic tropical polynomial [**64**, Corollary 5.1.5].

For the rest of this section we assume that $\Gamma(A)$ is strongly connected and $\lambda(A)$ equals zero. In this case it is easy to describe the tropical eigenvectors of A. Namely, we can form the *Kleene star*, which is the matrix

$$A^* = I \oplus A \oplus A^{\odot 2} \oplus \cdots \oplus A^{\odot(n-1)}, \tag{2.4}$$

where $A^{\odot k}$ is the kth tropical matrix power, and $I = A^{\odot 0}$ is the tropical identity matrix, with zero diagonal and all off-diagonal entries equal to ∞. Then we have $A \odot A^* = A^*$. That is to say that each column of A^* is a tropical eigenvector of A

with respect to the eigenvalue zero; see [**48**, Proposition 3.39]. More generally, the *tropical eigenspace* is the set

$$\operatorname{Eig}(A) \;=\; \{x \in \mathbb{T}^n \setminus \{\infty\mathbb{1}\} \mid A \odot x = x\}\,.$$

For $x, y \in \operatorname{Eig}(A)$ and $\lambda, \mu \in \mathbb{R}$ we compute

$$A \odot (\lambda \odot x \oplus \mu \odot y) \;=\; \lambda \odot (A \odot x) \oplus \mu \odot (A \odot y) \;=\; \lambda \odot x \oplus \mu \odot y,$$

and this says that the tropical eigenspace $\operatorname{Eig}(A)$ is a tropical cone. In fact, we have $\operatorname{Eig}(A) = \operatorname{tconv}(A^*)$ and, by construction, $\operatorname{tconv}(A^*) \subseteq \operatorname{tconv}(A)$.

EXAMPLE 10. The matrix V from Example 8 has minimum cycle mean $\lambda(V)=0$, and the Kleene star reads

$$V^* \;=\; \begin{pmatrix} 0 & 1 & 0 \\ 0 & 0 & 0 \\ 1 & 1 & 0 \end{pmatrix}.$$

The directed graph $\Gamma(V)$ is strongly connected, and each column of V^* lies in a cycle of weight zero. The tropical eigenspace is the tropical cone in $\mathbb{TP}^{n-1}$ spanned by the columns of V^*. In Figure 4 the eigenspace $\operatorname{Eig}(V) = \operatorname{Eig}(V^*)$, or rather its image in $\mathbb{TP}^2$, is the unique triangular cell in the covector decomposition of $\operatorname{tconv}(V)$.

The tropical convex hulls of Kleene stars play a very special role in tropical convexity. We continue with our general assumptions that $A \in \mathbb{T}^{n\times n}$ is a matrix such that $\Gamma(A)$ is strongly connected. Then A^* has finite coefficients only, and $\operatorname{tconv}(A^*)/\mathbb{R}\mathbb{1}$ is a tropical polytope in the tropical projective torus. We can identify $\mathbb{R}^n/\mathbb{R}\mathbb{1}$ with $\mathbb{R}^{n-1}$ via $(x_1, x_2, \ldots, x_n) + \mathbb{R}\mathbb{1} \mapsto (x_2 - x_1, \ldots, x_n - x_1)$; this identification was already used for drawing our figures. The following statement is essentially equivalent to [**32**, Proposition 18]; see also [**48**, Theorem 6.38].

PROPOSITION 11. *The tropical polytope* $\operatorname{tconv}(A^*)$, *seen as a subset of* $\mathbb{R}^{n-1} = \mathbb{R}^n/\mathbb{R}\mathbb{1}$ *is an ordinary convex polytope. We have*

$$\operatorname{tconv}(A^*) = \left\{x \in \mathbb{R}^n/\mathbb{R}\mathbb{1} \mid x_i - x_j \le a^*_{ij}\right\}.$$

The tropical polytopes like $\operatorname{tconv}(A^*)$ are exactly those tropical polytopes which, seen as subsets of $\mathbb{R}^{n-1}$, are convex in the ordinary sense. In [**49**] they have been called *polytropes*; see also [**48**, §6.5]. The bounded regions of tropical hyperplane arrangements, i.e., the bounded cells of the covector decomposition from the structure theorem (Theorem 3) are exactly the polytropes. Further, the integral polytropes, i.e., the tropical convex hulls of Kleene stars with integer coefficients, are precisely the *alcoved polytopes* (of type A) of Lam and Postnikov [**57**]. The hexagonal shape of the tropical projective plane in Figure 4 refers to the equality

$$\mathbb{TP}^2 \;=\; \operatorname{tconv}\begin{pmatrix} 0 & \infty & \infty \\ \infty & 0 & \infty \\ \infty & \infty & 0 \end{pmatrix},$$

which means that we may view $\mathbb{TP}^{n-1}$ as an "infinitely large" tropical unit ball, which is a polytrope [**48**, Example 6.39].

The Kleene star $A^* = (a^*_{ij})$ contains relevant information about the directed graph $\Gamma(A)$. More precisely, the coefficient a^*_{ij} records the length (or weight) of a shortest path from i to j. Computing shortest paths is a basic problem in combinatorial optimization. Suitable algorithms are, e.g., at the core of common navigational devices. In [**52**] we studied a generalization, where the coefficients of the matrix A are tropical polynomials. Again the definition (2.4) of the Kleene star makes sense, and then A^* describes the lengths of shortest paths in a parametric setting. The original motivation for [**52**] was a new computational approach for classifying the combinatorial types of polytropes. This continued previous work of Sturmfels and Tran [**76**] and Tran [**77**]. The parametric shortest path algorithms proposed in [**52**] were used by Cleveland et al. [**23**] for developing delay tolerant networks, with the ultimate goal to construct the Solar System Internet (SSI).

2.6. Tropical halfspaces. As its main benefit the compactification into $\mathbb{TP}^{d-1}$ makes tropical convexity more symmetric in the following sense. The main theorem for ordinary cones [**81**, §1.3 and §1.4] says that each ordinary cone is the intersection of finitely many linear halfspaces and conversely. In the tropical projective space there is a corresponding result for tropical polytopes. To describe it we need to explain what the tropical analogs to linear halfspaces are.

Let $a^+, a^- \in \mathbb{T}^d \backslash \{\infty \mathbb{1}\}$. For conciseness we assume that $\operatorname{supp}(a^+) \cap \operatorname{supp}(a^-) = \emptyset$. Then the set $\{x \in \mathbb{T}^d \mid a^+ \odot x \leq a^- \odot x\}$ is a *tropical halfspace* in $\mathbb{T}^d$, and we use the same name for the image under the projection to $\mathbb{TP}^{d-1}$. The following was proved by Gaubert and Katz [**39**], and this corrects an error in [**47**].

THEOREM 12. *Every tropical polytope in $\mathbb{TP}^{d-1}$ is the intersection of finitely many tropical halfspaces. Conversely, every such intersection is a tropical polytope.*

While the above result suggests that tropical polytopes are similar to ordinary polytopes, there are remarkable distinctions.

EXAMPLE 13. The tropical triangle T from Example 8 is the intersection of five tropical halfspaces. More precisely, T is the set of points $x \in \mathbb{TP}^2$ satisfying the homogeneous tropical matrix inequality

$$\begin{pmatrix} \infty & 0 & 0 \\ 0 & \infty & 1 \\ 0 & -1 & \infty \\ \infty & \infty & -2 \\ -2 & \infty & \infty \end{pmatrix} \odot x \leq \begin{pmatrix} 0 & \infty & \infty \\ \infty & 0 & \infty \\ \infty & \infty & 0 \\ 0 & 0 & \infty \\ \infty & -1 & 0 \end{pmatrix} \odot x. \tag{2.5}$$

Each pair of corresponding rows of the two matrices to the left and to the right defines one tropical halfspace. The inequality description (2.5) is irredundant. The tropical triangle T is depicted in Figure 4; yet the visualization in $\mathbb{TP}^2$ is not canonical in the sense that the boundary is deliberately drawn at some finite distance from the origin. Consequently, in such a picture some tropically convex sets would look nonlinear near the boundary; e.g., see [**48**, Figure 6.5].

In ordinary polytope theory McMullen's upper bound theorem [**66**] dictates how many inequalities are necessary at most to describe a polytope; see also [**81**, Theorem 8.23]. Allamigeon, Gaubert, and Katz [**8**] proved an analog for tropical polytopes; see also [**48**, Theorem 8.10]. The precise bounds differ slightly between the ordinary and the tropical setting, since nonnegativity constraints play a special role. So it may happen that a tropical triangle, i.e., the tropical convex hull of three points, like the one in Example 8, requires five tropical linear inequalities.

Arrangements of tropical halfspaces can also be studied combinatorially. This leads to yet another variation of the structure theorem (Theorem 3); see [**48**, §7.4].

2.7. Real Puiseux series. After the discussion above it is natural ask: how exactly are ordinary and tropical polytopes related? One way to answer this question goes through the tropicalization of ordinary polytopes defined over suitable ordered fields. A *(real) Puiseux series* is a formal power series of the form

$$\gamma(t) \;=\; \sum_{k=\ell}^{\infty} c_k t^{k/N},$$

where $\ell \in \mathbb{Z}$, $N \in \mathbb{N}$, $c_k \in \mathbb{R}$, and $c_\ell \neq 0$. With the coefficient-wise addition and the usual product of formal power series the real Puiseux series form a field, which we denote by $\mathbb{R}\{\{t\}\}$. Moreover, this field is ordered: $\gamma(t)$ inherits the sign from its leading coefficient c_ℓ. So a real Puiseux series is nonnegative if it is zero or its leading coefficient is positive. The ordered field $\mathbb{R}\{\{t\}\}$ admits an additional structure, which becomes visible by considering the *order map* $\operatorname{ord} : \mathbb{R}\{\{t\}\} \to \mathbb{Q} \cup \{\infty\}$, which sends $\gamma(t)$ to its leading exponent ℓ/N; and we set $\operatorname{ord}(0) = \infty$. The order map forms a discrete valuation. We have

$$\operatorname{ord}(\gamma+\delta) = \min\bigl(\operatorname{ord}(\gamma), \operatorname{ord}(\delta)\bigr) \quad \text{and} \quad \operatorname{ord}(\gamma \cdot \delta) = \operatorname{ord}(\gamma) + \operatorname{ord}(\delta),$$

provided that γ and δ both are strictly positive. This can be summarized as follows.

LEMMA 14. *Restricting the order map to nonnegative real Puiseux series yields a homomorphism* $\operatorname{ord} : \mathbb{R}\{\{t\}\}_{\geq 0} \to \mathbb{T}$ *of semirings.*

A minor technical nuisance is that the order map is not surjective onto $\mathbb{T}$. This can be remedied in several ways, for instance by passing to a suitable extension field which is large enough. One choice is a construction suggested by Markwig [**65**]; see also [**48**, §2.7]. In the sequel we assume that $\mathbb{L}$ is some real closed field [**18**], equipped with a valuation map, ord, which is surjective onto $\mathbb{T}$. For such fields Develin and Yu [**33**] proved the following; see also [**48**, Proposition 5.8]. Note that cones, polytopes, linear programming and such make sense over arbitrary ordered fields; e.g., see [44] and [**48**, §A.1].

THEOREM 15. *For any ordinary polyhedral cone, C, which is contained in the nonnegative orthant $\mathbb{L}^d_{\geq 0}$, the pointwise image* $\operatorname{ord}(C)$ *is a tropical cone in $\mathbb{T}^d$. Conversely, each tropical cone arises in this way.*

This implies a similar result for ordinary and tropical polytopes by considering the quotients $(\mathbb{L}^d_{\geq 0} \setminus \{0\})/\mathbb{L}_{>0}$ and $\mathbb{TP}^d = (\mathbb{T}^d \setminus \{\infty \mathbb{1}\})/\mathbb{R}\mathbb{1}$. The significance of Theorem 15 comes from the fact that $\mathbb{L}$ was assumed to be real closed. In this case the Tarski–Seidenberg transfer principle [**18**, Corollary 5.2.4] entails that the ordered fields $\mathbb{L}$ and $\mathbb{R}$ are elementary equivalent. In particular, the combinatorial types of polytopes over $\mathbb{L}$ and over $\mathbb{R}$ are the same; see [**14**, Proposition 5.12], [**51**] and [**48**, Theorem 8.36].

REMARK 16. For two real Puiseux series $\gamma, \delta \in \mathbb{R}\{\{t\}\}$ with $\gamma \leq \delta$ we have $\operatorname{ord}(\gamma) \geq \operatorname{ord}(\delta)$, which reverses the ordering.

2.8. Further aspects of tropical convexity and beyond. The connection between ordinary and tropical convexity can also be explored from a systematic point of view. For instance, Cohen, Gaubert, and Quadrat [**24**, **25**] investigated subsemimodules and convex subsets of general idempotent semimodules. To overcome the crucial nonnegativity assumption in Theorem 15, Loho and Végh proposed a version of tropical convexity in the signed tropical semiring [**63**]. Motivated by aspects of game theory Akian, Gaubert, and Vanucci introduced the concept of "ambitropical convexity" [**4**]. As a vast generalization in a different direction, Alessandrini [**5**] explored the tropicalization of arbitrary semialgebraic sets; Allamigeon, Gaubert, and Skomra studied the special case of tropical spectrahedra [**9**].

3. Tropical linear spaces

Speyer and Sturmfels [**74**] investigated the tropicalization of ordinary linear spaces and their parameter spaces. This leads to tropical linear spaces, which arise as certain tropical polytopes in a tropical projective space. We fix d and n such that $d \leq n$. A *finite tropical Plücker vector* is a map $\pi : \binom{[n]}{d} \to \mathbb{R}$ such that the minimum of the three numbers

$$\pi(\rho + ij) + \pi(\rho + k\ell)\,,\ \pi(\rho + ik) + \pi(\rho + j\ell)\,,\ \pi(\rho + i\ell) + \pi(\rho + jk)$$

is attained at least twice for any $(d-2)$-subset $\rho \subset [n]$ and pairwise distinct $i, j, k, \ell \in [n] - \rho$. Here we use the shortcut $\rho + ij$ for the set $\rho \cup \{i, j\}$; more generally, we employ this notation for the disjoint union of sets where the second set is written as a list of its elements. Let $\sigma \subset [n]$ have cardinality $d - 1$. Then $\sigma*$ is defined as the vector in $\mathbb{T}^n$ whose ith coordinate is $\pi(\sigma + i)$ if $i \notin \sigma$ and ∞ otherwise. The vector $\sigma*$ is a *cocircuit* of π. Now the tropical convex hull in $\mathbb{TP}^{n-1}$ of all cocircuits of π is the *uniform tropical linear space* $\mathcal{L}_\pi$.

EXAMPLE 17. With $d = 2$ and $n = 5$ we consider the $2{\times}5$-matrix

$$M = \begin{pmatrix} 1 & 0 & t & t^2 & t^3 \\ 0 & 1 & 1 & t-1 & t^{-4} \end{pmatrix},$$

whose entries are Puiseux series. Denoting the $2{\times}2$ submatrix formed by columns i and j by $M^{(ij)}$, we get a map $\pi : \binom{[5]}{2} \to \mathbb{R}$ by setting $\pi(ij) = \operatorname{ord}(\det(M^{(ij)}))$. This reads

$$12 \mapsto 0\,,\ 13 \mapsto 0\,,\ 14 \mapsto 0\,,\ 15 \mapsto -4\,,\ 23 \mapsto 1\,,$$
$$24 \mapsto 2\,,\ 25 \mapsto 3\,,\ 34 \mapsto 1\,,\ 35 \mapsto -3\,,\ 45 \mapsto -2.$$

Since the maximal minors of a matrix satisfy the Plücker relations, it follows that π is a tropical Plücker vector; see [**64**, §4.3]. The five cocircuits are the rows (or columns) of the symmetric matrix

$$\begin{pmatrix} \infty & 0 & 0 & 0 & -4 \\ 0 & \infty & 1 & 2 & 3 \\ 0 & 1 & \infty & 1 & -3 \\ 0 & 2 & 1 & \infty & -2 \\ -4 & 3 & -3 & -2 & \infty \end{pmatrix}. \tag{3.1}$$

Their tropical convex hull $\mathcal{L}_\pi$ in $\mathbb{TP}^4$, intersected with $\mathbb{R}^5/\mathbb{R}\mathbb{1}$, is the stratum $\mathcal{L} \cap \mathbb{TP}(\emptyset)$; it is a contractible one-dimensional polyhedral complex with two bounded edges and five unbounded ones. They form a trivalent tree with five leaves, one for each cocircuit; the leaves are the generators, and they sit in the boundary

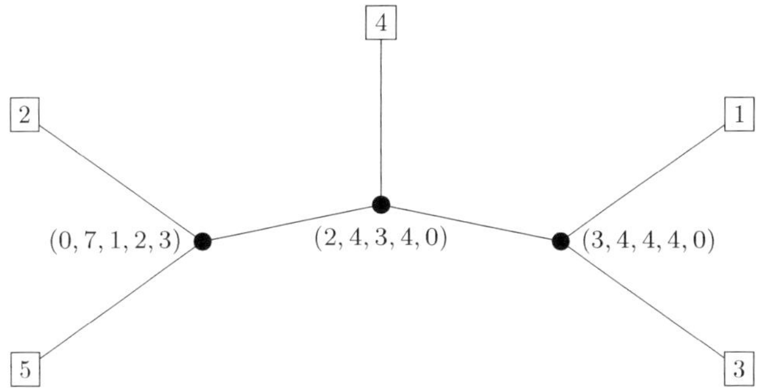

FIGURE 5. A tropicalized linear space in $\mathbb{TP}^4$. Leaf labels correspond to row/column indices of the cocircuit matrix (3.1).

$\partial\mathbb{TP}^4 = \mathbb{TP}^4 \setminus \mathbb{R}^5/\mathbb{R}\mathbb{1}$. The stratum $\mathcal{L}_\pi \cap \mathbb{TP}(\{i\})$ comprises the ith cocircuit only, and all other strata are empty. The tree is shown in Figure 5, with homogeneous coordinates in $\mathbb{R}^5/\mathbb{R}\mathbb{1}$ for the three interior nodes.

A tropical linear space is called a *tropicalized linear space* if it arises from the procedure followed in Example 17. In this case we start out with a matrix whose coefficients are complex Puiseux series, and the resulting tropical linear space is the tropicalization of the row space of that matrix. Classically, the row spaces of $d\times n$-matrices of full row rank d parameterize the d-dimensional linear subspaces in n-space, and they form the Grassmannian $\mathrm{Gr}(d,n)$. The Grassmannian is an algebraic variety, defined over the field of complex Puiseux series, cut out by the Plücker relations. However, since the coefficients of the Plücker relations are integers, i.e., complex Puiseux series which are constants, it suffices to view $\mathrm{Gr}(d,n)$ as a complex variety. Tropicalizing $\mathrm{Gr}(d,n)$, we arrive at the tropical variety $\mathrm{TGr}(d,n)$, which is called the *tropical Grassmannian*; see [**74**] or [**64**, §4.3]. The tropical Grassmannian $\mathrm{TGr}(d,n)$ is a polyhedral fan in

$$\mathbb{R}^{\binom{n}{d}}.$$

That fan is pure, and all its maximal cones share the same dimension $(n-d)d+1$. After modding out linealities and passing to the intersection with the unit sphere we arrive at a spherical polytopal complex of dimension $nd-n-d^2$.

THEOREM 18 (Speyer and Sturmfels [**74**]). *The points in* $\mathrm{TGr}(d,n)$ *are precisely the tropical Plücker vectors of tropicalized linear spaces.*

We have $\mathrm{TGr}(d,n) \cong \mathrm{TGr}(n-d,n)$ as polyhedral fans. Hence, for the rest of this section, we assume $n \geq 2d$. For $d=2$ the tropical Grassmannians $\mathrm{TGr}(2,n)$ are spaces of phylogenetic trees on n labeled leaves. In this case each tropical linear space is tropicalized. However, for $d \geq 3$ and $n \geq 7$ this is no longer true. In these cases the *Dressian* $\mathrm{Dr}(d,n)$, which is the polyhedral fan comprising all tropical Plücker vectors, is strictly larger than $\mathrm{TGr}(d,n)$.

REMARK 19. The tropical Grassmannians $\mathrm{TGr}(3,n)$ and the Dressians $\mathrm{Dr}(3,n)$ have explicitly been determined for the first few values of n: $n=6$ already occurs in [**74**], the case $n=7$ was resolved in [**41**]. The current state of the art is marked by $n=8$. Despite that the Dressian is (much) larger than the tropical Grassmannian it is easier to compute; this was accomplished already in [**42**]. Computing $\mathrm{TGr}(3,8)$ is a very recent achievement, which required sophisticated technology and nontrivial parallel algorithms in computer algebra [**15**]. About $\operatorname{TGr}(4,8)$ and $\operatorname{Dr}(4,8)$ and their even bigger friends we only have very limited partial information.

4. Applications

Recent years saw an increasing number of applications of tropical geometry, and tropical convexity especially, to many fields of mathematics and beyond. Here we want to highlight one application to topics in computational complexity. We close this section with a brief collection of further references.

4.1. Linear programming. An ordinary linear program is an optimization problem which seeks to minimize (or maximize) a linear function over an ordinary polyhedron. Classically this has been studied over the reals, but the concept makes sense over any ordered field, in particular, the field of real Puiseux series studied in Section 2.7. Before we look at a tropical analog, we want to re-examine our setup.

By Remark 16 the valuation map ord, which sends a real Puiseux series to its lowest degree, reverses the order. If we want to relate ordinary and tropical inequalities, this change of orientation can be confusing. To this end, we can instead take dual Puiseux series, which we denote by $\mathbb{R}\{\{t\}\}^*$, and whose exponents form a descending sequence. Then the *dual valuation*, ord^*, takes a dual Puiseux series to its highest degree. As in Lemma 14 we obtain a semiring homomorphism

$$\operatorname{ord}^* : \mathbb{R}\{\{t\}\}^*_{\geq 0} \to \mathbb{T}_{\max}, \quad \text{where} \quad \mathbb{T}_{\max} = (\mathbb{R} \cup \{-\infty\}, \max, +)$$

is the max-tropical semiring. However, ord^* preserves the ordering.

So we consider two matrices $A^+, A^- \in \mathbb{T}_{\max}^{m\times n}$, two vectors $b^+, b^- \in \mathbb{T}_{\max}^m$ and another vector $c \in \mathbb{T}_{\max}^n$. Often we abbreviate $A = (A^+, A^-)$ and $b = (b^+, b^-)$. Equipped with this input, the *tropical linear program* $\operatorname{TLP}(A, b, c)$ is the optimization problem

$$\begin{aligned} &\text{minimize} && \langle c, x\rangle_{\text{trop}} \\ &\text{subject to} && A^+ \mathbin{\overline{\odot}} x \mathbin{\overline{\oplus}} b^+ \geq A^- \mathbin{\overline{\odot}} x \mathbin{\overline{\oplus}} b^- \\ & && x \in \mathbb{T}_{\max}^n . \end{aligned} \tag{4.1}$$

Each row of the constraints corresponds to one tropical halfspace as in Theorem 12, but there are two minor differences. First, as pointed out above, here we consider tropical arithmetic with respect to $\oplus = \overline{\oplus} = \max$ as the tropical addition. This entails a tropical matrix multiplication which uses that version of tropical addition; to stress this we write the latter as "$\overline{\odot}$". Second, by taking the vectors b^+ and b^- into account we arrive at inhomogeneous inequalities, leading to *tropical polyhedra* rather than tropical cones. This ties in better with ordinary linear optimization; this is the reason for the latter. Note that the vector $-\infty\mathbb{1} \in \mathbb{T}_{\max}^n$ trivially satisfies any tropical linear inequality.

The following result shows why tropical linear programming is interesting. It is concerned with the computational complexity of deciding if some $x \neq -\infty\mathbb{1}$ exists which satisfies all the constraints. This is the tropical analog to the feasibility problem of linear programming.

THEOREM 20 (Akian, Gaubert, and Guterman [**3**]). *Deciding whether a tropical linear program has a nontrivial feasible solution lies in the complexity class* $\mathrm{NP} \cap \text{co-NP}$.

The complexity class NP comprises those decision problems which admit a nondeterministic polynomial algorithm which can settle the issue in the affirmative. Its "mirror image" is the complexity class co-NP which is about nondeterministic polynomial algorithms to certify the negation. Most crucially, there is no deterministic polynomial algorithm known for tropical linear programming feasibility. It is a major open problem in theoretical computer science whether or not the complexity classes NP ∩ co-NP and P (deterministic polynomial algorithms) agree.

Yet there is another aspect. By Theorem 15 we may view tropical cones and tropical polytopes as projections of ordinary cones and polytopes over real Puiseux series under the (dual) valuation map. In other words, each tropical polytope or tropical polyhedron admits a *lift* to their ordinary counterparts, which are defined over real Puiseux series. The same holds true for tropical linear programs: tropical linear programs arise as projections of ordinary linear programs over Puiseux series.

EXAMPLE 21. For a parameter $r \geq 1$ we define a linear program over $\mathbb{R}\{\{t\}\}^*$ with $2r$ variables and $3r + 1$ constraints:

$$
\begin{aligned}
&\text{minimize} && x_1 \\
&\text{subject to} && x_1 \leq t^2 \\
& && x_2 \leq t \\
& && \left.\begin{aligned} &x_{2j+1} \leq t\, x_{2j-1}\,, \quad x_{2j+1} \leq t\, x_{2j} \\ &x_{2j+2} \leq t^{1-1/2^j}(x_{2j-1} + x_{2j}) \end{aligned}\right] \quad 1 \leq j < r \\
& && x_{2r-1} \geq 0\,, \quad x_{2r} \geq 0.
\end{aligned} \tag{4.2}
$$

Since all the terms occurring in (4.2) are nonnegative, the tropicalization is a merely syntactic procedure, term by term. This yields:

$$
\begin{aligned}
&\text{minimize} && X_1 \\
&\text{subject to} && X_1 \leq 2 \\
& && X_2 \leq 1 \\
& && \left.\begin{aligned} &X_{2j+1} \leq 1 + X_{2j-1}\,, \quad X_{2j+1} \leq 1 + X_{2j} \\ &X_{2j+2} \leq 1 - 1/2^j + \max(X_{2j-1}, X_{2j}) \end{aligned}\right] \quad 1 \leq j < r.
\end{aligned} \tag{4.3}
$$

We write capital letters for the variables now to mark the distinction from the ordinary linear program (4.2). The "built-in positivity" of the tropical semiring was mentioned before; and this is responsible for not needing nonnegativity constraints in (4.3).

While we started out by viewing (4.2) as a linear program over Puiseux series, we may also consider t a suitable (positive) real number. In this way, we obtain a family of ordinary linear programs, over the reals, which depend on two parameters, r and t. However, r and t play very different roles. The idea is that the limit for t approaching infinity is defined, and it is very close to the tropicalization, e.g., measured in the Hausdorff metric. The precise relationship between tropicalization and this limiting process is known as *Maslov dequantization* [**60**].

As its key property the tropical linear program (4.1) exhibits convoluted combinatorics. While the details are beyond the scope of this survey, we wish to give a sketch. Let P denote the tropical polyhedron which forms the feasible region of (4.1). For a fixed real number λ we can consider the *tropical sublevel set* $S_\lambda := \{x \in P \mid x_1 \leq \lambda\}$. The latter set is tropically convex, whence, e.g., $u, v \in S_\lambda$

implies $u \overline{\odot} v \in S_\lambda$. It follows that the coefficientwise maximum in the set S_λ is well-defined and unique; let us call this point $p(\lambda)$. The map $\lambda \mapsto p(\lambda)$ is piecewise linear, and its image is the *tropical central path* of (4.1). One way to express the "convoluted combinatorics" is to say that, for a certain discrete curvature measure, the tropical central path has very large total curvature.

A standard class of algorithms for solving ordinary linear programs employs the *interior point method*. It starts at a certain point in the interior of the feasible region, called the *analytic center*, and it follows a branch of a real algebraic curve, known as the *central path* to approach the optimum. The interior point method is an iterative procedure which employs auxiliary optimization problems at each step; these auxiliary problems are then solved via Newton's method. There is a host of choices for everything, the precise form of the auxiliary problems, the step sizes and so forth; see Renegar [**71**] for the details. In [**6**, Theorem 15] it is shown that for a fairly general class of interior point algorithms the central path (of any linear program over dual real Puiseux series) is mapped to the tropical central path of the tropicalization. This can be exploited to derive information about the computational complexity of interior point algorithms. To obtain a precise statement, in the following *log-barrier interior point method* is short for "primal-dual path-following interior point algorithm with a log-barrier function which iterates in the wide neighborhood of the central path". The next result is [**6**, Theorem B]; see also [**7**].

THEOREM 22. *The number of iterations required by any log-barrier interior point method, from the analytic center to some point of duality measure* ≤ 1, *is exponential in* r *on the family of linear programs* (4.2) *over the reals, provided that* $t \gg 1$ *is sufficiently large.*

This theorem is relevant because it is known from the work of Karmarkar [**54**] that some interior point algorithms solve linear programs in polynomial time, measured in the size of the input. And the log-barrier interior point methods form prominent members of that class. The *size* is a function of the dimensions m and n as well as the coding lengths of the coefficients of the constraints and the objective function. Now the linear programs (4.2) show that, in general, the running time of the log-barrier interior point methods depend on the coefficient size in a crucial way. Recently, Theorem 22 was generalized to other barrier functions by Allamigeon, Gaubert, and Vandame; see [**10**].

4.2. Further applications. To show how far the ideas of tropical convexity can be carried we briefly mention further connections, without any attempt at completeness. In particular, there are even more applications of tropical geometry, beyond tropical convexity, which we ignore here.

Matroids. In Section 3 we saw how tropical linear spaces arise as tropical convex hulls in some tropical projective space. When each coefficient of each cocircuit generator is either 0 or ∞, we are in the constant coefficient case; see [**48**, §1.5]. Then the intersection of the tropical linear space with the tropical projective torus is an ordinary polyhedral fan; in this case it is the *Bergman fan* of some matroid; see Feichtner and Sturmfels [**36**], [**64**, §4.2] and [**48**, §10.8]. Bergman fans occur in the work of Bergman [**16**] as logarithmic limit sets of algebraic varieties, and they are related to the De Concini–Procesi compactification of the complement of a hyperplane arrangement [**30**]. More recently, Bergman fans played a central role in groundbreaking work of Adiprasito, Huh and Katz [**2**], who solved several previously open problems in matroid theory.

Combinatorial commutative algebra. Block and Yu [**17**] used the covector decompositions of tropical polytopes for explicit constructions of resolutions of certain ideals in polynomial rings. That idea was carried further in articles by Dochtermann, Joswig, and Sanyal [**34**] and Almousa, Dochtermann, and Smith [**11**]. The resolutions discussed in these articles are examples of cellular resolutions in the sense of [**67**, Chapter 4]. For connections between monomial ideals and tropical convexity see [**50**, §5].

Bruhat–Tits buildings. Keel and Tevelev investigated compactifications of certain moduli spaces in algebraic geometry [**56**]. Fixing some field, $\mathbb{K}$, equipped with a discrete valuation, this lead them to studying equivalence classes of R-lattices, where R is the valuation ring. The classical setting are the Bruhat–Tits buildings arising from the special linear groups $\mathrm{SL}_d\,\mathbb{K}$; see [**1**] for an introduction to buildings. As a key new tool, Keel and Tevelev introduced *membranes*, which are certain sets of equivalence classes of R-lattices; see [**56**, §4]. The membranes turn out to agree with the sets of lattice points in certain tropical linear spaces; see [**53**] and [**48**, §10.9]. More recently, Zhang developed algorithms for computing membranes [**79**]. El Maazouz, Hahn, Nebe, Stanojkovski, and Sturmfels built on these ideas to study matrix algebras arising in representation theory [**35**].

Economics. Questions in economics which are amenable to methods of tropical convexity are often inherently game-theoretic. For details see the articles [**3, 4, 24, 63**], which were mentioned before, and their references. This line of research lead Shiozawa to interprete the classical Ricardian theory of trade [**73**] in terms of tropical convexity; see also [**48**, §9.5]. Shortly after Crowell and Tran characterized incentive compatible mechanisms [**28**]. Mechanism design is a branch of optimization bordering economics. The goal is to define auctions for selling items such that, e.g., the revenue of the seller is maximized.

Data analysis. A basic method for dimension reduction in data science is known as principal component analysis (PCA). Geometrically, a version of this amounts to projecting data points onto a (low-dimensional) affine linear subspace. Yoshida, Zhang, and Zhang proposed a tropical analog to PCA, i.e., a method to project onto a tropical linear space, and applied it to topics in phylogenetic analysis [**78**]; see also Semple and Steel[**72**] and [**48**, §10.9]. A key goal in phylogenetics is to aggregate data from several trees (which are considered as distinct approximations of one ground truth) into one tree, with some consistency requirements. This is the *consensus tree problem*; see Bryant [**20**] and Bryant, Francis, and Steel [**21**]. To this end Lin, Sturmfels, Tang, and Yoshida [**58**] and Lin and Yoshida [**59**] studied a tropical version of the Fermat–Weber problem. Recently, we found an asymmetric variant which gives stronger results [**26**]. The key idea of this article is to translate the consensus tree problem into a tropical convex hull problem, where an optimal solution can be found from one (polytropal) cell of the covector decomposition from Theorem 3. This was further generalized by Cox and Curiel [**27**].

Zhang, Naitzat, and Lim initiated the study of feedforward neural networks with ReLU activation via tropical geometry [**80**]. Later sharp complexity bounds were found by Montúfar, Ren, and Zhang [**68**] and Haase, Hertrich, and Loho [**40**].

Computational complexity and combinatorial optimization. Akian, Gaubert, and Guterman [**3**] showed that deciding the feasibility of a tropical linear program is equivalent to computing an optimal strategy in a mean payoff game; see also [**48**, Chapter 9]. The latter problem is of interest to researchers in theoretical computer science because it is unknown whether or not such a strategy can be found in polynomial time. This line of research led Loho to study combinatorial generalizations of tropical linear programming [**61**]. Methods from tropical convexity have also been applied to multicriteria optimization [**50**].

There are many connections between tropical convexity and combinatorial optimization; the shortest paths and minimum cycle means were already mentioned in Section 2.5. One more recent addition are results concerning the computation of periodic time tables. This is a notoriously difficult optimization problem with numerous applications, notably in public transport. Bortoletto, Lindner, and Masing developed new methods based on tropical convexity; their article [**19**] won a best paper award at the "22nd Symposium on Algorithmic Approaches for Transportation Modelling, Optimization, and Systems (ATMOS 2022)".

References

[1] P. Abramenko and K. S. Brown, *Buildings*, Graduate Texts in Mathematics, vol. 248, Springer, New York, 2008. Theory and applications, DOI 10.1007/978-0-387-78835-7. MR2439729

[2] K. Adiprasito, J. Huh, and E. Katz, *Hodge theory for combinatorial geometries*, Ann. of Math. (2) **188** (2018), no. 2, 381–452, DOI 10.4007/annals.2018.188.2.1. MR3862944

[3] M. Akian, S. Gaubert, and A. Guterman, *Tropical polyhedra are equivalent to mean payoff games*, Internat. J. Algebra Comput. **22** (2012), no. 1, 1250001, 43, DOI 10.1142/S0218196711006674. MR2900854

[4] M. Akian, S. Gaubert, and S. Vannucci, *Ambitropical geometry, hyperconvexity and zero-sum games*, Preprint, `arXiv:2108.07748`, 2023.

[5] D. Alessandrini, *Logarithmic limit sets of real semi-algebraic sets*, Adv. Geom. **13** (2013), no. 1, 155–190, DOI 10.1515/advgeom-2012-0020. MR3011539

[6] X. Allamigeon, P. Benchimol, S. Gaubert, and M. Joswig, *Log-barrier interior point methods are not strongly polynomial*, SIAM J. Appl. Algebra Geom. **2** (2018), no. 1, 140–178, DOI 10.1137/17M1142132. MR3772004

[7] X. Allamigeon, P. Benchimol, S. Gaubert, and M. Joswig, *What tropical geometry tells us about the complexity of linear programming*, SIAM Rev. **63** (2021), no. 1, 123–164, DOI 10.1137/20M1380211. MR4209657

[8] X. Allamigeon, S. Gaubert, and R. D. Katz, *The number of extreme points of tropical polyhedra*, J. Combin. Theory Ser. A **118** (2011), no. 1, 162–189, DOI 10.1016/j.jcta.2010.04.003. MR2737191

[9] X. Allamigeon, S. Gaubert, and M. Skomra, *Tropical spectrahedra*, Discrete Comput. Geom. **63** (2020), no. 3, 507–548, DOI 10.1007/s00454-020-00176-1. MR4074332

[10] X. Allamigeon, S. Gaubert, and N. Vandame, *No self-concordant barrier interior point method is strongly polynomial*, STOC '22—Proceedings of the 54th Annual ACM SIGACT Symposium on Theory of Computing, ACM, New York, [2022] ©2022, pp. 515–528. MR4490019

[11] A. Almousa, A. Dochtermann, and B. Smith, *Root polytopes, tropical types, and toric edge ideals*, Algebr. Comb. **8** (2025), no. 1, 59–99, DOI 10.5802/alco.404. MR4878041

[12] F. Ardila and M. Develin, *Tropical hyperplane arrangements and oriented matroids*, Math. Z. **262** (2009), no. 4, 795–816, DOI 10.1007/s00209-008-0400-z. MR2511751

[13] F. L. Baccelli, G. Cohen, G. J. Olsder, and J.-P. Quadrat, *Synchronization and linearity*, Wiley Series in Probability and Mathematical Statistics: Probability and Mathematical Statistics, John Wiley & Sons, Ltd., Chichester, 1992. An algebra for discrete event systems. MR1204266

[14] P. Benchimol, *Tropical aspects of linear programming*, PhD Thesis, École Polytechnique, December 2014.

[15] D. Bendle, J. Böhm, Y. Ren, and B. Schröter, *Massively parallel computation of tropical varieties, their positive part, and tropical Grassmannians*, J. Symbolic Comput. **120** (2024), Paper No. 102224, 28, DOI 10.1016/j.jsc.2023.102224. MR4583112

[16] G. M. Bergman, *The logarithmic limit-set of an algebraic variety*, Trans. Amer. Math. Soc. **157** (1971), 459–469, DOI 10.2307/1995858. MR280489

[17] F. Block and J. Yu, *Tropical convexity via cellular resolutions*, J. Algebraic Combin. **24** (2006), no. 1, 103–114, DOI 10.1007/s10801-006-9104-9. MR2245783

[18] J. Bochnak, M. Coste, and M.-F. Roy, *Real algebraic geometry*, Ergebnisse der Mathematik und ihrer Grenzgebiete (3) [Results in Mathematics and Related Areas (3)], vol. 36, Springer-Verlag, Berlin, 1998. Translated from the 1987 French original; Revised by the authors, DOI 10.1007/978-3-662-03718-8. MR1659509

[19] E. Bortoletto, N. Lindner, and B. Masing, *Tropical neighbourhood search: A new heuristic for periodic timetabling*, 22nd Symposium on Algorithmic Approaches for Transportation Modelling, Optimization, and Systems (ATMOS 2022) (Dagstuhl, Germany) (Mattia D'Emidio and Niels Lindner, eds.), Open Access Series in Informatics (OASIcs), vol. 106, Schloss Dagstuhl – Leibniz-Zentrum für Informatik, 2022, pp. 3:1–3:19.

[20] D. Bryant, *A classification of consensus methods for phylogenetics*, Bioconsensus (Piscataway, NJ, 2000/2001), DIMACS Ser. Discrete Math. Theoret. Comput. Sci., vol. 61, Amer. Math. Soc., Providence, RI, 2003, pp. 163–183, DOI 10.1090/dimacs/061/11. MR1982426

[21] D. Bryant, A. Francis, and M. Steel, *Can We "future-proof" consensus trees?*, Systematic Biology **66** (2017), no. 4, 611–619.

[22] P. Butkovič, *Max-linear systems: Theory and algorithms*, Springer Monographs in Mathematics, Springer-Verlag London, Ltd., London, 2010, DOI 10.1007/978-1-84996-299-5. MR2681232

[23] J. Cleveland, A. Hylton, R. Short, B. Mallery, R. Green, J. Curry, D. V. Dabke, and O. Freides, *Introducing tropical geometric approaches to delay tolerant networking optimization*, 2022 IEEE Aerospace Conference (AERO), 2022, pp. 1–11.

[24] G. Cohen, S. Gaubert, and J.-P. Quadrat, *Duality and separation theorems in idempotent semimodules*, Linear Algebra Appl. **379** (2004), 395–422, DOI 10.1016/j.laa.2003.08.010. Tenth Conference of the International Linear Algebra Society. MR2039751

[25] G. Cohen, S. Gaubert, J.-P. Quadrat, and I. Singer, *Max-plus convex sets and functions*, Idempotent mathematics and mathematical physics, Contemp. Math., vol. 377, Amer. Math. Soc., Providence, RI, 2005, pp. 105–129, DOI 10.1090/conm/377/06987. MR2149000

[26] A. Comăneci and M. Joswig, *Tropical medians by transportation*, Math. Program. **205** (2024), no. 1-2, Ser. A, 813–839, DOI 10.1007/s10107-023-01996-8. MR4728892

[27] S. Cox and M. Curiel, *The tropical polytope is the set of all weighted tropical Fermat–Weber points*, Preprint, `arXiv:2310.07732`, 2023.

[28] R. A. Crowell and N. M. Tran, *Tropical geometry and mechanism design*, Preprint, `arXiv:1606.04880`, 2016.

[29] R. Cuninghame-Green, *Minimax algebra*, Lecture Notes in Economics and Mathematical Systems, vol. 166, Springer-Verlag, Berlin-New York, 1979. MR580321

[30] C. De Concini and C. Procesi, *Wonderful models of subspace arrangements*, Selecta Math. (N.S.) **1** (1995), no. 3, 459–494, DOI 10.1007/BF01589496. MR1366622

[31] J. A. De Loera, J. Rambau, and F. Santos, *Triangulations*, Algorithms and Computation in Mathematics, vol. 25, Springer-Verlag, Berlin, 2010. Structures for algorithms and applications, DOI 10.1007/978-3-642-12971-1. MR2743368

[32] M. Develin and B. Sturmfels, *Tropical convexity*, Doc. Math. **9** (2004), 1–27. MR2054977

[33] M. Develin and J. Yu, *Tropical polytopes and cellular resolutions*, Experiment. Math. **16** (2007), no. 3, 277–291. MR2367318

[34] A. Dochtermann, M. Joswig, and R. Sanyal, *Tropical types and associated cellular resolutions*, J. Algebra **356** (2012), 304–324, DOI 10.1016/j.jalgebra.2011.12.028. MR2891135

[35] Y. El Maazouz, M. A. Hahn, G. Nebe, M. Stanojkovski, and B. Sturmfels, *Orders and polytropes: Matrix algebras from valuations*, Beitr. Algebra Geom. **63** (2022), no. 3, 515–531, DOI 10.1007/s13366-021-00600-4. MR4473915

[36] E. M. Feichtner and B. Sturmfels, *Matroid polytopes, nested sets and Bergman fans*, Port. Math. (N.S.) **62** (2005), no. 4, 437–468. MR2191630

[37] A. Fink and F. Rincón, *Stiefel tropical linear spaces*, J. Combin. Theory Ser. A **135** (2015), 291–331, DOI 10.1016/j.jcta.2015.06.001. MR3366480

[38] S. Gaubert, *Theory of linear systems over dioids*, PhD thesis, Ecole Nationale Supérieure des Mines de Paris, 1992.

[39] S. Gaubert and R. D. Katz, *Minimal half-spaces and external representation of tropical polyhedra*, J. Algebraic Combin. **33** (2011), no. 3, 325–348, DOI 10.1007/s10801-010-0246-4. MR2772536

[40] C. A. Haase, C. Hertrich, and G. Loho, *Lower bounds on the depth of integral ReLU neural networks via lattice polytopes*, The Eleventh International Conference on Learning Representations, 2023.

[41] S. Herrmann, A. Jensen, M. Joswig, and B. Sturmfels, *How to draw tropical planes*, Electron. J. Combin. **16** (2009), no. 2, Special volume in honor of Anders Björner, Research Paper 6, 26, DOI 10.37236/72. MR2515769

[42] S. Herrmann, M. Joswig, and D. E. Speyer, *Dressians, tropical Grassmannians, and their rays*, Forum Math. **26** (2014), no. 6, 1853–1881, DOI 10.1515/forum-2012-0030. MR3334049

[43] S. Horn, *A topological representation theorem for tropical oriented matroids* (English, with English and French summaries), 24th International Conference on Formal Power Series and Algebraic Combinatorics (FPSAC 2012), Discrete Math. Theor. Comput. Sci. Proc., AR, Assoc. Discrete Math. Theor. Comput. Sci., Nancy, 2012, pp. 135–146. MR2957992

[44] R. G. Jeroslow, *There cannot be any algorithm for integer programming with quadratic constraints*, Operations Res. **21** (1973), 221–224, DOI 10.1287/opre.21.1.221. Mathematical programming and its applications. MR354007

[45] C. Jordan, M. Joswig, and L. Kastner, *Parallel enumeration of triangulations*, Electron. J. Combin. **25** (2018), no. 3, Paper No. 3.6, 27, DOI 10.37236/7318. MR3829292

[46] C. Jordan, M. Joswig, and L. Kastner, *mptopcom, version 1.4*, 2024, Open source software for the parallel enumeration of triangulations, `http://www.polymake.org/mptopcom`.

[47] M. Joswig, *Tropical halfspaces*, Combinatorial and computational geometry, Math. Sci. Res. Inst. Publ., vol. 52, Cambridge Univ. Press, Cambridge, 2005, pp. 409–431. MR2178330

[48] M. Joswig, *Essentials of tropical combinatorics*, Graduate Studies in Mathematics, vol. 219, American Mathematical Society, Providence, RI, 2021, DOI 10.1090/gsm/219. MR4423372

[49] M. Joswig and K. Kulas, *Tropical and ordinary convexity combined*, Adv. Geom. **10** (2010), no. 2, 333–352, DOI 10.1515/ADVGEOM.2010.012. MR2629819

[50] M. Joswig and G. Loho, *Monomial tropical cones for multicriteria optimization*, SIAM J. Discrete Math. **34** (2020), no. 2, 1172–1191, DOI 10.1137/17M1153066. MR4101364

[51] M. Joswig, G. Loho, B. Lorenz, and B. Schröter, *Linear programs and convex hulls over fields of Puiseux fractions*, Mathematical aspects of computer and information sciences, Lecture Notes in Comput. Sci., vol. 9582, Springer, [Cham], 2016, pp. 429–445, DOI 10.1007/978-3-319-32859-1_37. MR3517052

[52] M. Joswig and B. Schröter, *Parametric shortest-path algorithms via tropical geometry*, Math. Oper. Res. **47** (2022), no. 3, 2065–2081, DOI 10.1287/moor.2021.1199. MR4506364

[53] M. Joswig, B. Sturmfels, and J. Yu, *Affine buildings and tropical convexity*, Albanian J. Math. **1** (2007), no. 4, 187–211. MR2367213

[54] N. Karmarkar, *A new polynomial-time algorithm for linear programming*, Combinatorica **4** (1984), no. 4, 373–395, DOI 10.1007/BF02579150. MR779900

[55] R. M. Karp, *A characterization of the minimum cycle mean in a digraph*, Discrete Math. **23** (1978), no. 3, 309–311, DOI 10.1016/0012-365X(78)90011-0. MR523080

[56] S. Keel and J. Tevelev, *Geometry of Chow quotients of Grassmannians*, Duke Math. J. **134** (2006), no. 2, 259–311, DOI 10.1215/S0012-7094-06-13422-1. MR2248832

[57] T. Lam and A. Postnikov, *Alcoved polytopes. I*, Discrete Comput. Geom. **38** (2007), no. 3, 453–478, DOI 10.1007/s00454-006-1294-3. MR2352704

[58] B. Lin, B. Sturmfels, X. Tang, and R. Yoshida, *Convexity in tree spaces*, SIAM J. Discrete Math. **31** (2017), no. 3, 2015–2038, DOI 10.1137/16M1079841. MR3693600

[59] B. Lin and R. Yoshida, *Tropical Fermat-Weber points*, SIAM J. Discrete Math. **32** (2018), no. 2, 1229–1245, DOI 10.1137/16M1071122. MR3810501

[60] G. L. Litvinov, V. P. Maslov, and G. B. Shpiz, *Idempotent functional analysis: An algebraic approach* (Russian, with Russian summary), Mat. Zametki **69** (2001), no. 5, 758–797, DOI 10.1023/A:1010266012029; English transl., Math. Notes **69** (2001), no. 5-6, 696–729. MR1846814

[61] G. Loho, *Abstract tropical linear programming*, Electron. J. Combin. **27** (2020), no. 2, Paper No. 2.51, 68, DOI 10.37236/7718. MR4245106

[62] G. Loho and L. A. Végh, *Signed tropical convexity*, 11th Innovations in Theoretical Computer Science Conference, LIPIcs. Leibniz Int. Proc. Inform., vol. 151, Schloss Dagstuhl. Leibniz-Zent. Inform., Wadern, 2020, pp. Art. No. 24, 35. MR4048127

[63] G. Loho and L. A. Végh, *Signed tropical convexity*, 11th Innovations in Theoretical Computer Science Conference, LIPIcs. Leibniz Int. Proc. Inform., vol. 151, Schloss Dagstuhl. Leibniz-Zent. Inform., Wadern, 2020, pp. Art. No. 24, 35. MR4048127

[64] D. Maclagan and B. Sturmfels, *Introduction to tropical geometry*, Graduate Studies in Mathematics, vol. 161, American Mathematical Society, Providence, RI, 2015, DOI 10.1090/gsm/161. MR3287221

[65] T. Markwig, *A field of generalised Puiseux series for tropical geometry*, Rend. Semin. Mat. Univ. Politec. Torino **68** (2010), no. 1, 79–92. MR2759691

[66] P. McMullen, *The maximum numbers of faces of a convex polytope*, Mathematika **17** (1970), 179–184, DOI 10.1112/S0025579300002850. MR283691

[67] E. Miller and B. Sturmfels, *Combinatorial commutative algebra*, Graduate Texts in Mathematics, vol. 227, Springer-Verlag, New York, 2005. MR2110098

[68] G. Montúfar, Y. Ren, and L. Zhang, *Sharp bounds for the number of regions of maxout networks and vertices of Minkowski sums*, SIAM J. Appl. Algebra Geom. **6** (2022), no. 4, 618–649, DOI 10.1137/21M1413699. MR4521653

[69] J. Rambau, *TOPCOM: Triangulations of point configurations and oriented matroids*, Mathematical software (Beijing, 2002), World Sci. Publ., River Edge, NJ, 2002, pp. 330–340. MR1932619

[70] J. Rambau, *TOPCOM, version 1.1.2*, Available at http://www.rambau.wm.uni-bayreuth.de/TOPCOM/, 2023.

[71] J. Renegar, *A mathematical view of interior-point methods in convex optimization*, MPS/SIAM Series on Optimization, Society for Industrial and Applied Mathematics (SIAM), Philadelphia, PA; Mathematical Programming Society (MPS), Philadelphia, PA, 2001, DOI 10.1137/1.9780898718812. MR1857706

[72] C. Semple and M. Steel, *Phylogenetics*, Oxford Lecture Series in Mathematics and its Applications, vol. 24, Oxford University Press, Oxford, 2003. MR2060009

[73] Y. Shiozawa, *International trade theory and exotic algebras*, Evolut. Inst. Econ. Rev. **12** (2015), no. 1, 177–212.

[74] D. Speyer and B. Sturmfels, *The tropical Grassmannian*, Adv. Geom. **4** (2004), no. 3, 389–411, DOI 10.1515/advg.2004.023. MR2071813

[75] D. Speyer and B. Sturmfels, *Tropical mathematics*, Math. Mag. **82** (2009), no. 3, 163–173, DOI 10.1080/0025570x.2009.11953615. MR2522909

[76] B. Sturmfels and N. M. Tran, *Combinatorial types of tropical eigenvectors*, Bull. Lond. Math. Soc. **45** (2013), no. 1, 27–36, DOI 10.1112/blms/bds058. MR3033951

[77] N. M. Tran, *Polytropes and tropical eigenspaces: cones of linearity*, Discrete Comput. Geom. **51** (2014), no. 3, 539–558, DOI 10.1007/s00454-014-9574-9. MR3201242

[78] R. Yoshida, L. Zhang, and X. Zhang, *Tropical principal component analysis and its application to phylogenetics*, Bull. Math. Biol. **81** (2019), no. 2, 568–597, DOI 10.1007/s11538-018-0493-4. MR3902912

[79] L. Zhang, *Computing min-convex hulls in the affine building of* SL_d, Discrete Comput. Geom. **65** (2021), no. 4, 1314–1336, DOI 10.1007/s00454-020-00223-x. MR4249906

[80] L. Zhang, G. Naitzat, and L.-H. Lim, *Tropical geometry of deep neural networks*, Proceedings of the 35th International Conference on Machine Learning, 9 pp., Proc. Mach. Learn. Res. (PMLR), **80**, arXiv:1805.07091, 2018.

[81] G. M. Ziegler, *Lectures on polytopes*, Graduate Texts in Mathematics, vol. 152, Springer-Verlag, New York, 1995, DOI 10.1007/978-1-4613-8431-1. MR1311028

Technische Universität Berlin, Chair of Discrete Mathematics/Geometry; and Max Planck Institute for Mathematics in the Sciences, Leipzig, Germany

Email address: joswig@math.tu-berlin.de

Proceedings of Symposia in Pure Mathematics
Volume 111, 2025
https://doi.org/10.1090/pspum/111/02045

From chemical reaction networks to algebraic and polyhedral geometry—and back again

Elisenda Feliu and Anne Shiu

Abstract. We describe the contributions by Bernd Sturmfels and his collaborators in harnessing algebraic and combinatorial methods for analyzing chemical reaction networks. Topics explored include the steady-state variety, counting steady states, and the global attractor conjecture. We also recount some personal stories that highlight Sturmfels's long-lasting impact on this research area.

Introduction

How did Bernd Sturmfels get into the topic of chemical reaction networks? The story begins at the Mathematical Sciences Research Institute (MSRI) in 2003, during a yearlong program on "commutative algebra".[1] Karin Gatermann was visiting and presented new ideas about sparse polynomial systems arising from chemical reaction networks [**Gat01, GH02**]. One day, Gatermann and Sturmfels were talking at a blackboard, when Alicia Dickenstein walked by. Sturmfels invited her to join the conversation, and, as they say, the rest is history.

History, however, is sometimes bittersweet. Within a few years of her visit to MSRI, Gatermann died of cancer. To honor her, friends and colleagues organized a special issue of the *Journal of Symbolic Computation*. Sturmfels proposed to submit a research article on chemical reaction networks to this issue, in part to carry on Gatermann's legacy and research in this area. This was co-authored with Dickenstein, Gheorghe Craciun, and a PhD student of Sturmfels—Anne Shiu, an author of this article [**CDSS09**].

Much of that first article on chemical reaction networks was developed at a 2007 Institute for Mathematics and its Applications (IMA) workshop on "Applications in Biology, Dynamics, and Statistics"; see the group photo in Figure 1. The four

2020 *Mathematics Subject Classification.* Primary 05B35, 12D10, 13P25, 37C25, 37N25, 52B11, 92E20.

Key words and phrases. reaction network, steady state, deficiency, siphon, persistence, Newton polytope, mixed volume, matroid, attainable region problem.

The first author was supported by the Independent Research Fund of Denmark.

The second author was supported by the NSF (DMS-1752672).

[1]In fact, some seeds were planted earlier: In the mid-1990s, Sturmfels had some initial discussions with Karin Gatermann, and separately heard from his computer science colleague Alistair Sinclair about discrete-time versions of chemical reaction systems [**RSW92**].

FIGURE 1. Group photo at "Applications in Biology, Dynamics, and Statistics" workshop at the IMA in March 2007. Sturmfels is seated in the first row, fifth from the left. Dickenstein is in the third row, seventh from the left; and Craciun is the unique person in the zeroth row. The authors acknowledge the IMA (Institute for Mathematics and its Applications) for permission to use this photo.

authors met daily in a cafe at 7:00 am—those who know Sturmfels know that he wakes up early!

In this article, we recount additional personal stories behind Sturmfels's research impact in chemical reaction networks. Indeed, while Sturmfels is an author of only four articles in this area [**CDSS09**, **SS10**, **CKLS19**, **GHRS16**], a true measure of his impact is through the many people he brought together in collaborations. This article also highlights the many ideas—often borrowed from adjacent research areas—that he was the first to bring into this field. In fact, a theme running through Sturmfels's research on chemical reaction networks is summarized in the stated aim of [**GHRS16**]: "to demonstrate how biology can lead to interesting questions in algebraic geometry and to apply state-of-the-art techniques from computational algebra to biology."

This article is organized as follows. In Section 1, we introduce chemical reaction networks and give an overview of the main questions in this research area. Section 2 concerns chemical reaction systems having a special type of steady state, namely, complex-balanced steady states, and their associated toric structures. We also discuss the global attractor conjecture, including how Sturmfels gave the name to this conjecture. Siphons and their relation to boundary steady states are the focus of Section 3, followed by steady-state invariants and their relation to matroids in

Section 4. Next, Section 5 describes polyhedral methods for assessing whether a network admits multiple steady states, and then Section 6 explains how the mixed volume and techniques from numerical algebraic geometry are used to count or bound the number of steady states. Section 7 concerns convex hulls of trajectories of chemical reaction systems, and we end with a discussion in Section 8.

1. Background

1.1. Reaction networks and associated models. In this section we briefly introduce the mathematical setting of the theory of reaction networks. This theory was initiated by chemical engineers who wanted to model the time-dependent evolution of chemical reactions.

1.1.1. *Reaction networks.* Reaction networks model the interactions among a finite set of objects, called *species*. In this article, we let $\mathcal{S} = \{X_1, \ldots, X_n\}$ denote the set of species. A *complex* is a finite linear combination of the elements in $\mathcal{S}$ with nonnegative integer coefficients. We identify each complex with the vector it defines in $\mathbb{Z}^n$, so that a complex can be written in two forms:

$$y = \sum_{i=1}^{n} \alpha_i X_i \qquad \text{or} \qquad y = (\alpha_1, \ldots, \alpha_n) \in \mathbb{Z}^n_{\geq 0}.$$

A *reaction network* $\mathcal{G} = (\mathcal{C}, \mathcal{R})$ is a digraph whose nodes are complexes. Edges are referred to as *reactions*. Throughout this article, the number of reactions is denoted by r (so, $|\mathcal{R}| = r$) and the number of complexes by m (so, $|\mathcal{C}| = m$).

For example, the following network is McKeithan's "kinetic proofreading" model of T-cell signal transduction [**McK95, Son01**]:

(1.1)

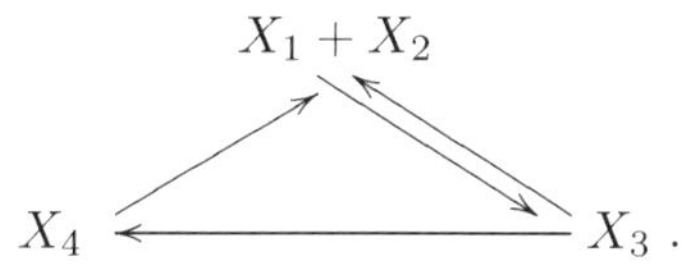

In this network, which we call the *McKeithan network*, X_1 represents the T-cell receptor, X_2 is the major histocompatibility complex (MHC) of an antigen-presenting cell, X_3 represents X_1 and X_2 bound together, and X_4 is the activated form of X_3. The set of complexes (nodes) and reactions (edges) are

(1.2) $$\mathcal{C} = \{X_1 + X_2,\ X_3,\ X_4\},$$
$$\mathcal{R} = \{X_1 + X_2 \longrightarrow X_3,\ X_3 \longrightarrow X_1 + X_2,\ X_3 \longrightarrow X_4,\ X_4 \longrightarrow X_1 + X_2\}.$$

Hence, there are $n = 4$ species, $m = 3$ complexes, and $r = 4$ reactions.

As mentioned earlier, reaction networks represent interactions among species. These species may arise in chemistry, biochemistry, or even in population biology or epidemiology. Therefore, species can be chemical compounds, proteins (as in the network in (1.1)), animals, classes of people, and so on. For instance, the classical Lotka–Volterra model of prey-predator interaction has the following underlying reaction network:

$$0 \longrightarrow P, \qquad P + U \longrightarrow 2U, \qquad U \longrightarrow 0,$$

where P is the prey and U the predator. As another example, the classical SIR model in epidemiology builds on the following reaction network:

$$S + I \longrightarrow 2I, \qquad I \longrightarrow R,$$

where S, I, R refer to susceptible, infected, and removed individuals.

1.1.2. *Dynamical systems.* Given a reaction network, one aims at modeling the time evolution of the species' abundances. This article focuses on deterministic (rather than stochastic) models with continuous (rather than discrete) time. Hence, we consider the concentrations of the species, and we model their time evolution using systems of ordinary differential equations (ODEs).

Formally, we let $x_i(t)$ denote the concentration of species X_i at time t, and we consider an ODE system of the form

$$\frac{dx_i}{dt} = \sum_{y \to y' \in \mathcal{R}} v_{y\to y'}\big(x(t)\big)(y'_i - y_i), \qquad i = 1, \dots, n. \tag{1.3}$$

Here $x(t) = (x_1(t), \dots, x_n(t))$, and $(y'_i - y_i)$ represents the net production of species X_i resulting from reaction $y \to y'$. Also, $v_{y\to y'}$ is a continuously differentiable function in $x_1, \dots, x_n$. A choice of functions $v_{y\to y'}$ for all reactions $y \to y'$ is referred to as a choice of *kinetics.* A widespread choice of kinetics is *mass-action kinetics*, in which the rate of each reaction is proportional to the product of the concentrations of the reactant species of the reaction. That is, each $v_{y\to y'}$ is a monomial of the following form:

$$v_{y\to y'}(x) = \kappa_{y\to y'}\, x_1^{y_1} \cdots x_n^{y_n}, \qquad \text{where } \kappa_{y\to y'} > 0. \tag{1.4}$$

The proportionality constant $\kappa_{y\to y'}$ is called the *reaction rate constant.* In this case, the right-hand side of each ODE in system (1.3) is a polynomial, and this is where the world of applied algebra comes into play.

The ODE system (1.3) admits a decomposition as the product of a matrix with the vector of kinetics, which we describe next. Choose an ordering of the reaction set $\mathcal{R}$, and construct the *stoichiometric matrix* $N \in \mathbb{R}^{n\times r}$, in which the ith column is the vector $y' - y$ if the ith reaction is $y \to y'$. Also, let $v(x)$ denote the length-r vector in which the entry corresponding to the reaction $y \to y'$ is $v_{y\to y'}(x)$. Now, after omitting reference to t, we rewrite (1.3) in matrix-vector form:

$$\frac{dx}{dt} = Nv(x). \tag{1.5}$$

Next, we describe other useful decompositions of the ODE system (1.3), for the case of mass-action kinetics. Choose an ordering of the set of complexes: $\mathcal{C} = \{y_1, \dots, y_m\}$. Let $Y \in \mathbb{R}^{n\times m}$ denote the matrix whose columns are the complexes, and let $x^Y = (x^{y_1}, \dots, x^{y_m})^\top \in \mathbb{R}^m$. (Recall that x^{y_i} denotes the monomial $x_1^{y_{i1}} \cdots x_m^{y_{im}}$). Let $C_{\mathcal{G}} \in \mathbb{R}^{m\times r}$ denote the incidence matrix of $\mathcal{G}$, and let $K_\kappa \in \mathbb{R}^{r\times m}$ denote the matrix whose (i,j)th entry is $\kappa_{y\to y'}$ if $y \to y'$ is the ith reaction and y the jth complex. Then $A_\kappa := C_{\mathcal{G}} K_\kappa \in \mathbb{R}^{m\times m}$ is the negative of the Laplacian matrix of the labeled digraph obtained from $\mathcal{G}$ by labeling each reaction $y \to y'$ with its reaction rate constant $\kappa_{y\to y'}$.

With this notation in place, and after observing that $N = YC_{\mathcal{G}}$, the right-hand side of (1.3) with mass-action kinetics admits the following two decompositions:

$$Nv(x) \;=\; YC_{\mathcal{G}}K_\kappa x^Y \;=\; YA_\kappa x^Y. \tag{1.6}$$

EXAMPLE 1.1. Let us find the ODE system under mass-action kinetics for the McKeithan network (1.1) using the objects just introduced. We choose the orderings of the complexes and reactions given in (1.2), and denote by κ_i the reaction rate constant of the ith reaction. This gives rise to the following objects:

$$N = \begin{bmatrix} -1 & 1 & 0 & 1 \\ -1 & 1 & 0 & 1 \\ 1 & -1 & -1 & 0 \\ 0 & 0 & 1 & -1 \end{bmatrix}, \qquad v(x) = (\kappa_1 x_1 x_2,\ \kappa_2 x_3,\ \kappa_3 x_3,\ \kappa_4 x_4)^\top,$$

$$Y = \begin{bmatrix} 1 & 0 & 0 \\ 1 & 0 & 0 \\ 0 & 1 & 0 \\ 0 & 0 & 1 \end{bmatrix}, \qquad C_{\mathcal{G}} = \begin{bmatrix} -1 & 1 & 0 & 1 \\ 1 & -1 & -1 & 0 \\ 0 & 0 & 1 & -1 \end{bmatrix},$$

$$K_\kappa = \begin{bmatrix} \kappa_1 & 0 & 0 \\ 0 & \kappa_2 & 0 \\ 0 & \kappa_3 & 0 \\ 0 & 0 & \kappa_4 \end{bmatrix}, \qquad A_\kappa = \begin{bmatrix} -\kappa_1 & \kappa_2 & \kappa_4 \\ \kappa_1 & -\kappa_2 - \kappa_3 & 0 \\ 0 & \kappa_3 & -\kappa_4 \end{bmatrix}.$$

Using (1.5) or (1.6), the ODE system describing the time evolution of the concentrations x_1, x_2, x_3, x_4 is as follows:

$$\begin{aligned} \tfrac{dx_1}{dt} &= -\kappa_1 x_1 x_2 + \kappa_2 x_3 + \kappa_4 x_4, & \tfrac{dx_3}{dt} &= \kappa_1 x_1 x_2 - \kappa_2 x_3 - \kappa_3 x_3, \\ \tfrac{dx_2}{dt} &= -\kappa_1 x_1 x_2 + \kappa_2 x_3 + \kappa_4 x_4, & \tfrac{dx_4}{dt} &= \kappa_3 x_3 - \kappa_4 x_4. \end{aligned} \tag{1.7}$$

From the ODEs (1.7) of the McKeithan network, we easily see that

$$\tfrac{dx_1}{dt} + \tfrac{dx_3}{dt} + \tfrac{dx_4}{dt} = 0, \tag{1.8}$$

and hence the sum $x_1(t) + x_3(t) + x_4(t)$ is a constant function in t for every solution of the ODE system. Its value equals the sum at the initial condition $(x_1(0), x_2(0), x_3(0), x_4(0))$.

The conservation relation (1.8) corresponds to the sum of the first, third, and fourth row of N being zero. Indeed, every vector in the left kernel of N gives such a relation. For the McKeithan network, one additional independent such relation exists, for instance, $x_2(t) + x_3(t) + x_4(t)$ is also constant.

In general, linear conservation relations arise as follows. From (1.5), we see that the derivatives $\frac{dx}{dt}$ belong to the vector subspace $\operatorname{Im}(N)$. Hence, given an initial condition $x(0) = z \in \mathbb{R}^n_{\geq 0}$, the trajectory is confined in the affine linear subspace $z + \operatorname{Im}(N)$. Additionally, both the positive and nonnegative real orthants, $\mathbb{R}^n_{>0}$ and $\mathbb{R}^n_{\geq 0}$, are forward invariant with respect to the mass-action ODE system (1.3) [**Vol72**]. A set is *forward invariant* if trajectories that begin in the set remain in the set for all positive time. From our discussion, we see that the following sets are also forward invariant:

$$\mathcal{P}_z = (z + \operatorname{Im}(N)) \cap \mathbb{R}^n_{\geq 0}, \qquad z \in \mathbb{R}^n_{\geq 0}, \tag{1.9}$$

and these sets are called *stoichiometric compatibility classes*. As they are the intersection of an affine linear subspace with the nonnegative orthant, they are polyhedra, and Sturmfels therefore calls these classes "invariant polyhedra". Equations defining $\mathcal{P}_z$ or, more precisely, the affine space $z + \operatorname{Im}(N)$, are *conservation laws*.

EXAMPLE 1.2 (Example 1.1, continued). We saw earlier that the McKeithan network has two linearly independent conservation laws:

$$x_1 + x_3 + x_4 = c_1, \qquad x_2 + x_3 + x_4 = c_2, \tag{1.10}$$

for some $c_1, c_2 \geq 0$ (which arise from some z as in (1.9)). One stoichiometric compatibility class of the McKeithan network, namely, the one containing the point

$z = \left(\frac{1}{2}, \frac{3}{2}, \frac{1}{4}, \frac{1}{4}\right)$, is as follows:

$$\mathcal{P}_z = \left\{(x_1, x_2, x_3, x_4) \in \mathbb{R}^4_{\geq 0} \mid x_1 + x_3 + x_4 = 1, \ x_2 + x_3 + x_4 = 2 \right\}.$$

This set $\mathcal{P}_z$ is the triangle depicted here.

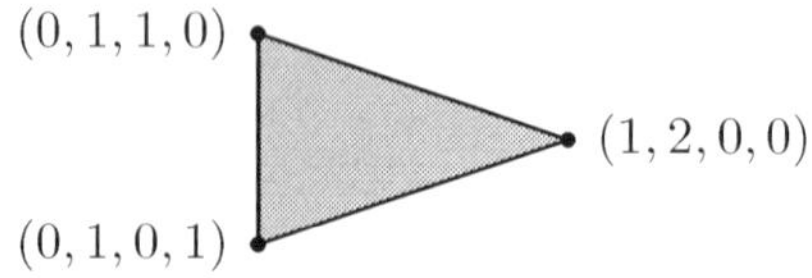

Recall that the boundary of a polyhedron is comprised of *faces*. For instance, $\mathcal{P}_z$ above has 3 *vertices* (0-dimensional faces) and 3 *edges* (1-dimensional faces). Similarly, a cube has 8 vertices, 12 edges, and 6 *facets* (maximal-dimension proper faces). Each face F of a stoichiometric compatibility class $\mathcal{P}$ is defined by the vanishing of a (possibly nonunique) subset of the variables x_i, with index set $W \subseteq \{1, \ldots, n\}$, as follows:

(1.11) $F = F_W = \mathcal{P} \cap Z_W$, where $Z_W := \{x \in \mathbb{R}^n \mid x_i = 0 \text{ for all } i \in W\}$.

For instance, for the invariant polyhedron shown earlier, we label two of the faces according to (1.11).

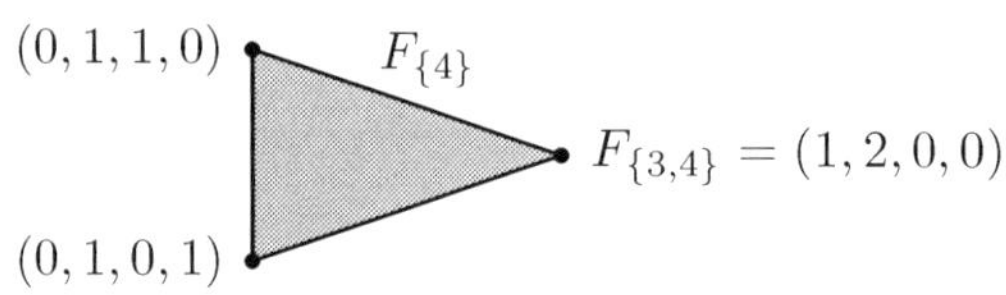

The final objects of interest are *steady states* (or *equilibrium points*). These are the solutions $x \in \mathbb{R}^n_{\geq 0}$ to the system $Nv(x) = 0$. In the case of mass-action kinetics, steady states are the nonnegative solutions to a system of polynomials. A steady state x is a *positive steady state* if $x \in \mathbb{R}^n_{>0}$.

1.2. Questions of interest and overview. The mathematical field pertaining to the study of reaction networks—which was initiated in the 1970s by Feinberg, Horn, Jackson, and others—focuses broadly on understanding the mathematical properties of the ODE system (1.3) for *unknown* parameter values. The rationale behind this arises from the fact that in many applications, parameter values are difficult to estimate, and, moreover, a certain degree of variability among individuals and systems exists.

The theory has two broad aims:

(1) identify characteristics of the reaction network that impose a certain behavior on the ODE system, mainly for mass-action kinetics, and
(2) develop tools to verify whether and where a property arises for some choice of parameter values.

Such properties of interest include the number of positive steady states (within a stoichiometric compatibility class), the existence of periodic solutions, the convergence of trajectories to a steady state, the existence of steady states at the boundary of the nonnegative orthant, and persistence. An ODE system is said to be persistent if no trajectory approaches the boundary. Moreover, one goal in this area is to obtain results on such properties that do not rely on simulations or numerical approaches.

In this research area, many mathematical techniques have found application, including ideas from applied algebra and polyhedral geometry. This article focuses on selected topics in this direction, specifically those where Sturmfels has contributed. These topics include the algebraic structure and convergence to steady state of complex-balanced systems [**CDSS09**]; siphons, persistence, and boundary steady states [**SS10**]; steady-state invariants via the study of matroids [**GHRS16**]; polyhedral methods and mixed volume [**GHRS16**]; and convex hulls of trajectories [**CKLS19**]. These topics are addressed in what follows, and we highlight Sturmfels's contribution in terms of both specific results and ideas he introduced.

2. Complex-balanced steady states and toric dynamical systems

2.1. The human story. In a few places in this article, we would like to share some personal stories involving Sturmfels. We begin with some recollections by Shiu:

As mentioned in the introduction, much progress on Bernd's first article on reaction networks happened at a March 2007 workshop at IMA. Before the workshop, Bernd met with me and fellow graduate student Jason Morton. Bernd went through the workshop's participant list and told Jason and me which researchers each of us had to meet. At the top of that list were "Feinberg and Fienberg"—Jason was to talk with Stephen Fienberg and I was to make sure to meet Martin Feinberg. Funny enough, Bernd is seated between Fienberg and Feinberg in the group photo at the workshop (see Figure 1*).*

My first night at the workshop, Bernd had me meet Alicia Dickenstein for dinner (I was nervous to meet such a prominent mathematician), and then, as mentioned earlier, the three of us, plus Gheorghe Craciun, met early each day. Gheorghe had been Feinberg's PhD student, and was already a prominent researcher in chemical reaction network theory. One morning, Bernd asked what was known about stability of complex-balanced steady states, and Gheorghe explained that these steady states are known to be local attractors but conjectured to be global attractors as well. "The global attractor conjecture," Bernd said, and from then on, the name stuck. More on this conjecture is in Section 2.3*.*

Another term Bernd coined in the article is (in the title!) "toric dynamical systems" (also known as complex-balanced systems). In June 2007, Bernd wrote the three of us an email, which began, "Dear all, this morning I picked up Jeremy Gunawardena's article 'Chemical reaction network theory for in-silico biologists', and I simply couldn't stop reading. This author is amazing!!" [**Gun03**][2] *He continued, "The main point of his paper is exactly what I was trying express all along: Deficiency zero is just a very special case of [toric dynamical systems], and it is the toric structure that really matters [...]".*

Then, following a detailed math description, he concluded, "So, now that I finally understand the literature, I would strongly urge that we use the term 'toric dynamical system' (TDS) [...]. This ensures in particular what I was trying to sell to you (Anne) on Wednesday in the car: deficiency zero implies TDS." Of course, these ideas had been known for some time (see Section 2.2*), but the algebraic*

[2]Gunawardena appears again later in another human story (Section 4.2).

packaging of the results, including a name that was inviting to algebraic geometers, was important for bringing newcomers into this area.

As for the car ride mentioned in the email, research meetings with Bernd held on modes of transportation were not uncommon! Once, he drove me and another PhD student (Peter Huggins) from a computational biology retreat near Lake Tahoe back to Berkeley, first with Peter in the front seat for a research chat, and then stopping halfway so Peter and I could switch and then it was my turn to have a math conversation. Efficient!

2.2. Steady-state structure. For a mass-action system (1.6), a *complex-balanced steady state* is a point $x^* \in \mathbb{R}^n_{>0}$ such that

$$A_\kappa (x^*)^Y = 0. \tag{2.1}$$

Clearly, by (1.6), every complex-balanced steady state is indeed a steady state. Complex-balanced steady states were introduced by Horn and Jackson [**HJ72**] and Feinberg [**Fei73**] in the 1970s, to account for thermodynamic constraints in a reaction network. It follows from (2.1) that complex-balanced steady states have the following characterization: For each complex y, the sum of the rates of all reactions with source y equals the sum of the rates of reactions with endpoint y.

Feinberg, Horn, and Jackson recognized that the existence of complex-balanced steady states requires the reaction network to be *weakly reversible*, that is, each connected component must be strongly connected [**Hor72**]. In this case, the image of $C_\mathcal{G}$ and A_κ coincide for all $\kappa \in \mathbb{R}^r_{>0}$ [**FH77**].

Being weakly reversible, however, is not sufficient for the existence of complex-balanced steady states. A well-known and celebrated result due to Feinberg is the *deficiency zero theorem* [**Fei87**]. The *deficiency* of a reaction network is defined as follows:

$$\delta = \dim(\ker(Y) \cap \operatorname{Im}(C_\mathcal{G})).$$

Hence, $\delta \geq 0$. Also, it can be shown that

$$\delta = m - \ell - s,$$

where ℓ is the number of connected components of $\mathcal{G}$, and s the rank of the stoichiometric matrix N. Clearly, if $\delta = 0$, then all steady states satisfy (2.1) and hence are complex-balanced. This is the case for the McKeithan network (1.1), as $\ell = 1$, $s = 2$ and hence

$$\delta = 3 - 1 - 2 = 0. \tag{2.2}$$

Therefore, every positive steady state x^* of the McKeithan network ODEs (1.7) is a solution to the system (2.1), which translates to:

$$-\kappa_1 x_1^* x_2^* + \kappa_2 x_3^* + \kappa_4 x_4^* = 0, \qquad \kappa_3 x_3^* - \kappa_4 x_4^* = 0$$

(one equation is removed as it is in the linear span of the above two). By adding the second equation to the first, we obtain an equivalent system

$$-\kappa_1 x_1^* x_2^* + (\kappa_2 + \kappa_3) x_3^* = 0, \qquad \kappa_3 x_3^* - \kappa_4 x_4^* = 0, \tag{2.3}$$

which now consists of binomial equations. For fixed κ, the intersection of the variety of (2.3) with $\mathbb{R}^4_{>0}$ is the positive part of a toric variety. Hence, it is irreducible and admits a parametrization. Such remarkable properties hold for all networks with $\delta = 0$, and are part of why complex-balanced steady states have received special mathematical attention.

For networks with $\delta > 0$, it is no longer true that there are complex-balanced steady states for all parameters κ. The set Σ of parameters that do generate complex-balanced steady states was shown in the 1970s to be cut out by δ equations [**Hor72**]. Building on that work, Craciun, Dickenstein, Shiu, and Sturmfels further elucidated the algebraic structure of the set Σ and also the set of complex-balanced steady states [**CDSS09**] (Theorems 2.1, 2.2, and 2.5). These results are phrased in the language of toric geometry, as the equations (2.1) (it turns out) define a binomial ideal. This is why, in [**CDSS09**], an ODE system (1.3) for which all steady states are complex-balanced is called a *toric dynamical system.*

Setup. The setup for the remainder of this subsection is as follows. Let $\mathcal{G}$ be a weakly reversible network with m complexes. For each $i \in \{1, \dots, m\}$, let I_i denote the subset of $\{1, \dots, m\}$ corresponding to the complexes in the same connected component as the complex y_i. Consider the submatrix A_{κ, I_i} of A_κ with rows and columns in I_i, and define $K_i \in \mathbb{Q}[\kappa]$ as $(-1)^{\#I_i - 1}$ times the minor of A_{κ, I_i} obtained by removing the row and column corresponding to y_i. Every coefficient of the polynomial K_i is positive. Also, by [**CDSS09**, Lemma 5], these polynomials K_i are algebraically independent.

Next, consider the following ideal obtained by saturation:[3]

$$T_{\mathcal{G}} = \langle K_i x^{y_j} - K_j x^{y_i} \mid y_i \to y_j \in \mathcal{R} \rangle : (x_1 \cdots x_n)^\infty \subseteq \mathbb{Q}[x, K] \subseteq \mathbb{Q}[x, \kappa].$$

This ideal gives a characterization of complex-balanced steady states, in the following sense [**CDSS09**].

THEOREM 2.1 (Complex-balanced steady states). *For $\kappa^* \in \mathbb{R}^r_{>0}$ and $x^* \in \mathbb{R}^n_{\geq 0}$, we have that x^* is a complex-balanced steady state of the mass-action system of $\mathcal{G}$ with rate constants κ^* if and only if (x^*, κ^*) lies is in the positive part of the toric variety defined by $T_{\mathcal{G}}$, denoted by $V_{>0}(T_{\mathcal{G}})$.*

To determine for which parameter values κ^* there are points of the form (x^*, κ^*) in $V_{>0}(T_{\mathcal{G}})$ (equivalently, which rate constants give rise to complex-balanced steady states), we consider the following elimination ideal:

$$M_{\mathcal{G}} = T_{\mathcal{G}} \cap \mathbb{Q}[\kappa].$$

The following is one of the main results in [**CDSS09**].

THEOREM 2.2 (Rate constants for complex balancing, part 1). *The mass-action system of a reaction network $\mathcal{G}$ with rate constants $\kappa^* \in \mathbb{R}^r_{>0}$ admits a complex-balanced steady state if and only if $\kappa^* \in V_{>0}(M_{\mathcal{G}})$. Furthermore, the codimension of $M_{\mathcal{G}}$ equals the deficiency δ of $\mathcal{G}$.*

Next, we illustrate Theorem 2.2 through two examples, one with $\delta = 0$ and one with $\delta > 0$.

EXAMPLE 2.3 (McKeithan network). Let us find $V_{>0}(M_{\mathcal{G}})$ for the McKeithan network. We must first compute the ideal $T_{\mathcal{G}}$. To this end, as there is one connected component and $m = 3$, we obtain the following using the matrix A_κ shown earlier in Example 1.1:

$$K_1 = (\kappa_2 + \kappa_3)\kappa_4, \qquad K_2 = \kappa_1 \kappa_4, \qquad K_3 = \kappa_1 \kappa_3. \tag{2.4}$$

The following `Macaulay2` code computes the ideal $T_{\mathcal{G}}$:

[3] Saturation of binomial ideals, especially in the context of algorithms, is an old idea that Bernd has been harnessing for many years; see the work of Bernd with Hoşten [**HS95**].

```
R = QQ[x1,x2,x3,x4, K1,K2,K3, MonomialOrder => Eliminate 4];
I = ideal(K1*x3-K2*x1*x2, K2*x4-K3*x3, K1*x4-K3*x1*x2);
TG = saturate(I, x1*x2*x3*x4);
groebnerBasis TG.
```

The output reveals that

$$T_{\mathcal{G}} = \langle K_1x_3 - K_2x_1x_2,\ K_2x_4 - K_3x_3,\ K_1x_4 - K_3x_1x_2\rangle.$$

No polynomials purely in the K_i's appear above as generators of $T_{\mathcal{G}}$, and so, as a monomial ordering that eliminates the x_i's was used, the elimination ideal $M_{\mathcal{G}}$ is the zero ideal. This is consistent with Theorem 2.2, as $\delta = 0$.

EXAMPLE 2.4 (Extended McKeithan network). We add the pair of reactions

$$2X_3 \underset{\kappa_6}{\overset{\kappa_5}{\rightleftharpoons}} 2X_4$$

to the McKeithan network. The resulting *Extended McKeithan network* has two connected components ($\ell = 2$) and $\delta = 1$. We let $y_4 = 2X_3$ and $y_5 = 2X_4$, and so we add $K_4 = \kappa_6$ and $K_5 = \kappa_5$ to (2.4). A computation like the one in Example 2.3 reveals the following ideal (where the generators form a Gröbner basis with respect to a monomial ordering for eliminating x_1, x_2, x_3, x_4):

$$T_{\mathcal{G}} = \langle K_2^2K_5 - K_3^2K_4,\ K_2x_4 - K_3x_3,\ K_4x_4^2 - K_5x_3^2,\ K_2x_1x_2 - K_1x_3,\\ K_3x_1x_2 - K_1x_4,\ K_3K_4x_4 - K_2K_5x_3\rangle.$$

We conclude:

$$M_{\mathcal{G}} = \langle K_2^2K_5 - K_3^2K_4\rangle = \langle \kappa_1^2\kappa_4^2\kappa_5 - \kappa_1^2\kappa_3^2\kappa_6\rangle. \tag{2.5}$$

Hence, if κ satisfies $\kappa_1^2\kappa_4^2\kappa_5 = \kappa_1^2\kappa_3^2\kappa_6$, then there are complex-balanced steady states. Otherwise, there are none.

In Examples 2.3 and 2.4, we determined which rate constants yield complex-balanced steady states by computing the ideals $T_{\mathcal{G}}$ and $M_{\mathcal{G}}$. Next, we explain how to avoid having to compute these ideals. Namely, Theorem 2.5 below gives explicit equations defining $V_{>0}(M_{\mathcal{G}})$, and they are always binomial in the K_i's.

Reorder the complexes so that complexes in the same connected component are together. Let $Y_1, \ldots, Y_\ell$ be the submatrices of Y corresponding to connected components. Then the *Cayley matrix* is defined as

$$\mathrm{Cay}_{\mathcal{G}}(Y) = \begin{bmatrix} Y_1 & Y_2 & \ldots & Y_\ell \\ \mathbf{1} & 0 & \ldots & 0 \\ 0 & \mathbf{1} & \ldots & 0 \\ \vdots & \vdots & \ddots & \vdots \\ 0 & 0 & \ldots & \mathbf{1} \end{bmatrix} \in \mathbb{R}^{(n+\ell)\times m}, \tag{2.6}$$

where $\mathbf{1}$ is a row vector of 1's of the appropriate size. With this in place, we can state the next result [**CDSS09**].

THEOREM 2.5 (Rate constants for complex-balancing, part 2). *For a reaction network $\mathcal{G}$, we have that $\kappa^* \in V_{>0}(M_{\mathcal{G}})$ if and only if $K^u = 1$ for all $u \in \mathrm{Ker}(\mathrm{Cay}_{\mathcal{G}}(Y)) \cap \mathbb{Z}^n$.*

EXAMPLE 2.6 (Example 2.4, continued). In the extended McKeithan network, we have

$$\mathrm{Cay}_{\mathcal{G}}(Y) = \begin{bmatrix} 1 & 0 & 0 & 0 & 0 \\ 1 & 0 & 0 & 0 & 0 \\ 0 & 1 & 0 & 2 & 0 \\ 0 & 0 & 1 & 0 & 2 \\ 1 & 1 & 1 & 0 & 0 \\ 0 & 0 & 0 & 1 & 1 \end{bmatrix}, \qquad \mathrm{Ker}(\mathrm{Cay}_{\mathcal{G}}(Y)) = \mathrm{span}\left\{ \begin{pmatrix} 0 \\ 2 \\ -2 \\ -1 \\ 1 \end{pmatrix} \right\}.$$

Hence, $\kappa \in V_{>0}(M_{\mathcal{G}})$ if and only if the equation $K_1^0 K_2^2 K_3^{-2} K_4^{-1} K_5 = 1$ holds, or, equivalently,

$$K_2^2 K_5 - K_3^2 K_4 = 0.$$

As it should, this binomial equation agrees with the one in (2.5) we obtained earlier.

We end this subsection by expanding on Sturmfels's contribution to two of the ideas mentioned above. The first concerns the Cayley matrix (2.6), which is so-named because it refers to the Cayley trick in elimination theory [**CDSS09**]. Notably, the Cayley trick was first developed by Sturmfels [**Stu94**], and later generalized [**HRS00**].

Another notable contribution by Sturmfels concerns the polynomials K_i, which we saw arise from minors of A_κ. These polynomials admit a graphical interpretation as labels of spanning forests of the digraph $\mathcal{G}$ labeled with the reaction rate constants. This is the *matrix-tree theorem* of Tutte, which concerns minors of a graph's Laplacian matrix [**Sta99, Tut48**]. Applications of this theorem appear frequently in the field of reaction networks, for example in [**CS15, DPM11, FW13c, FW12b, SWF17, FW13b, SFW18, Joh14**], and of course in [**CDSS09**]. In fact, the classical King–Altman method in enzyme kinetics follows from the matrix-tree theorem [**KA56**]. Gunawardena acknowledges having learned about the matrix-tree theorem from Sturmfels in [**TG09a**], and Feliu learned about it from Gunawardena's article. So it is probably fair to say that it was Sturmfels who brought it into the field!

2.3. Global attractor conjecture. A mass-action system that has a complex-balanced steady state is usually called a complex-balanced system, or toric dynamical system. Horn and Jackson proved that for these systems, each stoichiometric compatibility class $\mathcal{P}$, as in (1.9), has a unique positive steady state x^*, and that x^* is in fact complex-balanced [**HJ72**]. Sturmfels gave the name *Birch point* to x^*, to emphasize the connection to Birch's theorem in (algebraic) statistics.[4]

Next, given a Birch point x^* of a complex-balanced system, the following is a strict Lyapunov function (with respect to the stoichiometric compatibility class $\mathcal{P} = \mathcal{P}_{x^*}$):

$$\sum_{i=1}^{n} \big(x_i \log(x_i) \; - \; x_i \log(x_i^*) \; - \; x_i \; + x_i^* \big).$$

[4]The existence and uniqueness of positive steady states in toric dynamical systems is essentially equivalent to the existence and uniqueness of maximum likelihood estimators of toric, i.e., log-linear, statistical models. Extensions of Birch's theorem from the study of chemical reaction networks are found in [**GMS14, CMPY19**].

This Lyapunov function ensures that trajectories in $\mathcal{P}$ that start near the Birch point x^* converge to x^* [**HJ72**]; in other words, x^* is a local attractor (relative to $\mathcal{P}$). The following conjecture states that x^* is also a global attractor.

CONJECTURE 2.7 (Global attractor conjecture). *Consider a complex-balanced system (a toric dynamical system). Let $\mathcal{P}$ be a stoichiometric compatibility class of the system, and let x^* denote the Birch point (the unique positive steady state in $\mathcal{P}$). Then every trajectory beginning in the relative interior of $\mathcal{P}$ converges to x^*.*

As mentioned earlier, the name of this now-famous conjecture was coined by Sturmfels, and this attention revived interest in the conjecture, even though it had been stated much earlier, by Horn, in 1974 [**Hor74**]. Another contribution of Sturmfels and co-authors was to reframe what needs to be proven via the language of polytopes, as we explain next.

The Lyapunov function ensures that every trajectory in $\mathcal{P}$ either converges to x^* or converges to a point on the boundary of $\mathcal{P}$ [**SC94**]. Such a point on the boundary is necessarily a (boundary) steady state; this fact was shown in a special case by Sontag [**Son01**], and the general result follows in a similar manner. In summary, the global attractor conjecture is equivalent to the following assertion: *for a toric dynamical system, no trajectory beginning in the positive orthant converges to a boundary steady state.* It follows that if $\mathcal{P}$ has no boundary steady states, then the conjecture holds for this $\mathcal{P}$ [**SM00**].

As noted earlier, Sturmfels often calls stoichiometric compatibility classes "invariant polyhedra," and in [**CDSS09**] (and, independently, by Anderson [**And08**]), it is shown that vertices of such polyhedra cannot be points of convergence. This result set the stage for future progress on the conjecture—to rule out faces of polyhedra according to their dimension. Indeed, subsequently, relative-interior points of two types of faces were ruled out: facets (i.e., top-dimension, proper faces) [**AS10**] and "weakly dynamically nonemptiable" faces [**JS11**]. A related family of ideas contributed to a resolution of the conjecture for more networks: those having only one connected component [**And11**] and those for which the stoichiometric compatibility classes $\mathcal{P}$ have dimension at most 3 [**CNP13, Pan12**].

Furthermore, the vocabulary of polytopes led researchers to define "endotactic" [**CNP13**] and "strongly endotactic" networks [**GMS13**]; see Figure 2. Such networks, roughly speaking, have reaction arrows that point inward, rather than outward ("endo" means "inward", and "tactic" refers to movement as in "chemotaxis"). Geometrically, this means that arrows do not point "out" of the Newton polytope defined by the right-hand sides of the mass-action ODEs; Newton polytopes are discussed in Section 5. Intuitively, these inward-pointing arrows should guarantee that species concentrations avoid going to 0 or infinity—that is, trajectories avoid converging to the boundary. Indeed, for strongly endotactic networks, the global attractor conjecture has been proven [**GMS14**]. This conjecture, however, is still open for endotactic networks.

To summarize, the remaining cases for Conjecture 2.7 are toric dynamical systems of dimension at least 4 for which the boundary steady states lie on faces of dimension at least 1 and codimension at least 2. See Table 1.

The most recent progress on Conjecture 2.7 is due to Gheorghe Craciun, who proposed an outline of a proof in 2015 [**Cra15**]. One crucial idea is again to show how the boundary "repels" trajectories, but getting everything to fit together is very subtle. As this conjecture was the foremost open problem in the research area,

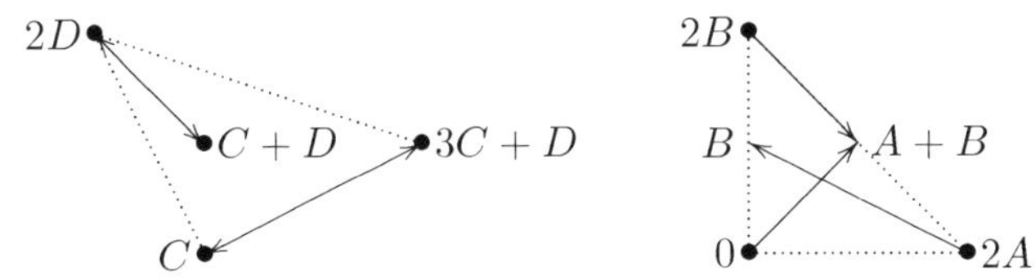

FIGURE 2. The network on the left is endotactic, while the one on the right is strongly endotactic. Both networks are endotactic, because no reaction arrows point out of the Newton polytopes (the triangles) unless they first pass through the respective triangle. The network on the right is strongly endotactic, because every proper face of the triangle (that is, each edge and vertex) has a reaction arrow that starts on that face and exits the face. In contrast, the network on the left is not strongly endotactic: the edge of the triangle from C to $3C + D$ has no reaction arrow that leaves that edge.

TABLE 1. Properties that guarantee that the global attractor conjecture holds.

Property	Reference
No boundary steady states	[**SM00**]
No relevant siphons (see Section 3)	[**ADS07**]
Strongly endotactic	[**GMS14**]
Dimension at most 3	[**Pan12**]
Each boundary steady state is a vertex or interior point of a facet	[**And08, AS10, CDSS09**]

Sturmfels initiated a workshop at San José State University in 2016 on Craciun's proposed proof [**GJM16**]. Participants—including members of Sturmfels's research group—were assigned parts of the paper to present. And, at Sturmfels's request, Craciun himself was in attendance but not allowed to speak during the presentations! The first installment of Craciun's work is now published [**Cra19**], and our community eagerly awaits the next one.

Finally, we note that ideas pertaining to being endotactic have grown outward ("exotactically," perhaps?) into new domains. For instance, the concept of "strongly endotactic" was used to analyze stochastic systems [**ADE18b, ADE18a, ACKN20**]. Also, results involving strongly endotactic networks (and pertaining to *persistence*, a topic of the next section) have been transferred to networks related to some "origin of life" models [**CDJY22**]. Bernd's influence has therefore come full circle: he brought new ideas to the reaction network community, and now ideas developed in this area are in turn proving to be useful in new domains.

3. Siphons, persistence, and boundary steady states

A siphon of a reaction network is a subset of species whose absence is forward-invariant with respect to the dynamics. In this section, we explain how siphons are related to persistence and boundary steady states, and elucidate Sturmfels's contribution to exploring the algebraic and computational aspects of siphons.

Let us describe the motivation behind siphons. In the McKeithan network (1.1), all reactions are "off" if and only if either (i) the concentrations of X_1, X_3, and X_4 are all zero, or (ii) the concentrations of X_2, X_3, and X_4 are all zero. Accordingly, we say that $\{X_1, X_3, X_4\}$ and $\{X_2, X_3, X_4\}$ (or, more accurately, $\{1,3,4\}$ and $\{2,3,4\}$) are (minimal) *siphons*.

More precisely, a *siphon* of a network is a nonempty subset $W \subseteq \{1, \dots, n\}$ of the set of species indices, such that if the product complex of a reaction contains an element of W, then so does the reactant complex. For example, if $X_1 + X_2 \to X_3$ is a reaction of a network, then $3 \in W$ implies that $1 \in W$ or $2 \in W$. A siphon of a network $\mathcal{G}$ is *minimal* if it is minimal with respect to inclusion among all siphons of $\mathcal{G}$.

The connection between siphons, faces of $\mathcal{P}$, and boundary steady states is the following result [**And08, ADS07**], where we recall that the face F_W induced by W was defined in (1.11):

PROPOSITION 3.1. *If there is a steady state in the relative interior of some face F_W of $\mathcal{P}$, then W is a siphon.*

This proposition—and our interest in the global attractor conjecture and persistence of mass-action systems—motivate the need for efficient tools for computing siphons. A mass-action system is *persistent* if no trajectory beginning in the relative interior of a stoichoimetric compatibility class $\mathcal{P}$ has an accumulation point, or *ω-limit point*, on the boundary of $\mathcal{P}$. In this context, Shiu and Sturmfels proved the following two results [**SS10**].

PROPOSITION 3.2. *If a reaction network is strongly connected, then its minimal siphons correspond exactly to minimal associated primes of the ideal of complex monomials x^{y_i}.*

For instance, the ideal of complex monomials for the McKeithan network (1.1) is $\langle x_1x_2, x_3, x_4 \rangle$ in $\mathbb{Q}[x_1, x_2, x_3, x_4]$, and its minimal associated primes are $\langle x_1, x_3, x_4 \rangle$ and $\langle x_2, x_3, x_4 \rangle$. These primes correspond exactly to the minimal siphons found earlier.

To state the more general result, which allows for networks that are not strongly connected, consider the following ideal in the quotient ring $\mathbb{Q}[x_1, \dots, x_n]/\langle x_1x_2 \cdots x_n \rangle$:

$$\mathfrak{I}_{\mathcal{G}} \;=\; \big\langle\, x^{y_i} \cdot (x^{y_j} - x^{y_i}) \;:\; y_i \to y_j \text{ is a reaction of } \mathcal{G} \,\big\rangle.$$

THEOREM 3.3 (Minimal siphons). *Let $\mathcal{G}$ be a reaction network. For each minimal prime $\mathfrak{P}$ of $\mathfrak{I}_{\mathcal{G}}$, consider the set:*

$$\{i \in \{1, \dots, n\} \mid x_i \in \mathfrak{P}\}. \tag{3.1}$$

The inclusion-minimal sets of the form (3.1) are precisely the minimal siphons of $\mathcal{G}$.

EXAMPLE 3.4. Consider the following network (its biological significance will be explained in Section 4.1):

$$\begin{aligned} X_1 + X_3 \to X_6 \to X_1 + X_4 \to X_7 \to X_1 + X_5, \\ X_2 + X_5 \to X_8 \to X_2 + X_4 \to X_9 \to X_2 + X_3. \end{aligned} \tag{3.2}$$

As in [**SS10**], we use `Macaulay 2` [**GS**] and, in particular, the command `decompose` to obtain minimal primes:

```
R = QQ[x_1 .. x_9];
I = ideal (x_1*x_3*(x_6-x_1*x_3),  x_6*(x_1*x_4-x_6),
x_1*x_4*(x_7-x_1*x_4), x_7*(x_1*x_5-x_7), x_2*x_5*(x_8-x_2*x_5),
x_8*(x_2*x_4-x_8), x_2*x_4*(x_9-x_2*x_4),  x_9*(x_2*x_3-x_9));
decompose (I + ideal product gens R)
```

The output reveals that there are 12 minimal primes:

$$\begin{gathered}
\langle x_1,\ x_2,\ x_6,\ x_7,\ x_8,\ x_9\rangle, \\
\langle x_3,\ x_4,\ x_5,\ x_6,\ x_7,\ x_8,\ x_9\rangle, \\
\langle x_1,\ x_4,\ x_5,\ x_6,\ x_7,\ x_8,\ x_9\rangle, \\
\langle x_2,\ x_3,\ x_4,\ x_6,\ x_7,\ x_8,\ x_9\rangle, \\
\langle x_2,\ x_3,\ x_4,\ x_6,\ x_8,\ x_9,\ x_1x_5 - x_7\rangle, \\
\langle x_2,\ x_3,\ x_6,\ x_8,\ x_9,\ x_4 - x_5,\ x_1x_5 - x_7\rangle, \\
\langle x_2,\ x_3,\ x_8,\ x_9,\ x_6 - x_7,\ x_4 - x_5,\ x_1x_5 - x_7\rangle, \\
\langle x_2,\ x_8,\ x_9,\ x_3 - x_5,\ x_4 - x_5,\ x_6 - x_7,\ x_1x_5 - x_7\rangle, \\
\langle x_1,\ x_4,\ x_5,\ x_6,\ x_7,\ x_8,\ x_2x_3 - x_9\rangle, \\
\langle x_1,\ x_5,\ x_6,\ x_7,\ x_8,\ x_3 - x_4,\ x_2x_4 - x_9\rangle, \\
\langle x_1,\ x_5,\ x_6,\ x_7,\ x_3 - x_4,\ x_8 - x_9,\ x_2x_4 - x_9\rangle, \\
\langle x_1,\ x_6,\ x_7,\ x_3 - x_5,\ x_4 - x_5,\ x_8 - x_9,\ x_2x_5 - x_9\rangle.
\end{gathered}$$

The last prime ideal contains x_1, x_6, x_7. Additionally, for all other prime ideals, the set of all monomials of the form x_i appearing in the ideal is *not* a proper subset of $\{x_1, x_6, x_7\}$. Thus, $\{1, 6, 7\}$ is a minimal prime. Similarly, there are two more minimal primes: $\{2, 8, 9\}$ and $\{3, 4, 5, 6, 7, 8, 9\}$.

The next result involves a certain square-free monomial ideal $\mathfrak{B}$, which is constructed from the stoichiometric subspace (see [**SS10**] for details). The result concerns relevant siphons; a siphon W is *relevant* if it defines a nonempty face of some stoichiometric compatibility class $\mathcal{P}$ (i.e., the face F_W of $\mathcal{P}$ is nonempty). Shiu and Sturmfels proved the following result, which extends a result of Angeli, De Leenheer, and Sontag [**ADS07**].

THEOREM 3.5. *For a reaction network $\mathcal{G}$, the following are equivalent:*

(1) *For every siphon W of $\mathcal{G}$, there exists a nonnegative, nonzero,*[5] *linear conservation law whose support is a subset of W.*
(2) *$\mathcal{G}$ has no relevant siphons.*
(3) $\mathrm{Sat}(\mathfrak{I}_{\mathcal{G}}, \mathfrak{B}) = \langle 1 \rangle$.

[5]The fact that "nonzero" is missing from the statement of this result in [**SS10**] was pointed out by Gilles Gnacadja (Amgen), whose connection to our community was initiated by an email from Sturmfels in 2007 and then a visit to Berkeley.

If the above (equivalent) conditions hold, then there are no boundary steady states in any stoichiometric compatibility class, and in particular the global attractor conjecture holds for every toric dynamical system arising from $\mathcal{G}$.

EXAMPLE 3.6. Recall that the minimal siphons of the McKeithan network are $\{1,3,4\}$ and $\{2,3,4\}$. Each of these contains the support of a nonnegative, nonzero, linear conservation law, as in (1.10), namely, $x_1 + x_3 + x_4 - c_1 = 0$ and $x_2 + x_3 + x_4 - c_2 = 0$, respectively. So, by Theorem 3.5, no boundary steady states exist in any stoichiometric compatibility class.

We note that Sturmfels, motivated in part by computations involving siphons, directly encouraged Thomas Kahle to develop theoretical and computation tools for binomial primary decomposition [**Kah10, KM14**] (which of course built on Sturmfels's foundational work with Eisenbud [**ES96**]). Much of Kahle's work was in collaboration with Ezra Miller, and the two of them were first introduced by one of Bernd's famous "I think you should talk" emails.

Indeed, Bernd truly succeeded in bringing algebraically inclined people into the reaction network community: Kahle later collaborated with Carsten Conradi (who first learned about algebraic aspects of this area from Karin Gatermann) on detecting whether an ideal (especially a steady-state ideal) is binomial [**CK15**]. Subsequently, Kahle and Conradi, with their jointly advised PhD student, Alexandru Iosif, analyzed multistationarity for systems having a monomial steady-state parametrization [**CIK19**]. Also, Ezra Miller went on to prove, with Manoj Gopalkrishnan and Anne Shiu, the strongly endotactic case of the global attractor conjecture [**GMS14, GMS13**] (recall Section 2.3).

We end this section by describing one more aspect of Sturmfels's work that pertains to boundary steady states. When studying boundary steady states, we often want to know the possible configurations of boundary faces F_W in the various stoichiometric compatibility classes. These classes (1.9) are subsets of $\mathbb{R}^n_{\geq 0}$ and can be defined by means of an equation $Zx = c$, where Z is a $d \times n$-matrix. In many applications, the image of $\mathbb{R}^n_{\geq 0}$ under Z is $\mathbb{R}^d_{\geq 0}$, and so the stoichiometric compatibility classes are in bijection with $\mathbb{R}^d_{\geq 0}$. In this setting, Sturmfels and collaborators introduced the *chamber complex* of Z. This complex is the partition of $\mathbb{R}^d_{\geq 0}$ into equivalence classes, where $c, c' \in \mathbb{R}^d_{\geq 0}$ are equivalent if their associated stoichiometric compatibility classes (which are polyhedra) have the same normal fan [**GHRS16**] and in particular are combinatorially isomorphic, with the same arrangement of boundary faces. The authors of [**GHRS16**] give a description of the chamber complex for a specific biological model: the Wnt shuttle network discussed later in Section 6.1. Coming full circle, a related approach to studying this chamber complex, which involves the ideal $\mathfrak{B}$ mentioned above, was previously described by Sturmfels in the context of siphons [**SS10**].

4. Invariants, matroids, and model selection

We now leave complex-balanced steady states and boundary steady states to enter a new topic, namely how steady-state invariants can be used for model discrimination (Section 4.1) and then discuss how Sturmfels introduced the idea of using matroids to help with this approach (Sections 4.2 and 4.3).

4.1. The idea behind invariants. In 2007, Jeremy Gunawardena wrote an influential article on how algebraic invariants can be used for nonparametric model selection [**Gun07**] (see also [**MG08**]). The idea can be formulated as follows. Given a reaction network $\mathcal{G}$, consider the *steady-state ideal* I, which is the ideal in $\mathbb{C}(\kappa)[x]$ generated by the polynomials on the right-hand side of the mass-action ODEs (1.6). If x^* is a steady state of a mass-action system defined by $\mathcal{G}$ and a choice of rate constants κ^*, then essentially every polynomial in I vanishes when evaluated at x^* and specialized at κ^*. Here, "essentially" means that we avoid polynomials in which some coefficient has a zero denominator when specialized at κ^*. In this context, an element of I is called a *(steady-state) invariant.*

Model selection aims at determining which of several competing models best represents a mechanism under study. Gunawardena's crucial insight is that by comparing the monomials appearing in the steady-state invariants of two competing models (networks), we can use experimental data at steady state to reject a model or support its validity, *without needing to know (or estimate) parameter values.*

To illustrate the idea, we present a slight modification of the models Gunawardena analyzed in [**Gun07**]. We consider a substrate S that exists in three states S_0, S_1, S_2. Such states arise for example when S admits the addition of a phosphate group (a process called *phosphorylation*) at two different sites: S_0 has no phosphate groups attached, S_1 has one phosphate, and S_2 has two. Phosphorylation is mediated by an enzyme E, and the reverse process, called *dephosphorylation*, is mediated by the enzyme F. One possible mechanism is the following (called *distributive*), where an independent encounter of S with the enzyme E or F is required for the phosphorylation or dephosphorylation of each site:

$$\begin{aligned}(4.1)\qquad \mathcal{G}_1:\qquad & E+S_0 \xrightarrow{\kappa_1} Y_1 \xrightarrow{\kappa_2} E+S_1 \xrightarrow{\kappa_3} Y_2 \xrightarrow{\kappa_4} E+S_2,\\ & F+S_2 \xrightarrow{\kappa_5} Y_3 \xrightarrow{\kappa_6} F+S_1 \xrightarrow{\kappa_7} Y_4 \xrightarrow{\kappa_8} F+S_0.\end{aligned}$$

An alternative mechanism (called *processive*) allows an encounter with the enzyme E to result in the phosphorylation of one or two sites.[6] This gives the following *mixed-mechanism* reaction network (notice that the dephosphorylation mechanism remains distributive):

$$\begin{aligned}\mathcal{G}_2:\qquad & E+S_0 \xrightarrow{\kappa_1} Y_1 \xrightarrow{\kappa_2} E+S_1 \qquad Y_1 \xrightarrow{\kappa_4} E+S_2,\\ & F+S_2 \xrightarrow{\kappa_5} Y_3 \xrightarrow{\kappa_6} F+S_1 \xrightarrow{\kappa_7} Y_4 \xrightarrow{\kappa_8} F+S_0.\end{aligned}$$

We rename the species as

$$X_1 = E, X_2 = F, X_3 = S_0, X_4 = S_1, X_5 = S_2, X_6 = Y_1, X_7 = Y_2, X_8 = Y_3, X_9 = Y_4.$$

The relabeled network $\mathcal{G}_1$ is the network (3.2) from earlier. Then, using mass-action kinetics, we form the steady-state ideals I_1 and I_2 of (respectively) $\mathcal{G}_1$ and $\mathcal{G}_2$:

$$\begin{aligned}I_1 = \langle & \kappa_3x_1x_4 - \kappa_4x_7,\ -\kappa_3x_1x_4 - \kappa_7x_2x_4 + \kappa_2x_6 + \kappa_6x_8,\ -\kappa_5x_2x_5 + \kappa_4x_7,\\ & \kappa_1x_1x_3 - \kappa_2x_6,\ \kappa_5x_2x_5 - \kappa_6x_8,\ \kappa_7x_2x_4 - \kappa_8x_9\rangle,\end{aligned}$$

$$\begin{aligned}I_2 = \langle & -\kappa_7x_2x_4 + \kappa_2x_6 + \kappa_6x_8,\ -\kappa_5x_2x_5 + \kappa_4x_6,\ \kappa_1x_1x_3 - \kappa_2x_6 - \kappa_4x_6,\\ & \kappa_5x_2x_5 - \kappa_6x_8,\ \kappa_7x_2x_4 - \kappa_8x_9\rangle.\end{aligned}$$

[6]The processive mechanism shown here differs slightly from some in the literature, e.g., [**CS15**].

In the networks $\mathcal{G}_1$ and $\mathcal{G}_2$, the species Y_i are called *intermediate species*, and their concentrations are typically difficult to measure experimentally. We therefore eliminate the corresponding variables from the steady-state ideals, as follows:

$$I_1 \cap \mathbb{C}(\kappa)[x_1, \ldots, x_5] = \langle \kappa_1\kappa_5 x_1 x_3 x_5 - \kappa_3\kappa_7 x_1 x_4^2 \rangle, \tag{4.2}$$

$$I_2 \cap \mathbb{C}(\kappa)[x_1, \ldots, x_5] = \langle -\kappa_4\kappa_7 x_2 x_4 + \kappa_5(\kappa_2 + \kappa_4) x_2 x_5,\ \kappa_1 x_1 x_3 - \kappa_7 x_2 x_4 \rangle. \tag{4.3}$$

In these elimination ideals, we focus on x_3, x_4, x_5 (corresponding to species S_0, S_1, S_2). By factoring x_1 from the generator in (4.2),[7] we find that, for all parameter values, at steady state (specifically, at positive steady states),

$$x_3 x_5 \text{ and } x_4^2 \text{ lie on a line.} \tag{4.4}$$

However, the same does not hold for $\mathcal{G}_2$. For this second network, we factor x_2 from the first polynomial, and obtain that at steady state,

$$x_4 \text{ and } x_5 \text{ lie on a line.} \tag{4.5}$$

The implications for model selection are as follows. By obtaining several sets of experimental data at steady state for x_3, x_4, x_5 (arising from distinct initial conditions $x(0)$), we can assess which of the pairs (x_4, x_5) or (x_3x_5, x_4^2) is more likely to have come from a line. We can then determine whether $\mathcal{G}_1$ or $\mathcal{G}_2$ is a more accurate model, or perhaps both models need to be rejected.

Of course, this is an idealistic approach, because experimental data contain errors, which are magnified when data values are multiplied (via monomials like x_3x_5), and the model itself is an approximation of reality. Accordingly, this approach requires a proper statistical treatment. Some ideas in this direction were given in [**HHTS12**], and related approaches to nonparametric model selection have been employed for example in [**DYHB13**].

Despite these challenges, Gunawardena's basic idea is very appealing, as it circumvents the need to know or estimate the parameters of a candidate model. Indeed, identifying suitable invariants—or at least the subsets of variables that are related by a polynomial in the steady-state ideal (more on this topic is in the next subsections)—can help with experimental design. More precisely, researchers can construct experiments to measure relevant subsets of variables so that polynomial relations can be tested.

4.2. The human story. In March 2013, Dickenstein, Gunawardena, and Shiu organized a workshop, "Mathematical Problems Arising from Biochemical Reaction Networks", at the American Institute for Mathematics (AIM). Sturmfels came to visit, and in the afternoon, a big audience gathered around him while he was standing at a whiteboard talking about *matroids* and their relation to the problem of steady-state invariants. The word "matroid" was foreign to most of the workshop participants, but the idea stuck. Some of us went back home and started reading on the topic.

Heather Harrington was present that day. Subsequently, she and collaborators applied ideas Sturmfels initiated—using matroids for finding steady-state invariants—to the problem of model selection for an important biological process, the Wnt signaling pathway [**MRBH15**], and later explored this topic further with Sturmfels and others [**GHRS16**]. These works constituted part of the thesis of

[7] We could instead avoid this factoring by first saturating the steady-state ideal with respect to $x_1 \cdots x_9$.

Sturmfels's PhD student Zvi Rosen, which was entitled "Algebraic matroids in applications" [**Ros15**].

4.3. Algebraic matroids. So, what is an algebraic matroid? Or well... what is a matroid? As it is not the aim of this article to give a proper introduction to matroids, we will simply say that it is a pair of objects $(X, \mathcal{I})$ such that X is a finite set and $\mathcal{I}$ is a set of subsets of X. The pair must satisfy some independence axioms, and so the elements of $\mathcal{I}$ are called *independent sets* [**Oxl11**]. The basic example of a matroid (a linear matroid) comes from a finite set of vectors $X = \{v_1, \ldots, v_n\}$ from a vector space. Here $\mathcal{I}$ consists of all linearly independent subsets of X.

Matroids have the following *downward-closed* property: Every subset of an independent set is an independent set (if $U \subseteq V \in \mathcal{I}$, then $U \in \mathcal{I}$). It thus makes sense to define the *circuits* of the matroid as the minimal sets (with respect to inclusion) not contained in $\mathcal{I}$.

Algebraic matroids arise in the context of irreducible algebraic varieties, and independent sets consist of variables that are *not* the support of any polynomial in the defining prime ideal. Specifically, consider a prime ideal I in a polynomial ring $K[x]$ where $x = (x_1, \ldots, x_n)$, and let $\mathcal{I}$ denote the following subset of the power set of $\{x_1, \ldots, x_n\}$:

$$\mathcal{I} = \{C \subseteq \{x_1, \ldots, x_n\} \mid I \cap K[C] = \{0\}\}.$$

Hence, circuits are the minimal subsets of variables for which there is a polynomial in the ideal involving only the variables in the subset. Additionally, for a circuit C, the ideal $I \cap K[C]$ is a principal ideal and the generator involves all variables in C. Several strategies can be employed to find the algebraic matroid [**Ros15**].

In the context of reaction networks, we are interested in algebraic matroids arising from the defining prime ideal I (in $\mathbb{Q}(\kappa)[x]$) of some irreducible component of the *steady-state variety* (that is, the zero set defined by the right-hand sides of the ODEs). When the set of positive steady states admits a rational parametrization (as in the examples below), we will instead consider the prime ideal I of the closure of the image of the parametrization.

Let us find the circuits of the algebraic matroids of the networks $\mathcal{G}_1$ and $\mathcal{G}_2$ above. For the distributive mechanism $\mathcal{G}_1$, from (4.1), we use symbolic software to solve the equations obtained by setting to zero the right-hand sides of the ODEs for $i = 4, \ldots, 9$, obtaining:

$$x_4 = \tfrac{\kappa_1}{\kappa_7} x_1 x_2^{-1} x_3, \qquad x_5 = \tfrac{\kappa_1\kappa_3}{\kappa_5\kappa_7} x_1^2 x_2^{-2} x_3, \qquad x_6 = \tfrac{\kappa_1}{\kappa_2} x_1 x_3,$$
$$x_7 = \tfrac{\kappa_1\kappa_3}{\kappa_7\kappa_4} x_1^2 x_2^{-1} x_3, \qquad x_8 = \tfrac{\kappa_1\kappa_3}{\kappa_6\kappa_7} x_1^2 x_2^{-1} x_3, \qquad x_9 = \tfrac{\kappa_1}{\kappa_8} x_1 x_3.$$

This yields a *steady-state parametrization*: for a given κ, the set of positive steady states is the image of the map $\mathbb{R}^3_{>0} \to \mathbb{R}^9_{>0}$ sending (x_1, x_2, x_3) to $(x_1, \ldots, x_9)$, where $x_4, \ldots, x_9$ come from the above expressions. Using implicitization [**CLO07**, §3.3], we find that the corresponding prime ideal is:

$$\widetilde{I}_1 = \big\langle \kappa_4 x_7 - \kappa_6 x_8, \kappa_2 x_6 - \kappa_8 x_9, -\kappa_5\kappa_8 x_5 x_9 + \kappa_7\kappa_6 x_4 x_8, \kappa_1\kappa_6 x_3 x_8 - \kappa_3\kappa_8 x_4 x_9,$$
$$\kappa_1\kappa_5 x_3 x_5 - \kappa_3\kappa_7 x_4^2, \kappa_5 x_2 x_5 - \kappa_6 x_8, \kappa_7 x_2 x_4 - \kappa_8 x_9,$$
$$\kappa_3\kappa_8 x_1 x_9 - \kappa_6\kappa_7 x_2 x_8, \kappa_3 x_1 x_4 - \kappa_6 x_8, \kappa_1 x_1 x_3 - \kappa_8 x_9\big\rangle.$$

Using Gröbner bases and elimination, we find that the minimal sets of variables $\{x_{i_1}, \ldots, x_{i_q}\}$ for which there is a polynomial in $\widetilde{I}_1 \cap \mathbb{Q}(\kappa)[x_{i_1}, \ldots, x_{i_q}]$ in those variables are the following:

$q = 2$: $\{6,9\},\{7,8\}$,

$q = 3$: $\{1,3,6\},\{1,3,9\},\{1,4,7\},\{1,4,8\},\{2,4,6\},\{2,4,9\},\{2,5,7\},\{2,5,8\},\{3,4,5\}$,

$q = 4$: $\{1,2,3,4\},\{1,2,3,5\},\{1,2,3,7\},\{1,2,3,8\},\{1,2,4,5\},\{1,2,5,6\},\{1,2,5,9\},$ $\{1,2,6,7\},\{1,2,6,8\},\{1,2,7,9\},\{1,2,8,9\},\{1,3,5,7\},\{1,3,5,8\},\{1,4,5,6\},$ $\{1,4,5,9\},\{1,5,6,7\},\{1,5,6,8\},\{1,5,7,9\},\{1,5,8,9\},\{2,3,4,7\},\{2,3,4,8\},$ $\{2,3,5,6\},\{2,3,5,9\},\{2,3,6,7\},\{2,3,6,8\},\{2,3,7,9\},\{2,3,8,9\},\{3,4,6,7\},$ $\{3,4,6,8\},\{3,4,7,9\},\{3,4,8,9\},\{3,5,6,7\},\{3,5,6,8\},\{3,5,7,9\},\{3,5,8,9\},$ $\{4,5,6,7\},\{4,5,6,8\},\{4,5,7,9\},\{4,5,8,9\}$.

These sets are the circuits of the matroid. We recognize the circuit $\{3,4,5\}$ as coming from a steady-state invariant we found earlier (recall (4.4)). As we started with an irreducible component intersecting $\mathbb{R}^9_{>0}$, factorization of x_1 is not necessary in this approach.

We repeat with the mixed-mechanism network $\mathcal{G}_2$ and find a steady-state parametrization given by:

$$x_4 = \tfrac{\kappa_1}{\kappa_7}\, x_1 x_2^{-1} x_3, \qquad x_5 = \tfrac{\kappa_1\kappa_4}{(\kappa_2+\kappa_4)\kappa_5}\, x_1 x_2^{-1} x_3, \qquad x_6 = \tfrac{\kappa_1}{\kappa_2+\kappa_4}\, x_1 x_3,$$
$$x_8 = \tfrac{\kappa_1\kappa_4}{(\kappa_2+\kappa_4)\kappa_6}\, x_1 x_3, \qquad x_9 = \tfrac{\kappa_1}{\kappa_8}\, x_1 x_3,$$

where we recall that x_7 is missing simply because $\mathcal{G}_2$ does not contain the species X_7. The corresponding prime ideal is as follows:

$$\widetilde{I}_2 = \big\langle (\kappa_2\kappa_6 + \kappa_6\kappa_4)x_8 - \kappa_4\kappa_8 x_9, x_6(\kappa_2+\kappa_4) - \kappa_8 x_9, (-\kappa_2\kappa_5 - \kappa_4\kappa_5)x_5 + \kappa_4\kappa_7 x_4,$$
$$x_2 x_5(\kappa_2\kappa_5 + \kappa_4\kappa_5) - \kappa_4\kappa_8 x_9, x_3 x_1 \kappa_1 - \kappa_8 x_9 \big\rangle.$$

The circuits of the corresponding algebraic matroid are the following:

$q = 2$: $\{4,5\},\{6,8\},\{6,9\},\{8,9\}$,

$q = 3$: $\{1,3,6\},\{1,3,8\},\{1,3,9\},\{2,4,6\},\{2,4,8\},\{2,4,9\},\{2,5,6\},\{2,5,8\},\{2,5,9\}$,

$q = 4$: $\{1,2,3,4\},\{1,2,3,5\}$.

As expected, $\{4,5\}$ is a circuit (recall the steady-state invariant coming from (4.5)).

By comparing the two algebraic matroids, we see that the set $\{1,2,3,5\}$ is a circuit of both. So, we consider the elimination ideals $\widetilde{I}_j \cap \mathbb{Q}(\kappa)[x_1,x_2,x_3,x_5]$, for $j = 1,2$, and examine the respective generators:

$$\kappa_1\kappa_3 x_1^2 x_3 - \kappa_5\kappa_7 x_2^2 x_5 \text{ (for } \mathcal{G}_1) \quad \text{and} \quad \kappa_1\kappa_4 x_1 x_3 - \kappa_5(\kappa_2+\kappa_4)x_2 x_5 \text{ (for } \mathcal{G}_2).$$

The monomials in these two polynomials differ, so we conclude that steady-state measurements of X_1, X_2, X_3, X_5 could lead to model discrimination.

4.4. Steady-state parametrizations. We end this section by discussing steady-state parametrizations and their relation to matroids.

In the previous subsection, we saw steady-state parametrizations for $\mathcal{G}_1$ and $\mathcal{G}_2$. These parametrizations consist of monomials in the x_i's, so the positive part of the steady-state variety is toric (recall that the ideals $\widetilde{I}_1$ and $\widetilde{I}_2$ are binomial). As another example, we revisit the McKeithan network. The steady-state equations (2.3) are linear in x_3 and x_4, so we solve for those variables to obtain a steady-state parametrization consisting of monomials:

$$x_3 = \tfrac{\kappa_1}{\kappa_2+\kappa_3} x_1 x_2, \qquad x_4 = \tfrac{\kappa_1\kappa_3}{(\kappa_2+\kappa_3)\kappa_4} x_1 x_2. \tag{4.6}$$

Of course, there is no guarantee that a steady-state parametrization exists, but it happens surprisingly often in networks arising in biological applications. This is because polynomial systems arising from reaction networks typically have only linear or quadratic terms, as they represent interactions of proteins. This is often enough to enable the *linear elimination* of variables at steady state. Furthemore, under favorable conditions, the matrix-tree theorem (our old friend!) guarantees that the expressions found after elimination are rational functions with positive coefficients—and hence define the positive part of the steady-state variety; see, e.g., [**FW13c, FW12b, SFW18, Gun12, TG09a**].

Steady-state parametrizations can greatly aid in the analysis of a reaction system [**CS18**], so we are interested in more strategies for finding them. One such strategy uses tools from toric geometry, in particular when the set of positive steady states define a toric variety, e.g., [**PDSC12, JMP18, PMD18**]. Another strategy, employed by Sturmfels and collaborators, uses the bases of degree 1 of the algebraic matroid [**GHRS16**, §5].

We end this section by noting a final application of matroids to reaction networks: Sturmfels and co-authors used matroids to assess *identifiability* of parameters, that is, the question of whether parameters—here, rate constants—can be recovered from data [**GHRS16**, §7].

5. Polyhedral methods and multistationarity

As the reader might have noticed so far, algebraic and polyhedral geometric tools go hand in hand in the study of steady states of reaction networks. We will see more instances of this phenomenon in this section (for assessing whether there are multiple steady states) and in Section 6 (for counting steady states).

5.1. Multistationarity and injectivity. Here we consider the important problem of deciding, for a given network, whether there exist parameter values for which there is more than one positive steady state in some stoichiometric compatibility class. When that happens, we say that the network is *multistationary*. Multistationarity has implications in cellular decision-making, and therefore answering this question has been the focus of much research.

Concretely, we consider a network $\mathcal{G}$, and let $Z \in \mathbb{R}^{d\times n}$ be a matrix such that the stoichiometric compatibility classes are of the form $Zx = c$ for $c \in \mathbb{R}^d$ (so $d = n - s$, and s is the rank of the stoichiometric matrix N). Then, using the notation in the mass-action equation (1.6), the network $\mathcal{G}$ is multistationary if and only if the system

$$YA_\kappa x^Y = 0, \qquad Zx = c \tag{5.1}$$

has at least two solutions $x \in \mathbb{R}^n_{>0}$, for some $\kappa \in \mathbb{R}^r_{>0}$ and $c \in \mathbb{R}^d$.

There are several methods to evaluate multistationarity quite effectively (see, for instance, [**BC10, BP18, CFMW17, PKC12, DPMST19, JS15, WS08**] and implementations [**EFJK12, DBMP14**]). One such approach, which precludes multistationarity, is to show that the function defined by the left-hand side of the equations in (5.1) is injective for all parameter values $\kappa \in \mathbb{R}^r_{>0}$ and $c \in \mathbb{R}^d$. If so, we say that the network is *injective*.

This idea was introduced by Craciun and Feinberg [**CF05**] and then further developed by numerous authors, e.g., in [**BC09, BC10, BDB07, Gna12, CF06, CF10, MR14, PKC12, SF12, MFR$^+$16, FW13a, FW12a, WF13, Fel14, JS12**].

For the purpose of this article, we highlight a formulation given in [**FW12a**] (see also [**MFR$^+$16**]). We begin by removing d linearly dependent rows from the stoichiometric matrix N, to form a new (full rank) matrix $N' \in \mathbb{R}^{s \times r}$. Let $B \in \mathbb{R}^{n \times r}$ be the matrix of mass-action exponents, that is, the i-th column is y if the i-th reaction is $y \to y'$. Then, for $\kappa = (\kappa_1, \ldots, \kappa_r)$ and $\lambda = (\lambda_1, \ldots, \lambda_n)$, we consider the following matrix:

$$M(\kappa, \lambda) = \begin{bmatrix} N' \operatorname{diag}(\kappa) B^\top \operatorname{diag}(\lambda) \\ Z \end{bmatrix} \in \mathbb{R}[\kappa, \lambda]^{n \times n}.$$

With this in place, we have the following result [**FW12a**].

THEOREM 5.1 (Injectivity criterion). *The network $\mathcal{G}$ is injective (and hence $\mathcal{G}$ is nonmultistationary) if and only if all coefficients of the polynomial* $\det(M(\kappa, \lambda))$ *have the same sign, and the polynomial is not the zero polynomial.*

Theorem 5.1 is shown by first establishing that the network is injective if and only if $\det(M(\kappa, \lambda))$ never vanishes, and then showing that, unless $\det(M(\kappa, \lambda))$ is the zero polynomial, this nonvanishing holds if and only if all coefficients of $\det(M(\kappa, \lambda))$ have the same sign. This second equivalence is in turn proven by showing that $\det(M(\kappa, \lambda))$ is homogeneous in κ, λ and is *multiaffine*, that is, each exponent is either 0 or 1.

5.2. The human story. Feliu recounts the following:

Several approaches related to injectivity of networks were unified in [**MFR$^+$16**]. *This happened after several of us—including Alicia Dickenstein—realized at the Dagstuhl Seminar on "Symbolic methods for chemical reaction networks" held in November 2012 at Schloss Dagstuhl, Germany, that several of our independent works had the same core techniques and principles.*

A few years later, while I was visiting Alicia at ICTP in Trieste in March 2015 to work on a common book project, Alicia said something on these lines (which pertains to ideas in the proof of Theorem 5.1 *mentioned above): "Bernd told me that the fact that* $\det(M(\kappa, \lambda))$ *does not vanish if and only all its coefficients have the same sign, is simply because the Newton polytope of* $\det(M(\kappa, \lambda))$ *is an n-dimensional cube and hence all points are vertices." At that time, I was not familiar with the relationship between signs of polynomials and properties of the Newton polytope, although later on Timo de Wolff referred to such a statement as "folklore in real algebraic geometry," and indeed, it goes back at least to Reznick* [**Rez78**]. *I was working on another project on multistationarity* [**CFMW17**], *which required deciding when a polynomial, which has both positive and negative coefficients, attains a negative value over the positive orthant. A bit of digging led me to understand this relation (Proposition* 5.2*), and since then, that result and related ideas are behind many of my works* [**CFMW17, TF21, CFM20, TF22, FKdWY22**].

5.3. Signs, Newton polytope, multistationarity, and dynamics. We now consider multivariate polynomials

$$p(x) = \sum_{\alpha \in \sigma(p)} a_\alpha x^\alpha \in \mathbb{R}[x_1, \ldots, x_n],$$

where $\sigma(p)$ is a finite subset of $\mathbb{Z}^n_{\geq 0}$ and $a_\alpha \neq 0$ (for all α). The *Newton polytope* of p is the convex hull of the exponent vectors of p:

$$\operatorname{New}(p) := \operatorname{Conv}\big(\{\alpha \in \sigma(p)\}\big) \subseteq \mathbb{R}^n. \tag{5.2}$$

Given a face F of the Newton polytope New(p), the polynomial p_F is obtained by considering only the exponents in F:

$$p_F(x) := \sum_{\alpha \in F \cap \sigma(p)} a_\alpha x^\alpha .$$

With this notation, the result recounted in Section 5.2 says the following.

PROPOSITION 5.2. *If there exists $x^* \in \mathbb{R}^n_{>0}$ such that $p_F(x^*) < 0$, then there also exists $y \in \mathbb{R}^n_{>0}$ such that $p(y) < 0$.*

Hence, if F is a vertex α, then $p_F(x) = a_\alpha x^\alpha$, and Proposition 5.2 establishes that if the coefficient of x^α is negative, then p attains some negative values in the positive orthant.

The study of the signs that a polynomial attains on the positive orthant leads to answering difficult questions pertaining to reaction networks. One of the first results in this direction arose when Craciun, Koeppl, and Pantea analyzed quadratic systems with fixed rate constants κ using the injectivity approach to precluding multistatinonarity described earlier (recall Theorem 5.1) [**PKC12**].

Another setting in which Proposition 5.2 can be harnessed arises from networks with no boundary steady states in stoichiometric compatibility classes that intersect $\mathbb{R}^n_{>0}$, and which satisfy an additional condition. For systems arising from such networks (and some choice of rate constants κ), the authors of [**CFMW17**] applied degree theory to construct a polynomial $p_\kappa \in \mathbb{R}[\xi_1, \ldots, \xi_q]$ such that[8]

- if $p_\kappa(\xi) > 0$ for *all* $\xi \in \mathbb{R}^q_{>0}$, then every stoichiometric compatibility class has exactly one positive steady state;
- if $p_\kappa(\xi) < 0$ for *some* $\xi \in \mathbb{R}^q_{>0}$, then there exists a stoichiometric compatibility class with at least two positive steady states.

It follows that the values of κ for which the polynomial p_κ attains negative values on $\mathbb{R}^q_{>0}$ are precisely those κ that enable multistationarity. The only subtlety arises when, for some κ, the polynomial p_κ takes both positive and zero values; but this situation typically does not arise in applications. Conditions for multistationarity can therefore be found by quantifier elimination, but this is infeasible in practice. Indeed, the polynomial p_κ (which we now view as a polynomial in both κ and ξ) is typically large. Instead, it is often profitable to analyze the coefficients of p_κ, and to apply Proposition 5.2.

However, Proposition 5.2 is silent when all negative coefficients arise from non-vertices of the Newton polytope. This situation happens, for instance, with the distributive phosphorylation mechanism $\mathcal{G}_1$ in (4.1) considered above. Feliu recalls how this problem was eventually resolved: *Timo de Wolff and I had talked for a while about using circuit numbers and tools from SONC (sums of nonnegative circuit functions) to determine the parameter region of multistationarity for $\mathcal{G}_1$. Bernd encouraged and brought Timo and me together with his Ph.D. student Nidhi Kaihnsa—plus Timo's PhD student Oğuzhan Yürük—to address this problem. Our results are now published in* [**FKdWY22**] *and further developed in* [**FKdWY23**].

Another context in which Proposition 5.2 is helpful arises when studying bistability and Hopf bifurcations, as follows. Let $J(x)$ denote the Jacobian matrix of $YA_\kappa x^Y$ in (1.6). For a steady state x^*, if all eigenvalues of $J(x^*)$ have negative

[8]The polynomial p_κ refers to the polynomial denoted by $(-1)^s a(\hat{x})$ in [**CFMW17**].

real part, then x^* is *asymptotically stable*, and trajectories starting nearby converge to x^* [**Per01**]. If all eigenvalues of $J(x^*)$ have negative real part except for a complex-conjugate pair of purely imaginary eigenvalues—and an additional nondegeneracy condition holds with respect to moving one parameter value—then *Hopf bifurcations* arise, and as a consequence, periodic solutions exist [**Per01**]. Periodic solutions can be biologically significant.

Finding eigenvalues of matrices with symbolic entries, like $J(x)$, is not possible in general. However, an alternative approach is to consider the Hurwitz matrix H associated with the characteristic polynomial of $J(x)$, and then to use the Routh-Hurwitz criterion [**Bar71**] to characterize the number of eigenvalues with positive vs. negative real part via the signs of the principal leading minors of H. Hence, Proposition 5.2 can again be used to assess stability and Hopf bifurcations; see for instance, [**EEG$^+$15, Yan02, Liu94, EKW00**]. This approach has been explored for phosphorylation networks and other signaling networks [**TF21, OSTT19, CMS19, CFM20**].

Finally, another algebraic approach for finding Hopf bifurcations, based on resultants and the Bezout matrix, was given by Guckenheimer, Myers, and Sturmfels in 1997 [**GMS97**].

6. Counting steady states, mixed volume, and numerical algebraic geometry

As we saw in the previous section, various results and tools are available for assessing whether a given network or family of networks is multistationary. Once a network is known to be multistationary, the next problem is to compute—or at least to bound—the maximum number of positive steady states (and, ideally, the maximum number of those that are stable). This problem of counting (or bounding) the number of positive steady states is an important and active research topic.

In this section, we highlight two of Sturmfels's contributions in this direction. Specifically, his work used for the first time (in this area) two concepts from adjacent areas—mixed volume (Section 6.1) and numerical algebraic geometry (Section 6.2). Both approaches are now quite active areas of research. At the end of the section, we highlight additional methods for counting steady states (Section 6.3).

6.1. Mixed volume. One way to bound the number of positive steady states, that is, the number of positive roots of the corresponding parameterized polynomial system, is to determine the maximum number of roots in the complex torus $\mathbb{C}^*$. One is then led to the concept of *mixed volume* from convex geometry [**Bet92**]. We do not define this precisely here, but we note that the mixed volume of polytopes is readily computable using computer algebra software, and its key property is the content of Bernshtein's theorem [**Ber75**], as follows.

PROPOSITION 6.1. *Let $g_1, \dots, g_n \in \mathbb{R}[x_1, \dots, x_n]$. Then, counted with multiplicity, the number of isolated solutions in $(\mathbb{C}^*)^n$ of the system $g_1(x) = \cdots = g_n(x) = 0$ is at most the mixed volume of the Newton polytopes* $\mathrm{New}(g_1), \dots, \mathrm{New}(g_n)$.

Therefore, an upper bound on the number of positive steady states can be obtained from the mixed volume of an appropriate set of Newton polytopes. Several ways to build such polytopes from the right-hand sides of the ODEs are described below.

In [**GHRS16**], Sturmfels and collaborators were the first to put related ideas into action. However, it is important to note that *the mixed volumes they define were* not *constructed to be upper bounds on the number of positive steady states*; their motivation was instead to assess which of the many parametrizations arising from bases of an algebraic matroid, as mentioned in Section 4.4, are the "best".[9] Nevertheless, these mixed volumes often appear to be good (but not tight) upper bounds. For instance, these mixed volumes range from 5 to 45 for a certain *Wnt shuttle network* (a model for an important biological signaling pathway, with 19 species and 36 reactions), as shown in Table 2, while the conjectured maximum number of positive steady states is 3 [**GHRS16**].

TABLE 2. Three reaction networks, their maximum number of positive steady states, matroidal mixed volume, augmented-system mixed volume, and references for these results. Both "≥ 3" entries are conjectured to be "3" [**GHRS16, OSTT19**]. The entry "$5 - 45$" refers to the fact that the mixed volumes obtained from various choices of parametrizations range from 5 to 45. The "?" entries were not computed, due to the complicated nature of matMV that makes it difficult to apply to general networks.

Network	Max # positive steady states	matMV	augMV	Reference
Wnt	≥ 3	$5 - 45$	56	[**GHRS16, Oba21**]
ERK	≥ 3	?	7	[**OSTT19**]
Fully irreversible ERK	1	?	3	[**OSTT19**]

With the door opened to mixed-volume computations, other formulations of mixed volumes for networks (that is, using other associated polytopes) were proposed [**GH21, OSTT19, Oba21, OSS20**]. These mixed volumes are compared in the recent dissertation of Obatake [**Oba21**], some of which we summarize here. Sturmfels's original mixed volume, which Obatake calls the *matroidal mixed volume* (matMV), is somewhat complicated. Indeed, it involves minimizing over choices of generators of a saturation ideal coming from the basis of an algebraic matroid, and this minimum depends on the choice of the basis (see Table 2).

In contrast, the mixed volume of Obatake, Shiu, Tang, and Torres [**OSTT19**]—the *augmented-system mixed volume* (augMV)—is straightforward to define: it is simply the mixed volume of the Newton polytopes of the polynomials obtained from the right-hand sides of the mass-action ODEs after linearly redundant ODEs are replaced using conservation laws. The mixed volume of Gross and Hill [**GH21**] is similar in spirit, and even lends itself to combinatorial arguments, thereby enabling mixed-volume computations for infinite families of networks [**GH21**]. Moreover, both augMV and the Gross–Hill mixed volume fulfill the inequality, mentioned earlier, that motivates our interest in the mixed volume [**Oba21**]:

$$\text{Max \# positive steady states} \quad \leq \quad \text{augMV} \quad \leq \quad \text{Gross–Hill MV} \tag{6.1}$$

[9]The idea is that parametrizations giving rise to mixed volumes that are equal to the true number of solutions in $(\mathbb{C}^*)^n$ generate good start systems for applying numerical homotopy-continuation methods.

In Table 2, we compare the matroidal and augmented-system mixed volumes for the Wnt shuttle network. Also shown in that table is data on the *ERK network* (and also an irreversible version) from [**RMBS16, OSTT19**]. For the ERK network, the mixed-volume bound is 7, while the maximum number of positive steady states is conjectured to be 3 [**OSTT19**].

We now consider a fourth type of mixed volume, which Obatake calls the steady-state-parametrization mixed volume (sspMV) [**Oba21**]. This mixed volume, which can be viewed as a saturation-free version of Sturmfels's matMV, is defined for networks having a steady-state parametrization. Also, like augMV and the Gross–Hill mixed volume, it bounds the number of positive steady states:

$$\text{Max \# positive steady states} \quad \leq \quad \text{sspMV}. \tag{6.2}$$

The procedure for computing sspMV (given some parametrization) is to substitute the parametrization into the conservation laws, clear denominators, and then compute the mixed volume of the resulting Newton polytopes. We illustrate this procedure in the following example.

EXAMPLE 6.2. We return to the McKeithan network (1.1) and the positive steady-state parametrization in (4.6). We plug this parametrization into the following conservation laws, as in (1.10): $x_1+x_3+x_4-c_1=0$ and $x_2+x_3+x_4-c_2=0$, which yields the following equations in x_1, x_2:

$$x_1+\tfrac{\kappa_1}{\kappa_2+\kappa_3}x_1x_2+\tfrac{\kappa_1\kappa_3}{(\kappa_2+\kappa_3)\kappa_4}x_1x_2-c_1=0, \quad x_2+\tfrac{\kappa_1}{\kappa_2+\kappa_3}x_1x_2+\tfrac{\kappa_1\kappa_3}{(\kappa_2+\kappa_3)\kappa_4}x_1x_2-c_2=0.$$

For these two polynomials, the Newton polytopes are (respectively) the following triangles, which we denote by P_1 and P_2.

The mixed volume $MV(P_1, P_2)$ equals 2, and so sspMV $= 2$ (for the parametrization we used). For this model, the mixed-volume bound (6.2) is not tight (recall that there is always a unique positive steady state—and it is complex-balanced—as, from (2.2), the deficiency is 0).

In light of the bounds (6.1) and (6.2), it is natural to compare augMV and sspMV. In fact, we can have augMV $\leq$ sspMV for some steady-state parametrization, while augMV $>$ sspMV for some "better" ones (and so sspMV improves the bound) [**Oba21**]. Our ideas have therefore come full circle: finding the "best" parametrizations was the original motivation for Sturmfels to propose mixed-volume computations for reaction networks!

6.2. Numerical algebraic geometry. As noted above, the Wnt shuttle network is conjectured to admit at most three positive steady states. In the first application of *numerical algebraic geometry* to reaction networks, Sturmfels and collaborators gave numerical evidence for this conjecture [**GHRS16**]. Specifically, they used the software `Bertini` [**BHSW**] to sample 10,000 pairs of parameter and total-amount values, and checked that the number of positive steady states was never more than 3. They also performed a similar analysis to explore how robust multistationarity of this network is to perturbations in the parameter and total-amount values.

Numerical algebraic geometry continues to play an important role in investigations of biochemical reaction networks [**GDH$^+$16**, **NGA$^+$20**, **HMBH20**]. For details on what numerical algebraic geometry is all about, we refer the reader to another article in this volume, "Numerical nonlinear algebra" [**BBC$^+$23**].

6.3. Other approaches to counting steady states. Here we describe additional approaches to counting or bounding the number of positive steady states. For specific families of biochemical reaction networks, several authors established unlimited multistability, that is, these families have no upper bound on the number of stable, positive steady states [**KFCS15**, **TG09b**, **FRW20**]. Next, the methods of Dickenstein, in collaboration with Frédéric Bihan and Magalí Giaroli, rely on finding "positively decorated" simplices from within the convex hull of all monomials appearing in the mass-action ODEs [**BDG20**, **GBD18**]. Finally, Nida Obatake and Elise Walker go beyond the mixed volume to volumes of "Newton–Okonkov bodies" to establish improved bounds on the number of steady states [**OW22**].

7. Convex hulls of trajectories and the attainable region problem

7.1. The human story. Shiu recalls the following:

The story begins in the early 1990s, when Martin Feinberg reached out to Victor Klee, who was one of Bernd's PhD advisors, for some help proving a result concerning boundary points of convex sets. Years later, in 2009, Martin mentioned these interactions to some of us, including Bernd, at a conference, and subsequently sent some notes and references on what is called the "attainable region problem."

This problem is of significant industrial interest and also inherently involves optimization and convex geometry. Bernd therefore hoped to generate interest in this problem among the convex/real algebraic geometers at Berkeley at the time, such as graduate student Cynthia Vinzant, so he had me give an informal talk on the topic.

Nothing came of it until years later. As Bernd wrote in an email to one of his PhD students, Nidhi Kaihnsa: "I had a very exciting discussion with a famous Chemical Engineer [Larry Biegler] yesterday, along with Cynthia Vinzant. The outcome is a very nice research theme I would like to suggest to you as [a] possible dissertation topic. The scientific context is 'attainable region theory'. Please google this. It looks very nonmathematical, right? But, I am absolutely convinced that this is a 'goldmine' for someone like you, who is open to the natural sciences. It's all about convex algebraic geometry." Of course, Bernd's assessment proved spot-on, and the results Nidhi and Bernd achieved are described below.

7.2. The attainable region problem. *Attainable region theory* is an important research area in chemical engineering; it pertains to the mathematics of understanding two related sets, the *attainable region* and the convex hull of trajectories. Here, trajectories $\{\phi(t) \mid t \geq 0\}$ are images of solutions $\phi(t)$ to (1.3). The convex hull of a trajectory is the smallest convex set containing the trajectory, and the *attainable region of a trajectory* (or the initial condition of the trajectory) is the (generally larger) set that is the smallest convex set that not only contains the trajectory but also is forward-closed with respect to the dynamics.

More generally, an *attainable region* is a convex set that describes all possible concentration vectors that can be obtained by basic chemical processes (mixing, reacting, and inflows) for a given reaction network and inflow rates. The goal, in chemical engineering applications, is to characterize (the boundary of) the attainable region, in order to optimize over this region. For instance, one might wish to maximize the production of a desired chemical [**MGH$^+$16**]. As recounted by Feinberg [**Fei02**], the attainable region problem was initiated by Horn in 1964 [**Hor64**]. Significant results were elucidated by Feinberg, Crowe, Glasser, and Hildebrandt from the 1980s onward [**FH97, HGC90**].

7.3. Contributions of Sturmfels and collaborators. The main results of Sturmfels and co-authors are as follows. First, Kaihnsa proved that for chemical reaction networks that are "linear", that is, the ODEs are given by linear polynomials, the two regions of interest, the attainable region and the convex hull of a trajectory, are always equal [**Kai18**]. Moreover, under some hypotheses, this region is a "spectrahedral shadow" (a convex set that can be described by linear matrix inequalities). Finally, Kaihnsa conducted some numerical experiments and conjectured that for a certain class of networks, the attainable region and the convex hull of a trajectory are again equal.

This conjecture was resolved (in the negative) in the follow-up paper of Kaihnsa and Sturmfels, with Ciripoi and Löhne [**CKLS19**]. The counterexample, and many other related examples, were enabled by an algorithm that the authors developed and implemented. Specifically, the algorithm focuses on decomposing into "patches" the boundary of the convex hull of a trajectory. Moreover, the mathematical foundation underlying these results is motivated by an important problem concerning convex bodies, which also is of general interest in the area of convex algebraic geometry: approximating convex hulls by convex polytopes. Indeed, in this context, the idea of patches was further developed by Plaumann, Sinn, and Wesner [**PSW23**].

8. Closing stories

In this article we did not manage to cover all works in the field of reaction networks that arose from matchmakings engineered by Sturmfels. For instance, recent works of Brustenga, Craciun, and Sorea [**CJS25, BiMCS22**] began while all were at MPI Leipzig, where Sturmfels is a director, and Sturmfels is acknowledged for bringing the team together. Another example involves Martin Helmer, who was Sturmfels's postdoc when he attended the global attractor conjecture workshop mentioned in Section 2.3, and subsequently collaborated with Michael Adamer on model testing and multistationarity for systems with toric steady states [**AH19, AH20**] before joining Feliu's research team [**FH19**].

Feliu recounts the following: *Bernd's ability to match people and integrate newcomers has also had an important effect on me. In 2010, I was entering this field following articles of Gunawardena, completely unaware of all the work of an algebraic nature that was already being developed. But Alicia Dickenstein, Mercedes Pérez-Millán, and Anne Shiu found out about my work with Carsten Wiuf, and, having learned from the master, Anne wrote to us in August 2011 making us aware of the SIAM conference in applied algebraic geometry in North Carolina, and expressing interest in meeting. I took to this large conference, without knowing anyone, but I was extremely well welcomed. One evening, when a big group were walking the streets of Raleigh, probably trying to find a restaurant, Bernd approached me with an unexpected small conversation many will recognize, something like: "So, I hear you started in pure maths and now work with reaction networks. Tell me more about you." That conference and the encounters made there marked the start of my journey in the very-welcoming applied algebraic geometry community.*

Feliu and Shiu share some final thoughts: *Both authors have benefited from the support and advice of Bernd, and, had it not been for this, our research careers would look quite different. We are extremely thankful for that. We thought we would honor Bernd by writing the article in his style (more or less, as we do need to sleep...). We worked intensively on the first draft for a week, taking advantage of the time zone difference such that the article never slept! And we must admit that it worked quite well.*

We hope that with this article we managed to convey our appreciation and admiration of Bernd and our recognition of his considerable influence on our research field and on all of us in this area. We definitely enjoyed writing it and digging into the past.

Happy birthday, Bernd!

Acknowledgments

The authors thank Tim Bates, Aatmun Baxi, Cashous Bortner, Crystal Farris, Seth Gerberding, Nidhi Kaihnsa, Tung Nguyen, Nida Obatake, Alexander Ruys de Perez, and Máté Telek for helpful comments on an earlier version of this article.

The authors also acknowledge Elizabeth Gross for useful discussions, and several people who shared personal recollections with us: Alicia Dickenstein, Thomas Kahle, and Nidhi Kaihnsa. Finally, the authors are grateful to three anonymous reviewers, whose detailed comments improved this work.

References

[ACKN20] D. F. Anderson, D. Cappelletti, J. Kim, and T. D. Nguyen, *Tier structure of strongly endotactic reaction networks*, Stochastic Process. Appl. **130** (2020), no. 12, 7218–7259, DOI 10.1016/j.spa.2020.07.012. MR4167205

[ADE18a] A. Agazzi, A. Dembo, and J.-P. Eckmann, *On the geometry of chemical reaction networks: Lyapunov function and large deviations*, J. Stat. Phys. **172** (2018), no. 2, 321–352, DOI 10.1007/s10955-018-2035-8. MR3824940

[ADE18b] A. Agazzi, A. Dembo, and J.-P. Eckmann, *Large deviations theory for Markov jump models of chemical reaction networks*, Ann. Appl. Probab. **28** (2018), no. 3, 1821–1855, DOI 10.1214/17-AAP1344. MR3809478

[ADS07] D. Angeli, P. De Leenheer, and E. Sontag, *A Petri net approach to persistence analysis in chemical reaction networks*, Biology and control theory: current challenges, Lect. Notes Control Inf. Sci., vol. 357, Springer, Berlin, 2007, pp. 181–216, DOI 10.1007/978-3-540-71988-5_9. MR2352231

[AH19] M. F. Adamer and M. Helmer, *Complexity of model testing for dynamical systems with toric steady states*, Adv. in Appl. Math. **110** (2019), 42–75, DOI 10.1016/j.aam.2019.06.001. MR3964190

[AH20] M. F. Adamer and M. Helmer, *Families of toric chemical reaction networks*, J. Math. Chem. **58** (2020), no. 9, 2061–2093, DOI 10.1007/s10910-020-01162-x. MR4152911

[And08] D. F. Anderson, *Global asymptotic stability for a class of nonlinear chemical equations*, SIAM J. Appl. Math. **68** (2008), no. 5, 1464–1476, DOI 10.1137/070698282. MR2407133

[And11] D. F. Anderson, *A proof of the global attractor conjecture in the single linkage class case*, SIAM J. Appl. Math. **71** (2011), no. 4, 1487–1508, DOI 10.1137/11082631X. MR2835244

[AS10] D. F. Anderson and A. Shiu, *The dynamics of weakly reversible population processes near facets*, SIAM J. Appl. Math. **70** (2010), no. 6, 1840–1858, DOI 10.1137/090764098. MR2596504

[Bar71] S. Barnett, *A new formulation of the theorems of Hurwitz, Routh and Sturm*, J. Inst. Math. Appl. **8** (1971), 240–250. MR288103

[BBC+23] D. J. Bates, P. Breiding, T. Chen, J. D. Hauenstein, A. Leykin, and F. Sottile, *Numerical nonlinear algebra*, Preprint, arXiv:2302.08585, (2023).

[BC09] M. Banaji and G. Craciun, *Graph-theoretic approaches to injectivity and multiple equilibria in systems of interacting elements*, Commun. Math. Sci. **7** (2009), no. 4, 867–900, DOI 10.4310/cms.2009.v7.n4.a4. MR2604624

[BC10] M. Banaji and G. Craciun, *Graph-theoretic criteria for injectivity and unique equilibria in general chemical reaction systems*, Adv. in Appl. Math. **44** (2010), no. 2, 168–184, DOI 10.1016/j.aam.2009.07.003. MR2576846

[BDB07] M. Banaji, P. Donnell, and S. Baigent, *P matrix properties, injectivity, and stability in chemical reaction systems*, SIAM J. Appl. Math. **67** (2007), no. 6, 1523–1547, DOI 10.1137/060673412. MR2349996

[BDG20] F. Bihan, A. Dickenstein, and M. Giaroli, *Lower bounds for positive roots and regions of multistationarity in chemical reaction networks*, J. Algebra **542** (2020), 367–411, DOI 10.1016/j.jalgebra.2019.10.002. MR4031126

[Ber75] D. N. Bernstein, *The number of roots of a system of equations* (Russian), Funkcional. Anal. i Priložen. **9** (1975), no. 3, 1–4. MR435072

[Bet92] U. Betke, *Mixed volumes of polytopes*, Arch. Math. (Basel) **58** (1992), no. 4, 388–391, DOI 10.1007/BF01189930. MR1152628

[BHSW] D. J. Bates, J. D. Hauenstein, A. J. Sommese, and C. W. Wampler, *Bertini: Software for numerical algebraic geometry*, Available at bertini.nd.edu with permanent DOI: dx.doi.org/10.7274/R0H41PB5.

[BiMCS22] L. Brustenga i Moncusí, G. Craciun, and M.-Ş. Sorea, *Disguised toric dynamical systems*, J. Pure Appl. Algebra **226** (2022), no. 8, Paper No. 107035, 24, DOI 10.1016/j.jpaa.2022.107035. MR4379914

[BP18] M. Banaji and C. Pantea, *The inheritance of nondegenerate multistationarity in chemical reaction networks*, SIAM J. Appl. Math. **78** (2018), no. 2, 1105–1130, DOI 10.1137/16M1103506. MR3784124

[CDJY22] G. Craciun, A. Deshpande, B. Joshi, and P. Y. Yu, *Autocatalytic recombination systems: A reaction network perspective*, Math. Biosci. **345** (2022), Paper No. 108784, 13, DOI 10.1016/j.mbs.2022.108784. MR4385892

[CDSS09] G. Craciun, A. Dickenstein, A. Shiu, and B. Sturmfels, *Toric dynamical systems*, J. Symbolic Comput. **44** (2009), no. 11, 1551–1565, DOI 10.1016/j.jsc.2008.08.006. MR2561288

[CF05] G. Craciun and M. Feinberg, *Multiple equilibria in complex chemical reaction networks. I. The injectivity property*, SIAM J. Appl. Math. **65** (2005), no. 5, 1526–1546, DOI 10.1137/S0036139904440278. MR2177713

[CF06] G. Craciun and M. Feinberg, *Multiple equilibria in complex chemical reaction networks. II. The species-reaction graph*, SIAM J. Appl. Math. **66** (2006), no. 4, 1321–1338, DOI 10.1137/050634177. MR2246058

[CF10] G. Craciun and M. Feinberg, *Multiple equilibria in complex chemical reaction networks: Semiopen mass action systems*, SIAM J. Appl. Math. **70** (2010), no. 6, 1859–1877, DOI 10.1137/090756387. MR2596505

[CFM20] C. Conradi, E. Feliu, and M. Mincheva, *On the existence of Hopf bifurcations in the sequential and distributive double phosphorylation cycle*, Math. Biosci. Eng. **17** (2020), no. 1, 494–513, DOI 10.3934/mbe.2020027. MR4045014

[CFMW17] C. Conradi, E. Feliu, M. Mincheva, and C. Wiuf, *Identifying parameter regions for multistationarity*, PLoS Comput. Biol. **13** (2017), no. 10, e1005751.

[CIK19] C. Conradi, A. Iosif, and T. Kahle, *Multistationarity in the space of total concentrations for systems that admit a monomial parametrization*, Bull. Math. Biol. **81** (2019), no. 10, 4174–4209, DOI 10.1007/s11538-019-00639-4. MR4012471

[CJS25] G. Craciun, J. Jin, and M.-Ş. Sorea, *The structure of the toric locus of a reaction network*, Nonlinearity **38** (2025), no. 1, Paper No. 015023, 28, DOI 10.1088/1361-6544/ad9c0b. MR4844581

[CK15] C. Conradi and T. Kahle, *Detecting binomiality*, Adv. in Appl. Math. **71** (2015), 52–67, DOI 10.1016/j.aam.2015.08.004. MR3406958

[CKLS19] D. Ciripoi, N. Kaihnsa, A. Löhne, and B. Sturmfels, *Computing convex hulls of trajectories*, Rev. Un. Mat. Argentina **60** (2019), no. 2, 637–662, DOI 10.33044/revuma.v60n2a22. MR4049807

[CLO07] D. Cox, J. Little, and D. O'Shea, *Ideals, varieties, and algorithms*, 3rd ed., Undergraduate Texts in Mathematics, Springer, New York, 2007. An introduction to computational algebraic geometry and commutative algebra, DOI 10.1007/978-0-387-35651-8. MR2290010

[CMPY19] G. Craciun, S. Müller, C. Pantea, and P. Y. Yu, *A generalization of Birch's theorem and vertex-balanced steady states for generalized mass-action systems*, Math. Biosci. Eng. **16** (2019), no. 6, 8243–8267, DOI 10.3934/mbe.2019417. MR4034297

[CMS19] C. Conradi, M. Mincheva, and A. Shiu, *Emergence of oscillations in a mixed-mechanism phosphorylation system*, Bull. Math. Biol. **81** (2019), no. 6, 1829–1852, DOI 10.1007/s11538-019-00580-6. MR3945912

[CNP13] G. Craciun, F. Nazarov, and C. Pantea, *Persistence and permanence of mass-action and power-law dynamical systems*, SIAM J. Appl. Math. **73** (2013), no. 1, 305–329, DOI 10.1137/100812355. MR3033151

[Cra15] G. Craciun, *Toric differential inclusions and a proof of the global attractor conjecture*, Preprint, `arXiv:1501.02860`, (2015).

[Cra19] G. Craciun, *Polynomial dynamical systems, reaction networks, and toric differential inclusions*, SIAM J. Appl. Algebra Geom. **3** (2019), no. 1, 87–106, DOI 10.1137/17M1129076. MR3920470

[CS15] C. Conradi and A. Shiu, *A global convergence result for processive multisite phosphorylation systems*, Bull. Math. Biol. **77** (2015), no. 1, 126–155, DOI 10.1007/s11538-014-0054-4. MR3303108

[CS18] C. Conradi and A. Shiu, *Dynamics of post-translational modification systems: recent progress and future challenges*, Biophys. J. **114** (2018), no. 3, 507–515.

[DBMP14] P. Donnell, M. Banaji, A. Marginean, and C. Pantea, *Control: An open source framework for the analysis of chemical reaction networks*, Bioinformatics **30** (2014), no. 11.

[DPM11] A. Dickenstein and M. Pérez Millán, *How far is complex balancing from detailed balancing?*, Bull. Math. Biol. **73** (2011), no. 4, 811–828, DOI 10.1007/s11538-010-9611-7. MR2785146

[DPMST19] A. Dickenstein, M. Pérez Millán, A. Shiu, and X. Tang, *Multistationarity in structured reaction networks*, Bull. Math. Biol. **81** (2019), no. 5, 1527–1581, DOI 10.1007/s11538-019-00572-6. MR3935185

[DYHB13] C. A. Dodson, S. Yeoh, T. Haq, and R. Bayliss, *A kinetic test characterizes kinase intramolecular and intermolecular autophosphorylation mechanisms*, Sci. Signal. **6** (2013), no. 282, ra54.

[EEG+15] H. Errami, M. Eiswirth, D. Grigoriev, W. M. Seiler, T. Sturm, and A. Weber, *Detection of Hopf bifurcations in chemical reaction networks using convex coordinates*, J. Comput. Phys. **291** (2015), 279–302, DOI 10.1016/j.jcp.2015.02.050. MR3327984

[EFJK12] P. Ellison, M. Feinberg, H. Ji, and D. Knight, *Chemical reaction network toolbox, version 2.2*, Available online at `http://www.crnt.osu.edu/CRNTWin`, 2012.

[EKW00] M. El Kahoui and A. Weber, *Deciding Hopf bifurcations by quantifier elimination in a software-component architecture*, J. Symbolic Comput. **30** (2000), no. 2, 161–179, DOI 10.1006/jsco.1999.0353. MR1777170

[ES96] D. Eisenbud and B. Sturmfels, *Binomial ideals*, Duke Math. J. **84** (1996), no. 1, 1–45, DOI 10.1215/S0012-7094-96-08401-X. MR1394747

[Fei87] M. Feinberg, *Foundations of chemical reaction network theory*, Applied Mathematical Sciences, vol. 202, Springer, Cham, 2019. MR3890056

[Fei02] M. Feinberg, *Toward a theory of process synthesis*, Industrial & engineering chemistry research **41** (2002), no. 16, 3751–3761.

[Fei73] M. Feinberg, *Complex balancing in general kinetic systems*, Arch. Rational Mech. Anal. **49** (1972/73), 187–194, DOI 10.1007/BF00255665. MR413930

[FH77] M. Feinberg and F. J. M. Horn, *Chemical mechanism structure and the coincidence of the stoichiometric and kinetic subspaces*, Arch. Rational Mech. Anal. **66** (1977), no. 1, 83–97, DOI 10.1007/BF00250853. MR446109

[FH97] M. Feinberg and D. Hildebrandt, *Optimal reactor design from a geometric viewpoint–I. Universal properties of the attainable region*, Chem. Eng. Sci. **52** (1997), no. 10, 1637–1665.

[Fel14] E. Feliu, *Injectivity, multiple zeros and multistationarity in reaction networks*, Proc. A. **471** (2015), no. 2173, 20140530, 18, DOI 10.1098/rspa.2014.0530. MR3298836

[FH19] E. Feliu and M. Helmer, *Multistationarity and bistability for Fewnomial chemical reaction networks*, Bull. Math. Biol. **81** (2019), no. 4, 1089–1121, DOI 10.1007/s11538-018-00555-z. MR3918779

[FKdWY22] E. Feliu, N. Kaihnsa, T. de Wolff, and O. Yürük, *The kinetic space of multistationarity in dual phosphorylation*, J. Dynam. Differential Equations **34** (2022), no. 2, 825–852, DOI 10.1007/s10884-020-09889-6. MR4413372

[FKdWY23] E. Feliu, N. Kaihnsa, T. de Wolff, and O. Yürük, *Parameter region for multistationarity in n-site phosphorylation networks*, SIAM J. Appl. Dyn. Syst. **22** (2023), no. 3, 2024–2053, DOI 10.1137/22M1504548. MR4623278

[FRW20] E. Feliu, A. D. Rendall, and C. Wiuf, *A proof of unlimited multistability for phosphorylation cycles*, Nonlinearity **33** (2020), no. 11, 5629–5658, DOI 10.1088/1361-6544/ab9a1e. MR4157864

[FW12a] E. Feliu and C. Wiuf, *Preclusion of switch behavior in networks with mass-action kinetics*, Appl. Math. Comput. **219** (2012), no. 4, 1449–1467, DOI 10.1016/j.amc.2012.07.048. MR2983855

[FW12b] E. Feliu and C. Wiuf, *Variable elimination in chemical reaction networks with mass-action kinetics*, SIAM J. Appl. Math. **72** (2012), no. 4, 959–981, DOI 10.1137/110847305. MR2968758

[FW13a] E. Feliu and C. Wiuf, *A computational method to preclude multistationarity in networks of interacting species*, Bioinformatics **29** (2013), 2327–2334.

[FW13b] E. Feliu and C. Wiuf, *Simplifying biochemical models with intermediate species*, J. R. Soc. Interface **10** (2013), no. 87.

[FW13c] E. Feliu and C. Wiuf, *Variable elimination in post-translational modification reaction networks with mass-action kinetics*, J. Math. Biol. **66** (2013), no. 1-2, 281–310, DOI 10.1007/s00285-012-0510-4. MR3004790

[Gat01] K. Gatermann, *Counting stable solutions of sparse polynomial systems in chemistry*, Symbolic computation: Solving equations in algebra, geometry, and engineering (South Hadley, MA, 2000), Contemp. Math., vol. 286, Amer. Math. Soc., Providence, RI, 2001, pp. 53–69, DOI 10.1090/conm/286/04754. MR1874271

[GBD18] M. Giaroli, F. Bihan, and A. Dickenstein, *Regions of multistationarity in cascades of Goldbeter-Koshland loops*, J. Math. Biol. **78** (2019), no. 4, 1115–1145, DOI 10.1007/s00285-018-1304-0. MR3922967

[GDH$^+$16] E. Gross, B. Davis, K. L. Ho, D. J. Bates, and H. A. Harrington, *Numerical algebraic geometry for model selection and its application to the life sciences*, J. R. Soc. Interface **13** (2016), no. 123.

[GH02] K. Gatermann and B. Huber, *A family of sparse polynomial systems arising in chemical reaction systems*, J. Symbolic Comput. **33** (2002), no. 3, 275–305, DOI 10.1006/jsco.2001.0512. MR1882230

[GH21] E. Gross and C. Hill, *The steady-state degree and mixed volume of a chemical reaction network*, Adv. in Appl. Math. **131** (2021), Paper No. 102254, 27, DOI 10.1016/j.aam.2021.102254. MR4294205

[GHRS16] E. Gross, H. A. Harrington, Z. Rosen, and B. Sturmfels, *Algebraic systems biology: A case study for the Wnt pathway*, Bull. Math. Biol. **78** (2016), no. 1, 21–51, DOI 10.1007/s11538-015-0125-1. MR3452313

[GJM16] E. Gross, M. Johnston, and N. Meshkat, *Discussing the proof of the global attractor conjecture*, SIAM news: A publication of Society for Industrial and Applied Mathematics (2016).

[GMS97] J. Guckenheimer, M. Myers, and B. Sturmfels, *Computing Hopf bifurcations. I*, SIAM J. Numer. Anal. **34** (1997), no. 1, 1–21, DOI 10.1137/S0036142993253461. MR1445727

[GMS13] M. Gopalkrishnan, E. Miller, and A. Shiu, *A projection argument for differential inclusions, with applications to persistence of mass-action kinetics*, SIGMA Symmetry Integrability Geom. Methods Appl. **9** (2013), Paper 025, 25, DOI 10.3842/SIGMA.2013.025. MR3056169

[GMS14] M. Gopalkrishnan, E. Miller, and A. Shiu, *A geometric approach to the global attractor conjecture*, SIAM J. Appl. Dyn. Syst. **13** (2014), no. 2, 758–797, DOI 10.1137/130928170. MR3199409

[Gna12] G. Gnacadja, *A Jacobian criterion for the simultaneous injectivity on positive variables of linearly parameterized polynomial maps*, Linear Algebra Appl. **437** (2012), no. 2, 612–622, DOI 10.1016/j.laa.2012.03.014. MR2921721

[GS] D. R. Grayson and M. E. Stillman, *Macaulay2, a software system for research in algebraic geometry*, Available at `http://www.math.uiuc.edu/Macaulay2/`.

[Gun03] J. Gunawardena, *Chemical reaction network theory for in-silico biologists*, Technical report, available at `https://vcp.upf.edu/papers/crnt.pdf`, 2003.

[Gun07] J. Gunawardena, *Distributivity and processivity in multisite phosphorylation can be distinguished through steady-state invariants*, Biophys. J. **93** (2007), 3828–3834.

[Gun12] J. Gunawardena, *A linear framework for time-scale separation in nonlinear biochemical systems*, PLOS ONE **7** (2012), no. 5, e36321.

[HGC90] D. Hildebrandt, D. Glasser, and C. M. Crowe, *Geometry of the attainable region generated by reaction and mixing: with and without constraints*, Ind. Eng. Chem. Res. **29** (1990), no. 1, 49–58.

[HHTS12] H. A. Harrington, K. L. Ho, T. Thorne, and M. P. H. Stumpf, *Parameter-free model discrimination criterion based on steady-state coplanarity.*, Proc. Natl. Acad. Sci. **109** (2012), 15746–15751.

[HJ72] F. Horn and R. Jackson, *General mass action kinetics*, Arch. Rational Mech. Anal. **47** (1972), 81–116, DOI 10.1007/BF00251225. MR400923

[HMBH20] H. A. Harrington, D. Mehta, H. M. Byrne, and J. D. Hauenstein, *Decomposing the parameter space of biological networks via a numerical discriminant approach*, Maple in Mathematics Education and Research (Cham) (Jürgen Gerhard and Ilias Kotsireas, eds.), Springer International Publishing, 2020, pp. 114–131.

[Hor64] F. Horn, *Attainable and nonattainable regions in chemical reaction technique*, Third European symposium on chemical reaction engineering, Pergamon Press London, 1964, pp. 1–10.

[Hor72] F. Horn, *Necessary and sufficient conditions for complex balancing in chemical kinetics*, Arch. Rational Mech. Anal. **49** (1972/73), 172–186, DOI 10.1007/BF00255664. MR413929

[Hor74] F. Horn, *The dynamics of open reaction systems*, Mathematical aspects of chemical and biochemical problems and quantum chemistry (Proc. SIAM-AMS Sympos. Appl. Math., New York, 1974), SIAM-AMS Proc., Vol. VIII, Amer. Math. Soc., Providence, RI, 1974, pp. 125–137. MR468667

[HRS00] B. Huber, J. Rambau, and F. Santos, *The Cayley trick, lifting subdivisions and the Bohne-Dress theorem on zonotopal tilings*, J. Eur. Math. Soc. (JEMS) **2** (2000), no. 2, 179–198, DOI 10.1007/s100970050003. MR1763304

[HS95] S. Hoşten and B. Sturmfels, *GRIN: an implementation of Gröbner bases for integer programming*, Integer programming and combinatorial optimization (Copenhagen, 1995), Lecture Notes in Comput. Sci., vol. 920, Springer, Berlin, 1995, pp. 267–276, DOI 10.1007/3-540-59408-6_57. MR1367987

[JMP18] M. D. Johnston, S. Müller, and C. Pantea, *A deficiency-based approach to parametrizing positive equilibria of biochemical reaction systems*, Bull. Math. Biol. **81** (2019), no. 4, 1143–1172, DOI 10.1007/s11538-018-00562-0. MR3918781

[Joh14] M. D. Johnston, *Translated chemical reaction networks*, Bull. Math. Biol. **76** (2014), no. 5, 1081–1116, DOI 10.1007/s11538-014-9947-5. MR3201336

[JS11] M. D. Johnston and D. Siegel, *Weak dynamical nonemptiability and persistence of chemical kinetics systems*, SIAM J. Appl. Math. **71** (2011), no. 4, 1263–1279, DOI 10.1137/100807557. MR2823502

[JS12] B. Joshi and A. Shiu, *Simplifying the Jacobian criterion for precluding multistationarity in chemical reaction networks*, SIAM J. Appl. Math. **72** (2012), no. 3, 857–876, DOI 10.1137/110837206. MR2968753

[JS15] B. Joshi and A. Shiu, *A survey of methods for deciding whether a reaction network is multistationary*, Math. Model. Nat. Phenom. **10** (2015), no. 5, 47–67, DOI 10.1051/mmnp/201510504. MR3395742

[KA56] E. L. King and C. Altman, *A schematic method of deriving the rate laws for enzyme-catalyzed reactions*, J. Phys. Chem. **60** (1956), no. 10, 1375–1378.

[Kah10] T. Kahle, *Decompositions of binomial ideals*, Ann. Inst. Statist. Math. **62** (2010), no. 4, 727–745, DOI 10.1007/s10463-010-0290-9. MR2652314

[Kai18] N. Kaihnsa, *Attainable regions of dynamical systems*, Preprint, arXiv:1802.07298, (2018).

[KFCS15] V. B. Kothamachu, E. Feliu, L. Cardelli, and O. S. Soyer, *Unlimited multistability and boolean logic in microbial signalling*, J. R. Soc. Interface **12** (2015), no. 108, 20150234.

[KM14] T. Kahle and E. Miller, *Decompositions of commutative monoid congruences and binomial ideals*, Algebra Number Theory **8** (2014), no. 6, 1297–1364, DOI 10.2140/ant.2014.8.1297. MR3267140

[Liu94] W. M. Liu, *Criterion of Hopf bifurcations without using eigenvalues*, J. Math. Anal. Appl. **182** (1994), no. 1, 250–256, DOI 10.1006/jmaa.1994.1079. MR1265895

[McK95] T. McKeithan, *Kinetic proofreading in T-cell receptor signal transduction*, Proc. Natl. Acad. Sci. USA **92** (1995), no. 11, 5042–5046.

[MFR+16] S. Müller, E. Feliu, G. Regensburger, C. Conradi, A. Shiu, and A. Dickenstein, *Sign conditions for injectivity of generalized polynomial maps with applications to chemical reaction networks and real algebraic geometry*, Found. Comput. Math. **16** (2016), no. 1, 69–97, DOI 10.1007/s10208-014-9239-3. MR3451424

[MG08] A. K. Manrai and J. Gunawardena, *The geometry of multisite phosphorylation*, Biophys. J. **95** (2008), 5533–5543.

[MGH+16] D. Ming, D. Glasser, D. Hildebrandt, B. Glasser, and M. Metgzer, *Attainable region theory: An introduction to choosing an optimal reactor*, John Wiley & Sons, 2016.

[MR14] V. P. Gerdt, W. Koepf, W. M. Seiler, and E. V. Vorozhtsov (eds.), *Computer algebra in scientific computing*, Lecture Notes in Computer Science, vol. 8660, Springer, Cham, 2014, DOI 10.1007/978-3-319-10515-4. MR3279659

[MRBH15] A. L. MacLean, Z. Rosen, H. M. Byrne, and H. A. Harrington, *Parameter-free methods distinguish Wnt pathway models and guide design of experiments*, Proc Nat Acad Sci. **9** (2015), no. 112, 2652–2657.

[NGA+20] K.-M. Nam, B. M. Gyori, S. V. Amethyst, D. J. Bates, and J. Gunawardena, *Robustness and parameter geography in post-translational modification systems*, PLOS Comput. Biol. **16** (2020), no. 5, 1–50.

[Oba21] N. K. Obatake, *On the Stability and Number of Steady States of Chemical Reaction Networks*, ProQuest LLC, Ann Arbor, MI, 2021. Thesis (Ph.D.)–Texas A&M University. MR4435713

[OSS20] N. Obatake, A. Shiu, and D. Sofia, *Mixed volume of small reaction networks*, Involve **13** (2020), no. 5, 845–860, DOI 10.2140/involve.2020.13.845. MR4190442
Math. Biol. **79** (2019), no. 4, 1515–1549.

[OSTT19] N. Obatake, A. Shiu, X. Tang, and A. Torres, *Oscillations and bistability in a model of ERK regulation*, J. Math. Biol. **79** (2019), no. 4, 1515–1549, DOI 10.1007/s00285-019-01402-y. MR4019931

[OW22] N. K. Obatake and E. Walker, *Newton-Okounkov bodies of chemical reaction systems*, Adv. in Appl. Math. **155** (2024), Paper No. 102672, 27, DOI 10.1016/j.aam.2024.102672. MR4700058

[Oxl11] J. Oxley, *Matroid theory*, 2nd ed., Oxford Graduate Texts in Mathematics, vol. 21, Oxford University Press, Oxford, 2011, DOI 10.1093/acprof:oso/9780198566946.001.0001. MR2849819

[Pan12] C. Pantea, *On the persistence and global stability of mass-action systems*, SIAM J. Math. Anal. **44** (2012), no. 3, 1636–1673, DOI 10.1137/110840509. MR2982726

[PDSC12] M. Pérez Millán, A. Dickenstein, A. Shiu, and C. Conradi, *Chemical reaction systems with toric steady states*, Bull. Math. Biol. **74** (2012), no. 5, 1027–1065, DOI 10.1007/s11538-011-9685-x. MR2909119

[Per01] L. Perko, *Differential equations and dynamical systems*, 3rd ed., Texts in Applied Mathematics, vol. 7, Springer-Verlag, New York, 2001, DOI 10.1007/978-1-4613-0003-8. MR1801796

[PKC12] C. Pantea, H. Koeppl, and G. Craciun, *Global injectivity and multiple equilibria in uni- and bi-molecular reaction networks*, Discrete Contin. Dyn. Syst. Ser. B **17** (2012), no. 6, 2153–2170, DOI 10.3934/dcdsb.2012.17.2153. MR2924455

[PMD18] M. Pérez Millán and A. Dickenstein, *The structure of MESSI biological systems*, SIAM J. Appl. Dyn. Syst. **17** (2018), no. 2, 1650–1682, DOI 10.1137/17M1113722. MR3811780

[PSW23] D. Plaumann, R. Sinn, and J. L. Wesner, *Families of faces and the normal cycle of a convex semi-algebraic set*, Beitr. Algebra Geom. **64** (2023), no. 3, 851–875, DOI 10.1007/s13366-022-00657-9. MR4615497

[Rez78] B. Reznick, *Extremal PSD forms with few terms*, Duke Math. J. **45** (1978), no. 2, 363–374. MR480338

[RMBS16] B. Y. Rubinstein, H. H. Mattingly, A. M. Berezhkovskii, and S. Y. Shvartsman, *Long-term dynamics of multisite phosphorylation*, Mol. Biol. Cell **27** (2016), no. 14, 2331–2340.

[Ros15] Z. H. Rosen, *Algebraic Matroids in Applications*, ProQuest LLC, Ann Arbor, MI, 2015. Thesis (Ph.D.)–University of California, Berkeley. MR3450261

[RSW92] Y. Rabinovich, A. Sinclair, and A. Wigderson, *Quadratic dynamical systems (preliminary version)*, Proceedings., 33rd Annual Symposium on Foundations of Computer Science, 1992, pp. 304–313.

[SC94] D. Siegel and Y. F. Chen, *Global stability of deficiency zero chemical networks*, Canad. Appl. Math. Quart. **2** (1994), no. 3, 413–434. MR1318551

[SF12] G. Shinar and M. Feinberg, *Concordant chemical reaction networks*, Math. Biosci. **240** (2012), no. 2, 92–113, DOI 10.1016/j.mbs.2012.05.004. MR3000363

[SFW18] M. Sáez, E. Feliu, and C. Wiuf, *Graphical criteria for positive solutions to linear systems*, Linear Algebra Appl. **552** (2018), 166–193, DOI 10.1016/j.laa.2018.04.014. MR3804483

[SM00] D. Siegel and D. MacLean, *Global stability of complex balanced mechanisms*, J. Math. Chem. **27** (2000), no. 1-2, 89–110, DOI 10.1023/A:1019183206064. MR1801004

[Son01] E. D. Sontag, *Structure and stability of certain chemical networks and applications to the kinetic proofreading model of T-cell receptor signal transduction*, IEEE Trans. Automat. Control **46** (2001), no. 7, 1028–1047, DOI 10.1109/9.935056. MR1842137

[SS10] A. Shiu and B. Sturmfels, *Siphons in chemical reaction networks*, Bull. Math. Biol. **72** (2010), no. 6, 1448–1463, DOI 10.1007/s11538-010-9502-y. MR2671581

[Sta99] R. P. Stanley, *Enumerative combinatorics. Vol. 2*, Cambridge Studies in Advanced Mathematics, vol. 62, Cambridge University Press, Cambridge, 1999. With a foreword by Gian-Carlo Rota and Appendix 1 by Sergey Fomin, DOI 10.1017/CBO9780511609589. MR1676282

[Stu94] B. Sturmfels, *On the Newton polytope of the resultant*, J. Algebraic Combin. **3** (1994), no. 2, 207–236, DOI 10.1023/A:1022497624378. MR1268576

[SWF17] M. Sáez, C. Wiuf, and E. Feliu, *Graphical reduction of reaction networks by linear elimination of species*, J. Math. Biol. **74** (2017), no. 1-2, 195–237, DOI 10.1007/s00285-016-1028-y. MR3590681

[TF21] A. Torres and E. Feliu, *Symbolic proof of bistability in reaction networks*, SIAM J. Appl. Dyn. Syst. **20** (2021), no. 1, 1–37, DOI 10.1137/20M1326672. MR4194953

[TF22] E. Feliu and M. L. Telek, *On generalizing Descartes' rule of signs to hypersurfaces*. part A, Adv. Math. **408** (2022), no. part A, Paper No. 108582, 29, DOI 10.1016/j.aim.2022.108582. MR4458911

[TG09a] M. Thomson and J. Gunawardena, *The rational parameterisation theorem for multisite post-translational modification systems*, J. Theoret. Biol. **261** (2009), no. 4, 626–636, DOI 10.1016/j.jtbi.2009.09.003. MR2974153

[TG09b] M. Thomson and J. Gunawardena, *Unlimited multistability in multisite phosphorylation systems*, Nature **460** (2009), no. 7252, 274–277.

[Tut48] W. T. Tutte, *The dissection of equilateral triangles into equilateral triangles*, Proc. Cambridge Philos. Soc. **44** (1948), 463–482, DOI 10.1017/s030500410002449x. MR27521

[Vol72] A. I. Vol′pert, *Differential equations on graphs* (Russian), Mat. Sb. (N.S.) **88(130)** (1972), 578–588. MR316796

[WF13] C. Wiuf and E. Feliu, *Power-law kinetics and determinant criteria for the preclusion of multistationarity in networks of interacting species*, SIAM J. Appl. Dyn. Syst. **12** (2013), no. 4, 1685–1721, DOI 10.1137/120873388. MR3116636

[WS08] L. Wang and E. D. Sontag, *On the number of steady states in a multiple futile cycle*, J. Math. Biol. **57** (2008), no. 1, 29–52, DOI 10.1007/s00285-007-0145-z. MR2393207

[Yan02] X. Yang, *Generalized form of Hurwitz-Routh criterion and Hopf bifurcation of higher order*, Appl. Math. Lett. **15** (2002), no. 5, 615–621, DOI 10.1016/S0893-9659(02)80014-3. MR1889512

Department of Mathematical Sciences, University of Copenhagen, Universitetsparken 5, 2100 Copenhagen, Denmark
Email address: efeliu@math.ku.dk

Department of Mathematics, Texas A&M University, College Station, Texas 77843-3368
Email address: annejls@tamu.edu

Proceedings of Symposia in Pure Mathematics
Volume **111**, 2025
https://doi.org/10.1090/pspum/111/02046

Numerical nonlinear algebra

Daniel J. Bates, Paul Breiding, Tianran Chen, Jonathan D. Hauenstein, Anton Leykin, and Frank Sottile

In honor of Bernd Sturmfels on his sixtieth birthday

Abstract. Numerical nonlinear algebra is a computational paradigm that uses numerical analysis to study polynomial equations. Its origins were methods to solve systems of polynomial equations based on the classical theorem of Bézout. This was decisively linked to modern developments in algebraic geometry by the polyhedral homotopy algorithm of Huber and Sturmfels, which exploits the combinatorial structure of the equations and led to efficient software for solving polynomial equations.

Subsequent growth of numerical nonlinear algebra continues to be informed by algebraic geometry and its applications. These include new approaches to solving, algorithms for studying positive-dimensional varieties, certification, and a range of applications both within mathematics and from other disciplines. With new implementations, numerical nonlinear algebra is now a fundamental computational tool for algebraic geometry and its applications. We survey some of these innovations and some recent applications.

1. Introduction

Bernd Sturmfels has a knack for neologisms, minting memorable mathematical terms that pithily portray their essence and pass into general use. Nonlinear algebra [**73**] is a Sturmfelsian neologism expressing the focus on computation in applications of algebraic geometry, the objects that appear in applications, and the theoretical underpinnings this inquiry requires. Numerical nonlinear algebra is numerical computation supporting nonlinear algebra. It is complementary to symbolic computation (also a key input to nonlinear algebra), and its development has opened up new vistas to explore and challenges to overcome.

2020 *Mathematics Subject Classification.* 65H14, 65H10, 14-04, 14Q65.

The views expressed in this article are those of the authors and do not reflect the official policy or position of the Department of Defense or the U.S. Government.

The research of D. B. was partially supported by the Naval Academy Research Council and a Kinnear Fellowship. The research of P. B. was supported by Deutsche Forschungsgemeinschaft (DFG) – Projektnr. 445466444. The research of T. C. was partially supported by AMS-Simons Research Enhancement Grant for Primarily Undergraduate Institution Faculty program and by NSF DMS award 1923099. The research of J. H. was partially supported by Robert and Sara Lumpkins Collegiate Professorship, National Science Foundation grant CCF 2331400, and Simons Foundation SFM-00005696. The research of A. L. was partially supported by NSF DMS award 2001267 and the Simons Fellows program. The research of F. S. was partially supported by NSF grant DMS-2201005.

Sturmfels did not create this field, but his work with Huber introducing the polyhedral homotopy algorithm [**56**] catalyzed it. This algorithm exemplifies Sturmfels's mathematical contributions, exploiting links between algebra and geometric combinatorics to address problems in other areas of mathematics, in this case the ancient problem of solving equations. He was also important for its development with his encouragement of researchers, early decisive use of its methods [**8, 101**], and by popularizing it [**13**].

We survey some of the main tools in this field, focusing on homotopy continuation. Our goal is to provide a reasonably comprehensive introduction to some of the main ideas in numerical nonlinear algebra. In Section 2 we describe polynomial homotopy continuation and its basic use to solve systems of polynomial equations. We develop the background and present some details of the polyhedral homotopy algorithm in Section 3. Numerical algebraic geometry, which uses these tools to represent and manipulate algebraic varieties on a computer, is presented in Section 4, along with new methods for solving equations that this perspective affords. A welcome and perhaps surprising feature is that there are often methods to certify the approximate solutions these algorithms provide, which is sketched in Section 5. We close this survey in Section 6 by presenting three domains in which numerical nonlinear algebra has recently been applied.

2. What is polynomial homotopy continuation?

Polynomial homotopy continuation is a numerical method for computing complex solutions to systems of polynomial equations,

$$F(\mathbf{x}) \;=\; F(x_1,\dots,x_n) \;=\; \begin{bmatrix} f_1(x_1,\dots,x_n) \\ \vdots \\ f_m(x_1,\dots,x_n) \end{bmatrix} \;=\; \mathbf{0}, \tag{2.1}$$

where $f_i(x_1,\dots,x_n) \in \mathbb{C}[x_1,\dots,x_n]$ for $1 \le i \le m$. A point $\mathbf{z} \in \mathbb{C}^n$ is a *zero*, or a *solution*, of F if $F(\mathbf{z}) = 0$. A solution is *regular* if the Jacobian matrix of the partial derivatives of F evaluated at $\mathbf{z}$ has rank n. Necessarily, $m \ge n$. When $m = n$, we refer to the system as *square*. Otherwise, the system is *overdetermined*.

The underlying idea of polynomial homotopy continuation is simple: to solve $F(\mathbf{x}) = 0$, we construct another system $G(\mathbf{x}) = 0$ of polynomial equations with known zeros, together with a *homotopy*. This is a family of polynomial systems $H(\mathbf{x}, t)$ that depend algebraically on $t \in \mathbb{C}$ and which interpolates between F and G in that $H(\mathbf{x}, 0) = F(\mathbf{x})$ and $H(\mathbf{x}, 1) = G(\mathbf{x})$. Considering one zero, $\mathbf{y}$, of G and restricting to $t \in [0, 1]$, the equation $H(\mathbf{x}, t) = 0$ defines a *solution path* $\mathbf{x}(t) \subset \mathbb{C}^n$ such that $H(\mathbf{x}(t), t) = 0$ for $t \in [0, 1]$ and $\mathbf{x}(1) = \mathbf{y}$. The path is followed (see Section 2.4) from $t = 1$ to $t = 0$ to compute an approximation of the solution $\mathbf{z} = \mathbf{x}(0)$ to F. Following this solution path is equivalent to solving the initial value problem

$$\frac{\partial}{\partial \mathbf{x}} H(\mathbf{x}, t)\left(\frac{\mathrm{d}}{\mathrm{d}t}\mathbf{x}(t)\right) + \frac{\partial}{\partial t} H(\mathbf{x}, t) \;=\; 0, \qquad \mathbf{x}(1) \;=\; \mathbf{y}. \tag{2.2}$$

This *Davidenko differential equation* [**27, 28**] is typically solved using a predictor-corrector scheme (see Section 2.4). We say that $\mathbf{x}(1) = \mathbf{y}$ gets *tracked* towards $\mathbf{x}(0)$. For this predictor-corrector scheme to successfully compute $\mathbf{x}(0)$, $\mathbf{x}(t)$ must be a regular zero of $H(\mathbf{x}, t) = 0$ for every $t \in (0, 1]$. Nonregular solutions at $t = 0$ are treated using specialized numerical methods called *endgames* [**80**].

So far, there is nothing specific about polynomials. When F is a system of polynomials, we can construct a *start system* G with known zeros and a homotopy H such that for every isolated zero $\mathbf{z}$ of F, there is at least one zero of G that can be tracked towards $\mathbf{z}$. That is, we can compute approximations of all isolated zeros of F.

The ideas behind this method were developed over many years, and by a number of people. García and Zangwill [**43**] proposed polynomial homotopy continuation and a classic reference is Morgan's book [**77**]. The textbook by Sommese and Wampler [**94**] is now a standard reference. Decades of conceptual advances have spawned several implementations of the core algorithms. Historically, the first implementation with wide use was `PHCpack` [**105**], followed a decade later by `Bertini` [**4**], which is also widely used. Later came the `HOM4PS` family [**23, 66**], `NAG4M2` [**67**], and `HomotopyContinuation.jl` [**15**]. The package `NumericalAlgebraicGeometry` of `Macaulay2` [**44**] (referred to as `NAG4M2`) includes interfaces to many of the alternatives listed above; for example, see [**45**].

We next explain polynomial homotopy continuation with more care and in more detail. We first discuss what is meant by numerical method and approximate zero of a system.

2.1. The solution to a system of polynomial equations. A solution to the system (2.1) is a point $\mathbf{z} \in \mathbb{C}^n$ satisfying (2.1). The collection of all such points is an *algebraic variety*,

$$V \;:=\; \{\mathbf{z} \in \mathbb{C}^n \mid f_1(\mathbf{z}) = \cdots = f_m(\mathbf{z}) = 0\}. \tag{2.3}$$

This defines solutions $\mathbf{z}$ *implicitly*, using just the definition of F. It is hard to extract any information other than "$\mathbf{z}$ is a solution to F" from this representation. A more useful representation of V is given by a *Gröbner basis* [**25, 101, 102**].

Consider a simple example of two polynomial equations in two variables,

$$x^2 + y^2 - 1 \;=\; 0,\; x^2 - y^3 - y - 1 \;=\; 0, \tag{2.4}$$

describing the intersection of two plane curves. A Gröbner basis is $\{y^3+y^2+y,\ x^2+y^2-1\}$. Its triangularity facilitates solving. The first equation, $y^3 + y^2 + y = 0$, has the three solutions $0, (-1 \pm \sqrt{-3})/2$. Substituting each into the second gives two solutions, for six solutions altogether. While these equations can be solved exactly, one cannot do this in general. A Gröbner basis is an equivalent implicit representation of V from which we may transparently extract numerical invariants such as the number of solutions or the dimension and degree of V. Finer questions about individual solutions may require computing them numerically.

Numerical methods only compute numerical approximations of solutions to a system (2.1). Thus $(1.271 + .341\sqrt{-1},\ -.500 + .866\sqrt{-1})$ is an approximation of a solution to (2.4). A numerical approximation of a point $\mathbf{z} \in V$ is any point $\mathbf{y} \in \mathbb{C}^n$ which is in some sense close to $\mathbf{z}$. For example, we could require that $\mathbf{y}$ is within some tolerance $\epsilon > 0$ of $\mathbf{z}$, i.e., $|\mathbf{y} - \mathbf{z}| < \epsilon$. Consequently, the concept of zero of (or solution to) a polynomial system is replaced by an *approximate zero* (defined in Section 2.4). This is fundamentally different than using exact methods like Gröbner bases, where the goal is to work with a representation of the true exact zeros of polynomial systems. As an approximate zero is not a true zero, a numerical computation does not yield all the information obtained in an exact computation. On the other hand, a numerical computation is often less costly than a symbolic computation. Other advantages are that the architecture of modern computers is optimized for floating point arithmetic and that numerical continuation is readily parallelized (see Remark 2.4). Another advantage is that approximate zeroes are easily refined using Newton's method (Section 2.4).

Despite not containing all the information of true zeros, in Section 5 we discuss how to use approximate zeros to obtain precise and provable results.

2.2. The parameter continuation theorem. Our discussion of polynomial homotopy continuation assumed that solution paths exist. The *parameter continuation theorem* by Morgan and Sommese [**79**] asserts this when the homotopy arises from a path in *parameter* space.

Suppose that the system of polynomials (2.1) depends on parameters $\mathbf{p} = (p_1, \ldots, p_k) \in \mathbb{C}^k$. Write $F(\mathbf{x}; \mathbf{p})$ for the polynomial system corresponding to a particular choice of $\mathbf{p}$, and further suppose that the map $\mathbf{p} \mapsto F(\mathbf{x}; \mathbf{p})$ is algebraic in that the coefficients of monomials in F are polynomial or rational functions of $\mathbf{p}$. For example, the parameters may be the coefficients in F. Consider the incidence variety

$$Z \;:=\; \{(\mathbf{x}, \mathbf{p}) \in \mathbb{C}^n \times \mathbb{C}^k \mid F(\mathbf{x}; \mathbf{p}) = 0\} \;\subseteq\; \mathbb{C}^n \times \mathbb{C}^k. \tag{2.5}$$

Let $\pi_1 \colon Z \to \mathbb{C}^n$ and $\pi_2 \colon Z \to \mathbb{C}^k$ be the projections onto the first and second factors. The map π_1 identifies points in the fiber $\pi_2^{-1}(\mathbf{p})$ with solutions to $F(\mathbf{x}; \mathbf{p}) = 0$. Standard arguments in algebraic geometry imply the existence of a dense open set $U \subset \mathbb{C}^k$ and a number d such that for all $\mathbf{p} \in U$, the fiber $\pi_2^{-1}(\mathbf{p})$ has dimension d. Furthermore, when this typical dimension d is zero, there is a number N such that for $\mathbf{p} \in U$, then number of regular solutions in the fiber $\pi_2^{-1}(\mathbf{p})$ is N.

THEOREM 2.1 (Parameter continuation theorem). *With these definitions and when $d = 0$, suppose that $\gamma(t) \colon [0, 1] \to \mathbb{C}^k$ is a continuous path.*

(a) *If $\gamma([0, 1]) \subset U$, then the homotopy $H(\mathbf{x}, t) = F(\mathbf{x}; \gamma(t))$ defines N continuous and isolated solution paths $\mathbf{x}(t)$.*
(b) *If $\gamma((0, 1]) \subset U$, then as $t \to 0$, the limits of the solution paths, if they exist, include all the isolated solutions to $F(\mathbf{x}; \gamma(0)) = 0$. This includes both regular solutions and solutions with multiplicity greater than one.*

At points $t \in [0, 1]$ with $\gamma(t) \in U$ where γ is differentiable, $\mathbf{x}(t)$ is differentiable.

A short proof for this theorem using Gröbner bases can be found in [**10**].

Strictly speaking, for differentiability at the endpoints, we want γ to be defined on some neighborhood of $[0, 1] \subset \mathbb{R}$. The point of the theorem is that any path satisfying $\gamma((0, 1]) \subset U$ can be used for polynomial homotopy continuation, with start system $G(\mathbf{x}) = F(\mathbf{x}; \gamma(1))$. Since the *branch locus* $B := \mathbb{C}^k \smallsetminus U$ is a subvariety, it has real codimension at least two and typical paths in the parameter space $\mathbb{C}^k$ do not meet B. When the typical fiber dimension d is zero, we call $\pi_2 \colon Z \to \mathbb{C}^k$ a *branched cover*. Theorem 2.1 can be generalized, replacing the parameter space $\mathbb{C}^k$ by an irreducible variety [**94**, Theorem 7.1.4]. Similarly, polynomials in F may be replaced by Laurent polynomials, as we shall encounter in Section 3. A *parameter homotopy* is one arising from a path γ such as in Theorem 2.1(b).

The parameter continuation theorem follows from Bertini's theorem, other standard results in algebraic geometry, and the implicit function theorem. A proof is given in [**79**].

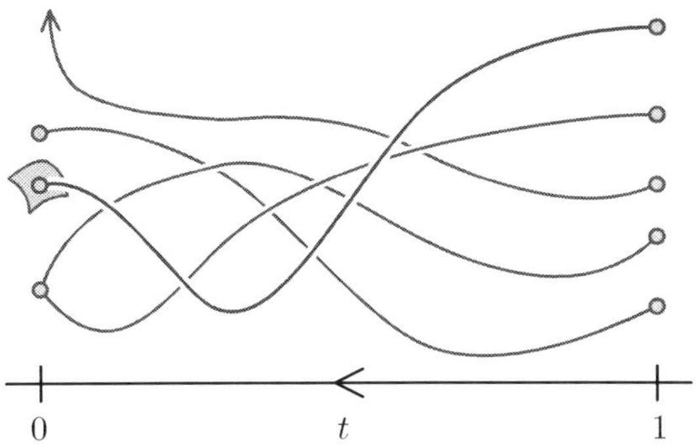

FIGURE 1. Homotopy paths

EXAMPLE 2.2. Figure 1 shows some of the possibilities for homotopy paths $\mathbf{x}(t)$, when Theorem 2.1(b) holds. The start system $G(\mathbf{x})$ at $t = 1$ has $N = 5$ regular zeros, and each lies on a unique path $\mathbf{x}(t)$ for $t \in (0, 1]$. One path has no finite limit as $t \to 0$, while the other four have limits. Two paths have the same limit, and their common endpoint is an isolated zero of $F(\mathbf{x})$ of multiplicity two. The endpoint of another path is not an isolated zero of $F(\mathbf{x})$, indicated by a small surface patch. The endpoint of the other path is the regular zero of the target system $F(\mathbf{x})$. ◇

2.3. The total degree homotopy. To reach *all* isolated zeros of the target system, the start system must have at least as many zeros as the target system. Thus, an upper bound on the number of isolated zeros is often needed to choose a homotopy. One such upper bound is provided by Bézout's theorem: The number of isolated zeros of the system $F = (f_1, \ldots, f_n)$ is at most $d_1 d_2 \cdots d_n$, where $d_i = \deg f_i$ for $i = 1, \ldots, n$. It inspires start systems of the form

$$\begin{bmatrix} b_0 x_1^{d_1} - b_1 \\ \vdots \\ b_0 x_n^{d_n} - b_n \end{bmatrix}. \tag{2.6}$$

For nonzero complex numbers $b_0, b_1, \ldots, b_n \in \mathbb{C}^\times = \mathbb{C} \setminus \{0\}$, this start system is outside the branch locus B and it has $d = d_1 d_2 \cdots d_n$ solutions, which are all are easily computed. This gives the *total degree homotopy*,

$$H(x_1, \ldots, x_n, t) := t \cdot \begin{bmatrix} b_0 x_1^{d_1} - b_1 \\ \vdots \\ b_0 x_n^{d_n} - b_n \end{bmatrix} + (1-t) \cdot \begin{bmatrix} f_1(x_1, \ldots, x_n) \\ \vdots \\ f_n(x_1, \ldots, x_n) \end{bmatrix}. \tag{2.7}$$

Such a convex combination of two similar systems is called a *straight-line homotopy*, which is a particular case of a parameter homotopy. For general choices of the parameters b_i the smoothness conditions of Theorem 2.1(b) hold [**94**, Theorem 8.4.1].

EXAMPLE 2.3. The total degree homotopy for the system (2.4) has the form

$$H(x, y, t) := t \cdot \begin{bmatrix} b_0 x^2 - b_1 \\ b_0 y^3 - b_2 \end{bmatrix} + (1-t) \cdot \begin{bmatrix} x^2 + y^2 - 1 \\ x^2 - y^3 - y - 1 \end{bmatrix}. \tag{2.8}$$

For $b_0, b_1, b_2 \in \mathbb{C}^\times$, the start system $H(x, y, 1) = (b_0x^2 - b_1, b_0y^3 - b_2)$ has six distinct complex zeros, all of which are regular, and the zero set of $H(x, y, t)$ consists of six paths in $\mathbb{C}^2 \times [0, 1]$, each smoothly parameterized by t for *almost all* choices of b_0, b_1, b_2. The parameter b_0 is used to avoid cancellation of the highest degree terms for all $t \in (0, 1]$. ◇

The phrase "almost all" in this example is because the set of choices of b_i for which the paths are singular lies in the complement of the dense open subset U of Theorem 2.1, and therefore has measure zero. Such situations occur frequently in this field, often referred to as *probability one*.

2.4. Path-tracking. Path-tracking is the numerical core of polynomial homotopy continuation.

Suppose that $H(\mathbf{x}, t)$ for $\mathbf{x} \in \mathbb{C}^n$ and $t \in \mathbb{C}$ is a homotopy with target system $F(\mathbf{x}) = H(\mathbf{x}, 0)$ and start system $G(\mathbf{x}) = H(\mathbf{x}, 1)$. Further suppose that for $t \in (0, 1]$, $H(\mathbf{x}, t) = 0$ defines smooth paths $\mathbf{x}(t)\colon (0, 1] \to \mathbb{C}^n$ such that each isolated solution to F is connected to at least one regular solution to G through some path, as in Theorem 2.1(b). By the implicit function theorem, each isolated solution to G is the endpoint $\mathbf{x}(1)$ of a unique path $\mathbf{x}(t)$. Lastly, we assume that all regular solutions to G are known.

Given this, the isolated solutions to F may be computed as follows: For each regular solution $\mathbf{x}(1)$ to G, track the path $\mathbf{x}(t)$ from $t = 1$ towards $t = 0$. If it converges, then $\mathbf{x}(0) = \lim_{t\to 0} \mathbf{x}(t)$ satisfies $F(\mathbf{x}(0)) = 0$, and this will find all isolated solutions to F.

The path $\mathbf{x}(t)$ satisfies the Davidenko differential equation, and thus we may compute values $\mathbf{x}(t)$ by solving the initial value problem (2.2). Consequently, we may use any numerical scheme for solving initial value problems. This is not satisfactory for solving nonlinear polynomial systems due to the propagation of error.

As the solution paths $\mathbf{x}(t)$ are defined implicitly, there are standard methods to mitigate error propagation. Let E be a system of n polynomials in n variables. Given a point $\mathbf{z}_0$ where the Jacobian matrix $JE = \frac{\partial E}{\partial \mathbf{x}}$ of E is invertible, we may apply the *Newton operator* N_E to $\mathbf{z}_0$, obtaining $\mathbf{z}_1$,

$$\mathbf{z}_1 \;:=\; N_E(\mathbf{z}_0) \;:=\; \mathbf{z}_0 - (JE(\mathbf{z}_0))^{-1}\, E(\mathbf{z}_0). \tag{2.9}$$

We explain this: if we approximate the graph of the function E by its tangent plane at $(\mathbf{z}_0, E(\mathbf{z}_0))$, then $\mathbf{z}_1 \in \mathbb{C}^n$ is the unique zero of this linear approximation. There exists a constant $0 < c < 1$ such that when $\mathbf{z}_0$ is sufficiently close to a regular zero $\mathbf{z}$ of E, we have *quadratic convergence* in that

$$\|\mathbf{z}_1 - \mathbf{z}\| \;\le\; c\|\mathbf{z}_0 - \mathbf{z}\|^2.$$

This is because $\mathbf{z}$ is a fixed point of N_E at which the derivative of N_E vanishes. The inequality follows from standard error estimates from Taylor's theorem for N_E in a neighborhood of $\mathbf{z}$. A consequence is that when $\mathbf{z}_0$ is sufficiently close to a regular zero $\mathbf{z}$, each Newton iterate starting from $\mathbf{z}_0$ doubles the number of accurate digits. Such a point $\mathbf{z}_0$ is an *approximate zero* of F. This leads to algorithms to certify numerical output as explained in Section 5.

Predictor-corrector algorithms for solving the initial value problem for homotopy paths take a discretization $1 = t_0 > t_1 > \cdots > t_m = 0$ of the interval $[0, 1]$ and iteratively compute approximations $\mathbf{x}(1) = \mathbf{x}(t_0), \mathbf{x}(t_1), \ldots, \mathbf{x}(t_m) = \mathbf{x}(0)$ to points on the solution path $\mathbf{x}(t)$. This requires an initial approximation $\mathbf{x}_0$ to $\mathbf{x}(t_0) = \mathbf{x}(1)$.

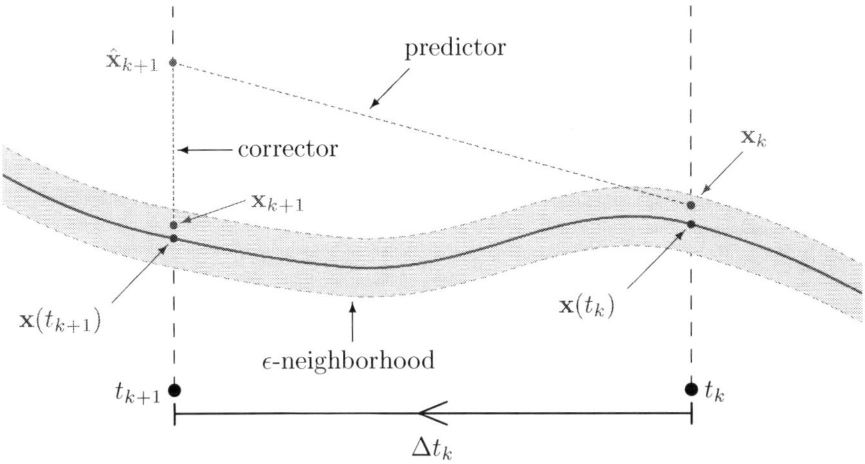

FIGURE 2. Euler prediction followed by Newton corrections. The image is adapted from [**14**] (we thank Sascha Timme for allowing us to use his figure).

Then, for $k = 0, \ldots, m-1$, given an approximation $\mathbf{x}_k$ to $\mathbf{x}(t_k)$, a prediction $\hat{\mathbf{x}}_{k+1}$ for $\mathbf{x}(t_{k+1})$ is computed. This typically uses one step in an iterative method for solving the initial value problem (a local solver). This is the *predictor step*. Next, one or more Newton iterations N_E for $E(\mathbf{x}) = H(\mathbf{x}, t_{k+1})$ are applied to $\hat{\mathbf{x}}_{k+1}$ to obtain a new approximation $\mathbf{x}_{k+1}$ to $\mathbf{x}(t_{k+1})$. This is the *corrector step*. Predictor steps generally cause us to leave the proximity of the path being tracked; corrector steps bring us back. The process repeats until $k = m-1$.

There are a number of efficient local solvers for solving initial value problems. They typically use approximations to the Taylor series for the trajectory $\mathbf{x}(t)$ for t near t_k. For example, the Euler predictor uses the tangent line approximation,

$$\hat{\mathbf{x}}_{k+1} := \mathbf{x}_k + \Delta t_k \Delta \mathbf{x}_k \qquad \text{where} \qquad \frac{\partial H}{\partial \mathbf{x}}(\mathbf{x}_k, t_k) \cdot \Delta \mathbf{x}_k + \frac{\partial H}{\partial t}(\mathbf{x}_k, t_k) = 0.$$

Here, $\Delta t_k = t_{k+1} - t_k$ and $(\Delta x_k, 1)$ spans the kernel of the Jacobian $JH(\mathbf{x}_k, t_k)$.

Figure 2 illustrates an Euler prediction followed by Newton corrections. It suggests a stopping criterion for Newton iterations based on a fixed tolerance ϵ. Another is to apply Newton iterations until quadratic convergence is observed.

REMARK 2.4. Since each solution path defined by $H(\mathbf{x}, t) = 0$ may be tracked independently, path-tracking is (in the words of Jan Verschelde) pleasingly parallelizable [**107**], which is a strength of polynomial homotopy continuation. $\diamond$

We have hardly mentioned the state of the art in path-tracking methods. There is a significant literature on other predictor-corrector schemes (see [**18**, Sections 15–18] for an overview), practical path-tracking heuristics [**5, 103**], and endgames for dealing with issues that arise near $t = 0$, such as divergent [**78**] or singular paths [**80**]. Indeed, the methods we describe would suffice only for the most well-conditioned paths ending at regular solutions at $t = 0$.

2.5. Squaring up. It is common to need to solve *overdetermined systems*, which have more equations than variables. While the version of the parameter continuation theorem we presented holds for overdetermined systems, this presents a challenge as both the total degree homotopy of Section 2.3 and the polyhedral homotopy from the next section enable us to find all isolated solutions to a *square* system $F(\mathbf{x}) = 0$ of polynomial equations. Let us discuss an example, and then explain the method of squaring up, which reduces the problem of solving overdetermined systems to that of solving square systems.

EXAMPLE 2.5. Let A, B, C be the following 2×3 matrices,

$$A := \begin{pmatrix} 1 & 3 & 5 \\ 2 & 4 & 6 \end{pmatrix}, \qquad B := \begin{pmatrix} 2 & 3 & 7 \\ 2 & 5 & -11 \end{pmatrix}, \qquad \text{and} \qquad C := \begin{pmatrix} 1 & -1 & 1 \\ -2 & 3 & -7 \end{pmatrix}.$$

We ask for the matrices of the form $D(x,y) := A + Bx + Cy$ that have rank 1. This is given by the vanishing of the three 2×2 minors of $D(x,y)$. This overdetermined system of equations in x, y has three solutions:

$$(2.10) \qquad (-\tfrac{4}{5}, \tfrac{3}{5}),\ (-0.15019, 0.16729),\ (-0.95120, 2.8373).$$

We may find these using the total degree homotopy as follows. The determinants of the first two columns and the last two columns of $D(x,y)$ give a square subsystem of the system of three minors, and these have $4 = 2 \cdot 2$ solutions (the Bézout number). In addition to the three solutions in (2.10), the fourth is $(-13/14, 3/14)$. Let $f(x,y)$ be the remaining minor. Then $f(-13/14, 3/14) = -963/98$, while the three solutions in (2.10) evaluate (close to) zero. This is a simplification of the general scheme. ◇

Let F be an overdetermined system consisting of m polynomials in n variables, where $m > n$. *Squaring up* F replaces it by a square system $G(\mathbf{x}) := MF(\mathbf{x})$ as follows: Let M be a (randomly chosen) $n \times m$ complex matrix, so that $G(\mathbf{x})$ consists of n polynomials, each of which is a linear combination of polynomials in $F(\mathbf{x})$. Next, find all isolated solutions to $G(\mathbf{x}) = 0$. Since the solutions of $F(\mathbf{x}) = 0$ are among those of $G(\mathbf{x}) = 0$, we need only to determine the zeros of G which are not zeros of F. A simple way is to evaluate F at each of the zeros of G and discard those that do not evaluate to zero (according to some heuristic). It is numerically more stable to apply the Newton operator for the overdetermined system F [**29**] to the zeros of G and retain those which converge quadratically. Example 2.5 is a simplification, using a very specific matrix M rather than a randomly chosen matrix.

REMARK 2.6. Suppose that the overdetermined system $F(\mathbf{x})$ depends on a parameter $\mathbf{p} \in \mathbb{C}^k$, i.e., we have $F(\mathbf{x}) = F(\mathbf{x}; \mathbf{p})$, and that for a general parameter $\mathbf{p} \in \mathbb{C}^k$ the system of equations $F(\mathbf{x}; \mathbf{p})$ has $N > 0$ isolated solutions (as in the parameter continuation theorem 2.1). Suppose further that we have already computed all the solutions of $F(\mathbf{x}; \mathbf{p}_0) = 0$ for a fixed parameter $\mathbf{p}_0 \in \mathbb{C}^k$ (either by squaring up or by using another method). Then, we can use the Newton operator for overdetermined systems from [**29**] for polynomial homotopy continuation along the path $F(\mathbf{x}; t\mathbf{p}_0 + (1-t)\mathbf{p})$ for any other parameter $\mathbf{p}$. ◇

3. Polyhedral homotopy

The total degree homotopy from Section 2.3 may be used to compute all isolated zeros to any system of polynomial equations. Its main flaw is that it is based on the Bézout bound. Many polynomial systems arising in nature have Bézout bound dramatically larger than their number of zeros; for these, the total degree homotopy will track many excess paths.

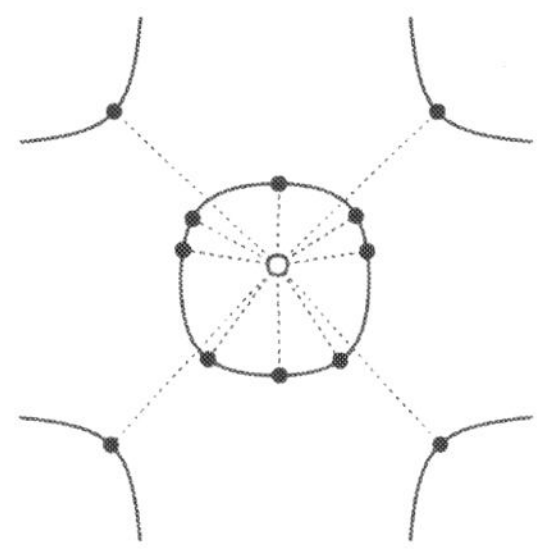

FIGURE 3. Twelve critical points of the distance function to a biquadratic.

EXAMPLE 3.1. To illustrate this potential inefficiency, consider the following problem, posed in [**30**]: Find the distance from a point $\mathbf{x}^* \in \mathbb{R}^d$ to a hypersurface given by the vanishing of a single polynomial f. A first step is to compute all critical points of the distance function $\|\mathbf{x} - \mathbf{x}^*\|$ for $f(\mathbf{x}) = 0$. We formulate this using a Lagrange multiplier λ,

$$f(\mathbf{x}) = 0 \qquad \text{and} \qquad \lambda(\mathbf{x} - \mathbf{x}^*) = \nabla f(\mathbf{x}).$$

When $d = 2$, $f = 5 - 3x_2^2 - 3x_1^2 + x_1^2x_2^2$, and $x^* = (0.025, 0.2)$, these equations become

$$5 - 3x_2^2 - 3x_1^2 + x_1^2x_2^2 = 0 \quad \text{and} \quad \lambda\begin{matrix} x_1 - 0.025 \\ x_2 - 0.2 \end{matrix} = \begin{matrix} -6x_1 + 2x_1x_2^2 \\ -6x_2 + 2x_1^2x_2 \end{matrix}, \tag{3.1}$$

which are polynomials in x_1, x_2, λ of degrees $4, 3, 3$, respectively. The system (3.1) has 12 solutions. We show the corresponding critical points in Figure 3. Note that a total degree homotopy for solving (3.1) follows $36 > 12$ homotopy paths. ◇

The discrepancy between the Bézout bound of 36 and the actual number of 12 solutions in this example is not due to the coefficients in the polynomials, but rather the monomials which appear in them. A polynomial of degree d in which not all terms of degree at most d occur is called *sparse*.

An alternative general-purpose homotopy is the *polyhedral homotopy* of Sturmfels and Huber [**56**] for solving systems of sparse polynomials. It is based on *Bernstein's bound* [**6**, Theorem A], which is at most the Bézout bound and at least the actual number of isolated zeros. Bernstein's bound is often significantly smaller than the Bézout bound, which makes the polyhedral homotopy an efficient tool for polynomial homotopy continuation because it produces fewer excess paths to track. Oftentimes it is optimal in that there are no excess paths to track [**11**, **70**].

The *polyhedral homotopy algorithm* is summarized in Algorithm 3.1. It is implemented in `PHCpack` [**105**], `HomotopyContinuation.jl` [**15**], and the `HOM4PS` family [**23**, **66**]. To understand how it works we first develop some theory. We begin with Bernstein's bound.

3.1. Bernstein's bound. The polyhedral homotopy takes place on the complex torus $(\mathbb{C}^\times)^n$, where $\mathbb{C}^\times := \mathbb{C} \setminus \{0\}$ is the set of invertible complex numbers. Each integer vector $\mathbf{a} \in \mathbb{Z}^n$ gives a *Laurent monomial* $\mathbf{x}^\mathbf{a} := x_1^{a_1} \cdots x_n^{a_n}$, which is a function on the torus $(\mathbb{C}^\times)^n$. A linear combination of Laurent monomials,

$$f := \sum_{\mathbf{a} \in \mathcal{A}} c_\mathbf{a} \mathbf{x}^\mathbf{a} \qquad c_\mathbf{a} \in \mathbb{C}^\times,$$

is a *Laurent polynomial.* The (finite) index set $\mathcal{A} \subset \mathbb{Z}^n$ is the *support* of f. The convex hull of $\mathcal{A}$ is the *Newton polytope* of f. The support of the polynomial f in Example 3.1 is the set containing the columns of the matrix $\left(\begin{smallmatrix} 0 & 0 & 2 & 2 \\ 0 & 2 & 0 & 2 \end{smallmatrix}\right)$, and its Newton polytope is the 2×2 square, $[0, 2] \times [0, 2]$.

Bernstein's bound concerns square systems of Laurent polynomials, and it is in terms of mixed volume [**36**, pp. 116–118]. The *Minkowski sum* of polytopes P and Q in $\mathbb{R}^n$ is

$$P + Q := \{\mathbf{x} + \mathbf{y} \mid \mathbf{x} \in P \text{ and } \mathbf{y} \in Q\}.$$

Given polytopes $P_1, \dots, P_n$ in $\mathbb{R}^n$ and positive scalars $t_1, \dots, t_n$, Minkowski proved that the volume $\text{vol}(t_1 P_1 + \cdots + t_n P_n)$ is a homogeneous polynomial in $t_1, \dots, t_n$ of degree n. He defined the *mixed volume* $\text{MV}(P_1, \dots, P_n)$ to be the coefficient of $t_1 \cdots t_n$ in that polynomial. While mixed volume is in general hard to compute, when $n = 2$, we have the formula

$$\text{MV}(P, Q) = \text{vol}(P + Q) - \text{vol}(P) - \text{vol}(Q). \tag{3.2}$$

This formula, reminiscent of the inclusion-exclusion formula, and its generalizations to $n > 2$ are the *polarization identities.*

Consider (3.2) for the 2×2 square and triangle below.

(3.3)

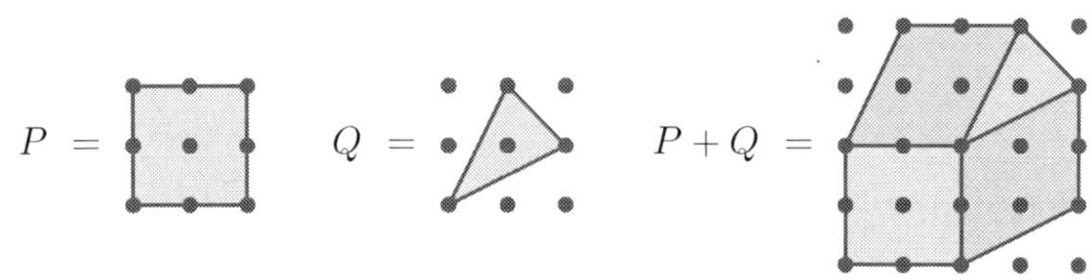

The final shape is the Minkowski sum $P + Q$, and the mixed volume $MV(P, Q)$ is the sum of the areas of the two (non-square) parallelograms, which is 8.

We give a version of Bernstein's theorem [**6**, Theorem A].

THEOREM 3.2 (Bernstein). *Let $f_1, \dots, f_n$ be Laurent polynomials and $P_1, \dots, P_n$ be their respective Newton polytopes. The number of isolated solutions to $F = (f_1, \dots, f_n)$ in $(\mathbb{C}^\times)^n$ is at most* $\text{MV}(P_1, \dots, P_n)$.

EXAMPLE 3.3. We show the supports and Newton polytopes of the three polynomials from Example 3.1, f, $\lambda(x_1 - x_1^*) - \partial f / \partial x_1$, and $\lambda(x_2 - x_2^*) - \partial f / \partial x_2$ (see Equation (3.1)).

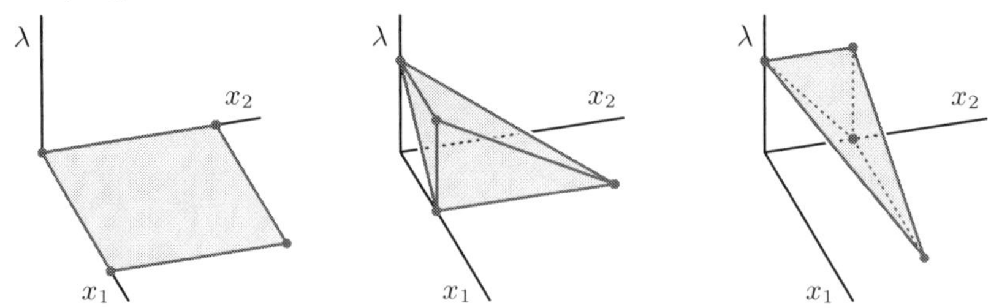

(3.4)

Their mixed volume is twelve, so the system (3.1) achieves the Bernstein bound. ◇

Bernstein proved that this bound is typically achieved in the following sense: For each $i = 1, \dots, n$, let $\mathcal{A}_i$ be the support of polynomial f_i and P_i its Newton polytope. The set of polynomial systems $G = (g_1, \dots, g_n)$ where each g_i has support a subset of $\mathcal{A}_i$ is a vector space V of dimension $|\mathcal{A}_1| + \cdots + |\mathcal{A}_n|$ whose coordinates are given by the coefficients of the polynomials g_i. Bernstein showed that there is a nonempty Zariski open subset $U \subset V$ consisting of systems G with exactly $\mathrm{MV}(P_1, \ldots, P_n)$ regular zeros. Also, if U' is the (larger) set of systems G with exactly $\mathrm{MV}(P_1, \ldots, P_n)$ solutions, counted with multiplicity, then U' is open, and Bernstein [**6**, Theorem B] gave an effective criterion for when $G \in V \setminus U'$.

We remark that Bernstein's bound and Bernstein's theorem are often called the Bernstein–Kushnirenko–Khovanskii (BKK) bound and BKK theorem due to their joint paper [**7**] and the circle of closely related work [**6, 59, 62**].

3.2. Polyhedral homotopy of Huber and Sturmfels. In their seminal work [**56**], Huber and Sturmfels developed a homotopy continuation method for solving systems of Laurent polynomials that takes advantage of the polyhedral structure of the system and Bernstein's bound in that it tracks only $\mathrm{MV}(P_1, \ldots, P_n)$ paths. Their work also provided a new interpretation for mixed volume in terms of mixed cells (the parallelograms in (3.3)) and a new proof of Bernstein's theorem [**6**, Theorem A]. We sketch its main ideas in some detail. This is also described in Sturmfels's award-winning article in *American Mathematical Monthly* [**100**].

REMARK 3.4. It is worth noting that around the same time, Verschelde, Verlinden, and Cools independently developed an alternative homotopy for solving systems of sparse polynomials that takes advantage of their combinatorial structure and Bernstein's bound [**106**]. ◇

Suppose that $F = (f_1, \ldots, f_n)$ is a system of Laurent polynomials and that $\mathcal{A}_i$ is the support of f_i for $i = 1, \ldots, n$, so that

$$f_i \;=\; \sum_{\mathbf{a} \in \mathcal{A}_i} c_{i,\mathbf{a}} \mathbf{x}^{\mathbf{a}}, \quad \text{with } c_{i,\mathbf{a}} \in \mathbb{C}^{\times}\,. \tag{3.5}$$

For now, assume that F is sufficiently generic, in that F has $\mathrm{MV}(P_1, \ldots, P_n)$ regular zeros.

In the polyhedral homotopy, the continuation parameter t appears in a very different way than in the total degree homotopy (2.7). First, the start system is at $t = 0$ and the target system F is at $t = 1$, but this is not a substantive difference. The polyhedral homotopy depends upon a choice of *lifting functions*, $\omega_i \colon \mathcal{A}_i \to \mathbb{Z}$, for $i = 1, \ldots, n$. That is, a choice of an integer $\omega_i(\mathbf{a})$ for each monomial $\mathbf{a}$ in $\mathcal{A}_i$. We address this choice in Section 3.3.

Given lifting functions, define the homotopy $H(\mathbf{x}, t) := (h_1, \ldots, h_n)$ by

$$h_i(\mathbf{x}, t) \;:=\; \sum_{\mathbf{a} \in \mathcal{A}_i} c_{i,\mathbf{a}}\, \mathbf{x}^{\mathbf{a}}\, t^{\omega_i(\mathbf{a})}.$$

By the parameter continuation theorem (Theorem 2.1) and our genericity assumption on F, over $t \in (0, 1]$ the system of equations $H(\mathbf{x}, t) = 0$ defines $\mathrm{MV}(P_1, \ldots, P_n)$ smooth paths. It is however not at all clear what happens as $t \to 0$. For example, $H(\mathbf{x}, 0)$ is undefined if some $\omega_i(\mathbf{a}) < 0$, and if $\omega_i(\mathbf{a}) > 0$ for all i and $\mathbf{a}$, then $H(\mathbf{x}, 0)$ is identically zero.

The key idea is to use an invertible change of coordinates to study the homotopy paths as $t \to 0$. This coordinate change depends upon a *weight* $\boldsymbol{\alpha} \in \mathbb{Z}^n$ and a positive integer r. Together, they describe the order of a series expansion of a subset of homotopy paths near $t = 0$. The weight gives a *cocharacter* of the torus, for $s \in \mathbb{C}^{\times}$, $s^{\boldsymbol{\alpha}} := (s^{\alpha_1}, \ldots, s^{\alpha_n})$. Set

$$\mathbf{y} \;=\; \mathbf{x} \circ s^{-\boldsymbol{\alpha}} \;:=\; (x_1 s^{-\alpha_1}, \ldots, x_n s^{-\alpha_n}).$$

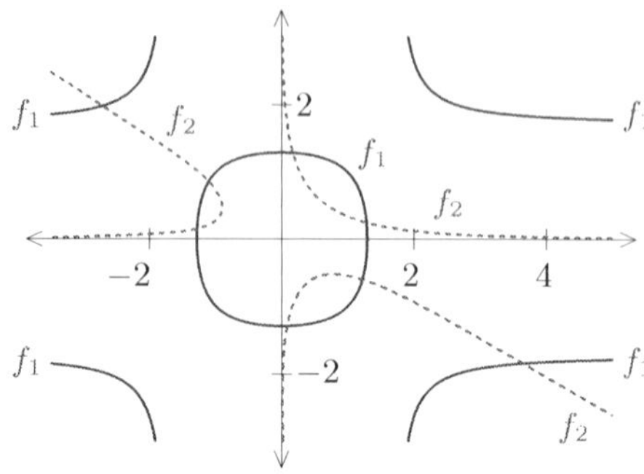

FIGURE 4. The zero sets of f_1 (solid lines) and f_2 (dotted lines) together with their intersections.

Then $\mathbf{x}=\mathbf{y}\circ s^{\boldsymbol{\alpha}}$, and we define $H^{(\boldsymbol{\alpha})}(\mathbf{y},s):=(h_1^{(\boldsymbol{\alpha})},\dots,h_n^{(\boldsymbol{\alpha})})$, where for $i=1,\dots,n$,

$$h_i^{(\boldsymbol{\alpha})(\mathbf{y},s)} \;:=\; s^{-\beta_i}h_i(\mathbf{y}\circ s^{\boldsymbol{\alpha}},s^r) \;=\; \sum_{\mathbf{a}\in\mathcal{A}_i} c_{i,\mathbf{a}}\,\mathbf{y}^{\mathbf{a}}\,s^{\langle\boldsymbol{\alpha},\mathbf{a}\rangle+r\omega_i(\mathbf{a})-\beta_i}, \tag{3.6}$$

where $\beta_i := \min\{\langle\boldsymbol{\alpha},\mathbf{a}\rangle + r\omega_i(\mathbf{a}) \mid \mathbf{a}\in\mathcal{A}_i\}$. The purpose of β_i is to ensure that s appears in $h_i^{(\boldsymbol{\alpha})}(\mathbf{y},s)$ with only non-negative exponents, and that $h_i^{(\boldsymbol{\alpha})}(\mathbf{y},0)$ is defined and not identically zero. Specifically, if $\mathcal{A}_i^{(\boldsymbol{\alpha})} := \{\mathbf{a}\in\mathcal{A}_i \mid \langle\boldsymbol{\alpha},\mathbf{a}\rangle + r\omega_i(\mathbf{a}) = \beta_i\}$, then

$$h_i^{(\boldsymbol{\alpha})}(\mathbf{y},0) \;=\; \sum_{\mathbf{a}\in\mathcal{A}_i^{(\boldsymbol{\alpha})}} c_{i,\mathbf{a}}\mathbf{y}^{\mathbf{a}} \qquad \text{for } i=1,\dots,n. \tag{3.7}$$

The purpose of the positive integer r is to keep the exponents integral. As $r>0$, we have that for $s\in[0,1]$, $t=s^r\to 0$ if and only if $s\to 0$. Thus the role of r and s is to parameterize the homotopy path through a series expansion. We remark on this later in the sketch of the proof for Algorithm 3.1.

We will see that for almost all $(\boldsymbol{\alpha},r)$, the system $H^{(\boldsymbol{\alpha})}(\mathbf{y},0)$ has no zeros in $(\mathbb{C}^*)^n$, but for appropriately chosen $(\boldsymbol{\alpha},r)$, the system $H^{(\boldsymbol{\alpha})}(\mathbf{y},0)$ has easily computed zeros, each defining a homotopy path to a solution of $H^{(\boldsymbol{\alpha})}(\mathbf{y},1)=H(\mathbf{x},1)$. For such an $(\boldsymbol{\alpha},r)$, $H^{(\boldsymbol{\alpha})}(\mathbf{y},0)$ is a *start subsystem*. The polyhedral homotopy algorithm consists of determining those $(\boldsymbol{\alpha},r)$, solving the start subsystems $H^{(\boldsymbol{\alpha})}(\mathbf{y},0)$, and then tracking the homotopy paths from $t=0$ to $t=1$.

Before discussing this in more detail, including the role of the choices of lifting functions ω_i, cocharacter $\boldsymbol{\alpha}$, and positive integer r, let us consider an example.

EXAMPLE 3.5. Let f_1 be the biquadratic from Example 3.1 and let $f_2 = 1 + 2x_1x_2 - 5x_1x_2^2 - 3x_1^2x_2$. Here, $\mathcal{A}_1 = \left(\begin{smallmatrix}0&0&2&2\\0&2&0&2\end{smallmatrix}\right)$ and $\mathcal{A}_2 = \left(\begin{smallmatrix}0&1&1&2\\0&1&2&1\end{smallmatrix}\right)$. Their Newton polygons are the square P and triangle Q in (3.3). Figure 4 shows their $8 = \mathrm{MV}(P,Q)$ common zeros.

Define ω_1 to be identically zero and set $\omega_2\left(\begin{smallmatrix}a\\b\end{smallmatrix}\right) := a+b$. Then $h_1(\mathbf{x},t)=f_1(\mathbf{x})$ and

$$h_2(\mathbf{x},t) \;=\; 1 + 2x_1x_2t^2 - 5x_1x_2^2t^3 - 3x_1^2x_2t^3.$$

Let $\boldsymbol{\alpha}=(0,-3)$ and $r=2$, so that $x_1=y_1$ and $x_2=y_2t^{-3}$. Then we may check that $\beta_1=-6$ and $\beta_2=\min\{0-0,4-3,6-6,6-3\}=0$, so that

$$h_1^{(\boldsymbol{\alpha})}(\mathbf{y},s) = \underline{-3y_2^2 + y_1^2y_2^2} \;+\; s^6(5-3y_1^2),$$

$$h_2^{(\boldsymbol{\alpha})}(\mathbf{y},s) = \underline{1 - 5y_1y_2^2} \ + \ s(2y_1y_2 - 3y_1^2y_2s^2).$$

The underlined terms are the binomials $h_1^{(\boldsymbol{\alpha})}(\mathbf{y},0)$ and $h_2^{(\boldsymbol{\alpha})}(\mathbf{y},0)$. They have four solutions,

$$(\sqrt{3}, 75^{-\frac{1}{4}}),\ (\sqrt{3}, -75^{-\frac{1}{4}}),\ (-\sqrt{3}, 75^{-\frac{1}{4}}),\ (-\sqrt{3}, -75^{-\frac{1}{4}}).$$

These four solutions lead to four homotopy paths, which is fewer than the $8 = \mathrm{MV}(P,Q)$ paths defined by $H(\mathbf{x},t) = 0$ over $t \in (0,1]$.

If we let $\boldsymbol{\gamma} = (-3,0)$ and $r = 2$, then

$$h_1^{(\boldsymbol{\gamma})}(\mathbf{y},s) = \underline{-3y_1^2 + y_1^2y_2^2} \ + \ s^6(5 - 3y_2^2),$$
$$h_2^{(\boldsymbol{\gamma})}(\mathbf{y},s) = \underline{1 - 3y_1^2y_2} \ + \ s(2y_1y_2 - 5y_1y_2^2s^2),$$

so that $h_1^{(\boldsymbol{\gamma})}(\mathbf{y},0)$ and $h_2^{(\boldsymbol{\gamma})}(\mathbf{y},0)$ are again binomials and they have four solutions

$$(27^{-\frac{1}{4}}, \sqrt{3}),\ (-27^{-\frac{1}{4}}, \sqrt{3}),\ (27^{-\frac{1}{4}}, -\sqrt{3}),\ (-27^{-\frac{1}{4}}, -\sqrt{3}).$$

These lead to the other four homotopy paths. The partition $4+4=8$ is seen in the decomposition of $P+Q$ in (3.3). The weight $\boldsymbol{\alpha}$ corresponds to the parallelogram on the upper left, which is the Minkowski sum of the supports of the components $h_1^{(\boldsymbol{\alpha})}(\mathbf{y},0)$ and $h_2^{(\boldsymbol{\alpha})}(\mathbf{y},0)$ of the start subsystem, and the weight $\boldsymbol{\gamma}$ corresponds to the parallelogram on the lower right. The only weights and positive integers for which the start subsystem has solutions are positive multiples of $(\boldsymbol{\alpha},2)$ and $(\boldsymbol{\gamma},2)$. $\diamond$

3.3. Mixed cells. In Example 3.5 only two choices of $(\boldsymbol{\alpha}, r)$ gave start subsystems $H^{(\boldsymbol{\alpha})}(\mathbf{y},0)$ with solutions. We now address the problem of computing the pairs $(\boldsymbol{\alpha},r)$, such that $H^{(\boldsymbol{\alpha})}(\mathbf{y},0)$ has solutions. This leads to an algorithm that computes these pairs given the start system $F = (f_1,\ldots,f_n)$. We will show that the pairs $(\boldsymbol{\alpha},r)$ which give zeros at $t=0$ correspond to certain *mixed cells* in a decomposition of the Minkowski sum $P_1+\cdots+P_n$, where P_i is the Newton polytope of f_i.

We examine the geometric combinatorics of the lifting functions ω_i, weights $\boldsymbol{\alpha}$, and positive integer r. Let $P \subset \mathbb{R}^{n+1}$ be a polytope. If P has the same dimension as its projection to $\mathbb{R}^n$, then it and all of its faces are *lower faces*. Otherwise, replace $\mathbb{R}^{n+1}$ by the affine span of P and assume that P has dimension $n+1$. A *lower facet* of P is a facet Q of P ($\dim Q = n$) whose inward-pointing normal vector has positive last coordinate. A *lower face* of P is any face lying in a lower facet. The union of lower faces forms the *lower hull* of P.

Let $\mathcal{A} \subset \mathbb{Z}^n$ be a finite set and $\omega\colon \mathcal{A} \to \mathbb{Z}$ be a lifting function. The *lift* of $\mathcal{A}$ is the set

$$\widehat{\mathcal{A}} \ := \ \{(\mathbf{a}, \omega(\mathbf{a})) \mid \mathbf{a} \in \mathcal{A}\} \subset \mathbb{Z}^{n+1}.$$

Let $\widehat{P} := \mathrm{conv}(\widehat{\mathcal{A}})$ be its convex hull. Given a lower face Q of $\widehat{P}$, the projection to $\mathbb{Z}^n$ of $Q \cap \widehat{\mathcal{A}}$ is a subset $\mathcal{C}(Q)$ of $\mathcal{A}$ whose convex hull is the projection to $\mathbb{R}^n$ of Q. If $(\boldsymbol{\alpha},r)$ is upward-pointing ($r>0$) and $\mathbf{a} \mapsto \langle \boldsymbol{\alpha}, \mathbf{a}\rangle + r\omega(\mathbf{a})$ achieves its minimum on Q, then we may observe that $\mathcal{C}(Q) = \mathcal{A}^{(\boldsymbol{\alpha})}$.

For each $i = 1,\ldots,n$, let $\mathcal{A}_i \subset \mathbb{Z}^n$ be a finite set, $\omega_i\colon \mathcal{A}_i \to \mathbb{Z}$ be a lifting function, and set $\widehat{P}_i := \mathrm{conv}(\widehat{\mathcal{A}}_i)$. Let $\widehat{P} := \widehat{P}_1 + \cdots + \widehat{P}_n$ be their Minkowski sum. As $\widehat{P}$ is a Minkowski sum, if Q is a lower face of $\widehat{P}$, for each $i=1,\ldots,n$ there is a lower face Q_i of $\widehat{P}_i$ with

$$Q \ = \ Q_1 + \cdots + Q_n. \tag{3.8}$$

DEFINITION 3.6. Lifting functions $\omega_i\colon \mathcal{A}_i \to \mathbb{Z}$ for $i = 1, \ldots, n$ are *generic* if for each lower facet $Q = Q_1 + \cdots + Q_n$ of $\widehat{P}$, with Q_i being a lower face of $\widehat{P}_i$ for $i = 1, \ldots, n$, we have that

$$\dim Q \quad = \quad n \quad = \quad \dim Q_1 + \cdots + \dim Q_n, \tag{3.9}$$

and $\dim Q_i = \#(Q_i \cap \widehat{\mathcal{A}}_i) - 1$.

A lower facet Q for which every Q_i in (3.9) has dimension 1 (and thus $\#Q_i \cap \widehat{\mathcal{A}}_i = 2$) is a *mixed facet* and its projection to $\mathbb{R}^n$ is a *mixed cell*. Mixed facets and mixed cells are parallelepipeds (Minkowski sums of independent line segments). ◇

Huber and Sturmfels show that almost all real lifting functions are generic and the density of rational numbers implies that there exist generic integral lifting functions. Setting $P_i := \operatorname{conv}(\mathcal{A}_i)$ for $i = 1, \ldots, n$, then the projection to $\mathbb{R}^n$ of the lower faces of $\widehat{P}$ forms a polyhedral subdivision of the Minkowski sum $P_1 + \cdots + P_n$, called a *mixed decomposition*.

This leads to a new interpretation for mixed volume.

THEOREM 3.7 (Huber–Sturmfels). *Suppose that $\omega_i\colon \mathcal{A}_i \to \mathbb{Z}$ for $i = 1, \ldots, n$ are generic lifting functions. Then the mixed volume $\mathrm{MV}(P_1, \ldots, P_n)$ is the sum of the volumes of the mixed cells in the induced polyhedral decomposition of the Minkowski sum $P_1 + \cdots + P_n$.*

PROOF. These constructions—the lifts $\widehat{P}_i$, lower faces, and the mixed subdivision—scale multilinearly with positive $t_1, \ldots, t_n \in \mathbb{R}$. For example, a lower face $Q = Q_1 + \cdots + Q_n$ (3.8) of $\widehat{P}_1 + \cdots + \widehat{P}_n$ corresponds to a lower face $t_1 Q_1 + \cdots + t_n Q_n$ of $t_1 \widehat{P}_1 + \cdots + t_n \widehat{P}_n$. Let $\pi\colon \mathbb{R}^{n+1} \to \mathbb{R}^n$ be the projection. This shows

$$\operatorname{vol}(t_1 P_1 + \cdots + t_n P_n) \quad = \quad \sum_Q \operatorname{vol}(\pi(t_1 Q_1 + \cdots + t_n Q_n)),$$

the sum over all lower facets Q. By Condition (3.9), $n = \dim(Q_1) + \cdots + \dim(Q_n)$, and thus

$$\operatorname{vol}(\pi(t_1 Q_1 + \cdots + t_n Q_n)) \quad = \quad t_1^{\dim(Q_1)} \cdots t_n^{\dim(Q_n)} \operatorname{vol}(\pi(Q)).$$

Hence the coefficient of $t_1 \cdots t_n$ in $\operatorname{vol}(t_1 P_1 + \cdots + t_n P_n)$ is the sum of the volumes of the mixed cells. □

EXAMPLE 3.8. Let us consider this construction on our running example, using the lifts from Example 3.5. Figure 5 shows two views of the lower hull of the Minkowski sum $\widehat{P} + \widehat{Q}$, along with the mixed decomposition.

Note that $\widehat{P} = P$ as the lifting function is 0 and $\widehat{Q}$ is affinely equivalent to Q. There are two mixed lower facets, whose corresponding mixed cells are the parallelograms of (3.3), showing them to be mixed cells of the mixed subdivision induced by these lifts. The dot product with $(\boldsymbol{\alpha}, 2) = (0, -3, 2)$ is minimized along the mixed lower facet

$$\operatorname{conv}\{(0,2,0), (2,2,0), (3,4,3), (1,4,3)\}$$

with minimal value -6 and the dot product with $(\boldsymbol{\gamma}, 2) = (-3, 0, 2)$ is minimized along the mixed lower facet

$$\operatorname{conv}\{(2,0,0), (2,2,0), (4,3,3), (4,1,3)\}$$

with minimal value -6. ◇

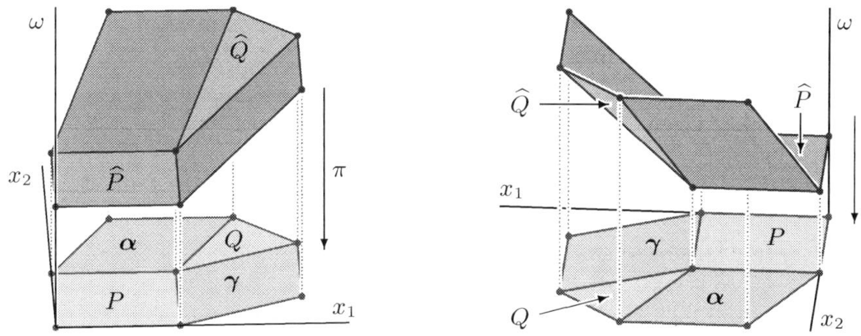

FIGURE 5. Two views of the lower hull of the lift and the induced mixed subdivision with mixed cells labeled by corresponding cocharacters.

Keeping Example 3.8 in mind, we return to our problem of studying the homotopy given by generic lifting functions $\omega_i\colon \mathcal{A}_i \to \mathbb{Z}$ for $i = 1, \ldots, n$, for the supports of our target system (3.5). Let $\widehat{P} := \widehat{P}_1 + \cdots + \widehat{P}_n$ be the Minkowski sum of the convex hulls $\widehat{P}_i$ of the lifted supports $\widehat{\mathcal{A}}_i$. A vector $(\boldsymbol{\alpha}, r) \in \mathbb{Z}^{n+1}$ with $r > 0$ is *upward-pointing*, and the linear function $\langle(\boldsymbol{\alpha}, r), -\rangle$ it defines achieves its minimum on $\widehat{P}$ along a lower face Q—the lower face of $\widehat{P}$ *exposed* by $(\boldsymbol{\alpha}, r)$. When Q has the form (3.8), then for each $i = 1, \ldots, n$, Q_i is the lower face of $\widehat{P}_i$ exposed by $(\boldsymbol{\alpha}, r)$, and the minimum value of $\langle(\boldsymbol{\alpha}, r), -\rangle$ along Q_i is

$$\min\{\langle(\boldsymbol{\alpha}, r), (\mathbf{a}, \omega(\mathbf{a}))\rangle = \langle\boldsymbol{\alpha}, \mathbf{a}\rangle + r\omega(\mathbf{a}) \mid \mathbf{a} \in \mathcal{A}_i\} \;=\; \beta_i, \tag{3.10}$$

which explains the geometric significance of $(\boldsymbol{\alpha}, r)$ and of β_i. When Q is a facet, there is a unique primitive (components have no common factor) upward-pointing integer vector $(\boldsymbol{\alpha}, r)$ that exposes Q. In this case, $\mathcal{A}_i^{(\boldsymbol{\alpha})} = \pi(Q_i \cap \widehat{\mathcal{A}}_i) = \mathcal{C}(Q_i)$ is the support of $h^{(\boldsymbol{\alpha})}(\mathbf{y}, 0)$.

We explain the algebraic consequences of this geometric combinatorics. Suppose that $H^{(\boldsymbol{\alpha})}(\mathbf{y}, s)$ is the system of polynomials $h_i^{(\boldsymbol{\alpha})}(\mathbf{y}, s)$ defined by (3.6). Then $H^{(\boldsymbol{\alpha})}(\mathbf{y}, 0)$ is given by the polynomials $h_i^{(\boldsymbol{\alpha})}(\mathbf{y}, 0)$ of (3.7). If, for some i, $\#\mathcal{C}(Q_i) = 1$, so that $\dim Q_i = 0$, then $h_i^{(\boldsymbol{\alpha})}(\mathbf{y}, 0)$ is a monomial and therefore $H^{(\boldsymbol{\alpha})}(\mathbf{y}, 0)$ has no solutions in $(\mathbb{C}^\times)^n$.

Suppose that $H^{(\boldsymbol{\alpha})}(\mathbf{y}, 0)$ has solutions in $(\mathbb{C}^\times)^n$. Necessarily, $\dim Q_i \geq 1$ for all i. By (3.9), $\dim Q_i = 1$ and $\#\mathcal{C}(Q_i) = 2$ for all i, and thus Q is a mixed facet. Consequently, each $h_i^{(\boldsymbol{\alpha})}(\mathbf{y}, 0)$ is a binomial and $H^{(\boldsymbol{\alpha})}(\mathbf{y}, 0)$ is a system of independent binomials, which may be solved by inspection. Thus the only start subsystems $H^{(\boldsymbol{\alpha})}(\mathbf{y}, 0)$ with solutions are those for which $(\boldsymbol{\alpha}, r)$ exposes a mixed facet Q of $\widehat{P}$. The following proposition, whose proof is sketched in Section 3.5, records the number of solutions to such a mixed system.

PROPOSITION 3.9. *The number of solutions to the system of binomials* $H^{(\boldsymbol{\alpha})}(\mathbf{y}, 0)$ *is the volume of the mixed cell* $\pi(Q) = \mathrm{conv}(\mathcal{C}(Q_1) + \cdots + \mathcal{C}(Q_n))$.

3.4. The polyhedral homotopy algorithm. We sketch this algorithm and provide a brief argument about its correctness.

Algorithm 3.1: The polyhedral homotopy algorithm

1 **Input:** A system $F = (f_1, \dots, f_n)$ of n polynomials in n variables, where f_i has support $\mathcal{A}_i$ and Newton polytope P_i. The system F is assumed general and has $\mathrm{MV}(P_1, \dots, P_n)$ regular solutions in $(\mathbb{C}^*)^n$.
2 **Output:** All complex zeros of F.
3 Compute generic lifting functions $\omega_i \colon \mathcal{A}_i \to \mathbb{Z}$ (see Definition 3.6). They define mixed cells in the Minkowski sum $P = P_1 + \dots + P_n$;
4 **for** *each mixed cell* Q *of* P **do**
5 Compute the pair $(\boldsymbol{\alpha}, r)$ given as the primitive upward pointing normal of the mixed facet of $\widehat{P}$ that corresponds to Q.
6 Solve the start subsystem $H^{(\boldsymbol{\alpha})}(\mathbf{y}, 0)$ and then use homotopy continuation to track those solutions along the homotopy $H^{(\boldsymbol{\alpha})}(\mathbf{y}, s)$ from $s = 0$ to $s = 1$, giving solutions to $H^{(\boldsymbol{\alpha})}(\mathbf{y}, 1)$.
7 **end**
8 The solutions computed in Step 6 to $H^{(\boldsymbol{\alpha})}(\mathbf{y}, 1) = H(\mathbf{x}, 1) = F(\mathbf{x})$ for all mixed cells are all the solutions to $F(\mathbf{x})$.

Sketch of Proof of Correctness. The system of equations $H(\mathbf{x}, t) = 0$ defines an algebraic curve C in $(\mathbb{C}^\times)^n \times \mathbb{C}_t^\times$ whose projection onto $\mathbb{C}_t^\times$ has degree equal to $\mathrm{MV} := \mathrm{MV}(P_1, \dots, P_n)$ with the fiber over $t = 1$ having MV points. This curve has MV branches near $t = 0$, each of which is a point $\mathbf{z}(t)$ in $\mathbb{C}\{t\}^n$. Here, $\mathbb{C}\{t\}$ is the field of Puiseux series, which contains the algebraic closure of the field $\mathbb{C}(t)$ of rational functions in t [**88**, Section 2.5.3]. Elements of $\mathbb{C}\{t\}$ may be represented by fractional power series of the form

$$\sum_{m \geq N} b_m t^{m/r},$$

where $m, N, r \in \mathbb{Z}$ with $r > 0$, and $b_m \in \mathbb{C}$. Observe that both the exponents of t and the denominators in those exponents are bounded below.

Fix a branch $\mathbf{z}(t)$ of C and let r be the least common denominator of all exponents of coordinates of $\mathbf{z}(t)$. Consider the lowest order terms of the coordinates in $\mathbf{z}(t)$,

$$(c_1 t^{\alpha_1/r}, \dots, c_n t^{\alpha_n/r}),$$

where $\alpha_i \in \mathbb{Z}$ and $r \in \mathbb{N}$. Set $\boldsymbol{\alpha} := (\alpha_1, \dots, \alpha_n)$. The substitution $t = s^r$ clears the denominators, converting $\mathbf{z}(t)$ to a vector $\mathbf{z}(s^r)$ of Laurent series in s. The coordinate change $\mathbf{z}(s^r) \circ s^{-\boldsymbol{\alpha}}$ converts these Laurent series to ordinary power series with constant coefficients $\mathbf{c} := (c_1, \dots, c_n)$. Finally, $\mathbf{c}$ is a solution to the start subsystem $H^{(\boldsymbol{\alpha})}(\mathbf{y}, 0)$.

The point is that for each branch $\mathbf{z}(t)$ of C near $t = 0$, there is a weight $\boldsymbol{\alpha}$ and positive integer r such that the vector $\mathbf{c}$ of lowest order coefficients of $\mathbf{z}(t)$ is a solution to $H^{(\boldsymbol{\alpha})}(\mathbf{y}, 0)$. The discussion preceding the statement of the polyhedral homotopy algorithm shows that $(\boldsymbol{\alpha}, r)$ exposes a mixed lower facet Q of $\widehat{P}$, and that $H^{(\boldsymbol{\alpha})}(\mathbf{y}, 0)$ has $\mathrm{vol}(\pi(Q))$ solutions. Furthermore, each solution $\mathbf{c}$ to $H^{(\boldsymbol{\alpha})}(\mathbf{y}, 0)$ may be developed into a power series solution $\mathbf{y}(s)$ to $H^{(\boldsymbol{\alpha})}(\mathbf{y}, s)$. Reversing the coordinate changes and reparameterization, this gives a solution $\mathbf{z}(t)$ to $H(\mathbf{x}, t) = 0$ in $(\mathbb{C}\{t\})^n$ and thus a branch of the curve C near $t = 0$.

Thus the homotopy paths for $H(\mathbf{x},t)$ correspond to the MV distinct branches $\mathbf{z}(t)$ of C near $t=0$ and the solutions computed in (3) give all MV solutions solutions to $F(\mathbf{x})$. □

REMARK 3.10. The assumption that $F(\mathbf{x})$ is general in the polyhedral homotopy algorithm ensures that $F(\mathbf{x})$ has MV regular solutions *and* that $H(\mathbf{x},t)|_{t\in(0,1]}$ consists of MV smooth paths. Thus, to solve a given system $F(\mathbf{x})=(f_1,\dots,f_n)$ where f_i has support $\mathcal{A}_i$, one first generates a general system $G=(g_1,\dots,g_n)$ where g_i has support $\mathcal{A}_i$. In practice, this is done by choosing random complex numbers as coefficients, and then with probability one, $G(\mathbf{x})$ is general and satisfies the hypotheses of the polyhedral homotopy algorithm. The polyhedral homotopy algorithm is used to solve $G(\mathbf{x})=0$, and then a parameter homotopy with start system G and target system F is used to compute the solutions to $F(\mathbf{x})=0$. ◇

3.5. Solving binomial systems. To complete the discussion, we take a brief look at Step 6 in Algorithm 3.1. By construction, the subsystems $H^{(\boldsymbol{\alpha})}(\mathbf{y},0)$ in Step 6 are binomial systems. We explain how to solve such a system.

Suppose that $H(\mathbf{y})$ is a system of binomials

$$H(\mathbf{y})=\begin{bmatrix}p_1\mathbf{y}^{\mathbf{u}^{(1)}}-q_1\mathbf{y}^{\mathbf{v}^{(1)}} & \cdots & p_n\mathbf{y}^{\mathbf{u}^{(n)}}-q_n\mathbf{y}^{\mathbf{v}^{(n)}}\end{bmatrix}$$

where each $p_i,q_i\neq 0$ and $\mathbf{u}^{(1)}-\mathbf{v}^{(1)},\dots,\mathbf{u}^{(n)}-\mathbf{v}^{(n)}$ are linearly independent. This is equivalent to the assertion that the Minkowski sum of the supports of the binomials is a parallelepiped $\pi(Q)$ of dimension n. Then for $\mathbf{y}\in(\mathbb{C}^\times)^n$, $H(\mathbf{y})=0$ becomes

$$\mathbf{y}^{\mathbf{u}^{(i)}-\mathbf{v}^{(i)}} = q_i/p_i \qquad \text{for } i=1,\dots,n. \tag{3.11}$$

Let A be the $n\times n$ matrix with rows $\mathbf{u}^{(1)}-\mathbf{v}^{(1)},\dots,\mathbf{u}^{(n)}-\mathbf{v}^{(n)}$. Then $\det A=\operatorname{vol}(\pi(Q))$. The *Smith normal form* of A consists of unimodular integer matrices X,Y (integer matrices with determinant ±1) and a diagonal matrix $D=\operatorname{diag}(d_1,\dots,d_n)$ such that $XAY=D$ and thus $|\det A|=|\det D|=d_1\cdot d_2\cdots d_n$. The unimodular matrices X and Y give coordinate changes on $(\mathbb{C}^\times)^n$ which convert the system (3.11) into a diagonal system of the form

$$x_i^{d_i} = b_i \qquad \text{for } i=1,\dots,n.$$

All $d_1\cdots d_n$ solutions may be found by inspection, and then the coordinate changes may be reversed to obtain all solutions to the original system $H(\mathbf{y})$.

4. Numerical algebraic geometry

We have described methods to compute all isolated solutions to a system of polynomial equations. *Numerical algebraic geometry* uses this ability to compute zero-dimensional algebraic varieties to represent and manipulate higher-dimensional algebraic varieties on a computer. This is an essential component of numerical nonlinear algebra. Besides expanding the reach of numerical methods, the geometric ideas behind numerical algebraic geometry have led to new methods for solving systems of polynomial equations, including regeneration and monodromy. While the term was coined in [**91**], the fundamental ideas were developed in a series of papers including [**90, 92**], and a more thorough treatment is in [**94**, Part III].

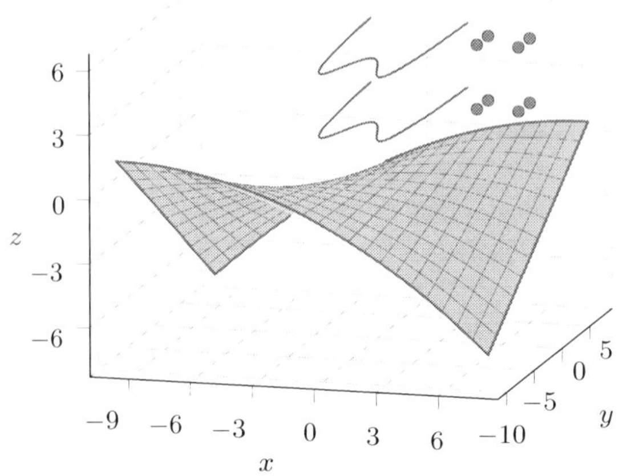

FIGURE 6. A reducible variety, defined implicitly by (4.1).

EXAMPLE 4.1. Consider the following square system of polynomials in the variables x, y, z:

$$F(x,y,z) = \begin{bmatrix} f(x,y,z)g(x,y,z)(x-4)(x-6) \\ f(x,y,z)g(x,y,z)(y-3)(y-5) \\ f(x,y,z)(z-2)(z-5) \end{bmatrix}, \tag{4.1}$$

where

$$f(x,y,z) = \tfrac{1}{40}(2xy - x^2) - z - 1 \qquad \text{and} \qquad g(x,y,z) = x^4 - 4x^2 - y - 1.$$

Figure 6 shows the real part of the variety V of $F(x,y,z) = 0$, consisting of a quadric (degree 2) surface, two quartic (degree 4) curves (at $z = 2$ and $z = 5$, respectively), and eight points. ◇

Given any system $F(\mathbf{x})$ defining a reducible variety V, implemented symbolic algorithms (primary decomposition and computing radicals) will decompose the variety V as follows. These methods will compute a list $I_1, \ldots, I_r$, where each I_i is the ideal of an irreducible component V_i of V. Each ideal I_i is represented by a Gröbner basis, which is a finite set of generators, and thus serves as a data structure encoding information about V_i. For example, the dimension and degree of a component V_i may be computed from the data I_i.

In numerical algebraic geometry, the data structure to represent a positive-dimensional component of a variety is a witness set, which we now describe. Suppose that $F(\mathbf{x})$ is a system of polynomials, and let V be an irreducible component of the variety defined by $F(\mathbf{x}) = 0$. A *witness set* W for the component V is a triple $W = (F, L, L \cap V)$, where L is a general (affine) linear subspace complimentary to V in that $\operatorname{codim}(L) = \dim(V)$ and $L \cap V$ consists of numerical approximations of the points in the intersection of L and V. Generality (see Section 4.1) ensures that the *linear slice* $L \cap V$ is transverse and consists of $\deg(V)$ points. In practice, L is represented by $\dim(V)$ randomly chosen polynomials of degree one.

The simple algorithm of *membership testing* illustrates the utility of this data structure. Given a witness set $W = (F, L, L \cap V)$ for an irreducible variety $V \subset \mathbb{C}^n$ and a point $\mathbf{x}_0 \in \mathbb{C}^n$, we would like to determine if $\mathbf{x}_0 \in V$. Evaluating F at $\mathbf{x}_0$

only implies that $\mathbf{x}_0$ lies near the variety defined by F, not that it lies near the irreducible component V. We instead choose a general linear subspace L' with the same codimension as L, but for which $\mathbf{x}_0 \in L'$ (that is, L' is otherwise general, given that $\mathbf{x}_0 \in L'$). Next, form the *linear slice homotopy*,

$$H(\mathbf{x}, t) := (F(\mathbf{x}), tL(\mathbf{x}) + (1-t)L'(\mathbf{x})), \tag{4.2}$$

and use it to track the points of $L \cap V$ from $t = 1$ to $t = 0$, obtaining the points of $L' \cap V$. As the intersection of V with the other components of the variety of F has lower dimension than V, the complement in V of the other components is path-connected, and thus $\mathbf{x}_0$ lies in $L' \cap V$ if and only if $\mathbf{x}_0 \in V$.

The core of this membership test reveals another algorithm involving witness sets. Given a witness set $W = (F, L, L \cap V)$ and a general linear subspace L' with the same codimension as L, the step of following the points of $L \cap V$ along the homotopy (4.2) to obtain the points $L' \cap V$ is called *moving a witness set*. This is because $W' = (F, L', L' \cap V)$ is a new witness set for V. This may also be considered to be an algorithm for sampling points of V.

The rest of this section discusses algorithms for computing a witness set and the corresponding numerical irreducible decomposition of a variety V. It concludes with a summary of regeneration and monodromy, two new methods for solving systems of polynomials.

REMARK 4.2. The set of points in the linear slice $L \cap V$ is considered a concrete version of André Weil's generic points of a variety [**108**]. We call it *witness point set*.

A witness point set is related to Chow groups from intersection theory [**42**]. Indeed, a witness set for an irreducible variety V may be interpreted as a specific way to represent the class of V in the Chow ring of $\mathbb{P}^n$. In [**96**] this point of view was used to extend witness sets to represent subvarieties of varieties other than $\mathbb{P}^n$.

4.1. More on linear slices. An irreducible algebraic subvariety V of affine or projective space has two fundamental invariants—its *dimension*, $\dim(V)$, and its *degree*, $\deg(V)$. The dimension of V is the dimension of its (dense subset of) smooth points, as a complex manifold. Equivalently, this is the dimension of its tangent space at any smooth point.

By Bertini's theorem [**88**, Theorem 2, §6.2], there is a dense Zariski-open subset of (affine) linear spaces L of codimension $\dim(V)$ such that the linear slice $L \cap V$ is transverse. Here, a codimension d linear subspace is defined by d independent degree one polynomials. The degree of V is the maximal number of points in such an intersection. By Bertini's theorem again, this maximum is achieved by linear spaces chosen from a Zariski open subset of the Grassmannian of linear subspaces of codimension $\dim(V)$.

In practice, L is represented by $\dim(V)$ random degree one polynomials (their coefficients are chosen randomly). By the nature of Zariski open sets, for most reasonable probability distributions on these coefficients, a suitably general L will be found with probability one.

When the variety V defined by the vanishing of $F(\mathbf{x})$ is reducible and the maximum dimension of an irreducible component is d, then a randomly chosen linear subspace L of codimension d will meet each irreducible component V' of V of dimension d in $\deg(V')$ points $L \cap V'$, and it will not intersect any components of V of dimension less than d. If V' is the unique component of dimension d, then $(F, L, L \cap V')$ is a witness set for V'.

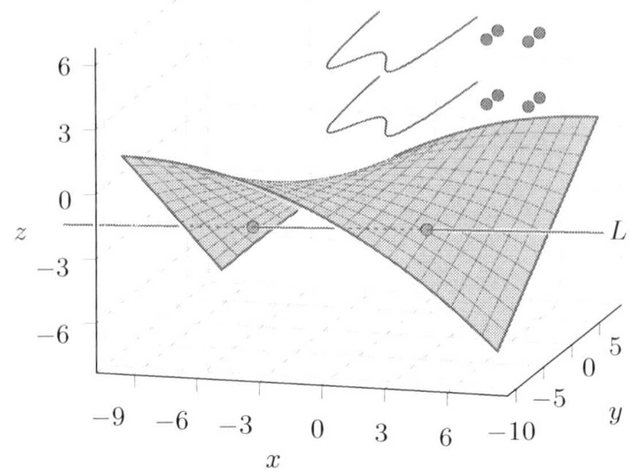

FIGURE 7. Slice of V by a line L.

EXAMPLE 4.3. We continue Example 4.1. To compute the linear slice $L \cap V$ with the line L parameterized by $(t, -t-2, -3+t/4)$, we add to F two degree one polynomials $x+y+2$ and $z+3-x/4$. The augmented system defines the intersection $L \cap V$. It has two solutions $(10/3, -16/3, -13/6)$ and $(-8, 6, -5)$. The line L is sufficiently general so that it only meets the two-dimensional surface defined by $f(x, y, z)$, and neither of the curves nor any isolated points. Figure 7 shows this configuration. ◇

While finding a witness point set for the top-dimensional component of V in Example 4.3 was straightforward, finding witness point sets for the other components is not as simple. To find points on curves, we intersect V with the vertical plane P defined by $x + y - 2 = 0$, finding eight isolated solutions. These come from the two curves of degree four, each contributing four points. This number eight does not yet tell us that there are two curves, there may be a single curve of degree eight or some other configuration. Furthermore, the plane intersects the surface in a curve C, and we may have found additional non-isolated points on C. This is displayed in Figure 8.

Methods to remove points on higher-dimensional components and to determine which points lie on which components of the same dimension are described in the next section.

4.2. Numerical irreducible decomposition. A system $F(\mathbf{x})$ of polynomials in n variables defines the algebraic variety $V := \{\mathbf{x} \in \mathbb{C}^n \mid F(\mathbf{x}) = 0\}$. Were V irreducible, a witness set would be an acceptable representation for V. An analog when V is reducible is a *numerical irreducible decomposition* of V. This data structure for representing V consists of a collection of witness sets $(F, L', L' \cap V')$, one for each irreducible component V' of V. We present a numerical irreducible

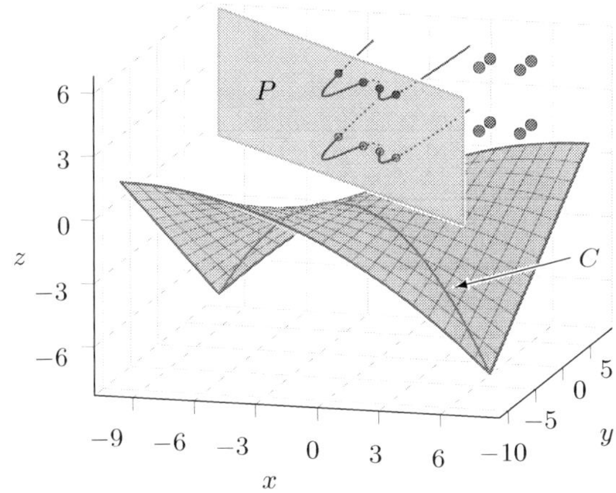

FIGURE 8. The plane P intersects V in eight isolated points and a curve C.

decomposition for our running example:

$$\big(F, [x+y+2, z+3-\tfrac{x}{4}], \{(\tfrac{10}{3}, -\tfrac{16}{3}, -\tfrac{13}{6}), (-8, 6, -5)\}\big)$$

$$(F, [x+y+2], \{(-2.06, 0.06, 2), (-0.40, -1.60, 2), (0.69, -2.69, 2), (1.76, -3.76, 2)\}),$$

$$(F, [x+y+2], \{(-2.06, 0.06, 5), (-0.40, -1.60, 5), (0.69, -2.69, 5), (1.76, -3.76, 5)\}),$$

$$(F, [], \{(4,3,2)\}), \ (F, [], \{(4,3,5)\}), \ (F, [], \{(4,5,2)\}), \ (F, [], \{(4,5,5)\}),$$

$$(F, [], \{(6,3,2)\}), \ (F, [], \{(6,3,5)\}), \ (F, [], \{(6,5,2)\}), \ (F, [], \{(6,5,5)\}).$$

We later present the *cascade algorithm* (Algorithm 4.2) to compute a numerical irreducible decomposition. We first explain its constituents.

4.2.1. *Witness point supersets.* A starting point is to compute, for each i, a set of points U_i in a linear slice $L \cap V$ of V with a codimension i linear space L, where U_i contains all witness point sets $L \cap V'$ for V' an irreducible component of dimension i. For this, let $\ell_1, \dots, \ell_{n-1}$ be randomly chosen (and hence independent) degree one polynomials. For each i, let L^i be defined by $\ell_1, \dots, \ell_i$, and let F_i be a subsystem of F consisting of $n-i$ randomly chosen linear combinations of elements of F. Then $(F_i, \ell_1, \dots, \ell_i)$ is a square subsystem of (F, L^i), and we may use it to compute points U_i that lie in $L^i \cap V$, as explained in Section 2.5. In the literature, these sets U_i are called witness point supersets.

By the generality of $\ell_1, \dots, \ell_{n-1}$, there will be no solutions to $(F_i, \ell_1, \dots, \ell_i)$ when i exceeds the dimension d of V—yet another application of Bertini's theorem. By the same generality, the set U_i contains witness point sets for each irreducible component of V of dimension i, and perhaps some points on irreducible components of V of larger dimension. The next two sections describe how to remove points of U_i that lie on components of V of dimension exceeding i, and then how to decompose such an equidimensional slice into witness point sets for the irreducible components of V of dimension i.

4.2.2. *Removing points on higher-dimensional components.* Suppose that we have computed witness point supersets $U_0, U_1, \dots, U_d$, where d is the dimension of V as in Section 4.2.1. By the generality of $\ell_1, \dots, \ell_d$, U_d is equal to the linear slice $L^d \cap V$, and thus is the union of witness point sets for the irreducible components of V of dimension d. For each $i = 0, \ldots, d$, let V_i be the union of all irreducible components of V of dimension i. Then $W_i := L^i \cap V_i \subset U_i$ consists of points in U_i lying on some component of V of dimension i. This union of the witness point sets of i-dimensional components of V is an *equidimensional slice*. Also, note that $W_d = U_d$.

Since points of $U_i \smallsetminus W_i$ lie on $V_{i+1}, \ldots, V_d$, downward induction on i and the membership test computes $U_i \smallsetminus W_i$ and thus W_i. This uses the observation that the membership test, starting with $W_j = L^j \cap V_j$, may be used to determine if a point $\mathbf{x}_0 \in U_i$ lies on V_j, for any $j > i$. This is invoked in Step 7 in Algorithm 4.1 for computing such equidimensional slices.

Algorithm 4.1: Computing equidimensional slices

1 **Input:** $F(\mathbf{x}), \ell_1, \ldots, \ell_d, U_1, \ldots, U_d$ as above.
2 **Output:** Equidimensional slices $W_0, \ldots, W_d$ of V.
3 Set $W_d := U_d$.
4 **for** *i from* $d - 1$ *down to* 0 **do**
5   Set $W_i := \{\}$.
6   **for** *each point* $\mathbf{x}_0 \in U_i$ **do**
7   If $\mathbf{x}_0 \notin V_{i+1} \cup \cdots \cup V_d$, then $W_i := W_i \cup \{\mathbf{x}_0\}$.
8   **end**
9 **end**
10 Return $W_d, \ldots, W_1, W_0$.

REMARK 4.4. An alternative to Algorithm 4.1 is a *local dimension test* [**3**], which can determine if a point $\mathbf{x}_0 \in U_i$ lies on a component of dimension exceeding i. ◇

4.2.3. *Decomposing equidimensional slices.* Suppose that we have the equidimensional slices $W_0, \ldots, W_d$ of V, where $W_i = L^i \cap V_i$ for each i, as computed in Algorithm 4.1. Fix i and suppose that the irreducible decomposition of V_i is

$$V_i \;=\; X_1 \cup X_2 \cup \cdots \cup X_r ,$$

so that $X_1, \ldots, X_r$ are all of the irreducible components of V of dimension i. Then

$$W_i \;=\; L^i \cap V_i \;=\; \left(L^i \cap X_1\right) \sqcup \left(L^i \cap X_2\right) \sqcup \cdots \sqcup \left(L^i \cap X_r\right). \tag{4.3}$$

This union is disjoint by the generality of $\ell_1, \ldots, \ell_d$, as the intersection of $X_j \cap X_k$ with $j \neq k$ has dimension less than i. Call (4.3) the *witness set partition* of the equidimensional slice W_i. Each part $L^i \cap X_j$ is a witness point set for X_j. Computing a witness set partition of W_i is tantamount to computing a numerical irreducible decomposition of V_i.

Suppose that $H(\mathbf{x}, t) := (F(\mathbf{x}), tL^i(\mathbf{x}) + (1-t)L'(\mathbf{x}))$ is a linear slice homotopy (4.2). As with moving a witness set, if we track a point $\mathbf{x} \in L^i \cap X_j$ to a point $\mathbf{x}' \in L' \cap V$, then all points of the homotopy path, including its endpoint $\mathbf{x}'$, lie on X_j.

Suppose that we combine linear slice homotopies together, moving points of $W_i = L^i \cap V_i$ to $L' \cap V_i$ on to $L'' \cap V_i$, and then back to $L^i \cap V_i$. The three convex combinations,

$$tL^i(\mathbf{x}) + (1-t)L'(\mathbf{x}), \quad tL'(\mathbf{x}) + (1-t)L''(\mathbf{x}), \quad \text{and} \quad tL''(\mathbf{x}) + (1-t)L^i(\mathbf{x}),$$

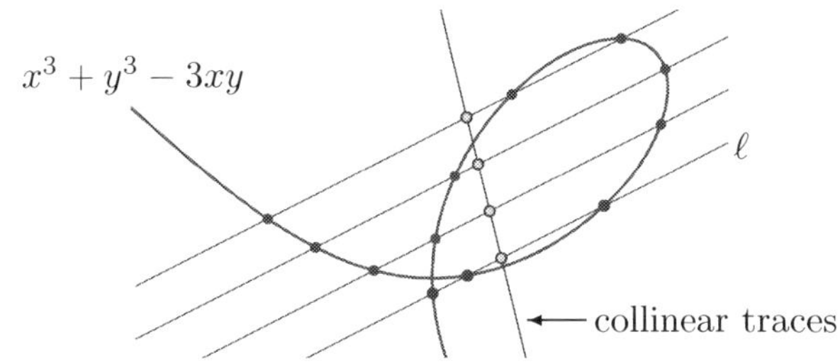

FIGURE 9. The trace test for the folium of Descartes.

for $t \in [0, 1]$ together form a based loop in the space of codimension i affine linear subspaces. Tracking each $\mathbf{x}_0 \in W_i$ along the three homotopies gives another point $\sigma(\mathbf{x}_0) \in W_i$. This computes a *monodromy permutation* σ of W_i. This has the property that the partition of W_i into the cycles of σ refines the witness set partition (4.3).

Following additional based loops may lead to other partitions of W_i into cycles of monodromy permutations. The common coarsening of these monodromy cycle partitions is an *empirical partition* of W_i. Every empirical partition is a refinement of the witness set partition. Since the smooth locus of $X_j \smallsetminus (\bigcup_{k \neq j} X_k)$ is path-connected, the common coarsening of all empirical partitions is the witness set partition. Thus computing monodromy permutations will eventually give the witness set partition.

The problem with this approach to numerical irreducible decomposition is that only when V_i is irreducible is there a stopping criterion. Namely, if we discover an empirical partition consisting of a single part, then we conclude that V_i is irreducible, and (F, L^i, W_i) is a numerical irreducible decomposition of V_i. All other cases lack a stopping criterion. That is, there is no proof that a witness set partition has been found.

A common heuristic for stopping is the trace test [**92**]. To begin, form a linear slice homotopy (4.2) using a linear subspace L' such that $L^i \cap L'$ has codimension $i+1$ in projective space. Then the convex combination $tL^i(\mathbf{x}) + (1-t)L'(\mathbf{x})$ forms a *pencil* of linear spaces. The *trace test* follows from the observation that while each homotopy path $\mathbf{x}(t)$ for $t \in [0, 1]$ tracked from a point $\mathbf{x} \in W_i$ is nonlinear, if we sum over all points $\mathbf{x} \in L^i \cap X_j$ in a single part of the witness set partition, then that sum or its average is an affine-linear function of the homotopy parameter t. Given a subset $S \subset W_i$, the average of the points tracked from S is the *trace* of S. The trace is an affine-linear function if and only if S is the full witness point set [**94**, Theorem 15.5.1].

EXAMPLE 4.5. Consider the folium of Descartes which is defined by $f = x^3 + y^3 - 3xy$. A general line ℓ meets the folium in three points W with the triple (f, ℓ, W) forming a witness set for the folium. Figure 9 shows these witness sets on four parallel lines, which lie in a pencil. Note that the four traces are collinear. $\diamond$

The collinearity of traces may be seen as a consequence of Vièta's formula that the sum of the roots of a monic polynomial of degree δ in y is the coefficient of $-y^{\delta-1}$ [**68**].

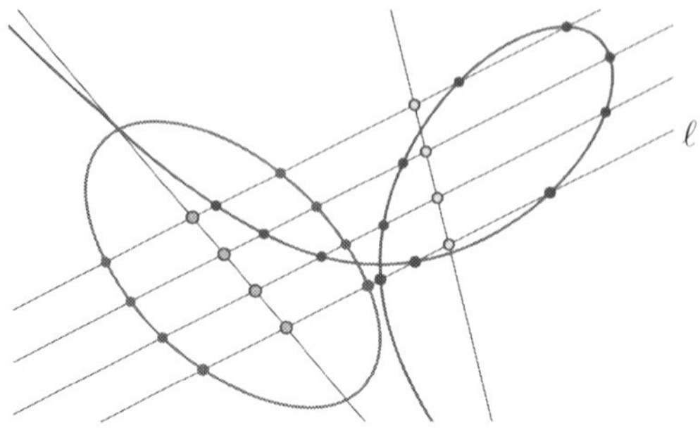

FIGURE 10. Numerical irreducible decomposition for the ellipse and folium.

This gives the following heuristic for stopping for computing a numerical irreducible decomposition. Given a part S of an empirical partition of W_i, track all points of S along a linear slice homotopy given by a pencil containing L^i. If the traces are collinear, then S is a witness point set for some component of V_i. Otherwise, either compute more monodromy permutations to coarsen the empirical partition or check the collinearity of the trace for the union of S with other parts of the empirical partition. This is called the *trace test*. The reason this is only a heuristic is that there is no way to certify collinearity of traces of approximate solutions. In [**16**] the authors develop a trace test specifically for systems of polynomials that are sparse in the sense of Section 3.

EXAMPLE 4.6. Suppose that V is the union of the ellipse $8(x+1)^2+3(2y+x+1)^2=8$ and the folium, as in Figure 10. A witness set for V consists of the five points $W=V\cap\ell$. Tracking points of W as ℓ varies over several loops in the space of lines gives an empirical partition of W into two sets, of cardinalities two and three, respectively. Applying the trace test to each subset verifies that each is a witness set of a component of V. ◇

The methods described in this and the previous sections combine to give the *cascade algorithm* for computing a numerical irreducible decomposition. This was introduced in [**90**] and is implemented in `PHCpack` [**105**], `Bertini` [**4**] and `HomotopyContinuation.jl` [**15**], and in `NAG4M2` [**67**] through its interfaces. We present a simplified form in Algorithm 4.2.

4.3. Advanced methods for solving. The perspective afforded by numerical algebraic geometry and its tools—witness sets and monodromy—lead to new algorithms for solving systems of equations. We describe two such algorithms. Regeneration [**51**] is a bootstrap method that constructs a numerical irreducible decomposition one equation at a time. Monodromy solving [**32**] exploits that it is often easier to find a system of equations for which a given point is a solution than to find a solution to a given system.

4.3.1. *Regeneration.* Let $F := (f_1, \ldots, f_m)$ be a system of polynomials in n variables. Rather than solve all equations at once as in (2.3), we instead consider the sequence of varieties

$$\mathbb{C}^n \;=\; V_0 \supset V_1 \supset V_2 \supset \cdots \supset V_m = V,$$

Algorithm 4.2: The cascade algorithm

Input: A system $F(\mathbf{x})$ of polynomials in n variables defining a variety V of dimension d.
Output: A numerical irreducible decomposition of V.
for *each dimension* i *from* d *down to* 0 **do**
  Choose a codimension i linear space L^i and compute $L^i \cap V$, yielding points U_i.
  Remove from U_i any points lying on higher-dimensional components as in Algorithm 4.1. Call the remaining points W_i.
  Compute the witness set partition $W_i = S_1 \sqcup \cdots \sqcup S_r$ using monodromy and the trace test as explained in Section 4.2.3.
  Return (F, L^i, S_j) for $j = 1, \ldots, r$. These are witness sets for the irreducible components of V of dimension i.
end

where V_i is defined by the first i polynomials in F. The approach of *equation-by-equation solvers* [**34, 51, 52, 93**] is to iteratively compute V_i given V_{i-1} for each $i = 1, \ldots, m$.

Let $X \subset V_{i-1}$ be an irreducible component of dimension d. Then

$$X \cap V_i \;=\; \{\mathbf{x} \in X \mid f_i(\mathbf{x}) = 0\}.$$

If f_i vanishes identically on X, then X is an irreducible component of V_i. Otherwise $X \cap V_i$, if nonempty, is a union of irreducible components of V_i of dimension $d-1$. We explain how to obtain a numerical irreducible decomposition of $X \cap V_i$ given a witness set for X.

Let $((f_1, \ldots, f_{i-1}), L^d, L^d \cap X)$ be a witness set for X. By the generality of L^d, with probability one we may conclude that a general polynomial f vanishes on X only if f vanishes at each point of $W := L^d \cap X$. *Regeneration* is a method to use W to compute a witness point superset for $X \cap V_i$. Let $\ell_1, \ldots, \ell_d$ be d linear polynomials defining L^d, and suppose that δ is the degree of f_i. Form the convex combination $\ell(t) := t\ell_d + (1-t)\ell'$ of ℓ_d with a new general degree one polynomial, ℓ'. Use the straight-line homotopy

$$(f_1, \ldots, f_{i-1},\ \ell(t),\ \ell_1, \ldots, \ell_{d-1})$$

to move the witness point set $W_1 = W = L^d \cap X$ at $t = 1$ to witness point sets $W_2, \ldots, W_\delta$ at distinct points $t = t_2, \ldots, t_\delta$, respectively.

Then the product $f := \ell_d \cdot \ell(t_2) \cdots \ell(t_\delta)$ has degree δ in $\mathbf{x}$ and we have

$$U' \;=\; W_1 \cup W_2 \cup \cdots \cup W_\delta \;=\; L^{d-1} \cap (X \cap \{\mathbf{x} \mid f(\mathbf{x}) = 0\}),$$

where L^{d-1} is defined by $\ell_1, \ldots, \ell_{d-1}$. Use the straight-line homotopy

$$(f_1, \ldots, f_{i-1},\ tf + (1-t)f_i,\ \ell_1, \ldots, \ell_{d-1})$$

to track the points of U' at $t = 1$ to the set U at $t = 0$. Then

$$U \;=\; L^{d-1} \,\cap\, \big(X \,\cap\, V_i\big)$$

is a witness point superset for $X \cap V_i$. Finally, use monodromy and the trace test of Section 4.2.3 to decompose U into witness point sets for the irreducible components of $X \cap V_i$.

As regeneration computes a numerical irreducible decomposition of the variety V of F, it will also compute all isolated solutions to F.

4.3.2. *Solving by monodromy.* Suppose that we wish to solve a system $F = F(\mathbf{x};\mathbf{p})$ of polynomials that lies in a parameter space of polynomial systems as in Section 2.2, and that evaluation at a general point $\mathbf{x}_0 \in \mathbb{C}^n$ gives n independent linear equations in the parameters $\mathbf{p} \in \mathbb{C}^k$. For example, $F(\mathbf{x};\mathbf{p})$ could be the family of all polynomial systems $f_1(\mathbf{x}),\dots,f_n(\mathbf{x})$ where the degree of f_i is d_i and $\mathbf{p} \in \mathbb{C}^k$ is the vector of coefficients. More generally, each $f_i(\mathbf{x})$ could be a sparse polynomial with support $\mathcal{A}_i$.

Consider the incidence variety (2.5) with projections π_1 to $\mathbb{C}^n$ and π_2 to $\mathbb{C}^k$.

$$
\begin{array}{ccl}
 & Z := & \{(\mathbf{x},\mathbf{p}) \in \mathbb{C}^n \times \mathbb{C}^k \mid F(\mathbf{x};\mathbf{p}) = 0\} \subseteq \mathbb{C}^n \times \mathbb{C}^k \\
 & {}^{\pi_1}\swarrow \quad \searrow^{\pi_2} & \\
 & \mathbb{C}^n \qquad \mathbb{C}^k &
\end{array}
\tag{4.4}
$$

For any parameters $\mathbf{p} \in \mathbb{C}^k$, $\pi_2^{-1}(\mathbf{p})$ is the set of solutions $\mathbf{x} \in \mathbb{C}^n$ to $F(\mathbf{x};\mathbf{p}) = 0$. On the other hand, if we fix a general $\mathbf{x}_0 \in \mathbb{C}^n$, then $\pi_1^{-1}(\mathbf{x}_0) \subset \mathbb{C}^k$ is defined by n independent linear equations on $\mathbb{C}^k$, and is thus a linear subspace of dimension $k-n$. (This implies that Z is irreducible and has dimension k, which explains why the general fiber $\pi_2^{-1}(\mathbf{p})$ is finite and $\pi_2 \colon Z \to \mathbb{C}^k$ is a branched cover.) Imposing $k-n$ additional general degree one equations on $\pi_1^{-1}(\mathbf{x}_0)$ gives a single parameter value $\mathbf{p}_0 \in \mathbb{C}^k$ such that $F(\mathbf{x}_0;\mathbf{p}_0) = 0$, that is, a system of polynomials $F(\mathbf{x};\mathbf{p}_0)$ for which $\mathbf{x}_0$ is a solution.

The underlying idea of monodromy solving [**32**] is to use monodromy to discover all solutions to $F(\mathbf{x};\mathbf{p}_0) = 0$ and then use a parameter homotopy to find solutions to any desired system of polynomials in the family. Similar to the description of monodromy in Section 4.2.3, if we choose general points $\mathbf{p}_1, \mathbf{p}_2 \in \mathbb{C}^k$, we may form the trio of parameter homotopies $F(\mathbf{x}; t\mathbf{p}_0 + (1-t)\mathbf{p}_1)$, $F(\mathbf{x}; t\mathbf{p}_1 + (1-t)\mathbf{p}_2)$, and $F(\mathbf{x}; t\mathbf{p}_2 + (1-t)\mathbf{p}_0)$. For $t \in [0,1]$, these form a loop in the parameter space based at $\mathbf{p}_0$, and we may track the point $\mathbf{x}_0$ along this loop to obtain a possibly new point $\mathbf{x}' \in \pi_2^{-1}(\mathbf{p}_0)$ so that $F(\mathbf{x}';\mathbf{p}_0) = 0$.

More generally, given a subset $S \subset \pi_2^{-1}(\mathbf{p}_0)$ of computed points, we may track it along a possibly new loop in $\mathbb{C}^k$ based at $\mathbf{p}_0$ to obtain a subset $S' \subset \pi_2^{-1}(\mathbf{p}_0)$. Thus we may discover additional solutions to $F(\mathbf{x};\mathbf{p}_0) = 0$.

When the number N of solutions to a general system in the family is known (e.g., via Bernstein's bound for the sparse systems of Section 3.1), this method has a stopping criterion. Otherwise, some heuristic may be used once sufficiently many solutions are known. The technique of solving using monodromy was introduced in [**32**], where a more complete description may be found. It is implemented in the package `MonodromySolver` of `Macaulay2` [**44**] and `HomotopyContinuation.jl` [**15**] and widely used, in particular when it is not necessary to compute all solutions to a given system.

Suppose that we have a branched cover $\pi \colon Z \to Y$ with Y rational (e.g., as in (4.4) where $Y = \mathbb{C}^k$), and we know all solutions $\pi^{-1}(\mathbf{y}_0)$ for a parameter value $\mathbf{y}_0$ not in the branch locus, B. As in Section 4.2.3, tracking all points in $\pi^{-1}(\mathbf{y})$ as $\mathbf{y}$ varies along a loop in $Y \smallsetminus B$ based at $\mathbf{y}_0$ gives a monodromy permutation σ of $\pi^{-1}(\mathbf{y}_0)$, which we regard as an element of the symmetric group S_N, where

$N = |\pi^{-1}(\mathbf{y}_0)|$. The set of all monodromy permutations forms the *monodromy group* of the branched cover Z.

This is in fact a Galois group [**47**, **58**, **97**, **104**]: Let $\mathbb{K} = \mathbb{C}(Y)$ be the field of rational functions on the parameter space Y and let $\mathbb{L} = \mathbb{C}(Z)$ be the function field of the incidence variety Z. As π is dominant, we may regard $\mathbb{K}$ as a subfield of $\mathbb{L}$ via π^{-1}, and $\mathbb{L}/\mathbb{K}$ is a field extension of degree N. The Galois group of the normal closure of $\mathbb{L}/\mathbb{K}$ is equal to the monodromy group of Z, and we call it the Galois group of the branched cover Z.

There are several approaches to computing Galois groups using methods from numerical nonlinear algebra. In [**69**], monodromy permutations were computed and used to show some Galois groups were the full symmetric group (see Section 6.2). Other approaches, including methods guaranteed to compute generators of Galois groups, were developed in [**50**]. Yahl [**110**] introduced a method to certify that a Galois group contains a simple transposition, using ideas from this section and from Section 5.

A Galois group that is imprimitive (preserves a partition of the solutions) is equivalent to the branched cover decomposing as a composition of branched covers, and this may be exploited for solving (computing points in a fiber). This is explained in [**1**, **17**].

5. Certification

Let F be a square system of polynomials and let $\mathbf{z}_0$ be a point presumed to be an approximation of a solution to F. We discuss methods that can give a computational proof that Newton iterates starting from $\mathbf{z}_0$ converge to a nearby regular zero $\mathbf{z}$ of F. Such methods *certify* the numerical solution $\mathbf{z}_0$ to F. Certification methods can also prove that two numerical solutions correspond to two distinct zeros, and are thus a useful tool in both theoretical and applied problems in numerical nonlinear algebra.

There are two main strategies to certify solutions to square polynomial systems, *Smale's α-theory* and *Krawczyk's method*. A difference is that Smale's α-theory uses exact arithmetic, while Krawczyk's method uses floating-point arithmetic, and is typically more efficient.

REMARK 5.1. There are other approaches. In [**31**] the authors develop methods to certify overdetermined systems which require global information. In [**49**], overdetermined systems are reformulated as square systems to enable certification. ◇

5.1. Smale's α-theory.

Smale's α-theory certifies approximate zeros of a square system F. An approximate zero of F is a data structure representing a solution to F. We recall its definition from Section 2.4.

DEFINITION 5.2. Let $F(\mathbf{x})$ be a square system of n polynomials in n variables. Writing $JF := \partial F/\partial \mathbf{x}$ for its Jacobian matrix, its Newton operator N_F (2.9) is $N_F(\mathbf{x}) := \mathbf{x} - (JF(\mathbf{x}))^{-1} F(\mathbf{x})$. A point $\mathbf{z}_0 \in \mathbb{C}^n$ is an *approximate zero* of F if there exists a regular zero $\mathbf{z} \in \mathbb{C}^n$ of F such that the sequence $\{\mathbf{z}_k \mid k \geq 0\}$ of Newton iterates defined by $\mathbf{z}_{k+1} = N_F(\mathbf{z}_k)$ for $k \geq 0$ converges quadratically to $\mathbf{z}$ in that

$$\|\mathbf{z}_{k+1} - \mathbf{z}\| \;\leq\; \frac{1}{2}\|\mathbf{z}_k - \mathbf{z}\|^2 \qquad \forall k \geq 0.$$

We call $\mathbf{z}$ the *associated zero* of $\mathbf{z}_0$. ◇

Smale's α-theory certifies that a point $\mathbf{x}$ is an approximate zero using only local information encoded in two functions of F and $\mathbf{x}$,

$$\beta(F,\mathbf{x}) \;:=\; \|JF(\mathbf{x})^{-1}F(\mathbf{x})\| \quad \text{and} \quad \gamma(F,\mathbf{x}) \;:=\; \max_{k\geq 2}\left\|\frac{1}{k!}\,JF(\mathbf{x})^{-1}\mathrm{D}^kF(\mathbf{x})\right\|^{\frac{1}{k-1}}.$$

Here, β is the size of a Newton step, $\mathrm{D}^kF(\mathbf{x})$ is the symmetric tensor of derivatives of order k at $\mathbf{x}$, and $JF(\mathbf{x})^{-1}\mathrm{D}^kF(\mathbf{x})$ is the corresponding multilinear map $(\mathbb{C}^n)^k \to \mathbb{C}^n$. The norm is the operator norm $\|A\| := \max_{\|v\|=1}\|A(v,\ldots,v)\|$.

Let $\alpha(F,\mathbf{x}) := \beta(F,\mathbf{x})\cdot\gamma(F,\mathbf{x})$ be the product of these two functions. We state two results of Smale [**9**, Theorem 4 and Remark 6 in Chapter 8].

THEOREM 5.3. *Let $\mathbf{x}\in\mathbb{C}^n$ and let F be a system of n polynomials in n variables.*

(a) *If $\alpha(F,\mathbf{x}) < \frac{13-3\sqrt{17}}{4} \approx 0.15767$, then $\mathbf{x}$ is an approximate zero of F whose associated zero $\mathbf{z}$ satisfies $\|\mathbf{x}-\mathbf{z}\| \leq 2\beta(F,\mathbf{x})$.*

(b) *If $\mathbf{x}$ is an approximate zero of F and $\mathbf{y}\in\mathbb{C}^n$ satisfies $\|\mathbf{x}-\mathbf{y}\| < \frac{1}{20\,\gamma(F,\mathbf{x})}$, then $\mathbf{y}$ is also an approximate zero of F with the same associated zero as $\mathbf{x}$.*

Shub and Smale [**89**] derived an upper bound for $\gamma(F,\mathbf{x})$ which can be computed using exact arithmetic, and thus one may decide algorithmically if $\mathbf{x}$ is an approximate zero of F, using only data of F and the point $\mathbf{x}$ itself.

The software `alphaCertified` [**53**] uses this theorem in an algorithm. An implementation is publicly available for download.[1] If the polynomial system F has only real coefficients, then `alphaCertified` can decide if an associated zero is real. The idea is as follows. We write $\overline{\mathbf{x}}$ for the complex conjugate of a point $\mathbf{x}\in\mathbb{C}^n$.

Suppose that $\mathbf{x}\in\mathbb{C}^n$ be an approximate zero of F with associated zero $\mathbf{z}$. Since the Newton operator has real coefficients, $N_F(\overline{\mathbf{x}}) = \overline{N_F(\mathbf{x})}$, we see that $\overline{\mathbf{x}}$ is an approximate zero of F with associated zero $\overline{\mathbf{z}}$. Consequently, if $\|\mathbf{x}-\overline{\mathbf{x}}\| < \frac{1}{20\,\gamma(F,\mathbf{x})}$, then $\mathbf{z}=\overline{\mathbf{z}}$ by Theorem 5.3(b).

5.2. Krawczyk's method. Interval arithmetic certifies computations using floating-point arithmetic. *Krawczyk's method* [**63**] adapts Newton's method to interval arithmetic and can certify zeros of analytic function $\mathbb{C}^n\to\mathbb{C}^n$. This is explained in [**84**].

Real interval arithmetic involves the set of compact real intervals,

$$\mathbb{IR} \;:=\; \{[x,y] \mid x,y\in\mathbb{R}, x\leq y\}.$$

For $X,Y\in\mathbb{IR}$ and $\circ\in\{+,-,\cdot,\div\}$, we define $X\circ Y := \{x\circ y \,|\, x\in X, y\in Y\}$. (For $\div$ we require that $0\notin Y$.) For intervals $I,J,K\in\mathbb{IR}$ we have $I\cdot(J+K)\subseteq I\cdot J+I\cdot K$, but the inclusion may be strict. Indeed,

$$[0,1]\cdot([-1,0]+[1,1]) \;=\; [0,1]\cdot[0,1] \;=\; [0,1],$$
$$\text{but}\quad [0,1]\cdot[-1,0]+[0,1]\cdot[1,1] \;=\; [-1,0]+[0,1] \;=\; [-1,1].$$

Thus there is no distributive law in interval arithmetic.

Complex intervals are rectangles in the complex plane of the form

$$X+\sqrt{-1}Y \;=\; \{x+\sqrt{-1}y \mid x\in X, y\in Y\}, \quad \text{where } X,Y\in\mathbb{IR}.$$

[1]`https://franksottile.github.io/research/stories/alphaCertified/index.html`

Let $\mathbb{IC}$ be the set of all complex intervals. Writing $\frac{X}{Y}$ for $X \div Y$, we define arithmetic for complex intervals $I = X + \sqrt{-1}\, Y$ and $J = W + \sqrt{-1} Z$ as follows:

$$\begin{aligned} I + J &:= (X + W) + \sqrt{-1}\,(Y + Z), \\ I - J &:= (X - W) + \sqrt{-1}\,(Y - Z), \\ I \cdot J &:= (X \cdot W - Y \cdot Z) + \sqrt{-1}\,(X \cdot Z + Y \cdot W), \\ \frac{I}{J} &:= \frac{X \cdot W + Y \cdot Z}{W \cdot W + Z \cdot Z} + \sqrt{-1}\,\frac{Y \cdot W - X \cdot Z}{W \cdot W + Z \cdot Z}. \end{aligned}$$

As before, for $\frac{I}{J}$ we assume that $0 \notin (W \cdot W + Z \cdot Z)$.

As with real intervals, there is no distributive law for complex intervals. Consequently, evaluating a polynomial at intervals is not well defined. Evaluation at intervals is well defined for expressions of a polynomial as a straight-line program, which is an evaluation of the polynomial via a sequence of arithmetic operations that does not involve distributivity.

EXAMPLE 5.4. Consider the polynomial $f(x, y, z) = x(y+z) = xy+xz$. These two expressions of the distributive law are different straight-line programs for f, and we have shown that they have distinct evaluations on the triple $([0, 1], [-1, 0], [1, 1])$. $\diamond$

We sidestep this issue with the notion of an interval enclosure.

DEFINITION 5.5. Let F be a system of m polynomials in n variables. We call a map

$$\square F \colon (\mathbb{IC})^n \to (\mathbb{IC})^m$$

such that $\{F(\mathbf{x}) \mid \mathbf{x} \in \mathbf{I}\} \subseteq \square F(\mathbf{I})$ for every $\mathbf{I} \in (\mathbb{IC})^n$ an *interval enclosure* of F. $\diamond$

Let $\square F$ be an interval enclosure of a square polynomial system F and let $\square JF$ be an interval enclosure of its Jacobian map $JF \colon \mathbb{C}^n \to \mathbb{C}^{n\times n}$. Furthermore, let $\mathbf{I} \in (\mathbb{IC})^n$, $\mathbf{x} \in \mathbb{C}^n$, and let $Y \in \mathbb{C}^{n\times n}$ be an invertible matrix. The *Krawczyk operator* these define is

$$K_{\mathbf{x},Y(\mathbf{I})} := \mathbf{x} - Y \cdot \square F(\mathbf{x}) + (\mathbf{1}_n - Y \cdot \square JF(I))(\mathbf{I} - \mathbf{x}),$$

where $\mathbf{1}_n$ is the $n \times n$ identity matrix. The norm of a matrix interval $A \in (\mathbb{IC})^{n\times n}$ is $\|A\|_\infty := \max_{B\in A} \max_{\mathbf{v}\in\mathbb{C}^n} \|B\mathbf{v}\|_\infty / \|\mathbf{v}\|_\infty$, where $\|(v_1, \ldots, v_n)\|_\infty = \max_{1\le i\le n} |v_i|$ for $\mathbf{v} \in \mathbb{C}^n$.

We state the main theorem underlying Krawczyk's method in the complex case that appeared in [**12**, **19**]. The approach stems from the real case analysis in [**75**] with a more modern presentation in [**76**]. See also [**84**] for an alternative version of the statement.

THEOREM 5.6. *Let $F = (f_1, \ldots, f_n)$ be a system of n polynomials in n variables, $\mathbf{I} \in (\mathbb{IC})^n$, $\mathbf{x} \in \mathbf{I}$, and let $Y \in \mathbb{C}^{n\times n}$ be invertible.*

(a) *If $K_{\mathbf{x},Y}(\mathbf{I}) \subset \mathbf{I}$, there is a zero of F in $\mathbf{I}$.*
(b) *If $\sqrt{2}\, \|\mathbf{1}_n - Y \cdot \square JF(\mathbf{I})\|_\infty < 1$, then F has a unique zero in $\mathbf{I}$.*

Several choices have to be made to implement Krawczyk's method. For instance, we have to choose interval enclosures of both F and its Jacobian JF. Example 5.4 shows that this is nontrivial as different straight-line programs for the same polynomial system can produce different results in interval arithmetic. Furthermore, choosing $\mathbf{I}$ in Theorem 5.6 too small might cause the true zero not to lie in $\mathbf{I}$, while choosing $\mathbf{I}$ too large can be an obstacle for the contraction property in Theorem 5.6(a). Heuristics are usually implemented to address these issues.

The key idea for combining interval arithmetic with floating point operations is that one can round *outwards*. Rouding outwards, an interval computed in floating point arithmetic contains the exact interval. More specifically, suppose that $X, Y \in \mathbb{IC}$ are intervals and we want to compute $X \circ Y$, $\circ \in \{+, -, *, /\}$. Let u denote the machine precision. On the computer we get the interval $\mathrm{fl}(X \circ Y) := \{(x \circ y)(1+\delta) \mid |\delta| \leq u, x \in X, y \in Y\}$, which *contains* $X \circ Y$. Consequently, it can happen that the Krawczyk operator $K_{\mathbf{x},Y}$ is a contraction for an interval I, but that machine arithmetic cannot verify this, because $\mathrm{fl}(X \circ Y)$ is too large.

Krawczyk's method is implemented in `HomotopyContinuation.jl` [**12, 15**], `Macaulay2` [**44**] package `NumericalCertification` [**19, 65**], and in the commercial `MATLAB` package `INTLAB` [**85**]. Krawczyk's method can also certify the reality of a zero: Assume that F has real coefficients. Suppose that we have found an interval $\mathbf{I} \in (\mathbb{IC})^n$ and a matrix $Y \in \mathbb{C}^{n\times n}$ such that $K_{\mathbf{x},Y}(\mathbf{I}) \subset \mathbf{I}$ and $\sqrt{2}\,\|\mathbf{1}_n - Y \cdot \square JF(\mathbf{I})\|_\infty < 1$. By Theorem 5.6, F has a unique zero $\mathbf{z}$ in $\mathbf{I}$. Since $\overline{\mathbf{z}}$ is also a zero of F, if $\{\overline{\mathbf{y}} \mid \mathbf{y} \in K_{\mathbf{x},Y}(\mathbf{I})\} \subset \mathbf{I}$, then $\mathbf{z} = \overline{\mathbf{z}}$.

6. Applications

While we have largely discussed the theory and many aspects, methods, and some implementations of numerical nonlinear algebra, these were all motivated by its applications to questions within and from outside of mathematics. Many of these are well known and may be found in other contributions in this volume. We present three applications here, involving synchronization of oscillators, enumerative geometry, and computer vision.

6.1. The Kuramoto model. In his 1673 treatise on pendulums and horology [**57**], Christiaan Huygens observed an "odd kind of sympathy" between pendulum clocks, which was one of the earliest observations of synchronization among coupled oscillators. Other examples range from pacemaker cells in the heart to the formation of circadian rhythm in the brain to synchronized flashing of fireflies. The Kuramoto model emerged from this study and has found many interesting applications.

A network of oscillators can be modeled as a swarm of points circling the origin which pull on each other. For weakly coupled and nearly identical oscillators, the natural separation of timescales [**64, 109**] allows a simple description of the long-term behavior in terms of phases of the oscillators. Kuramoto singled out the simplest case governed by equations

$$\dot{\theta}_i \;=\; \omega_i - \sum_{j\sim i} k_{ij}\sin(\theta_i - \theta_j) \qquad \text{for } i = 0, \ldots, n. \tag{6.1}$$

Here, $\theta_0, \ldots, \theta_n \in [0, 2\pi)$ are the phases of the oscillators, ω_i are their natural frequencies, $k_{ij} = k_{ji}$ are coupling coefficients, and $j \sim i$ is adjacency in the graph G underlying the network. This is the Kuramoto model [**64**]. It is simple enough to be analyzed yet it exhibits interesting emergent behaviors, and has initiated an active research field [**99**].

One core problem that can be studied algebraically is *frequency synchronization*. This occurs when the tendency of oscillators to relax to their limit cycles and the influences of their neighbors reach equilibrium, and the oscillators are all tuned to their mean frequency. Such synchronized configurations correspond to equilibria of (6.1), which are solutions to a nonlinear system of equations. Even though this system is derived from a simplification of the oscillator model, its utility extends far beyond this narrow setting. For example, in electrical engineering, it coincides with a special case of the power flow equations, derived from laws of alternating current circuits [**35**].

The equilibrium equations become algebraic after a change of variables. Numerical nonlinear algebra has been used to solve these and related families of equations finding synchronization configurations that cannot be found by simulations or symbolic computation. For example, the IEEE 14 bus system from electrical engineering is a well-studied test case, yet its full set of solutions remained unknown until it was computed using total degree and polyhedral homotopy methods by Mehta, Nguyen, and Turitsyn [**72**], using `Bertini` [**4**] and `HOM4PS-2.0` [**66**].

For rank one coupling, Coss, Hauenstein, Hong, Molzahn [**24**] showed that the equilibrium equation of (6.1) may be reformulated as a set of decoupled univariate radical equations, which are easy to solve.

Determining the number of complex equilibria (solutions to the equilibrium equations (6.1)) is another line of research that has used numerical nonlinear algebra. In the 1980s, Baillieul and Byrnes [**2**] showed that a complete network of three oscillators has at most six complex equilibria, and all may be real. For a complete network of four oscillators, they constructed 14 real equilibria. There are 20 complex equilibria. In the 2010s, Molzahn, Niemerg, Mehta, and Hauenstein [**74**] showed there could be 16 real equilibria; and in 2020, Lindberg, Zachariah, Boston, and Lesieutre [**71**] improved this to 18. It remains unknown if all 20 complex equilibria can be real.

We have a more complete answer for the enumeration of complex equilibria. Using the bihomogeneous Bézout bound of an algebraic formulation of the equilibrium equations of (6.1), Baillieul and Byrnes showed that a network of $n+1$ oscillators has at most $\binom{2n}{n}$ complex equilibria [**2**]. This upper bound is attained for generic parameters $\{\omega_i\}$ and $\{k_{ij}\}$ whose network is a complete graph.

For sparse networks whose underlying graph is not complete, the bihomogeneous Bézout bound is not sharp, as the equations are sparse in the sense of Section 3. Bernstein's theorem (Theorem 3.2) is used in [**72**] to give a bound that depends upon the underlying graph. This is elegantly expressed in terms of the normalized volumes of symmetric edge polytopes. To a connected graph G we associated its *symmetric edge polytope*, which is defined by

$$\Delta_G \ := \ \text{conv}\{e_i - e_j \mid i \sim j \text{ in } G\}. \tag{6.2}$$

For a network of $n+1$ oscillators this has dimension n. Figure 11 shows symmetric edge polytopes for connected graphs on three vertices. A result in [**22**] is that for a connected graph G and generic parameters, there are exactly $n!\text{vol}(\Delta_G)$ complex equilibria. We may see the bound of six from [**2**] for the network K_3 in Figure 11; the hexagon Δ_{K_3} is composed of six primitive triangles.

This symmetric edge polytope Δ_G is quite natural and has been independently studied in geometry and number theory [**26**]. Table 1 shows examples of the numbers of complex equilibria obtained through this connection. Finding exact formulae for other families of networks remains an active topic. For trees and cycle networks, it is possible for all complex equilibria to be real. It is still unknown if the same holds for other families of networks.

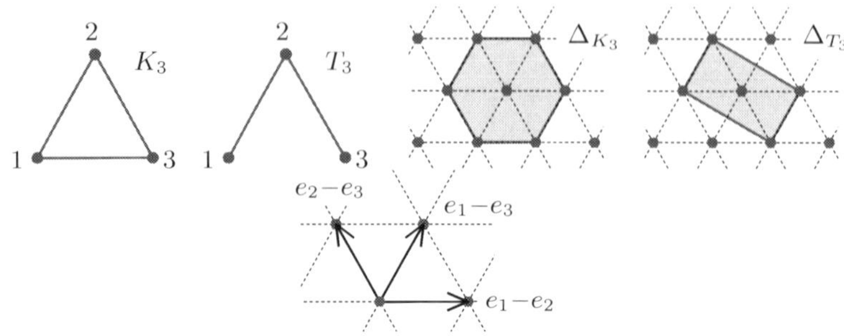

FIGURE 11. Connected graphs on three vertices, their symmetric edge polytopes, and the coordinates.

TABLE 1. Known results for the generic and maximum complex equilibria of (6.1).

A tree network with $n+1$ nodes [**21**]	2^n
A cycle network with $n+1$ nodes [**21**]	$(n+1)\binom{n}{\lfloor n/2 \rfloor}$
Cycles of lengths $2m_1, \ldots, 2m_k$ joined along an edge [**26**]	$\frac{1}{2^{k-1}} \prod_{i=1}^{k} m_i \binom{2m_i}{m_i}$
Cycles of lengths $2m_1+1$ and $2m_2+1$ joined along an edge [**26**]	$(m_1+m_2+2m_1m_2)\binom{2m_1}{m_1}\binom{2m_2}{m_2}$
A wheel graph with $n+1$ nodes for odd n [**26**]	$(1-\sqrt{3})^n + (1+\sqrt{3})^n$
A wheel graph with $n+1$ nodes for even n [**26**]	$(1-\sqrt{3})^n + (1+\sqrt{3})^n - 2$

6.2. Numerical nonlinear algebra in enumerative geometry. Paraphrasing Schubert [**86**], enumerative geometry is the art of counting geometric figures satisfying conditions imposed by other, fixed, geometric figures. Traditionally, these counting problems are solved by understanding the structure of the space of figures we are to count well enough to construct their cohomology or Chow rings [**42**], where the computations are carried out. Numerical nonlinear algebra allows us to actually compute the solutions to a given instance of an enumerative problem and then glean information about the problem that is not attainable by other means.

While the polyhedral homotopy of Section 3 based on Bernstein's theorem may be viewed as a numerical homotopy method to solve a class of enumerative problems, perhaps the first systematic foray in this direction was in [**55**] by Sturmfels and coauthors, who exploited structures in the Grassmannian to give three homotopy methods for solving simple Schubert problems. These are enumerative problems that ask for the k-planes in $\mathbb{C}^n$ that meet a collection of linear subspaces nontrivially, such as finding all (462) 3-planes in $\mathbb{C}^7$ that meet twelve 4-planes [**87**]. Their number may be computed using Pieri's formula. The Pieri homotopy algorithm from [**55**] was later used [**69**] to study Galois groups in Schubert calculus. This included showing that a particular Schubert problem with 17589 solutions had Galois group the full symmetric group S_{17589}.

One of the most famous and historically important enumerative problems is the problem of five conics: How many plane conics are simultaneously tangent

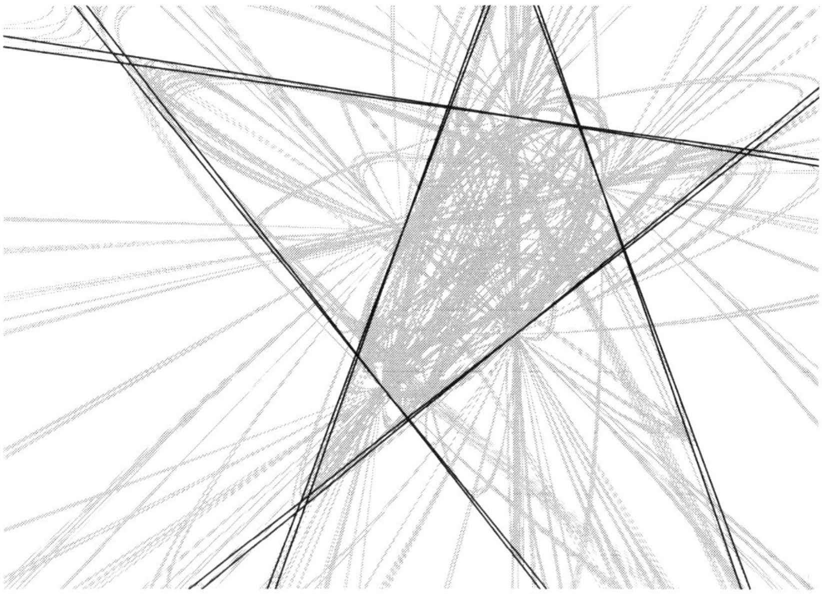

FIGURE 12. 3264 real conics tangent to five (nearly degenerate) conics.

to five given plane conics? This was posed by Steiner [**98**] who gave the answer 7776. Briefly, a conic $ax^2 + bxy + cy^2 + dxz + eyz + fz^2 = 0$ in $\mathbb{P}^2$ is given by the point $[a, b, c, d, e, f]$ in $\mathbb{P}^5$, and the condition to be tangent to a given conic is a sextic in $a, b, \ldots, f$. By Bézout's theorem, Steiner expected $6^5 = 7776$. The only problem with this approach is that every "doubled-line conic", one of the form $(\alpha x + \beta y + \gamma z)^2$, is tangent to every conic, and thus the Bézout count of 7776 includes a contribution from the doubled-line conics. Chasles [**20**] essentially introduced the Chow ring of smooth conics to give the correct answer of 3264 [**61**].

Fulton [**41**, p. 55] asked how many of the 3264 conics could be real, later determining that all can be real, but he did not publish his argument. His argument involves deforming an asymmetric pentagonal arrangement of lines and points to prove the *existence* of five real conics having all 3264 tangent conics real. Ronga, Tognoli, and Vust [**83**] published a different proof of existence via a delicate local computation near a symmetric arrangement that had 102 tangent conics, each of multiplicity 32. Fulton's argument is sketched in [**95**, Chapter 9] and Sturmfels with coauthors wrote a delightful article "3264 conics in a second" [**13**] in which they give an explicit example of five real conics with 3264 real tangent conics, together with a proof using certification as in Section 5 using numerical nonlinear algebra. Figure 12 shows a picture.

6.3. Computer vision. Computer vision is a field of artificial intelligence that trains computers to interpret and understand the visual world. Several types of problems in computer vision are amenable to algebraic computational methods. We shall focus on one type—minimal problems—and one method—homotopy continuation. Minimal problems are the backbone of the *structure from motion* pipeline that is used for three-dimensional (3D) reconstruction in applications from medical imaging to autonomous vehicles.

The article "Snapshot of algebraic vision" in this volume [**60**] treats other types of problems and other algebraic methods, including symbolic computation.

All problems amenable to this algebraic analysis share a purely geometric problem at their core. For computer vision, this often begins with basic projective geometry. We consider the projective space $\mathbb{P}^3$ as the 3D world, as it compactifies the Euclidean space $\mathbb{R}^3$. A mathematical model of a pin-hole camera C is a projective linear map given by a matrix

$$C := [R \mid t], \quad R \in \mathbb{R}^{3\times 3} \text{ and } t \in \mathbb{R}^{3\times 1},$$

which captures the images of world points in the *image plane* $\mathbb{P}^2$. A *calibrated* (a.k.a. *pinhole*) camera C has $R \in \mathrm{SO}(3)$. While this is formulated in the $\mathbb{P}^3$ compactifying $\mathbb{R}^3$, for computations we extend scalars to the complex numbers.

We may also interpret a calibrated camera as an element of the special Euclidean group $\mathrm{SE}(3)$ acting on $\mathbb{R}^3$, the rotation R followed by the translation t. It is convenient to operate in a fixed affine chart on $\mathbb{P}^3$ and consider the (affine) camera plane as a plane of points with the third coordinate equal to 1 in $\mathbb{R}^3$ (the local affine coordinates of the camera). The image of a point is obtained by intersecting the image plane with the line going through the point and the center of the camera. Figure 13 illustrates this as well as the definition of *depth*. The relative positions of the two cameras is encoded by an element $[R \mid t]$ of the Euclidean special group $\mathrm{SE}(3)$. The third coordinate of t is set equal to 1 to remove a scaling ambiguity. In the (3D) coordinate frame of the first camera the image of the tip of the little finger lies in the camera plane and the actual tip is the point obtained by scaling this by the *depth* λ.

6.3.1. *Minimal problems.* A reconstruction problem is *minimal* if it has a finite number of (complex) solutions for general values of the parameters. As in Sections 2.2 and 4.3.2, a minimal problem gives rise to a branched cover $\pi\colon \mathcal{M} \to \mathcal{P}$, where the base space $\mathcal{P}$ is called the *problem space* and the total space $\mathcal{M}$ (incidence variety) is called the *problem-solution* manifold. The number of the solutions, which is the degree of the branched cover, is the *degree* of the problem.

EXAMPLE 6.1. A classical minimal problem is computing the calibrated camera from three points in space and their image projections [**48**]. One classical formulation [**46**] of this problem is as a system of three quadratic polynomial equations

$$\begin{aligned}
\|X_1 - X_2\|^2 &= \|\lambda_1 x_1 - \lambda_2 x_2\|^2,\\
\|X_2 - X_3\|^2 &= \|\lambda_2 x_2 - \lambda_3 x_3\|^2,\\
\|X_3 - X_1\|^2 &= \|\lambda_3 x_3 - \lambda_1 x_1\|^2,
\end{aligned}$$

in three unknown *depths* $\lambda_1, \lambda_2, \lambda_3$. The parameters are the three ($i = 1, 2, 3$) world points $X_i \in \mathbb{R}^3$ and the points $x_i = (x_{i1}, x_{i2}, 1)^T \in \mathbb{R}^3$ representing three images in $\mathbb{P}^2$.

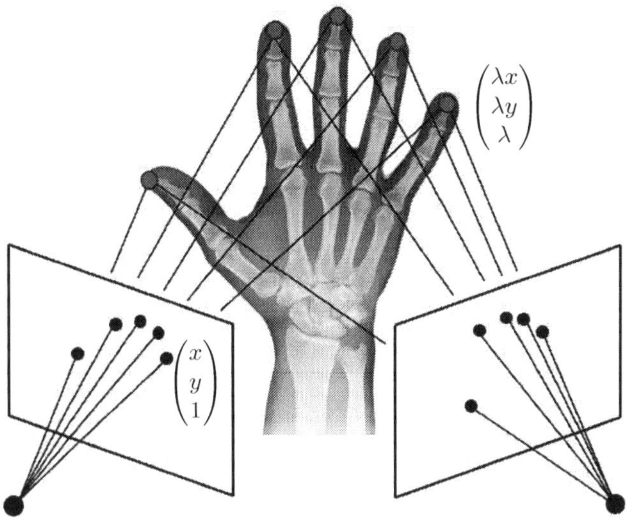

FIGURE 13. *Five point problem*: Given images of five points in two views, find the relative pose $[R \mid t]$ of the two cameras with $t = (t_1, t_2, 1)^T$.

This formulation implicitly uses that R is orthogonal and preserves the norm. Recovery of the camera $C = [R \mid t]$ from the depths is an exercise in linear algebra.

This problem has degree eight. That is, for generic X_i and x_i, $i = 1, 2, 3$, the system has eight complex solutions [**39**]. In practice, there are fewer solutions with positive depths λ_i. This gives a branched cover of degree eight over the problem manifold $\mathcal{P} \cong \mathbb{C}^{15}$. $\diamond$

We formulate perhaps the most consequential of all 3D reconstruction problems.

EXAMPLE 6.2. The *five point problem* of computing the relative pose of two calibrated cameras from five point correspondences in two images is featured in Figure 13.

Consider (paired) images $x_i = (x_{i1}, x_{i2}, 1)^T$, $y_i = (y_{i1}, y_{i2}, 1)^T$ and depths λ_i and μ_i, where $i = 1, \dots, 5$, in the first and second cameras, respectively. Write down all or sufficiently many of $\binom{5}{2}$ same-distance equations

$$\|\lambda_i x_i - \lambda_j x_j\|^2 = \|\mu_i y_i - \mu_j y_j\|^2, \quad (1 \le i < j \le 5),$$

between the image points, and one same-orientation equation

$$\det[\lambda_1 x_1 - \lambda_2 x_2 \mid \lambda_1 x_1 - \lambda_3 x_3 \mid \lambda_1 x_1 - \lambda_4 x_4] \\ = \det[\mu_1 y_1 - \mu_2 y_2 \mid \mu_1 y_1 - \mu_3 y_3 \mid \mu_1 y_1 - \mu_4 y_4].$$

These determinants are the signed volume of the same tetrahedron (formed by the world points $X_1, \dots, X_4$ in different coordinate frames). The equality of the volumes is implied by the same-distance equations but not the equality of signs. Fix one depth, $\lambda_1 = 1$, to fix the ambiguity in scaling. This gives a system of equations in the remaining nine unknown depths λ and μ.

The solution space is the space of vectors of non-fixed depths $\mathcal{P} = \mathbb{R}^9$. The projection from the incidence variety/problem-solution manifold to $\mathcal{P}$ gives a covering of degree 20. $\diamond$

As in Section 4.3.2, we may analyze the Galois group of the branched cover. Decomposing a monodromy group of a minimal problem as shown in [**33**] may lead to an easier 3D reconstruction. The classical *epipolar geometry* approach to the five point problem [**48**, Section 9] is realized in this way. This gives a two-stage procedure for the relative pose recovery with the essential stage being a problem of degree 10.

6.3.2. *Engineering meets mathematics.* The five point problem of Example 6.2 plays a practical role in solvers for geometric optimization problems in vision based on RANSAC [**40, 82**]. This problem has many practical solutions based on or inspired by Gröbner basis techniques that also use the epipolar geometry formulation [**81**].

Recently, homotopy continuation has found practical use for minimal problems whose degrees are too high for efficient symbolic computation. The first step toward practical fast computation was a special solver MINUS [**37**] based on `Macaulay2` [**44**] core C++ code for homotopy continuation and optimized for performance on modern hardware. Featured in [**38**], it is deployed on two minimal problems of degrees 216 and 312 involving point as well as line correspondences. This computes *all* complex solutions and uses postprocessing to filter out irrelevant solutions.

Unlike [**38**], the work in [**54**] combines a neural network classifier with homotopy continuation. Rather than compute all solutions and then pick a relevant one, it follows only one continuation path (over $\mathbb{R}$). That strategy finds the relevant solution with high probability in several practical scenarios. It is comparable to state-of-art algorithms for the five point problem and exceeds the performance for a four point problem. While matching four points in three calibrated views is not a minimal problem, there is a relaxation of degree 272 that is minimal, and the solution of the relaxation may be verified by using the original (overdetermined) formulation.

References

[1] C. Améndola, J. Lindberg, and J. I. Rodriguez, *Solving parameterized polynomial systems with decomposable projections*, Preprint, `arXiv:1612.08807`, 2016.

[2] J. Baillieul and C. I. Byrnes, *Geometric critical point analysis of lossless power system models*, IEEE Trans. Circuits and Systems **29** (1982), no. 11, 724–737, DOI 10.1109/TCS.1982.1085093. MR691229

[3] D. J. Bates, J. D. Hauenstein, C. Peterson, and A. J. Sommese, *A numerical local dimensions test for points on the solution set of a system of polynomial equations*, SIAM J. Numer. Anal. **47** (2009), no. 5, 3608–3623, DOI 10.1137/08073264X. MR2576513

[4] D. J. Bates, J. D. Hauenstein, A. J. Sommese, and C. W. Wampler, *Bertini: Software for Numerical Algebraic Geometry*, Available at `bertini.nd.edu` with permanent DOI: `dx.doi.org/10.7274/R0H41PB5`.

[5] D. J. Bates, J. D. Hauenstein, A. J. Sommese, and C. W. Wampler, *Numerically solving polynomial systems with Bertini*, SIAM, 2013.

[6] D. N. Bernstein, *The number of roots of a system of equations* (Russian), Funkcional. Anal. i Priložen. **9** (1975), no. 3, 1–4. MR435072

[7] D. N. Bernstein, A. G. Kušnirenko, and A. G. Hovanskiĭ, *Newton polyhedra* (Russian), Uspehi Mat. Nauk **31** (1976), no. 3(189), 201–202. MR492376

[8] G. Blekherman, J. Hauenstein, J. C. Ottem, K. Ranestad, and B. Sturmfels, *Algebraic boundaries of Hilbert's SOS cones*, Compos. Math. **148** (2012), no. 6, 1717–1735, DOI 10.1112/S0010437X12000437. MR2999301

[9] L. Blum, F. Cucker, M. Shub, and S. Smale, *Complexity and real computation*, Springer-Verlag, New York, 1998. With a foreword by Richard M. Karp, DOI 10.1007/978-1-4612-0701-6. MR1479636

[10] V. Borovik and P. Breiding, *A short proof for the parameter continuation theorem*, J. Symbolic Comput. **127** (2025), Paper No. 102373, 8, DOI 10.1016/j.jsc.2024.102373. MR4783587

[11] P. Breiding, F. Sottile, and J. Woodcock, *Euclidean distance degree and mixed volume*, Found. Comput. Math. **22** (2022), no. 6, 1743–1765, DOI 10.1007/s10208-021-09534-8. MR4509115

[12] P. Breiding, K. Rose, and S. Timme, *Certifying zeros of polynomial systems using interval arithmetic*, ACM Trans. Math. Software **49** (2023), no. 1, Art. 11, 14, DOI 10.1145/3580277. MR4567897

[13] P. Breiding, B. Sturmfels, and S. Timme, *3264 conics in a second*, Notices Amer. Math. Soc. **67** (2020), no. 1, 30–37. MR3970037

[14] Y. C. Lu, *Singularity theory and an introduction to catastrophe theory*, Universitext, Springer-Verlag, New York, 1976. With an introduction by Peter Hilton. MR461562

[15] Y. C. Lu, *HomotopyContinuation.jl: A Package for Homotopy Continuation in Julia*, Mathematical Software – ICMS 2018 (Cham), Springer International Publishing, 2018, pp. 458–465.

[16] T. Brysiewicz and M. Burr, *Sparse trace tests*, Math. Comp. **92** (2023), no. 344, 2893–2922, DOI 10.1090/mcom/3849. MR4628769

[17] T. Brysiewicz, J. I. Rodriguez, F. Sottile, and T. Yahl, *Solving decomposable sparse systems*, Numer. Algorithms **88** (2021), no. 1, 453–474, DOI 10.1007/s11075-020-01045-x. MR4297926

[18] P. Bürgisser and F. Cucker, *Condition: The geometry of numerical algorithms*, Grundlehren der mathematischen Wissenschaften [Fundamental Principles of Mathematical Sciences], vol. 349, Springer, Heidelberg, 2013, DOI 10.1007/978-3-642-38896-5. MR3098452

[19] M. Burr, K. Lee, and A. Leykin, *Effective certification of approximate solutions to systems of equations involving analytic functions*, ISSAC'19—Proceedings of the 2019 ACM International Symposium on Symbolic and Algebraic Computation, ACM, New York, 2019, pp. 267–274. MR4007469

[20] M. Chasles, *Construction des coniques qui satisfont à cinque conditions*, C. R. Acad. Sci. Paris **58** (1864), 297–308.

[21] T. Chen, R. Davis, and D. Mehta, *Counting equilibria of the Kuramoto model using birationally invariant intersection index*, SIAM J. Appl. Algebra Geom. **2** (2018), no. 4, 489–507, DOI 10.1137/17M1145665. MR3867607

[22] T. Chen, E. Korchevskaia, and J. Lindberg, *On the typical and atypical solutions to the Kuramoto equations*, 2022.

[23] T. Chen, T.-L. Lee, and T.-Y. Li, *Hom4PS-3: A parallel numerical solver for systems of polynomial equations based on polyhedral homotopy continuation methods*, Mathematical software—ICMS 2014, Lecture Notes in Comput. Sci., vol. 8592, Springer, Heidelberg, 2014, pp. 183–190, DOI 10.1007/978-3-662-44199-2_30. MR3334764

[24] O. Coss, J. D. Hauenstein, H. Hong, and D. K. Molzahn, *Locating and counting equilibria of the Kuramoto model with rank-one coupling*, SIAM J. Appl. Algebra Geom. **2** (2018), no. 1, 45–71, DOI 10.1137/17M1128198. MR3755653

[25] D. A. Cox, J. Little, and D. O'Shea, *Ideals, varieties, and algorithms*, 4th ed., Undergraduate Texts in Mathematics, Springer, Cham, 2015. An introduction to computational algebraic geometry and commutative algebra, DOI 10.1007/978-3-319-16721-3. MR3330490

[26] A. D'Alì, E. Delucchi, and M. Michałek, *Many faces of symmetric edge polytopes*, Electron. J. Combin. **29** (2022), no. 3, Paper No. 3.24, 42, DOI 10.37236/10387. MR4458147

[27] D. F. Davidenko, *On a new method of numerical solution of systems of nonlinear equations* (Russian), Doklady Akad. Nauk SSSR (N.S.) **88** (1953), 601–602. MR54339

[28] D. F. Davidenko, *On approximate solution of systems of nonlinear equations* (Russian), Ukrain. Mat. Žurnal **5** (1953), 196–206. MR57029

[29] J. P. Dedieu and M. Shub, *Newton's method for overdetermined systems of equations*, Math. Comp. **69** (2000), no. 231, 1099–1115, DOI 10.1090/S0025-5718-99-01115-1. MR1651750

[30] J. Draisma, E. Horobeţ, G. Ottaviani, B. Sturmfels, and R. R. Thomas, *The Euclidean distance degree of an algebraic variety*, Found. Comput. Math. **16** (2016), no. 1, 99–149, DOI 10.1007/s10208-014-9240-x. MR3451425

[31] T. Duff, N. Hein, and F. Sottile, *Certification for polynomial systems via square subsystems*, J. Symbolic Comput. **109** (2022), 367–385, DOI 10.1016/j.jsc.2020.07.010. MR4316045
[32] T. Duff, C. Hill, A. Jensen, K. Lee, A. Leykin, and J. Sommars, *Solving polynomial systems via homotopy continuation and monodromy*, IMA J. Numer. Anal. **39** (2019), no. 3, 1421–1446, DOI 10.1093/imanum/dry017. MR3984062
[33] T. Duff, V. Korotynskiy, T. Pajdla, and M. H. Regan, *Galois/monodromy groups for decomposing minimal problems in 3D reconstruction*, SIAM J. Appl. Algebra Geom. **6** (2022), no. 4, 740–772, DOI 10.1137/21M1422872. MR4522866
[34] T. Duff, A. Leykin, and J. I. Rodriguez, *u-generation: Solving systems of polynomials equation-by-equation*, Numer. Algorithms **95** (2024), no. 2, 813–838, DOI 10.1007/s11075-023-01590-1. MR4690003
[35] F. Dörfler and F. Bullo, *Synchronization in complex networks of phase oscillators: A survey*, Automatica J. IFAC **50** (2014), no. 6, 1539–1564, DOI 10.1016/j.automatica.2014.04.012. MR3214901
[36] G. Ewald, *Combinatorial convexity and algebraic geometry*, Graduate Texts in Mathematics, vol. 168, Springer-Verlag, New York, 1996, DOI 10.1007/978-1-4612-4044-0. MR1418400
[37] R. Fabbri, *MINUS: MInimial problem NUmerical continuation Solver*, `https://github.com/rfabbri/minus`.
[38] R. Fabbri, T. Duff, H. Fan, M. H. Regan, D. da Costa de Pinho, E. P. Tsigaridas, C. W. Wampler, J. D. Hauenstein, P. J. Giblin, B. B. Kimia, A. Leykin, and T. Pajdla, *TRPLP – Trifocal relative pose from lines at points*, 2020 IEEE/CVF Conference on Computer Vision and Pattern Recognition, CVPR 2020, Seattle, WA, USA, June 13-19, 2020, IEEE, 2020, pp. 12070–12080.
[39] J.-C. Faugère, G. Moroz, F. Rouillier, and M. Safey El Din, *Classification of the perspective-three-point problem, discriminant variety and real solving polynomial systems of inequalities*, ISSAC 2008, ACM, New York, 2008, pp. 79–86, DOI 10.1145/1390768.1390782. MR2500376
[40] M. A. Fischler and R. C. Bolles, *Random sample consensus: A paradigm for model fitting with applications to image analysis and automated cartography*, Comm. ACM **24** (1981), no. 6, 381–395, DOI 10.1145/358669.358692. MR618158
[41] W. Fulton, *Introduction to intersection theory in algebraic geometry*, CBMS Regional Conference Series in Mathematics, vol. 54, Published for the Conference Board of the Mathematical Sciences, Washington, DC; by the American Mathematical Society, Providence, RI, 1984, DOI 10.1090/cbms/054. MR735435
[42] W. Fulton, *Intersection theory*, 2nd ed., Ergebnisse der Mathematik und ihrer Grenzgebiete. 3. Folge. A Series of Modern Surveys in Mathematics [Results in Mathematics and Related Areas. 3rd Series. A Series of Modern Surveys in Mathematics], vol. 2, Springer-Verlag, Berlin, 1998, DOI 10.1007/978-1-4612-1700-8. MR1644323
[43] C. B. García and W. I. Zangwill, *Finding all solutions to polynomial systems and other systems of equations*, Math. Programming **16** (1979), no. 2, 159–176, DOI 10.1007/BF01582106. MR527572
[44] D. R. Grayson and M. E. Stillman et al., *Macaulay2: A software system for research in algebraic geometry*, Available at `www.macaulay2.com`.
[45] E. Gross, S. Petrović, and J. Verschelde, *Interfacing with PHCpack*, J. Softw. Algebra Geom. **5** (2013), 20–25, DOI 10.2140/jsag.2013.5.20. MR3273856
[46] J. A. Grunert, *Das pothenotische Problem in erweiterter Gestalt nebst über seine Anwendungen in Geodäsie*, In Grunerts Archiv für Mathematik und Physik (1841), 238–248.
[47] J. Harris, *Galois groups of enumerative problems*, Duke Math. J. **46** (1979), no. 4, 685–724. MR552521
[48] R. Hartley and A. Zisserman, *Multiple view geometry in computer vision*, 2nd ed., Cambridge University Press, Cambridge, 2003. With a foreword by Olivier Faugeras. MR2059248
[49] J. D. Hauenstein, N. Hein, and F. Sottile, *A primal-dual formulation for certifiable computations in Schubert calculus*, Found. Comput. Math. **16** (2016), no. 4, 941–963, DOI 10.1007/s10208-015-9270-z. MR3529130
[50] J. D. Hauenstein, J. I. Rodriguez, and F. Sottile, *Numerical computation of Galois groups*, Found. Comput. Math. **18** (2018), no. 4, 867–890, DOI 10.1007/s10208-017-9356-x. MR3833644

[51] J. D. Hauenstein, A. J. Sommese, and C. W. Wampler, *Regeneration homotopies for solving systems of polynomials*, Math. Comp. **80** (2011), no. 273, 345–377, DOI 10.1090/S0025-5718-2010-02399-3. MR2728983

[52] J. D. Hauenstein, A. J. Sommese, and C. W. Wampler, *Regenerative cascade homotopies for solving polynomial systems*, Appl. Math. Comput. **218** (2011), no. 4, 1240–1246, DOI 10.1016/j.amc.2011.06.004. MR2831632

[53] J. D. Hauenstein and F. Sottile, *Algorithm 921: alphaCertified: Certifying solutions to polynomial systems*, ACM Trans. Math. Software **38** (2012), no. 4, Art. 28, 20, DOI 10.1145/2331130.2331136. MR2972672

[54] P. Hruby, T. Duff, A. Leykin, and T. Pajdla, *Learning to solve hard minimal problems*, Proceedings of the IEEE/CVF Conference on Computer Vision and Pattern Recognition, 2022, pp. 5532–5542.

[55] B. Huber, F. Sottile, and B. Sturmfels, *Numerical Schubert calculus*, J. Symbolic Comput. **26** (1998), no. 6, 767–788, DOI 10.1006/jsco.1998.0239. Symbolic numeric algebra for polynomials. MR1662035

[56] B. Huber and B. Sturmfels, *A polyhedral method for solving sparse polynomial systems*, Math. Comp. **64** (1995), no. 212, 1541–1555, DOI 10.2307/2153370. MR1297471

[57] C. Huygens, *Horologium oscillatorium: Sive de motu pendulorum ad horologia aptato demonstrationes geometricae*, F. Muguet, Paris, 1673.

[58] C. Jordan, *Traité des substitutions et des équations algébriques* (French), Les Grands Classiques Gauthier-Villars. [Gauthier-Villars Great Classics], Éditions Jacques Gabay, Sceaux, 1989. Reprint of the 1870 original. MR1188877

[59] A. G. Hovanskiĭ, *Newton polyhedra, and the genus of complete intersections* (Russian), Funktsional. Anal. i Prilozhen. **12** (1978), no. 1, 51–61. MR487230

[60] J. Kileel and K. Kohn, *Snapshot of algebraic vision*, Combinatorial, Computational, and Applied Algebraic Geometry: A Tribute to Bernd Sturmfels, Proc. Sympos. Pure Math., vol. 111, Amer. Math. Soc., Providence, RI ©2025.

[61] S. L. Kleiman, *Chasles's enumerative theory of conics: A historical introduction*, Studies in algebraic geometry, MAA Stud. Math., vol. 20, Math. Assoc. America, Washington, DC, 1980, pp. 117–138. MR589410

[62] A. G. Kouchnirenko, *Polyèdres de Newton et nombres de Milnor* (French), Invent. Math. **32** (1976), no. 1, 1–31, DOI 10.1007/BF01389769. MR419433

[63] R. Krawczyk, *Newton-Algorithmen zur Bestimmung von Nullstellen mit Fehlerschranken* (German, with English summary), Computing (Arch. Elektron. Rechnen) **4** (1969), 187–201, DOI 10.1007/bf02234767. MR255046

[64] Y. Kuramoto, *International Symposium on Mathematical Problems in Theoretical Physics, January 23–29, 1975, Kyoto University, Kyoto/Japan*, Lecture Notes in Physics (2005), 420–422.

[65] K. Lee, *Certifying approximate solutions to polynomial systems on* Macaulay2, ACM Commun. Comput. Algebra **53** (2019), no. 2, 45–48. MR4033384

[66] T. L. Lee, T. Y. Li, and C. H. Tsai, *HOM4PS-2.0: A software package for solving polynomial systems by the polyhedral homotopy continuation method*, Computing **83** (2008), no. 2-3, 109–133, DOI 10.1007/s00607-008-0015-6. MR2457354

[67] A. Leykin, *Numerical algebraic geometry*, J. Softw. Algebra Geom. **3** (2011), 5–10, DOI 10.2140/jsag.2011.3.5. MR2881262

[68] A. Leykin, J. I. Rodriguez, and F. Sottile, *Trace test*, Arnold Math. J. **4** (2018), no. 1, 113–125, DOI 10.1007/s40598-018-0084-3. MR3810571

[69] A. Leykin and F. Sottile, *Galois groups of Schubert problems via homotopy computation*, Math. Comp. **78** (2009), no. 267, 1749–1765, DOI 10.1090/S0025-5718-09-02239-X. MR2501073

[70] J. Lindberg, N. Nicholson, J. I. Rodriguez, and Z. Wang, *The maximum likelihood degree of sparse polynomial systems*, SIAM J. Appl. Algebra Geom. **7** (2023), no. 1, 159–171, DOI 10.1137/21M1422550. MR4566812

[71] J. Lindberg, A. Zachariah, N. Boston, and B. Lesieutre, *The distribution of the number of real solutions to the power flow equations*, IEEE Transactions on Power Systems (2022), 1–1.

[72] D. Mehta, H. D. Nguyen, and K. Turitsyn, *Numerical polynomial homotopy continuation method to locate all the power flow solutions*, IET Generation, Transmission & Distribution **10** (2016), no. 12, 2972–2980.

[73] M. Michałek and B. Sturmfels, *Invitation to nonlinear algebra*, Graduate Studies in Mathematics, vol. 211, American Mathematical Society, Providence, RI, [2021] ©2021, DOI 10.1090/gsm/211. MR4423369

[74] D. Molzahn, M. Niemerg, D. Mehta, and J. Hauenstein, *Investigating the maximum number of real solutions to the power flow equations: Analysis of lossless four-bus systems*, Preprint, `arXiv:1603.05908`, 2016.

[75] R. E. Moore, *Interval analysis*, Prentice-Hall, Inc., Englewood Cliffs, NJ, 1966. MR231516

[76] R. E. Moore, R. B. Kearfott, and M. J. Cloud, *Introduction to interval analysis*, Society for Industrial and Applied Mathematics (SIAM), Philadelphia, PA, 2009, DOI 10.1137/1.9780898717716. MR2482682

[77] A. Morgan, *Solving polynomial systems using continuation for engineering and scientific problems*, Classics in Applied Mathematics, vol. 57, Society for Industrial and Applied Mathematics (SIAM), Philadelphia, PA, 2009. Reprint of the 1987 original [MR1049872]; Pages 304–534: computer programs section, also available as a separate file online, DOI 10.1137/1.9780898719031.pt1. MR3396207

[78] A. P. Morgan, *A transformation to avoid solutions at infinity for polynomial systems*, Appl. Math. Comput. **18** (1986), no. 1, 77–86, DOI 10.1016/0096-3003(86)90029-9. MR815773

[79] A. P. Morgan and A. J. Sommese, *Coefficient-parameter polynomial continuation*, Appl. Math. Comput. **29** (1989), no. 2, 123–160, DOI 10.1016/0096-3003(89)90099-4. MR977815

[80] A. P. Morgan, A. J. Sommese, and C. W. Wampler, *Computing singular solutions to polynomial systems*, Adv. in Appl. Math. **13** (1992), no. 3, 305–327, DOI 10.1016/0196-8858(92)90014-N. MR1176579

[81] D. Nistér, *An efficient solution to the five-point relative pose problem*, IEEE Transactions on Pattern Analysis and Machine Intelligence **26** (2004), no. 6, 756–770.

[82] R. Raguram, O. Chum, M. Pollefeys, J. Matas, and J.-M. Frahm, *USAC: A universal framework for random sample consensus*, IEEE Transactions on Pattern Analysis Machine Intelligence **35** (2013), no. 8, 2022–2038.

[83] F. Ronga, A. Tognoli, and T. Vust, *The number of conics tangent to five given conics: the real case*, Rev. Mat. Univ. Complut. Madrid **10** (1997), no. 2, 391–421. MR1605670

[84] S. M. Rump, *Solving algebraic problems with high accuracy*, Parallel and large-scale computers: performance, architecture, applications (Montreal, Que., 1982), IMACS Trans. Sci. Comput., II, IMACS, New Brunswick, NJ, 1983, pp. 299–300. MR751813

[85] S. M. Rump, *INTLAB - INTerval LABoratory*, Developments in Reliable Computing, Kluwer Academic Publishers, 1999, pp. 77–104.

[86] H. Schubert, *Kalkül der abzählenden Geometrie* (German), Springer-Verlag, Berlin-New York, 1979. Reprint of the 1879 original; With an introduction by Steven L. Kleiman. MR555576

[87] H. Schubert, *Losüng des Charakteritiken-Problems für lineare Räume beliebiger Dimension*, Mittheil. Math. Ges. Hamburg (1886), 135–155, (dated 1885).

[88] I. R. Shafarevich, *Basic algebraic geometry. 1*, 2nd ed., Springer-Verlag, Berlin, 1994. Varieties in projective space; Translated from the 1988 Russian edition and with notes by Miles Reid. MR1328833

[89] M. Shub and S. Smale, *Complexity of Bézout's theorem. I. Geometric aspects*, J. Amer. Math. Soc. **6** (1993), no. 2, 459–501, DOI 10.2307/2152805. MR1175980

[90] A. J. Sommese and J. Verschelde, *Numerical homotopies to compute generic points on positive dimensional algebraic sets*, J. Complexity **16** (2000), no. 3, 572–602, DOI 10.1006/jcom.2000.0554. Complexity theory, real machines, and homotopy (Oxford, 1999). MR1787886

[91] A. J. Sommese and C. W. Wampler, *Numerical algebraic geometry*, The mathematics of numerical analysis (Park City, UT, 1995), Lectures in Appl. Math., vol. 32, Amer. Math. Soc., Providence, RI, 1996, pp. 749–763. MR1421365

[92] A. J. Sommese, J. Verschelde, and C. W. Wampler, *Symmetric functions applied to decomposing solution sets of polynomial systems*, SIAM J. Numer. Anal. **40** (2002), no. 6, 2026–2046 (2003), DOI 10.1137/S0036142901397101. MR1974173

[93] A. J. Sommese, J. Verschelde, and C. W. Wampler, *Solving polynomial systems equation by equation*, Algorithms in algebraic geometry, IMA Vol. Math. Appl., vol. 146, Springer, New York, 2008, pp. 133–152, DOI 10.1007/978-0-387-75155-9_8. MR2397941

[94] A. J. Sommese and C. W. Wampler II, *The numerical solution of systems of polynomials*, World Scientific Publishing Co. Pte. Ltd., Hackensack, NJ, 2005. Arising in engineering and science, DOI 10.1142/9789812567727. MR2160078

[95] F. Sottile, *Real solutions to equations from geometry*, University Lecture Series, vol. 57, American Mathematical Society, Providence, RI, 2011, DOI 10.1090/ulect/057. MR2830310

[96] F. Sottile, *General witness sets for numerical algebraic geometry*, ISSAC'20—Proceedings of the 45th International Symposium on Symbolic and Algebraic Computation, ACM, New York, [2020] ©2020, pp. 418–425, DOI 10.1145/3373207.3403995. MR4144068

[97] F. Sottile and T. Yahl, *Galois groups in enumerative geometry and applications*, Preprint, arXiv:2108.07905, 2021.

[98] J. Steiner, *9. Elementare Lösung einer geometrischen Aufgabe, und über einige damit in Beziehung stehende Eigenschaften der Kegelschnitte*, J. Reine Angew. Math. **37** (1848), 161–192b, DOI 10.1515/crll.1848.37.161. MR1581626

[99] S. H. Strogatz, *From Kuramoto to Crawford: Exploring the onset of synchronization in populations of coupled oscillators*, Phys. D **143** (2000), no. 1-4, 1–20, DOI 10.1016/S0167-2789(00)00094-4. Bifurcations, patterns and symmetry. MR1783382

[100] B. Sturmfels, *Polynomial equations and convex polytopes*, Amer. Math. Monthly **105** (1998), no. 10, 907–922, DOI 10.2307/2589283. MR1656923

[101] B. Sturmfels, *Solving Systems of Polynomial Equations*, CBMS Regional Conferences Series, no. 97, American Mathematical Society, 2002.

[102] B. Sturmfels, *What is ... a Gröbner basis?*, Notices Amer. Math. Soc. **52** (2005), no. 10, 1199–1200. MR2186903

[103] S. Timme, *Mixed precision path tracking for polynomial homotopy continuation*, Adv. Comput. Math. **47** (2021), no. 5, Paper No. 75, 23, DOI 10.1007/s10444-021-09899-y. MR4320841

[104] R. Vakil, *Schubert induction*, Ann. of Math. (2) **164** (2006), no. 2, 489–512, DOI 10.4007/annals.2006.164.489. MR2247966

[105] J. Verschelde, *Algorithm 795: PHCpack: A General-Purpose Solver for Polynomial Systems by Homotopy Continuation*, ACM Trans. Math. Softw. **25** (1999), no. 2, 251–276.

[106] J. Verschelde, P. Verlinden, and R. Cools, *Homotopies exploiting Newton polytopes for solving sparse polynomial systems*, SIAM J. Numer. Anal. **31** (1994), no. 3, 915–930, DOI 10.1137/0731049. MR1275120

[107] J. Verschelde and G. Yoffe, *Polynomial homotopies on multicore workstations*, Proceedings of the 4th International Workshop on Parallel and Symbolic Computation (New York, NY, USA), PASCO '10, Association for Computing Machinery, 2010, p. 131–140.

[108] A. Weil, *Foundations of algebraic geometry*, American Mathematical Society, Providence, RI, 1962. MR144898

[109] A. T. Winfree, *Biological rhythms and the behavior of populations of coupled oscillators*, Journal of theoretical biology **16** (1967), no. 1, 15–42.

[110] T. Yahl, *Computing Galois groups of Fano problems*, J. Symbolic Comput. **119** (2023), 81–89, DOI 10.1016/j.jsc.2023.02.004. MR4556920

Department of Mathematics, U.S. Naval Academy, 572C Holloway Road, Mail Stop 9E, Annapolis, Maryland 21402
Email address: dbates@usna.edu

Universität Osnabrück, FB Mathematik/Informatik, Albrechtstr. 28a, 49076 Osnabrück, Germany
Email address: pbreiding@uni-osnabrueck.de
URL: http://www.paulbreiding.org

Department of Mathematics, Auburn University at Montgomery, Montgomery, Alabama 36116,
Email address: ti@nranchen.org
URL: http://www.tianranchen.org

Department of Applied & Computational Mathematics & University of Notre Dame, Notre Dame, Indiana 46556
Email address: hauenstein@nd.edu
URL: http://www.nd.edu/~jhauenst

School of Mathematics, Georgia Institute of Technology, 686 Cherry Street, Atlanta, Georgia 30332-0160
Email address: leykin@math.gatech.edu

Department of Mathematics, Texas A&M University, College Station, Texas 77843
Email address: sottile@tamu.edu
URL: https://franksottile.github.io/

Proceedings of Symposia in Pure Mathematics
Volume **111**, 2025
https://doi.org/10.1090/pspum/111/02047

Sums of squares, spectrahedra, and Bernd

Grigoriy Blekherman

ABSTRACT. I will give a brief personal introduction to some aspects of nonnegative polynomials, sums of squares, spectrahedra, and polynomial optimization. In the process, I will highlight the influence of Bernd Sturmfels on the subject.

1. A long time ago in a galaxy far far away

We begin with a personal story. It was the spring of 2009 and I was just starting to do research again, after a gap of about a year and a half, and struggling with depression. After reading a recent preprint by Bruce Reznick (Bruce is another mathematician for whose support I am particularly indebted), I realized that there is a way of understanding the difference between the cones of nonnegative polynomials and sums of squares via dimensions of their faces. These initial ideas (with a lot of help from co-authors!) eventually resulted in [**7**], and were further developed in [**6**]. I emailed Bruce and he invited me to give a talk at an AMS Sectional meeting. It was an amazing opportunity that gave me a chance to re-enter the research community.

I contacted Frank Sottile looking for advice on how I may restart my career. Frank replied immediately and suggested that I contact Bernd. This was the first time I had ever contacted him by email. I received an enthusiastic reply early the next morning. Bernd suggested that I apply for a special semester postdoc at IPAM for the Fall 2010. For the intervening year Bernd helped me find a position at Virginia Bioinformatics Institute (VBI) working under the direction of Reinhard Laubenbacher. The year at VBI and the semester at IPAM were formative experiences for my career.

2020 *Mathematics Subject Classification.* Primary 14P99; Secondary 90C22, 52A20.

I am extremely grateful to the entire "Real Algebraic Geometry and Polynomial Optimization" community for their support throughout my career, and I owe a particularly large debt to Bernd Sturmfels.

I would like to thank Reinhard Laubenbacher for his amazing and positive mentoring, and the IPAM semester served to make new connections and broaden my research horizons. None of this would have been possible without Bernd's help and guidance. I would also like to acknowledge support from Saugata Basu, Pablo Parrilo, Vicki Powers, Rekha Thomas and many others. I feel very fortunate to be a part of this community.

2. Setting the table

I will now describe the context of my email exchange with Bernd. How can one decide whether a function, for instance a polynomial in one variable, takes only nonnegative values? As an example, consider $p(x) = 2 + 4x - 4x^3 + 2x^4$. It is not difficult to show that $p(x)$ is globally nonnegative, but it is not immediately apparent. However, if we find a decomposition of $p(x)$ as a sum of squares of polynomials:

$$p(x) = (x^2 - 1)^2 + (x^2 - 2x - 1)^2,$$

then its nonnegativity becomes obvious. This example raises an immediate question of whether such a decomposition, which certifies nonnegativity of $p(x)$, is possible for any nonnegative polynomial $p(x)$? Another important consideration is how to compute a sum of squares decomposition, if one exists.

We begin with the history of nonnegative polynomials and sums of squares. The story begins in 1885 at the PhD defense of Hermann Minkowski, where David Hilbert was one of the examiners. During his defense Minkowski claimed that there exist nonnegative polynomials that are not sums of squares, although he did not provide an example or a proof. Three years later Hilbert published a seminal paper [**25**] classifying all cases in terms of number of variables n and degree $2d$ for which all n-variate nonnegative polynomials of degree at most $2d$ are sums of squares.

We use $P_{n,\leq 2d}$ and $\Sigma_{n,\leq 2d}$ to respectively denote the sets of n-variate nonnegative polynomials and sums of squares of degree at most $2d$. A subset C of a finite-dimensional real vector space V is called a *convex cone* if C is closed under addition and multiplication by nonnegative scalars. We note that both $P_{n,\leq 2d}$ and $\Sigma_{n,\leq 2d}$ are convex cones in the vector space $\mathbb{R}[x]_{n,\leq 2d}$ of n-variate polynomials of degree at most $2d$, and $\dim \mathbb{R}[x]_{n,\leq 2d} = \binom{n+2d}{2d}$. We refer the reader to [**2**] for background on convex geometry.

EXERCISE 2.1. Show that $P_{n,\leq 2d}$ and $\Sigma_{n,\leq 2d}$ are closed full-dimensional convex cones in $\mathbb{R}[x]_{n,\leq 2d}$ not containing any lines.

We now state Hilbert's 1888 theorem.

THEOREM 2.2. $P_{n,\leq 2d} = \Sigma_{n,\leq 2d}$ *only in the following three cases:*
(1) $n = 1$, *the case of univariate polynomials,*
(2) $2d = 2$, *the case of quadratic polynomials,*
(3) $n = 2$, $2d = 4$, *the exceptional case of ternary quartics.*

In all other cases there exist nonnegative polynomials that are not sums of squares.

The convexity structure of the cones $P_{n,\leq 2d}$ and $\Sigma_{n,\leq 2d}$ turns out to be very important in understanding their relationship. One of the great advantages of having convex sets is duality theory in convex geometry, which we now describe.

3. Enter duality

One can view a convex cone $C \subset \mathbb{R}^n$ in two ways: the first is simply as a collection of points in $\mathbb{R}^n$ which is closed under addition and multiplication by nonnegative scalars. The second, dual way, is that C is a subset of $\mathbb{R}^n$ which satisfies (potentially infinitely many) linear inequalities. Geometrically, we collect the normal vectors of all the hyperplanes H that isolate C, i.e., C lies completely in one of the two closed half-spaces that H divides $\mathbb{R}^n$ into. The normal vectors that we take always point into the halfspace containing C.

The normal vectors of the linear inequalities defining C themselves form a convex cone (!), which is called the dual cone of C and denoted by C^*. Duality is a powerful tool in convex geometry, and as observed by Rostalski and Sturmfels in [**47**], it is directly related to the duality theory in algebraic geometry.

The boundary of C is a hypersurface (set of codimension 1) in $\mathbb{R}^n$. The *algebraic boundary* $\partial_a C$ of C is the *Zariski closure of* the boundary ∂C of C, i.e., the smallest algebraic hypersurface containing ∂C. For an arbitrary convex cone C, it is possible that the algebraic boundary of C is equal to $\mathbb{R}^n$, i.e., there are no algebraic equations vanishing on the boundary of C. However, if C is a semialgebraic set (i.e., it is defined by finitely many polynomial inequalities), then the algebraic boundary of C is defined by the vanishing of a single polynomial. We now give a couple of illustrative examples.

EXAMPLE 3.1. Consider the cone $\mathcal{S}^n_+$ of positive semidefinite matrices. The boundary of C consists of positive semidefinite matrices which are singular. The algebraic boundary of $\mathcal{S}^n_+$ consists of all singular matrices, and it is defined by the vanishing of the determinant.

EXAMPLE 3.2. For the case of higher degrees consider the cone $P_{n,2d}$ of non-negative forms (homogeneous polynomials, i.e., each term has the same degree) of degree $2d$ in n variables. The boundary of $P_{n,2d}$ consists of all nonnegative forms $f \in P_{n,2d}$ which have a nontrivial zero in $\mathbb{R}^n$: there exists nonzero $x \in \mathbb{R}^n$ such that $f(x) = 0$. A real zero of a nonnegative polynomial is *a singular point*, i.e., if $f(x) = 0$, then its gradient $\nabla f(x)$ at x is also zero. This follows, since a zero of a nonnegative polynomial is a local minimum. The algebraic boundary of $P_{n,2d}$ consist of all singular forms of degree $2d$ and it is given by the *discriminant* [**38**]. We will see in Section 3.2 that understanding homogeneous nonnegative polynomials is equivalent to understanding non-homogeneous nonnegative polynomials in one less variable. However, studying homogenous polynomials is often more convenient. For instance, the constant polynomial 1 lies on the boundary of $P_{1,\leq 2d}$, but it has no zeroes in $\mathbb{R}$.

Bernd angle: The *dual variety* to the algebraic boundary $\partial_a C$ collects all vectors that are normal to the tangent space of $\partial_a C$ at a smooth point. Just like in convex duality, the normal vectors themselves form a variety, which is directly related to the extreme rays of the dual cone C^*. This connection was made by Rostalski and Sturmfels in [**47**].

The theory of duality provides a convenient way to see how the objective value depends on the input parameters of the problem. These considerations led Nie, Sturmfels and Ranestad to introduce the concept of *algebraic degree of a semidefinite program* in [**40**]. Roughly speaking, the algebraic degree measures the degree of the algebraic numbers that arise as solutions to a semidefinite program. We invite the reader to consult the chapter by Rostalski and Sturmfels in [**8**, Chapter 5] and the article "Algebraic degree in nonlinear models" by Ranestad in this volume for more details on both of these topics.

3.1. Examples of dual cones. We now look at some examples of dual cones to nonnegative polynomials. As we will see, the extreme rays of the dual cones are easy to understand, they are simply *point evaluations*.

EXAMPLE 3.3. Consider the cone $P_{1,\leq 2}$ of univariate polynomials of degree at most 2 nonnegative on $\mathbb{R}$. Extreme rays of the dual cone cone $P^*_{1,\leq 2}$ are point evaluations on the points of $\mathbb{R}$. More concretely, if we write quadratic polynomials as $a_2t^2 + a_1t + a_0$, then point evaluations are the image of $\mathbb{R}$ under *the monomial map* $\varphi : \mathbb{R} \to \mathbb{R}^3$, $\varphi(t) = (t^2, t, 1)$. If we represent a quadratic $q(t)$ by its vector of coefficients $v_q = (a_2, a_1, a_0)$, then the standard dot product $\langle v_q, \varphi(t)\rangle$ gives the value $q(t)$ of q at t.

There is an additional extreme ray of $P^*_{1,\leq 2d}$, which comes from considering $\varphi(t)$ as t approaches infinity. If we rescale $\varphi(t)$ by dividing all coordinates by t^2, then we get $\frac{\varphi(t)}{t^2} = (1, \frac{1}{t}, \frac{1}{t^2})$. As t approaches infinity we approach the point $v_\infty = (1, 0, 0)$, which can be thought of as *evaluation at infinity*. We see that $\langle v_q, \varphi(t)\rangle = a_2$, i.e., evaluation at infinity gives the coefficient of t^2.

EXAMPLE 3.4. Let C be the cone of univariate quadratic polynomials nonnegative on the interval $[0, 1]$. Extreme rays of the dual cone C^* are point evaluations on the points of $[0, 1]$. As in the previous example, if we write quadratic polynomials as $a_2t^2 + a_1t + a_0$, then point evaluations are the image of $[0, 1]$ under the same monomial map $\varphi : \mathbb{R} \to \mathbb{R}^3$, $\varphi(t) = (t^2, t, 1)$. Since the interval $[0, 1]$ is compact, we do not have any additional extreme rays, whereas for the whole of $\mathbb{R}$ we picked up an additional extreme ray coming from evaluation at infinity.

The image $\varphi([0, 1])$ is a (segment of a) *space curve*, and the dual cone C^* is the convex hull of a space curve. The boundary of C^* is swept out by lines passing through one of the endpoints $(0, 0, 0)$ or $(1, 1, 1)$ and an additional point of $\varphi([0, 1])$.

Bernd angle: The algebraic boundary of the convex hull of a compact space curve is a surface in $\mathbb{R}^3$. A systematic study of the algebraic properties of these surfaces was undertaken by Ranestad and Sturmfels in [**45**]. In particular, they expressed the degree of the defining equation of the algebraic boundary in terms of the degree, genus and singularities of the original curve. It should be pointed out that in Example 3.3 the space curve is not compact, and in Example 3.4 we are dealing with the segment of a curve.

3.2. Dual cones in depth. We first observe that questions about the relationship between nonnegative polynomials and sums of squares can be equivalently stated for homogeneous polynomials.

EXERCISE 3.5. Define the *homogenization* $\tilde{p}$ of a polynomial $p(x_1, \ldots, x_n)$ of degree d by introducing a new variable x_0 and multiplying all monomials in p by a power of x_0, so that all monomials have degree d. More formally,

$$\tilde{p} = x_0^d \cdot p\left(\frac{x_1}{x_0}, \ldots, \frac{x_n}{x_0}\right).$$

Show that p is nonnegative if and only if $\tilde{p}$ is, and the same is also true for sums of squares. Given a homogenous polynomial p we can also *dehomogenize* it by setting any one of its variables equal to 1.

In view of the above exercise, questions about global nonnegativity and sums of squares are often studied for homogeneous polynomials, which are also called forms. We will use $P_{n,2d}$ and $\Sigma_{n,2d}$ to respectively denote the sets of n-variate homogenous nonnegative polynomials and sums of squares of degree $2d$. The cones $P_{n,2d}$ and $P_{n-1,\leq 2d}$ are the same, and can be mapped to each other via the homogenization/dehomogenization maps. The same holds for $\Sigma_{n,2d}$ and $\Sigma_{n-1,\leq 2d}$.

From now on we work with homogeneous nonnegative polynomials and sums of squares. We need to define a (homogeneous) generalization of the univariate monomial map φ. This map is called the *Veronese map* and is denoted $\nu_{n,d}$, where n stands for the number of variables and d for the degree. In coordinates, $\nu_{n,d}$ sends a point $x = (x_1, \ldots, x_n) \in \mathbb{R}^n$ to all of the monomials of degree d in n variables. We give an example of $\nu_{3,2}$:

$$\nu_{3,2} : \mathbb{R}^3 \to \mathbb{R}^6, \quad \nu_{3,2}(x_1, x_2, x_3) = (x_1^2, x_1x_2, x_1x_3, x_2^2, x_2x_3, x_3^2).$$

The idea is the same as in Example 3.4: the point $\nu_{n,d}(x)$ corresponds to *a point evaluation* for n-variate polynomials of degree d.

Let's see why this is the case. Let $p = \sum c_\alpha x^\alpha$ be a homogeneous polynomial of degree d. We can represent p by the vector v_p of coefficients c_α. Then $p(x) = \langle v_p, \nu_{n,d}(x)\rangle$ where $\langle\, ,\, \rangle$ is the standard inner product.

To capture nonnegativity we do not need the whole image $\nu_{n,2d}(\mathbb{R}^n)$: since we deal with homogeneous polynomials it suffices to consider points x lying on the unit sphere $\mathbb{S}^{n-1}$. Therefore we obtain the following.

PROPOSITION 3.6. *The dual cone $P^*_{n,2d}$ is the conical hull of the image of the unit sphere under the Veronese map $\nu_{n,2d}(\mathbb{S}^{n-1})$.*

The following exercise is useful for establishing that a cone is closed.

EXERCISE 3.7. Let S be a compact subset of $\mathbb{R}^n$, such that the convex hull of S does not contain the origin. Show that the conical hull of S is closed. Give an example of a compact subset of $\mathbb{R}^2$ whose conical hull is not closed.

There is one more interesting layer to our story. The special orthogonal group $SO(n)$ acts on $\mathbb{R}^n$ via rotations. This action can be extended to forms of degree d via $Af(x) = f(A^{-1}x)$ for a matrix A in $SO(n)$ and a form f. The unit sphere is the orbit of one point under the action of $SO(n)$. We can identify the Veronese image $\nu_{n,2d}(\mathbb{S}^{n-1})$ as the orbit of one point under the extended action of $SO(n)$.

Bernd angle: A convex hull of an orbit of a point was called an *orbitope* by Sanyal, Sottile, and Sturmfels in [**48**]. The authors showed several interesting examples of orbitopes with semidefinite representations. For instance, the group $SO(3)$ in the vector space of 3×3 real matrices is tautologically an orbit of a single point (e.g. the identity matrix) under the action of $SO(3)$ by left multiplication. Therefore, the convex hull of $SO(3)$ is an orbitope, and the authors gave an explicit description of $\operatorname{conv} SO(3)$ as a spectrahedron, as well as examining the case of $\operatorname{conv} SO(n)$ in detail.

However, orbitopes are very complicated objects in general [**3**]. In a celebrated paper Scheiderer [**49**] showed that the convex hull of $\nu_{3,6}(\mathbb{S}^2)$ has no spectrahedral representation answering a question posed by Nemirovski [**24**] and Helton and Nie [**37**].

3.3. Dual cone to sums of squares. The cone of sums of squares is intimately related to positive semidefinite matrices. We begin with an illustrative example. Consider $p(x) = (1 + 2x + 3x^2)^2$. Let $\mathbf{x} = (1, x, x^2)^T$ be the vector of monomials, and let $c_p = (1, 2, 3)^T$ be the vector of coefficients. Then $p(x) = (\mathbf{x}^T c_p)(c_p^T \mathbf{x}) = \mathbf{x}^T c_p c_p^T \mathbf{x} = \mathbf{x}^T A \mathbf{x}$ where A is the rank one positive semidefinite matrix $c_p c_p^T$. This shows that squares are in one-to-one correspondence with rank 1 positive semidefinite matrices.

By taking sums, any sum of squares of polynomials of degree at most 2 can be written as $\mathbf{x}^T A\mathbf{x}$ where A is a positive semidefinite matrix. Since a positive semidefinite matrix is a sum of rank 1 matrices, the converse also holds: if $p(x) = \mathbf{x}^T A\mathbf{x}$, then $p(x)$ is a sum of squares.

We can write $\mathbf{x}^T A\mathbf{x}$ in a slightly different manner. Let $X = \mathbf{x}\mathbf{x}^T$ be the *Hankel matrix* of monomials. Then $\mathbf{x}^T A\mathbf{x} = \langle A, X\rangle$, where $\langle A, X\rangle = \text{trace}(AX)$ is the trace inner product.

Therefore the cone of sums of squares $\Sigma_{1,\leq 4}$ is the projection of the cone $\mathcal{S}^3_+$ of 3×3 positive semidefinite matrices via the map π which sends $A \in \mathcal{S}^3_+$ to the inner product $\langle A, X\rangle = \mathbf{x}^T A\mathbf{x}$ with the Hankel moment matrix X. Convex duality tells us that the dual cone $\Sigma^*_{1,\leq 2}$ is a *spectrahedral cone*: it is the intersection of $\mathcal{S}^3_+$ with the linear subspace of Hankel matrices, i.e., matrices of the form

$$\begin{pmatrix} y_0 & y_1 & y_2 \\ y_1 & y_2 & y_3 \\ y_2 & y_3 & y_4. \end{pmatrix}.$$

We illustrate this correspondence via one more example.

EXAMPLE 3.8. Consider the cone $\Sigma_{3,4}$ of ternary quartics which are sums of squares. Ternary quartics are sums of squares of ternary quadrics. Let $\mathbf{x} = (x^2, xy, xz, y^2, yz, z^2)^t$ be the vectors of all monomials appearing in ternary quadrics. Applying the same logic as above we se that $p \in \Sigma_{3,4}$ if and only if we can find a matrix $A \in \mathcal{S}^6_+$ such that

$$p = \mathbf{x}^T A\mathbf{x} = \langle A, X\rangle,$$

where $X = \mathbf{x}\mathbf{x}^T$. Therefore the cone $\Sigma_{3,4}$ is a projection of the cone $\mathcal{S}^6_+$, and by convex duality the dual cone $\Sigma^*_{3,4}$ is the section $\mathcal{S}^6_+$ with the following subspace of *generalized Hankel matrices:*

$$\begin{pmatrix} y_{400} & y_{310} & y_{301} & y_{220} & y_{211} & y_{202} \\ y_{310} & y_{220} & y_{211} & y_{130} & y_{121} & y_{112} \\ y_{301} & y_{211} & y_{202} & y_{121} & y_{112} & y_{013} \\ y_{220} & y_{130} & y_{121} & y_{040} & y_{031} & y_{022} \\ y_{211} & y_{121} & y_{112} & y_{031} & y_{022} & y_{013} \\ y_{202} & y_{112} & y_{103} & y_{022} & y_{013} & y_{004} \end{pmatrix},$$

where the subscripts correspond to exponents in x, y, and z, respectively.

Dual cones to sums of squares are special examples of spectrahedral cones; since we are slicing with the linear subspace of generalized Hankel matrices, they are called *Hankel spectrahedra.*

3.4. Back to the primal. As we saw in Example 3.2 the algebraic boundary of $P_{n,2d}$ is irreducible and given by the vanishing of the discriminant. The picture for the cone of sums of squares is more subtle. It can be shown that the discriminant hypersurface is an irreducible component of $\partial_a\Sigma_{n,2d}$, but when $\Sigma_{n,2d} \neq P_{n,2d}$ there are additional irreducible components in the algebraic boundary $\partial_a\Sigma_{n,2d}$.

Bernd angle: In [**5**] Blekherman, Hauenstein, Ottem, Ranestad, and Sturmfels fully described the algebraic boundary of the cones $\Sigma_{3,6}$ and $\Sigma_{4,4}$, which correspond to the boundary cases of Hilbert's theorem. For the case $(3, 6)$ the discriminant is a polynomial of degree 75 in 28 variables. The discriminant gives the algebraic boundary of $P_{3,6}$, and it forms an irreducible component of the boundary of $\Sigma_{3,6}$. There is a unique additional component to the algebraic boundary of $\Sigma_{3,6}$, and it is given by a polynomial of degree 83200 in 28 variables. For the case $(4, 4)$ the discriminant is a polynomial of degree 108 in 35 variables. The discriminant gives the algebraic boundary of $P_{4,4}$, and it forms an irreducible component of the boundary of $\Sigma_{4,4}$. There is a unique additional component to the algebraic boundary of $\Sigma_{4,4}$, and it is given by a polynomial of degree 38475 in 35 variables. Both numbers 83200 and 38475 are established by drawing a connection to between the algebraic boundary of the cone of sums of squares and Noether–Lefschetz divisors in moduli spaces of K3 surfaces.

We do not know a good description of the algebraic boundary of $\Sigma_{n,2d}$ outside of the cases of equality and boundary cases of Hilbert's theorem. In particular, even the number of irreducible components is not known.

PROBLEM 3.9. Describe the algebraic boundary of cones $\Sigma_{n,2d}$. In particular, find the number of irreducible components of $\Sigma_{n,2d}$ and their degrees.

4. Spectrahedra and spectrahedral cones

We have seen that the dual cones $\Sigma_{n,2d}^*$ are spectrahedral cones, i.e., they are slices of the cone of positive semidefinite matrices with a linear subspace. A *spectrahedron* is a slice of the cone of positive semidefinite matrices with an affine subspace. Spectrahedra are an important class of convex sets in optimization, since they are precisely the feasible regions of semidefinite programs.

EXAMPLE 4.1. Let $C = \mathcal{S}_+^n \cap L$ be a spectrahedral cone. The algebraic boundary $\partial_a C$ of C is contained in the variety defined by the vanishing of the determinant restricted to the subspace L. While the determinant is an irreducible polynomial, its restriction to L may factor. For instance, if L is a two-dimensional subspace passing through the interior of $\mathcal{S}_+^n$ then $\partial_a C$ factors as a union of two lines.

Bernd angle 1: We now consider the following basic problem: what kind of convex cones C can we get when we slice $\mathcal{S}_+^n$ with a 4-dimensional subspace L? The algebraic boundary $\partial_a C$ is contained in (and for sufficiently generic subspaces L equal to) the restriction of the determinant to the subspace L. The defining equation of the algebraic boundary is homogeneous since C is a cone, and thus we can view it as a special degree n surface in the projective space $\mathbb{P}^3$. Such surfaces are classically called *symmetroids* and much is known about, for instance, their singularity structure. For $n = 4$ these cones were called *quartic spectrahedra* and investigated in detail in [**41**]. We refer the reader to [**41**] and follow up work in [**23**], [**14**] about rational and quintic spectrahedra.

Bernd angle 2: As we saw in Example 4.1 the algebraic boundary of a spectrhedral cone may factor. Projections of spectrahedral cones are known as *spectrahedral shadows* or *semidefinitely representable cones*. By convex duality, spectrahedral cones and spectrahedral shadows are dual to each other. In [**50**] Sinn and Sturmfels considered the algebraic boundary of semidefinitely representable cones. They found descriptions of the number of irreducible components and their degrees if the slicing subspace L and the projection π are sufficiently generic.

5. History comes back

It should be noted that Hilbert did not provide an explicit nonnegative polynomial that is not a sum of squares of polynomials. The first explicit example appeared only much later and is due to Motzkin [**36**] (See Exercise 5.2). The main difficulty of constructing an example is certifying that a given polynomial is globally nonnegative without relying on a sums of squares decomposition. Motzkin's idea was to guarantee nonnegativity via classical inequalities! The Motzkin polynomial

$$M(x, y) = x^2y^4 + x^4y^2 + 1 - 3x^2y^2$$

is nonnegative by applying the arithmetic mean/geometric mean inequality, and an argument detailed in the exercises below shows that it is not a sum of squares. A historical aside: it seems that Motzkin did not realize the significance of his construction, and the fact that this was the first explicit example of a nonnegative polynomial that is not a sum of squares was pointed out to him by Olga Taussky-Todd [**46**].

EXERCISE 5.1. The *Newton polytope* N_p of a polynomial p is the convex hull of the vectors of monomial exponents that occur in p. For example, the Newton Polytope of $x^2 - 2xy + 1$ is the convex hull of vectors $(2,0), (1,1)$ and $(0,0)$. Show that if a polynomial $p = \sum_i q_i^2$ is a sum of squares, then the Newton Polytope of each q_i is contained in $\frac{1}{2}N_p$. This result originally appeared in [**15**].

EXERCISE 5.2. Use Exercise 5.1 to show that the Motzkin polynomial is not a sum of squares. Also show that the homogeneous polynomial $S(x, y, z) = x^4y^2 + y^4z^2 + z^4x^2 - 3x^2y^2z^2$, constructed by Choi and Lam, is nonnegative but not a sum of squares.

Since then many other examples of nonnegative polynomials that are not sums of squares have appeared, particularly in the work of Choi, Lam, and Reznick (see [**46**] for a nice overview).

To complete our historical detour we note that after proving his 1888 theorem, Hilbert showed in 1893 that any nonnegative bivariate polynomial is a sum of squares of rational functions. At first, writing

$$f = \sum \left(\frac{p_i}{q_i}\right)^2$$

may seem quite limiting, however observe that if we bring all of the rational functions to a common denominator r then we are writing

$$f \cdot r^2 = \sum s_i^2,$$

i.e., we hope to find a square multiplier r^2, such that $f \cdot r^2$ is a sum of squares of polynomials. This is clearly a generalization of writing f as a sum of squares of polynomials, where we take r to be the constant polynomial 1. In fact we can even allow sum of squares multipliers, instead of just a single square.

EXERCISE 5.3. Show that f is a sum of squares of rational functions if and only if there exists a sum of squares polynomial h such that $f \cdot h$ is a sum of squares of polynomials.

Hilbert's 17th problem asked whether a nonnegative polynomial in any number of variables can always be written a sum of squares of rational functions. E. Artin solved this problem in the affirmative in 1927. His original proof, which relied on the newly established Artin–Schreier theory of real closed fields, was purely existential. In particular, it did not provide any bounds on the degree of the multiplier h [**13**]. The search for a multiplier h of fixed degree such that $f \cdot h$ is a sum of squares can also be formulated as a semidefinite program, so it is important for computations to understand in what degree rational certificates exist.

To this day, our understanding of both upper and lower degree bounds is quite poor. For the best current upper bound see [**33**] and for the lower bounds see [**4**]. For the special case of nonnegative polynomials on curves see [**12**]. While philosophically, Artin's solution to Hilbert's 17th problem provides a quite satisfactory answer on the power of sums of squares to explain nonnegativity, computationally much remains unknown.

PROBLEM 5.4. Find better degree bounds (both upper and lower) for rational sums of squares representations in Hilbert's 17th problem, and more generally, for rational sums of squares representations on polynomials nonnegative on varieties.

6. Back to the future

A modern way of looking at the relationship between nonnegative polynomials and sums of squares is to consider nonnegative polynomials and sums of squares on projective varieties. We first reformulate Hilbert's theorem using the Veronese map $\nu_{n,d}$, which we revisit and describe from a projective point of view.

The dth Veronese map $\nu_{n+1,d}$ is a map from

$$\mathbb{P}^n \quad \text{to} \quad \mathbb{P}^{\binom{n+d}{d}-1},$$

which sends a point with projective coordinates $[x_0 : \cdots : x_n]$ to all monomials of degree d in variables $x_0, \ldots, x_n$:

$$\nu_{n+1,d} \colon \mathbb{P}^n \hookrightarrow \mathbb{P}^{\binom{n+d}{d}-1},$$

$$[x_0 : \cdots : x_n] \mapsto [\text{all monomials of degree } d \text{ in these variables}].$$

Here again is an explicit example of the second Veronese map from $\mathbb{P}^2$ to $\mathbb{P}^5$, this time stated projectively:

$$\nu_{3,2} \colon \mathbb{P}^2 \hookrightarrow \mathbb{P}^5, \; [x_0 : x_1 : x_2] \mapsto \left[x_0^2 : x_0x_1 : x_0x_2 : x_1^2 : x_1x_2 : x_2^2\right].$$

The map $\nu_{n+1,d}$ is very useful because it "linearizes" degree d forms: degree d forms are simply linear forms on

$$\nu_{n+1,d}(\mathbb{P}^n).$$

This follows by simply restricting a linear form on $\mathbb{P}^{\binom{n+d}{d}-1}$ to the image of the Veronese map $\nu_{n+1,d}$. Moreover, quadratic forms on $\nu_{n+1,d}(\mathbb{P}^n)$ are forms of degree $2d$.

We can make the following identifications: the cone $P_{n+1,2d}$ of globally nonnegative quadratic forms is the cone of quadratic forms nonnegative on $\nu_{n+1,d}(\mathbb{P}^n)$, and the cone $\Sigma_{n+1,2d}$ of sums of squares is the cone of quadratic forms on $\nu_{n+1,d}(\mathbb{P}^n)$ which can be written as a sum of squares of linear forms (modulo the ideal of $\nu_{n+1,d}(\mathbb{P}^n)$).

DEFINITION 6.1. Let P_X denote the convex cone of quadratic forms nonnegative on a real variety X. Let Σ_X denote the cone of quadratic forms that can we written as a sum of squares of linear forms modulo the ideal $I(X)$ of X.

Let us take a more detailed look at the cones P_X and Σ_X. Both these cones consist of functions on the variety X, and therefore they naturally live in the *coordinate ring* $\mathbb{R}(X)$ of X: $\mathbb{R}(X) = \mathbb{R}[x_0, \ldots, x_n]/I(X)$. Moreover, we only deal with quadratic functions, and therefore both cones are subsets of the degree 2 part of $\mathbb{R}(X)$, which is a real vector space:

$$\Sigma_X \subseteq P_X \subset \mathbb{R}(X)_2 = \mathbb{R}[x_0, \ldots, x_n]_2/I(X)_2.$$

EXERCISE 6.2. Suppose that $I(X)$ contains all quadratics vanishing on the real points of X. Show that P_X and Σ_X are closed, full-dimensional cones in $\mathbb{R}(X)_2$.

We can now see that Hilbert's theorem deals with cases of equality of P_X and Σ_X where $X = \nu_{n+1,d}(\mathbb{P}^n)$. We now generalize this to arbitrary real varieties X. We start with some remarks.

A real projective variety is a variety in complex projective space $\mathbb{CP}^n$ that is defined by real equations. The real points $X_{\mathbb{R}}$ of X are the points of X contained in the real projective space $\mathbb{RP}^n$. It is possible to have real varieties with no real points. For example, the variety defined by $x_0^2 + \cdots + x_n^2$ defines a real hypersurface in $\mathbb{CP}^n$ with no real points. We would like to rule out such examples. A projective variety is *totally real* if real points $X_{\mathbb{R}}$ of X are Zariski dense in X, i.e. any form vanishing on $X_{\mathbb{R}}$ must vanish on all of X. Equivalently we say that an irreducible variety X is totally real if it contains a smooth real point.

To understand all cases of equality between degree $2d$ nonnegative forms on a real projective variety $X \subseteq \mathbb{P}^n$ it suffices to understand all cases of equality between P_X and Σ_X. We can use the Veronese map in the same way to transfer the degree $2d$ equality question to the question of equality between $P_{\nu_{n+1,d}(X)}$ and $\Sigma_{\nu_{n+1,d}(X)}$.

EXERCISE 6.3. Let $X \subset \mathbb{CP}^n$ be an irreducible projective variety that linearly spans $\mathbb{CP}^n$. Show the following inequality between degree and codimension of X:

$$\deg X \geq \operatorname{codim} X + 1.$$

(Hint: Think about what happens to degree and codimension of a variety when taking hyperplane sections and apply Bertini's theorem.)

DEFINITION 6.4. An irreducible projective variety X is called a variety of minimal degree if

$$\deg X = \operatorname{codim} X + 1.$$

EXERCISE 6.5. Show that the cases of equality in Hilbert's theorem all correspond to equality $P_X = \Sigma_X$ where X is a variety of minimal degree. What happens in the border cases $n = 3$, $2d = 6$ and $n = 4$, $2d = 4$? (Hint: The degree of the Veronese variety can be computed via Bezout's theorem).

The classification of varieties of minimal is an important achievement in classical algebraic geometry, and it is due to del Pezzo and Bertini. For a modern account, see [**22**]. Hilbert's theorem can be generalized to irreducible real varieties in the following way.

THEOREM 6.6 ([**9**,**11**]). *Let X be a totally real irreducible variety in $\mathbb{P}^n$. Then the following equivalence holds:*

$$P_X = \Sigma_X \iff X \text{ is a variety of minimal degree.}$$

If $P_X = \Sigma_X$ holds, then we need at most $\dim X + 1$ *many squares for writing an element in P_X as a sum of squares modulo $I(X)$.*

If a variety X is reducible, then there is no geometric way of generalizing the concept of minimal degree. However, there is *homological generalization* that works for reducible varieties. The idea is to look at generators of the ideal of X and examine polynomial relations between generators, which are called *syzygies.* One then needs to consider *syzygies of syzygies* (second syzygies) and so on. The process will eventually terminate by Hilbert's syzygy theorem, i.e., at some point there will be no polynomial relations between kth syzygies. The data of considering syzygies and syzygies of syzygies and so on, can be packaged into a *minimal free resolution* of the ideal $I(X)$. We refer the reader to [**20**, **35**] for some background on free resolutions.

Varieties of minimal degree have very special syzygies: kth syzygies are given by linear polynomials for all k, in other words varieties of minimal degree have *Castelnuovo–Mumford regularity* 2 [**21**]. This can be extended to reducible varieties [**19**], and amazingly it provides the right generalization for equality between nonnegative polynomials and sums of squares!

THEOREM 6.7 ([**9**–**11**]). *Let X be a totally real variety in $\mathbb{P}^n$. Then the following equivalence holds:*

$$P_X = \Sigma_X \iff X \text{ has regularity } 2.$$

If $P_X = \Sigma_X$ holds, then we need at most $\dim X + 1$ *many squares for writing an element in P_X as a sum of squares modulo $I(X)$.*

7. Polynomial optimization

We end with giving a very brief introduction to the use of sums of squares in polynomial optimization. We invite the reader to consult [**8**, **31**, **32**] for more details. As we saw in the previous Section, we can use sums of squares to certify nonnegativity of polynomial functions on varieties. We do one more small explicit example.

EXAMPLE 7.1. Let $X = \{0,1\} \subset \mathbb{R}$ be the variety defined by the equation $x(1-x) = 0$. We can show that the function x is nonnegative on X by writing

$$x = x^2 + x(1-x).$$

We wrote x as a *sum of squares modulo the ideal of* X, just as we did in Section 6. This certifies nonnegativity on X, since a sum of squares is globally nonnegative and any polynomial in the ideal of X is identically zero on X.

Now that we see how to use sums of squares to certify nonnegativity, we make a connection to optimization. Suppose that we are given a problem of minimizing a polynomial function $f(x)$ on a variety X. The infimum γ^* of $f(x)$ on X is simply the best lower bound we have for $f(x)$:

$$\gamma^* = \min_{x \in X} f(x) = \max_{f(x)-\gamma \geq 0 \text{ on } X} \gamma.$$

We see that if we can understand the set of nonnegative functions on X algorithmically, then we can do optimization. Many difficult problems, e.g., MAXCUT in combinatorial optimization, can be written as optimization of a quadratic function on a variety defined by quadratics. Therefore, the set of nonnegative quadratics on a variety can be complicated. However, sums of squares give us a subset of *obviously nonnegative polynomials* on X. Therefore we may formulate the following *sum of squares relaxation*:

$$\gamma' = \max \gamma \quad | \quad f(x) - \gamma \text{ is a sum of squares on } X.$$

However, to make this algorithmically feasible we need to also limit the degree of sums of squares that we use. We get a *sequence of relaxations:*

$$\gamma'(d) = \max \gamma \quad | \quad f(x) - \gamma \text{ is a sum of squares of degree at most } d \text{ on } X.$$

These relaxations give us a lower bound for the minimum, and under mild conditions converge to γ^* [**30**, **39**]. In general, we have

$$\gamma'(1) \leq \gamma'(2) \leq \cdots \leq \gamma^*.$$

Bernd angle 1: In [**17**] Cifuentes, Harris, and Sturmfels studied the following problem: start with a real variety X defined by quadratics; describe quadratic functions f such that the sum of squares relaxation is exact in the first step, i.e.,

$$\gamma'(1) = \gamma^*.$$

The authors focused on the case of minimizng Euclidean distance to a variety, which built on the work of [**16**]. Here we can think of a point in $\mathbb{R}^n$ as defining a quadratic function, namely the distance to the point, that we wish to minimize on our variety X. The authors showed that the region of SDP-exactness is a bundle of spectrahedral shadows surrounding the variety.

Bernd angle 2: A different viewpoint to optimization on varieties was adopted in the influential paper [**18**] by Draisma, Horobeţ, Ottaviani, Sturmfels, and Thomas. Many interesting practical problems can be stated as finding the nearest point on an algebraic variety; or often a semialgebraic subset of a variety, as in applications a nonnegative solution is often desired. Local optimization methods, such as many forms of gradient descent, can reliably find a local minimum. However, it may be important to know how many local minima the problem has in total. For a generic point, the number of local minima is a constant, which the authors called the *Euclidean distance degree of the variety.*

8. Conclusion

The past twenty years have seen incredible activity at the interface between convex geometry, real algebraic geometry and optimization. I highlighted only some of the developments in the article above based on my personal taste and highlighting Bernd's influence on the subject. Sums of squares provide a very flexible optimization framework with applications in dynamical systems, control and robotics to name just a few [**28**, **34**, **42**, **51**]. Additionally, sums of squares play an increasingly prominent role in theoretical computer science, often proving best complexity algorithms for difficult computational problems [**26**, **29**, **43**, **44**]. There is also a fascinating connection between sums of squares algorithms and the *unique games conjecture* [**1**, **27**]. There are many interesting theorems in this area that are waiting to be discovered!

References

[1] B. Barak and D. Steurer, *Sum-of-squares proofs and the quest toward optimal algorithms*, Proceedings of the International Congress of Mathematicians—Seoul 2014. Vol. IV, Kyung Moon Sa, Seoul, 2014, pp. 509–533. MR3727623

[2] A. Barvinok, *A course in convexity*, Graduate Studies in Mathematics, vol. 54, American Mathematical Society, Providence, RI, 2002, DOI 10.1090/gsm/054. MR1940576

[3] A. Barvinok and G. Blekherman, *Convex geometry of orbits*, Combinatorial and computational geometry, Math. Sci. Res. Inst. Publ., vol. 52, Cambridge Univ. Press, Cambridge, 2005, pp. 51–77. MR2178312

[4] G. Blekherman, J. Gouveia, and J. Pfeiffer, *Sums of squares on the hypercube*, Math. Z. **284** (2016), no. 1-2, 41–54, DOI 10.1007/s00209-016-1644-7. MR3545483

[5] G. Blekherman, J. Hauenstein, J. C. Ottem, K. Ranestad, and B. Sturmfels, *Algebraic boundaries of Hilbert's SOS cones*, Compos. Math. **148** (2012), no. 6, 1717–1735, DOI 10.1112/S0010437X12000437. MR2999301

[6] G. Blekherman, S. Iliman, M. Juhnke-Kubitzke, and M. Velasco, *Gap vectors of real projective varieties*, Adv. Math. **283** (2015), 458–472, DOI 10.1016/j.aim.2015.07.015. MR3383808

[7] G. Blekherman, S. Iliman, and M. Kubitzke, *Dimensional differences between faces of the cones of nonnegative polynomials and sums of squares*, Int. Math. Res. Not. IMRN **18** (2015), 8437–8470, DOI 10.1093/imrn/rnu202. MR3417682

[8] G. Blekherman, P. A. Parrilo, and R. R. Thomas (eds.), *Semidefinite optimization and convex algebraic geometry*, MOS-SIAM Series on Optimization, vol. 13, Society for Industrial and Applied Mathematics (SIAM), Philadelphia, PA; Mathematical Optimization Society, Philadelphia, PA, 2013. MR3075433

[9] G. Blekherman, D. Plaumann, R. Sinn, and C. Vinzant, *Low-rank sum-of-squares representations on varieties of minimal degree*, Int. Math. Res. Not. IMRN **1** (2019), 33–54, DOI 10.1093/imrn/rnx113. MR4023754

[10] G. Blekherman, R. Sinn, and M. Velasco, *Do sums of squares dream of free resolutions?*, SIAM J. Appl. Algebra Geom. **1** (2017), no. 1, 175–199, DOI 10.1137/16M1084560. MR3633773

[11] G. Blekherman, G. G. Smith, and M. Velasco, *Sums of squares and varieties of minimal degree*, J. Amer. Math. Soc. **29** (2016), no. 3, 893–913, DOI 10.1090/jams/847. MR3486176

[12] G. Blekherman, G. G. Smith, and M. Velasco, *Sharp degree bounds for sum-of-squares certificates on projective curves* (English, with English and French summaries), J. Math. Pures Appl. (9) **129** (2019), 61–86, DOI 10.1016/j.matpur.2018.12.010. MR3998790

[13] J. Bochnak, M. Coste, and M.-F. Roy, *Real algebraic geometry*, Ergebnisse der Mathematik und ihrer Grenzgebiete (3) [Results in Mathematics and Related Areas (3)], vol. 36, Springer-Verlag, Berlin, 1998. Translated from the 1987 French original; Revised by the authors, DOI 10.1007/978-3-662-03718-8. MR1659509

[14] T. Brysiewicz, K. Kozhasov, and M. Kummer, *Nodes on quintic spectrahedra*, Matematiche (Catania) **76** (2021), no. 2, 415–430, DOI 10.4418/2021.76.2.8. MR4334896

[15] M. D. Choi, T. Y. Lam, and B. Reznick, *Sums of squares of real polynomials*, K-theory and algebraic geometry: Connections with quadratic forms and division algebras (Santa Barbara, CA, 1992), Proc. Sympos. Pure Math., vol. 58, Amer. Math. Soc., Providence, RI, 1995, pp. 103–126, DOI 10.1090/pspum/058.2/1327293. MR1327293

[16] D. Cifuentes, S. Agarwal, P. A. Parrilo, and R. R. Thomas, *On the local stability of semidefinite relaxations*, Math. Program. **193** (2022), no. 2, Ser. B, 629–663, DOI 10.1007/s10107-021-01696-1. MR4438380

[17] D. Cifuentes, C. Harris, and B. Sturmfels, *The geometry of SDP-exactness in quadratic optimization*, Math. Program. **182** (2020), no. 1-2, Ser. A, 399–428, DOI 10.1007/s10107-019-01399-8. MR4115654

[18] J. Draisma, E. Horobeţ, G. Ottaviani, B. Sturmfels, and R. R. Thomas, *The Euclidean distance degree of an algebraic variety*, Found. Comput. Math. **16** (2016), no. 1, 99–149, DOI 10.1007/s10208-014-9240-x. MR3451425

[19] D. Eisenbud, M. Green, K. Hulek, and S. Popescu, *Small schemes and varieties of minimal degree*, Amer. J. Math. **128** (2006), no. 6, 1363–1389. MR2275024

[20] D. Eisenbud, *Commutative algebra: With a view toward algebraic geometry*, Graduate Texts in Mathematics, vol. 150, Springer-Verlag, New York, 1995, DOI 10.1007/978-1-4612-5350-1. MR1322960

[21] D. Eisenbud and S. Goto, *Linear free resolutions and minimal multiplicity*, J. Algebra **88** (1984), no. 1, 89–133, DOI 10.1016/0021-8693(84)90092-9. MR741934

[22] D. Eisenbud and J. Harris, *On varieties of minimal degree (a centennial account)*, Algebraic geometry, Bowdoin, 1985 (Brunswick, Maine, 1985), Proc. Sympos. Pure Math., vol. 46, Amer. Math. Soc., Providence, RI, 1987, pp. 3–13, DOI 10.1090/pspum/046.1/927946. MR927946

[23] M. Helsø and K. Ranestad, *Rational quartic spectrahedra*, Math. Scand. **127** (2021), no. 1, 79–99, DOI 10.7146/math.scand.a-121456. MR4299228

[24] J. W. Helton and J. Nie, *Sufficient and necessary conditions for semidefinite representability of convex hulls and sets*, SIAM J. Optim. **20** (2009), no. 2, 759–791, DOI 10.1137/07070526X. MR2515796

[25] D. Hilbert, *Ueber die Darstellung definiter Formen als Summe von Formenquadraten* (German), Math. Ann. **32** (1888), no. 3, 342–350, DOI 10.1007/BF01443605. MR1510517

[26] S. B. Hopkins, P. Kothari, A. H. Potechin, P. Raghavendra, and T. Schramm, *On the integrality gap of degree-4 sum of squares for planted clique*, ACM Trans. Algorithms **14** (2018), no. 3, Art. 28, 31, DOI 10.1145/3178538. MR3841330

[27] S. Khot, *On the unique games conjecture*, 25th Annual IEEE Conference on Computational Complexity—CCC 2010, IEEE Computer Soc., Los Alamitos, CA, 2010, pp. 99–121. MR2932348

[28] M. Korda, D. Henrion, and I. Mezić, *Convex computation of extremal invariant measures of nonlinear dynamical systems and Markov processes*, J. Nonlinear Sci. **31** (2021), no. 1, Paper No. 14, 26, DOI 10.1007/s00332-020-09658-1. MR4197393

[29] P. K. Kothari, P. Manohar, and B. H. Zhang, *Polynomial-time sum-of-squares can robustly estimate mean and covariance of Gaussians optimally*, Algorithmic learning theory 2022, Proc. Mach. Learn. Res. (PMLR), vol. 167, Proceedings of Machine Learning Research PMLR, [place of publication not identified], [2022] ©2022, pp. 28 pp. [638–667]. MR4461172

[30] J. B. Lasserre, *Global optimization with polynomials and the problem of moments*, SIAM J. Optim. **11** (2000/01), no. 3, 796–817, DOI 10.1137/S1052623400366802. MR1814045

[31] J. B. Lasserre, *Moments, positive polynomials and their applications*, Imperial College Press Optimization Series, vol. 1, Imperial College Press, London, 2010. MR2589247

[32] M. Laurent, *Sums of squares, moment matrices and optimization over polynomials*, Emerging applications of algebraic geometry, IMA Vol. Math. Appl., vol. 149, Springer, New York, 2009, pp. 157–270, DOI 10.1007/978-0-387-09686-5_7. MR2500468

[33] H. Lombardi, D. Perrucci, and M.-F. Roy, *An elementary recursive bound for effective Positivstellensatz and Hilbert's 17th problem*, Mem. Amer. Math. Soc. **263** (2020), no. 1277, v+125, DOI 10.1090/memo/1277. MR4071235

[34] A. Majumdar, A. A. Ahmadi, and R. Tedrake, *Control design along trajectories with sums of squares programming*, 2013 IEEE International Conference on Robotics and Automation, IEEE, 2013, pp. 4054–4061.

[35] E. Miller and B. Sturmfels, *Combinatorial commutative algebra*, Graduate Texts in Mathematics, vol. 227, Springer-Verlag, New York, 2005. MR2110098

[36] T. S. Motzkin, *The arithmetic-geometric inequality*, Inequalities (Proc. Sympos. Wright-Patterson Air Force Base, Ohio, 1965), Academic Press, New York-London, 1967, pp. 205–224. MR223521

[37] A. Nemirovski, *Advances in convex optimization: Conic programming*, International Congress of Mathematicians. Vol. I, Eur. Math. Soc., Zürich, 2007, pp. 413–444, DOI 10.4171/022-1/17. MR2334199

[38] J. Nie, *Discriminants and nonnegative polynomials*, J. Symbolic Comput. **47** (2012), no. 2, 167–191, DOI 10.1016/j.jsc.2011.08.023. MR2854855

[39] J. Nie, *Optimality conditions and finite convergence of Lasserre's hierarchy*, Math. Program. **146** (2014), no. 1-2, Ser. A, 97–121, DOI 10.1007/s10107-013-0680-x. MR3232610

[40] J. Nie, K. Ranestad, and B. Sturmfels, *The algebraic degree of semidefinite programming*, Math. Program. **122** (2010), no. 2, Ser. A, 379–405, DOI 10.1007/s10107-008-0253-6. MR2546336

[41] J. C. Ottem, K. Ranestad, B. Sturmfels, and C. Vinzant, *Quartic spectrahedra*, Math. Program. **151** (2015), no. 2, Ser. B, 585–612, DOI 10.1007/s10107-014-0844-3. MR3348164

[42] M. Posa, M. Tobenkin, and R. Tedrake, *Lyapunov analysis of rigid body systems with impacts and friction via sums-of-squares*, HSCC'13—Hybrid systems: Computation and control, ACM, New York, 2013, pp. 63–72, DOI 10.1145/2461328.2461340. MR3199189

[43] P. Raghavendra, T. Schramm, and D. Steurer, *High dimensional estimation via sum-of-squares proofs*, Proceedings of the International Congress of Mathematicians—Rio de Janeiro 2018. Vol. IV. Invited lectures, World Sci. Publ., Hackensack, NJ, 2018, pp. 3389–3423. MR3966537

[44] P. Raghavendra and M. Yau, *List decodable learning via sum of squares*, Proceedings of the 2020 ACM-SIAM Symposium on Discrete Algorithms, SIAM, Philadelphia, PA, 2020, pp. 161–180. MR4141192

[45] K. Ranestad and B. Sturmfels, *On the convex hull of a space curve*, Adv. Geom. **12** (2012), no. 1, 157–178, DOI 10.1515/advgeom.2011.021. MR2911165

[46] B. Reznick, *Some concrete aspects of Hilbert's 17th Problem*, Real algebraic geometry and ordered structures (Baton Rouge, LA, 1996), Contemp. Math., vol. 253, Amer. Math. Soc., Providence, RI, 2000, pp. 251–272, DOI 10.1090/conm/253/03936. MR1747589

[47] P. Rostalski and B. Sturmfels, *Dualities in convex algebraic geometry*, Rend. Mat. Appl. (7) **30** (2010), no. 3-4, 285–327. MR2830307

[48] R. Sanyal, F. Sottile, and B. Sturmfels, *Orbitopes*, Mathematika **57** (2011), no. 2, 275–314, DOI 10.1112/S002557931100132X. MR2825238

[49] C. Scheiderer, *Spectrahedral shadows*, SIAM J. Appl. Algebra Geom. **2** (2018), no. 1, 26–44, DOI 10.1137/17M1118981. MR3755652

[50] R. Sinn and B. Sturmfels, *Generic spectrahedral shadows*, SIAM J. Optim. **25** (2015), no. 2, 1209–1220, DOI 10.1137/140978478. MR3359683

[51] J. Wang, C. Schlosser, M. Korda, and V. Magron, *Exploiting term sparsity in moment-SOS hierarchy for dynamical systems*, IEEE Trans. Automat. Control **68** (2023), no. 12, 8232–8237, DOI 10.1109/tac.2023.3293014. MR4699233

Email address: `greg@math.gatech.edu`

Proceedings of Symposia in Pure Mathematics
Volume 111, 2025
https://doi.org/10.1090/pspum/111/02048

Algebraic degree in nonlinear models

Kristian Ranestad

ABSTRACT. This note is a gentle introduction to algebraic degree as the cardinality of the set of critical points in a number of nonlinear algebraic models, in particular in models of polynomial optimization. We discuss algebraic degree in Euclidean distance, maximal likelihood and semidefinite programming, as introduced by Bernd Sturmfels.

1. Introduction

Algebraic methods are ubiquitous in mathematical modeling. Linear algebra has taken a dominating role, thanks to the precision and the effective computational tools it allows, but nonlinear algebraic methods are often more natural in modeling. They are more flexible, and may therefore give a more precise model. However, they come with a computational cost, and often also a cost of representation. The latter concerns the fact that solutions of higher degree equations are often not found in the field of the coefficients. Typically, equations with rational coefficients may have irrational or nonreal complex solutions. The algebraic degree is a measure of this cost.

A mathematical model has parameters and equations or inequalities that the parameters satisfy. An algebraic model has polynomial equations and inequalities. Here we restrict to equations. The set of solutions to the polynomial equations, or common zeros of polynomials, is then called an algebraic set. So our basic setup is a set of n parameters and a set of polynomials $f_1, \ldots, f_r$ with real coefficients, whose common zeros is an algebraic set $X \subset \mathbb{R}^n$. To find the parameters of points on X, one may, in principle, always eliminate parameters, which is equivalent to projecting away from parameters: $\pi : X \to \mathbb{R}^k$, with $k < n$. If X is a finite set of points, one may choose coordinates and project away from all but one coordinate ($k = 1$) such that X and $\pi(X)$ have the same cardinality and $\pi(X)$ is the zero set in $\mathbb{R}$ of one polynomial in one parameter. The real zeros of that polynomial are the values of that one parameter at the points on X. The appearence of non-real zeros means that the degree of the polynomial is larger than the cardinality of X. The degree of the polynomial is called the algebraic degree of X. It counts the number of complex zeros of the polynomial, counted with multiplicity. In particular, if $X_{\mathbb{C}}$ is the set of common complex solutions of $f_1, \ldots, f_r$, and this set is finite, then the algebraic degree of X is called the degree of $X_{\mathbb{C}}$.

2020 *Mathematics Subject Classification.* Primary 14C17.

Finding the algebraic degree of X therefore amounts to computing the degree of $X_{\mathbb{C}}$, when it is finite.

Stated this way, it is among the first central problems in algebraic geometry. Intersection theory is developed over time to answer this problem. Fulton's book [**19**] is the standard modern reference. Good textbook introductions that covers most of the relevant material in this note are given by Eisenbud and Harris [**17**] and by Michalek and Sturmfels [**36**].

Algebraic degree has been promoted by Bernd Sturmfels as a key concept in algebraic modeling. With collaborators he has mixed techniques from different mathematical fields to find formulas for algebraic degrees, exploiting and respecting the specific properties of the application; see [**7, 8, 16, 38, 42**]. His impetus has succeeded in initiating research into important aspects of nonlinear algebraic models and often unveiled new connections between different mathematical topics; see [**43**] for a nice overview in his own words and a sample of references,[**1–4, 10, 18, 21–24, 27–31, 33–35, 37, 39, 41, 44**]. The more recent book on *Metric Algebraic Geometry* [**5**] contains numerous examples, and exhibits a proliferation of models and methods that involve computing algebraic degrees.

The purpose of these notes is twofold. On the one hand it is to give a quick guide to classical formulas for the degree in terms of degrees and structures of the polynomial equations. On the other hand, it is to see a couple of important applications in polynomial optimization: The problem of finding the minimal distance from a point to a variety, and the problem of maximal likelihood estimation in statistical models.

2. The degree of a complex variety

Let $X \subset \mathbb{R}^n$ be a finite set, the common real zeros $X = V(f_1, \ldots, f_r)$ of polynomials $f_1, \ldots, f_r$, of degrees $d_1, \ldots, d_r$. Similarly let $X_{\mathbb{C}} \subset \mathbb{C}^n$ be the set of common complex zeros. A formula for the cardinality of X given in terms of the d_i does not exist in general. The typical example is a quadratic equation that have real roots only if the discriminant is nonnegative, and systems of linear equations that have common zeros only if the coefficient matrix and the augmented matrix have the same rank. The first issue may be circumvented by allowing complex solutions, while the latter issue is circumvented by extending the parameter space with points at infinity to the projective space. Thus we consider the common complex zeros $\bar{X} := \bar{X}_{\mathbb{C}} = V(F_1, \ldots, F_r) \subset \mathbb{P}^n_{\mathbb{C}}$ of the homogeneous polynomials, or forms, $F_1, \ldots, F_r$, where each form F_i is the homogenization of f_i with respect to a homogenous variable x_0. Two benefits of considering $\bar{X}$ is that the tool box of intersection theory is well suited for degree computations in $\mathbb{P}^n_{\mathbb{C}}$, and explains many of the known degree formulas. The cost is that the difference $\bar{X} \setminus X_{\mathbb{C}}$ may contain real or complex solutions to the homogeneous equations. We may call them irrelevant or *false solutions*. It is a relevant research question to find instances with a minimal number of false solutions for a given problem.

The notion of degree is defined using multiplicity. First, assume $\bar{X}$ is a finite set of points, then to each point in $x \in \bar{X}$ is attached a *multiplicity* $m_X(x)$, a nonnegative integer which equals one if there are n polynomials $F_{i_1}, \ldots, F_{i_n}$ among the F_i whose gradient vectors at the point x span $\mathbb{C}^n$ ([**19**, Chapter 7], [**17**, 1.29]). By the degree of $\bar{X}$ we mean

$$\deg \bar{X} = \sum_{x \in \bar{X}} m_X(x).$$

More generally, an algebraic set $\bar{X} \subset \mathbb{P}^n_{\mathbb{C}}$ decomposes as a union of irreducible closed subsets called varieties. When $\bar{X}$ is finite, these irreducible algebraic subsets are points. The dimension of a variety $\bar{X} \subset \mathbb{P}^n_{\mathbb{C}}$ is the largest number $m \geq 1$ such that the $\bar{X} \cap V(H_1, \ldots, H_m)$ is nonempty for any of set of linear forms $H_1, \ldots, H_m$. Since the common zeros, $V(H_1, \ldots, H_m)$, of the H_i is a linear space L of codimension at most m, this is just saying that $\bar{X} \cap L$ is nonempty for any linear space of dimension at least $n - m$. For general L of codimension $n - m$ the intersection $\bar{X} \cap L$ is then finite. An algebraic set $\bar{X}$ is equidimensional, if all the irreducible components have the same dimension. In this case the degree of $\bar{X}$ is the degree of $\bar{X} \cap L$, whenever the intersection is finite and L has codimension equal to the dimension of $\bar{X}$.

2.1. Degree formulas: Complete intersections and determinantal varieties. Here are a few examples of degree formulas.

2.1.1. *Bézout.*

THEOREM 2.1 (Bézout [**19**, Theorem 8.4] and [**17**, Theorem 2.4]). *Let $\bar{X}, \bar{Y} \subset \mathbb{P}^n_{\mathbb{C}}$ be equidimensional algebraic sets whose dimensions sum to n, then the degree of $\bar{X} \cap \bar{Y}$, if it is finite, is $\deg \bar{X} \cdot \deg \bar{Y}$.*

If F is a form of degree d and $H_1, \ldots, H_{n-1}$ are linear forms, such that $V(F, H_1, \ldots, H_{n-1})$ is finite, the latter is the zero-set of F restricted to the line $V(H_1, \ldots, H_{n-1})$, so

$$\deg V(F) \cdot \deg V(H_1, \ldots, H_{n-1}) = \deg V(F) = d.$$

Applying the theorem repeatedly to the intersection of r hypersurfaces $V(F_i), i = 1, .., s, s = 2, \ldots r$ and linear spaces $V(H_1, \ldots, H_{n-s})$ we get the following.

COROLLARY 2.2. *If $\bar{X} = V(F_1, \ldots, F_r)$ has codimension r and the forms F_i have degree d_i, then $\bar{X}$ is called a complete intersection and*

$$\deg \bar{X} = d_1 \cdots d_r.$$

2.1.2. *Rectangular determinantal.* If the equations that define $\bar{X}$ are all $(r \times r)$-minors of a $(r \times m)$-matrix M whose entries are homogeneous forms, then there are classical formulas for the degree of $\bar{X}$. Let

$$S_{m-r+1}(d_1, \ldots, d_r) = \sum_{i_1 + \cdots + i_r = m-r+1} d_1^{i_1} \cdots d_r^{i_r}.$$

be the complete symmetric function of degree $m - r + 1$ in $d_1, \ldots, d_r$.

THEOREM 2.3 (Giambelli, Thom, and Porteous, [**19**, Theorem 14.4]). *Let*

$$M(x) = \begin{pmatrix} a_{11} & a_{12} & \cdots & a_{1m} \\ a_{21} & a_{22} & \cdots & a_{2m} \\ \vdots & \vdots & & \vdots \\ a_{r1} & a_{r2} & \cdots & a_{rm} \end{pmatrix}$$

be a matrix whose entries a_{ij} are homogeneous forms of degree d_i in $\mathbb{C}[x_0, \ldots, x_n]$. Then the set

$$D_{r-1}(M) = \{x \in \mathbb{P}^n_{\mathbb{C}} \mid \operatorname{rank} M(x) \leq r - 1\}$$

is either empty, or it has codimension at most $m - r + 1$, and whenever the codimension equals $m - r + 1$, then $D_{r-1}(M)$ is a scheme of degree

$$\deg D_{r-1}(M) = S_{m-r+1}(d_1, \ldots, d_r).$$

There are similar degree formulas, when $\bar{X}$ is defined by $(s \times s)$-minors of M, with $s \leq r$, and there are integers $e_1, \ldots, e_r$ and $d_1, \ldots, d_m$ such that the entries of M have degrees $\deg a_{ij} = d_j - e_i$, see [**19**, Chapter 14] and [**17**, Chapter 12].

2.1.3. *Symmetric determinantal.*

THEOREM 2.4 ([**25**, Proposition 12]). . *Let $M(x)$ be a symmetric $(m \times m)$-matrix with linear entries. Then the set*

$$D_r(M) = \{x \in \mathbb{P}^n_{\mathbb{C}} \mid \operatorname{rank} M(x) \leq r\}$$

has codimension at most $\binom{m-r+1}{2}$, and whenever the codimension equals $\binom{m-r+1}{2}$, then $D_r(M)$ is a scheme of degree

$$\deg D_r(M) = \prod_{j=0}^{m-r-1} \frac{\binom{m+j}{m-r-j}}{\binom{2j+1}{j}}.$$

These degree formulas are very useful, but have limitations.

2.1.4. *Residual and excess intersection.* When $\bar{X} \subset \mathbb{P}^n_{\mathbb{C}}$ is an algebraic set and $\bar{Y} \subset \bar{X}$ is a proper algebraic subset of the same dimension, then there is a well defined *residual algebraic set* $\bar{Z} \subset \bar{X}$, such that $\bar{Z} \cup \bar{Y} = \bar{X}$ and

(1) if $\deg \bar{X} > \deg \bar{Y}$, then $\dim \bar{Z} = \dim \bar{X}$ and $\deg \bar{Z} = \deg \bar{X} - \deg Y$.
(2) if $\deg \bar{X} = \deg \bar{Y}$, then $\bar{Z}$ is either empty or $\dim \bar{Z} < \dim \bar{X}$.

Both cases occur naturally in models: Assume that the model Z_0 is contained in an algebraic set $\bar{X}$. If $\bar{Y} \subset \bar{X}$ is an algebraic subset that is disjoint from the model, then $Z_0 \subset Z$, where Z is an algebraic set residual to Y in X. If $\bar{Z}$ and $\bar{Y}$ have the same dimension, computing the degree of $\bar{Z}$ is a problem of *residual intersection*. When $\dim \bar{Z} < \dim \bar{Y}$, computing the degree of $\bar{Z}$ is an *excess intersection* problem. For general approaches to these problems see [**19**, Chapter 9] and [**17**, Chapter 13].

2.2. Application: Critical locus. The main issue in the applications we will discuss is to find the critical locus of a function on an algebraic set. There are two classical approaches to this problem: The method of Lagrange multipliers and the method of computing Euler characteristic with a pencil of divisors. By a pencil, we mean a set parameterized linearly by $\mathbb{P}^1_{\mathbb{C}}$. The level sets $\{f = \lambda\}_{\lambda \in \mathbb{C}}$ of a rational function $f = p/q : \mathbb{C}^n \to \mathbb{C}$ extend naturally to a pencil of hypersurfaces:

$$\{\mu p - \lambda q = 0\} \quad [\lambda : \mu] \in \mathbb{P}^1_{\mathbb{C}}.$$

By abuse of notation, we often write $\lambda \in \mathbb{P}^1_{\mathbb{C}}$ in stead of $[\lambda : \mu]$ for the parameter of a pencil.

2.2.1. *Critical locus: Lagrange multipliers.* The method of Lagrange multipliers from calculus may be generalized to cover many of the examples that we will discuss. Classically, this method finds the critical points on Y of a rational function $f = p/q : \mathbb{C}^n \to \mathbb{C}$ as the locus where the gradient of the level set of the function is in the linear span of the gradients of the forms defining Y. The pencil of level sets of f are the zeros of polynomials $\mu p - \lambda q, \quad [\lambda : \mu] \in \mathbb{P}^1_{\mathbb{C}}$. At each point $x \in \mathbb{C}^n$ these functions have a gradient vector, the normal vector to a hyperplane $H_{x,[\lambda:\mu]}$ through the point. In particular, at each point $x \in \mathbb{C}^n$ the pencil of level functions defines a pencil of hyperplanes $H_{x,[\lambda:\mu]}$. In a generalization of the method, we consider a pencil of hyperplanes in $\mathbb{P}^n_{\mathbb{C}}$ for each point $x \in \mathbb{P}^m_{\mathbb{C}}$, and a r-dimensional linear space space in $\mathbb{P}^n_{\mathbb{C}}$ for each point in a r-dimensional variety $\bar{Y} \subset \mathbb{P}^m_{\mathbb{C}}$.

Let $\bar{Y} \subset \mathbb{P}^m_{\mathbb{C}}$ be a r-dimensional projective variety. Let

$$\alpha : \bar{Y} \to \text{Grass}(n-r, (\mathbb{C}^{n+1})^*)$$

be a map that associates to each $x \in \bar{Y}$ an r-dimensional linear space $Y_x \subset \mathbb{P}^n_{\mathbb{C}}$ defined by $n-r$ linear forms, and let

$$\Lambda : \mathbb{P}^m_{\mathbb{C}} \to \text{Grass}(2, (\mathbb{C}^{n+1})^*)$$

be a map that associates to each point $x \in \mathbb{P}^m_{\mathbb{C}}$ a $n-2$-dimensional linear space Λ_x, that we identify with the pencil of hyperplanes in $\mathbb{P}^n_{\mathbb{C}}$ that contains Λ_x.

Consider the incidence

$$C_{\Lambda,\alpha}(\bar{Y}) = \{(x, H_x) \in \bar{Y} \times (\mathbb{P}^n_{\mathbb{C}})^* \mid Y_x \subset H_x, \Lambda_x \subset H_x\} \subset \bar{Y} \times (\mathbb{P}^n_{\mathbb{C}})^*,$$

and let

$$X_{\Lambda,\alpha}(\bar{Y}) = \pi_1(C_{\Lambda,\alpha}(\bar{Y})) \subset Y,$$

where $\pi_1 : Y \times (\mathbb{P}^n_{\mathbb{C}})^* \to \bar{Y}$ is the projection to the first factor.

DEFINITION 2.5. $X_{\Lambda,\alpha}(Y)$ is the *Lagrangian locus* of Λ with respect to α on Y.

EXAMPLE 2.6. Let $n = m$ and assume $\alpha : \bar{Y} \to G(n-r, (\mathbb{C}^{n+1})^*)$ is the Gauss map, so that $\alpha(x) = Y_x$ is the tangent space to $\bar{Y}$ at x. Assume that Λ is a constant map, a pencil of hyperplanes in $\mathbb{P}^n_{\mathbb{C}}$ independent of x. Then the Lagrangian locus $X_{\Lambda,\alpha}(\bar{Y})$ is the set of points $x \in \bar{Y}$ such that the tangent space Y_x is contained in some hyperplane H in the pencil Λ. The degree of the Lagrangian locus, if finite, counts with multiplicity the number of hyperplanes in Λ that are tangent to $\bar{Y}$. This is the degree of the dual variety $\bar{Y}^* \subset (\mathbb{P}^m_{\mathbb{C}})^*$ to $\bar{Y}$, the closure of the variety of hyperplanes tangent to $\bar{Y}$ at smooth points. Its degree is classically called the class of $\bar{Y}$.

EXAMPLE 2.7. Let $\bar{Y} \subset \mathbb{P}^m_{\mathbb{C}}$ be a r-dimensional projective variety.
Let

$$\alpha : Y \to \text{Grass}(n-r, (\mathbb{C}^{n+1})^*)$$

be defined by

$$x \mapsto \{\sum_{j=0}^{n} a_{ij} y_j = 0, \quad i = 1, \dots, n-r\},$$

where a_{ij} are forms in $\mathbb{C}[x_0, \dots, x_m]$ of degree d_i, and let Λ be the map

$$x \mapsto \{s(\sum_{j=0}^{n} b_j y_j) + t(\sum_{j=0}^{n} c_j y_j) = 0\}\}$$

that to each point $x \in \mathbb{P}^m_{\mathbb{C}}$ associates the pencil of hyperplanes generated by

$$\{\sum_{j=0}^{n} b_j y_j = 0\} \text{ and } \{\sum_{j=0}^{n} c_j y_j = 0\},$$

where b_i and c_i are forms in $\mathbb{C}[x_0,\dots,x_m]$ of degree e_1 and e_2, respectively. Then the matrix

$$M_{\Lambda,\alpha}(x) = \begin{pmatrix} b_0 & b_1 & \dots & b_n \\ c_0 & c_1 & \dots & c_n \\ a_{10} & a_{11} & \dots & a_{1n} \\ a_{20} & a_{21} & \dots & a_{2n} \\ \vdots & \vdots & \vdots & \vdots \\ a_{(n-r)0} & a_{(n-r)1} & \dots & a_{(n-r)n} \end{pmatrix}$$

of size $(n-r+2)\times(n+1)$ has rank at most $n-r+1$ in the Lagrangian locus

$$X_{\Lambda,\alpha}(Y) := D_{n-r+1}(M_{\Lambda,\alpha}) \cap Y.$$

The Lagrangian locus $X_{\Lambda,\alpha}(Y)$ is empty, or has codimension at most r in Y (see [**19**, Theorem 14.4], [**17**, Chapter 12]). If the codimension equals r, i.e., $X_{\Lambda,\alpha}(Y)$ is finite, then the degree of $X_{\Lambda,\alpha}(Y)$ is

$$S_r(d_1,\dots,d_{n-r},e_1,e_2)\cdot \deg Y.$$

2.2.2. *Critical locus: Euler characteristic.* A critical point p of a function f on a variety Y is a singular point on the variety $Y\cap\{f(x)=f(p)\}$, i.e., the divisor (codimension one subvariety) on Y defined by the level set of f through p. If Y is nonsingular and f is a sufficently general function, then general level set $\{f(x)=\lambda\}$ intersects Y in a nonsingular divisor $Y_{f,\lambda}=Y\cap\{f(x)=\lambda\}$. The topological Euler characteristic $\chi(Y_{f,\lambda})$ of $Y_{f,\lambda}$, whenever this is nonsingular, is independent of λ. It changes, however, when $Y_{f,\lambda}$ is singular. Also, the Euler characteristic of Y is related to the Euler characteristic of the $Y_{f,\lambda}$, since the level sets define a fibration of Y.

Let $\bar{Y}\subset\mathbb{P}^m_{\mathbb{C}}$ be a complex algebraic variety, and $\{F_\lambda=0\},\lambda\in\mathbb{P}^1_{\mathbb{C}}$ a pencil of hypersurfaces in $\mathbb{P}^m_{\mathbb{C}}$ that intersect $\bar{Y}$ in a pencil of divisors Y_λ. If the divisors have a common intersection on $\bar{Y}$, we consider the blowup $\tilde{Y}\to\bar{Y}$ in this locus. Then the strict transforms $\tilde{Y}_\lambda\subset\tilde{Y}$ are the fibers of a fibration $\pi:\tilde{Y}\to\mathbb{P}^1_{\mathbb{C}}$. Assume that there is a finite set $S\subset\mathbb{P}^1_{\mathbb{C}}$ of cardinality η, over which the fibers of π are singular. Then $\chi(\pi^{-1}(\mathbb{P}^1_{\mathbb{C}}\setminus S))=\chi(\tilde{Y}_{\lambda,\mathrm{ns}})\cdot\chi(\mathbb{P}^1_{\mathbb{C}}\setminus S)$ and

$$\chi(\tilde{Y})=(2-\eta)\chi(\tilde{Y}_{\lambda,\mathrm{ns}})+\sum_{\lambda'\in S}\chi(\tilde{Y}_{\lambda'})$$

where $\tilde{Y}_{\lambda,\mathrm{ns}}$ is nonsingular (see [**20**] and [**9**, Section 6]).

EXAMPLE 2.8. Let $\bar{Y}\subset\mathbb{P}^n_{\mathbb{C}}$ be a nonsingular curve of degree d and genus g, and let $f:\bar{Y}\to\mathbb{P}^1_{\mathbb{C}}$ be a general linear function. Then $\chi(Y_\lambda)=d$ for a general $\lambda\in\mathbb{P}^1_{\mathbb{C}}$ while $\chi(Y_\lambda)=d-1$ whenever Y_λ has a double point, i.e., f is ramified at some $y\in Y_\lambda$. Now $\chi(Y_\lambda)=2-2g$, so $2-2g=(2-\eta)d+\eta(d-1)$, i.e., $\eta=2d+2g-2$, which is the degree of the dual variety $Y^*\subset(\mathbb{P}^n_{\mathbb{C}})^*$, cf. Example 2.6.

3. Polynomial optimization

Let $Y\subset\mathbb{R}^n$ be the common zeros $V(f_1,\dots,f_s)$ of a set of polynomials f_1, $\dots$, f_s, and let g be a polynomial function of degree d on $\mathbb{R}^n$. Assume the restriction of g to Y is bounded. The set of points on Y at which g obtains its maximal value is a subset of the critical points $X\subset Y$ of g on Y. The set X is defined by polynomial equations. So as above, we may homogenize the equations and consider complex solutions. If the *Zariski closure*, $\bar{X} \subset \mathbb{P}^n_{\mathbb{C}}$, is finite, then the degree of $\bar{X}$, is *the algebraic degree* of the optimization problem.

The polynomial equations for $\bar{X}$ may be found using the method of Lagrange multipliers as in Section 2.2.1. Let $\bar{Y} \subset \mathbb{P}^n_{\mathbb{C}}$ be an r-dimensional variety, the Zariski closure $V(F_1, \ldots, F_s)$ of Y, and let

$$\alpha : \bar{Y} \to \mathrm{Grass}(n-r, (\mathbb{C}^{n+1})^*); \quad x \mapsto \mathrm{Jac}(F_1, \ldots, F_s)(x)$$

be the map that associates to each x the space of linear forms spanned by the gradients of the F_i at x. In coordinates, these are the row vectors of the Jacobian matrix. Let G be the homogenization of $g(x)$, and let

$$\Lambda : \mathbb{P}^n_{\mathbb{C}} \to \mathrm{Grass}(2, (\mathbb{C}^{n+1})^*); \quad x \mapsto \mathrm{Jac}(G, x_0^d).$$

Then the Zariski closure of the critical locus of g on Y is the Lagrangian locus $X_{\Lambda,\alpha}(\bar{Y})$ of Λ with respect to α on $\bar{Y}$.

In applications, $\bar{Y}$ does not always have a well defined tangent space at every point, i.e., the rank of the Jacobian matrix itself varies over $\bar{Y}$, or the Lagrangian locus of Λ is not finite. These are typical obstacles to find a degree formula.

We will see two cases: The first case is when the rank of the Jacobian matrix does not vary over $\bar{Y}$, i.e. when Y is nonsingular, and Λ is sufficiently general. Then the algebraic degree is an invariant of the dual variety to $\bar{Y}$. The second case are examples where the Jacobian matrix drops rank in a subvariety of $\bar{Y}$ while a formula for the degree still has been obtained.

We start with a standard example of a model and continue with some important optimization problems.

3.1. Complete intersection. Assume $Y \subset \mathbb{R}^n$ has dimension $n-r$ and is the common zeros of r polynomials $g_1, \ldots, g_r$ of degrees $d_1, .., d_r$, and let $f(x_1, \ldots, x_n)$ be a polynomial of degree d_0. The polynomial optimization problem is to find the point $p = (a_1, \ldots, a_n) \in Y$ where f attains its maximal value. The point p is a critical point for $f|_Y : Y \to \mathbb{R}$, i.e., the gradient of $f - f(p)$ at the point p lies in the span of the gradients of the g_i.

Homogenizing, we set

$$G_i(x_0, \ldots, x_n) = x_0^{d_i} g_i(x_1/x_0, \ldots, x_n/x_0),$$

$$F(x_0, \ldots, x_n) = x_0^{d_0} f(x_1/x_0, \ldots, x_n/x_0),$$

$$\text{and} \quad F_\lambda(x_0, \ldots, x_n) = F(x_0, \ldots, x_n) - \lambda x_0^{d_0}.$$

Then $\bar{Y} \subset \mathbb{P}^n_{\mathbb{C}}$ has dimension $n-r$ and is the common zeros of the r homogeneous forms $G_1, \ldots, G_r$ of degrees $d_1, .., d_r$, and $\bar{p} = (1 : a_1 : \cdots : a_n) \in \bar{Y}$. Let $\lambda = f(p)$. Then also $F_\lambda(\bar{p}) = 0$, and

$$\Big(\frac{\partial}{\partial x_1} F_\lambda(\bar{p}), \ldots, \frac{\partial}{\partial x_n} F_\lambda(\bar{p})\Big) = \Big(\frac{\partial}{\partial x_1} F(\bar{p}), \ldots, \frac{\partial}{\partial x_n} F(\bar{p})\Big).$$

In this case we may therefore simplify the computation in the method of Lagrange multipliers. The Zariski closure $\bar{X} \subset \bar{Y}$ of the critical locus of g on Y is the locus of points $y \in \bar{Y}$ such that the gradient $(\frac{\partial}{\partial x_1} F(y), \ldots, \frac{\partial}{\partial x_n} F(y))$ lies in the span of the gradients

$$\Big(\frac{\partial}{\partial x_1} G_i(y), \ldots, \frac{\partial}{\partial x_n} G_i(y)\Big), i = 1, \ldots, r.$$

PROPOSITION 3.1. *For a general set of polynomials* $g_1, .., g_r$ *and* f, *the set* $\bar{X}$ *is finite with cardinality*

$$\deg \bar{X} = d_1 d_2 \cdots d_r \sum_{i_0+i_1+\cdots+i_r=n-r} (d_0-1)^{i_0}\cdots(d_r-1)^{i_r}.$$

PROOF. The Zariski closure $\bar{X}$ of the critical locus is the set of points $y \in \bar{Y}$ such that the $(r+1 \times n)$-matrix

$$M(y) = \begin{pmatrix} \frac{\partial}{\partial x_1}F(y) & \frac{\partial}{\partial x_2}F(y) & \dots & \frac{\partial}{\partial x_n}F(y) \\ \frac{\partial}{\partial x_1}G_1(y) & \frac{\partial}{\partial x_2}G_1(y) & \dots & \frac{\partial}{\partial x_n}G_1(y) \\ \vdots & \vdots & & \vdots \\ \frac{\partial}{\partial x_1}G_r(y) & \frac{\partial}{\partial x_2}G_r(y) & \dots & \frac{\partial}{\partial x_n}G_r(y) \end{pmatrix}$$

has rank at most r. Now, $\bar{Y}$ has degree $d_1 d_2 \cdots d_r$ and dimension $n-r$, while the locus $D_r(M) \subset \mathbb{P}^n_{\mathbb{C}}$ where $M(x)$ has rank at most r has degree $S_{n-r}(d_0-1, d_1-1, \dots, d_r-1)$ and dimension r by Theorem 2.3, so the proposition follows from the Bertini and Bezout theorems (see [**17**, Theorem 0.3, Corollary 2.4]). □

We turn to some important applications.

3.2. Euclidean distance. Let $u = (u_1, \dots, u_n) \in \mathbb{R}^n$. Then $f_u(x) = (x_1 - u_1)^2 + \cdots + (x_n - u_n)^2$ is the squared Euclidean distance from u to $x = (x_1, \dots, x_n)$. Let $Y \in \mathbb{R}^n$ be an r-dimensional variety. If $y \in Y$ is a point on Y with minimal distance to u, then y is a critical point for f_u. Let

$$\alpha : \bar{Y} \to \mathrm{Grass}(n-r, (\mathbb{C}^{n+1})^*)$$

be the Gauss map, that associates to $y \in \bar{Y}$ the space of linear forms defining the tangent space to Y at y. Homogenizing, as above, we get a pencil of homogenous forms

$$F_\lambda(x) = F_u(x) - \lambda x_0^2 = (x_1 - u_1 x_0)^2 + \cdots + (x_n - u_n x_0)^2 - \lambda x_0^2.$$

Let

$$\Lambda_u : \mathbb{P}^n_{\mathbb{C}} \to \mathrm{Grass}(2, (\mathbb{C}^{n+1})^*); \quad y \mapsto \mathrm{Jac}(F_u, x_0^2).$$

Then the Lagrangian locus $X_{\Lambda_u,\alpha}(\bar{Y})$ is the Zariski closure of the critical locus of f on Y. The cardinality of $X_{\Lambda_u,\alpha}(\bar{Y})$ is the same for an open set of points $u \in \mathbb{P}^n_{\mathbb{C}}$.

DEFINITION 3.2. The Euclidean distance degree of Y, denoted EDdegree(Y), is the cardinality of $X_{\Lambda_u,\alpha}(\bar{Y})$ for a general $u \in \mathbb{P}^n_{\mathbb{C}}$.

When Y is a complete intersection defined by polynomials $(g_1, \dots, g_r)$ of degrees $d_1, \dots, d_r$, then, by the above formula, the Euclidean distance degree is

$$\begin{aligned} EDdegree(Y) &= d_1 d_2 \cdots d_r S_{n-r}(1, d_1 - 1, \dots, d_r - 1) \\ &= d_1 d_2 \cdots d_r \sum_{i_1+\cdots+i_r \le n-r} (d_1-1)^{i_1}\cdots(d_r-1)^{i_r}. \end{aligned}$$

When Y is defined by one polynomial equation $g(x) = 0$ of degree d, then this formula reduces to

$$EDdegree(Y) = dS_{n-1}(1, d-1) = d \sum_{i \le n-1} (d-1)^i = d\frac{(d-1)^n - 1}{d-2}.$$

The Euclidean distance degree has been computed for many classes of varieties; see [**1**, **8**, **16**, **35**, **41**].

3.3. Maximal likelihood for discrete models. Let $\Delta_n \subset \mathbb{R}^{n+1}$ be the simplex of points with positive coordinates summing to one. An algebraic subset $Y_0 \subset \Delta_n$ defined by homogeneous forms $F_1, \ldots, F_r \in \mathbb{C}[x_0, \ldots, x_n]$ of degrees $\deg F_i = d_i$, may be seen as a statistical model for discrete data, with $u \in \Delta_n$ a data point. The maximal likelihood estimate is a maximal point for the likelihood function

$$l_u(x) = \frac{x_0^{u_0} x_1^{u_1} \cdots x_n^{u_n}}{(x_0 + x_1 + \cdots + x_n)^{(u_0+u_1+\cdots+u_n)}}.$$

The critical points of this function is the *likelihood locus* on Y_0 with respect to u. Since the logarithm of l_u is well defined on Δ_n, the critical points of l_u and of $Log(l_u)$ on Y_0 coincide. The gradient of $Log(l_u)$ is the vector

$$\left(\frac{u_0}{x_0} - \frac{u_0 + u_1 + \cdots + u_n}{x_0 + x_1 + \cdots + x_n}, \ldots, \frac{u_n}{x_n} - \frac{u_0 + u_1 + \cdots + u_n}{x_0 + x_1 + \cdots + x_n}\right) = \left(\frac{u_0}{x_0} - \lambda, \ldots, \frac{u_n}{x_n} - \lambda\right)$$

with $\lambda = \frac{u_0+u_1+\cdots+u_n}{x_0+x_1+\cdots+x_n}$. Let $\bar{Y} \subset \mathbb{P}^n_{\mathbb{C}}$ be the Zariski closure of Y_0, and consider the critical locus of l_u on $\bar{Y}$. It includes the critical locus of l_u on Y_0 and can be found by the method of Lagrangian multipliers. Let

$$\mathcal{H} = \{x_0 \cdot x_1 \cdots x_n \cdot (x_0 + x_1 + \cdots + x_n) = 0\} \subset \mathbb{P}^n_{\mathbb{C}}.$$

The critical locus $\mathrm{ML}_u(\bar{Y})$ is the set of points $x \in \bar{Y} \setminus \mathcal{H}$ such that a vector $(\frac{u_0}{x_0} - \lambda, \ldots, \frac{u_n}{x_n} - \lambda)$ is in the linear span of the gradients of the F_i at x, for some $\lambda \in \mathbb{C}$.

When $\bar{Y}$ is nonsingular, the critical locus $\mathrm{ML}_u(\bar{Y})$ for a general point u is a finite set of cardinality independent of u.

DEFINITION 3.3. The *maximal likelihood degree* of Y_0, denoted $\mathrm{MLdegree}(Y_0)$, is the cardinality of $\mathrm{ML}_u(\bar{Y})$, for a general u, whenever it is finite.

When $\bar{Y}$ is a complete intersection, the set $\mathrm{ML}_u(\bar{Y})$ is the locus of points on $Y \setminus \mathcal{H}$, where the matrix

$$M_{\bar{Y},u}(x) = \begin{pmatrix} u_0/x_0 & u_1/x_1 & \ldots & u_n/x_n \\ 1 & 1 & \ldots & 1 \\ \partial F_1/\partial x_0 & \partial F_1/\partial x_1 & \ldots & \partial F_1/\partial x_n \\ \partial F_2/\partial x_0 & \partial F_2/\partial x_1 & \ldots & \partial F_2/\partial x_n \\ \vdots & \vdots & \vdots & \vdots \\ \partial F_r/\partial x_0 & \partial F_r/\partial x_1 & \ldots & \partial F_r/\partial x_n \end{pmatrix}$$

of size $(r+2) \times (n+1)$ has rank at most $n - r + 1$.

By multiplying column i in the matrix $M_{\bar{Y},u}(x)$ by x_i, we get

$$M'_{\bar{Y},u}(x) = \begin{pmatrix} u_0 & u_1 & \ldots & u_n \\ x_0 & x_1 & \ldots & x_n \\ x_0\partial F_1/\partial x_0 & x_1\partial F_1/\partial x_1 & \ldots & x_n\partial F_1/\partial x_n \\ x_0\partial F_2/\partial x_0 & x_1\partial F_2/\partial x_1 & \ldots & x_n\partial F_2/\partial x_n \\ \vdots & \vdots & \vdots & \vdots \\ x_0\partial F_r/\partial x_0 & x_1\partial F_r/\partial x_1 & \ldots & x_n\partial F_r/\partial x_n \end{pmatrix}$$

whose rank does not change outside $\mathcal{H}$. Notice that $M'_{\bar{Y},u}(x)$ has a well-defined rank at any point in $\mathbb{P}^n_{\mathbb{C}}$. Furthermore, for general F_i, it has maximal rank $n - r + 2$

on $\mathcal{H}$. So the set $\mathrm{ML}_u(\bar{Y})$ coincides with the locus $D_{r+1}(M'_{\bar{Y},u})$ and MLdegree(Y_0) coincides with the degree of this locus:

$$\mathrm{MLdegree}(Y_0) = d_1 \cdots d_r S_{r+1}(d_1, \ldots, d_r, 1, 0) = d_1 d_2 \cdots d_r \sum_{i_1+\cdots+i_r \leq n-r} d_1^{i_1} \cdots d_r^{i_r}.$$

3.4. Maximal likelihood, Gaussian distribution. The parameters for a multivariate Gaussian distribution on $\mathbb{R}^n$ are a vector $\mu \in \mathbb{R}^n$, the mean, and a positive definite $(n \times n)$- matrix Σ, the covariance. Its density function is

$$f_{\mu,\Sigma}(x) = \frac{(e^{-\frac{1}{2}(x-\mu)^T \Sigma^{-1}(x-\mu)}}{\sqrt{(2\pi)^n \det \Sigma)}}.$$

The interest is in a statistical model where Σ^{-1} belongs to some m-dimensional linear family $\mathcal{U}$ of symmetric $(n \times n)$-matrices. So Σ belongs to the family $\mathcal{U}^{-1}$ of inverses of matrices in $\mathcal{U}$.

Given sample vectors $x_1, .., x_s \in \mathbb{R}^n$ one estimates the mean μ as the mean of the sample vectors, and a sample covariance matrix S as the mean of the matrices $\Sigma_i = (x_i - \mu)(x_i - \mu)^T$. The maximal likelihood estimate is then the positive definite matrix Σ that maximizes the function

$$f_{\mu,\Sigma}(x_1) \cdots f_{\mu,\Sigma}(x_s).$$

Under mild generality assumptions, this maximum is the unique positive definite matrix $\Sigma \in \mathcal{U}^{-1}$ such that S is the image of Σ under the projection from $\mathcal{U}^\perp$. Here $\mathcal{U}^\perp$ is the orthogonal subspace to $\mathcal{U}$ under the *trace inner product*

$$\langle \Sigma_1, \Sigma_2 \rangle = \mathrm{trace}(\Sigma_1 \cdot \Sigma_2).$$

The ML-degree of the linear model $\mathcal{U}$ is then the degree of the projection map from $\mathcal{U}^\perp$ restricted to $\mathcal{U}^{-1}$. It is found by considering the Zariski closure

$$\bar{\mathcal{U}}^{-1} \subset \mathbb{P}_{\mathbb{C}}^{\binom{n+1}{2}-1}$$

and the projection center

$$\mathbb{P}(\mathcal{U}^\perp) \subset \mathbb{P}_{\mathbb{C}}^{\binom{n+1}{2}-1}.$$

If $\bar{\mathcal{U}}^{-1}$ does not intersect the projection center, then the projection is finite of degree equal to the ML-degree of $\mathcal{U}$.

This ML-degree of a linear model $\mathcal{U}$ is closely related to the algebraic degree in semidefinite programming (see below) with respect to a d-dimensional linear family $\mathcal{U}$ of symmetric matrices.

The ML-degree has been computed for a number of different statistical models, far beyond the ones described above. Further examples may be found in [**2, 7, 18, 24, 27, 42**]. Behind the computation of ML-degree is an appreciation for likelihood geometry, also promoted by Bernd Sturmfels. See [**3**] for an insightful presentation.

3.5. Semidefinite programming (SDP). Semidefinite programming concerns optimization in the space of positive semidefinite symmetric matrices. We consider a m-dimensional affine space $\mathcal{U}$ of symmetric $(n \times n)$-matrices, and the subset $\mathcal{U}_+$ of positive semidefinite matrices.

The semidefinite programming (SDP) problem is, given a symmetric matrix B, to maximize the linear function

$$f_B(X) = \text{trace}(B \cdot X)$$

on $\mathcal{U}_+$.

The set $\mathcal{U}_+$ is convex, so an optimal solution $\hat{X}$ is a matrix on the boundary of $\mathcal{U}_+$, i.e., a matrix of rank $r < n$. If $\mathcal{U}_r \subset \mathcal{U}$ is the locus of matrices rank $\leq r$, then the linear function $f_B(X)$ restricted to $\mathcal{U}_r$ has a critical point at $\hat{X}$. The algebraic degree in SDP for a general m-dimensional subspace $\mathcal{U}$ of symmetric $(n\times n)$-matrices at rank r is denoted $\delta(m, n, r)$. It is the cardinality of the Zariski closure of the critical locus of $f_B(X)$ restricted to $\mathcal{U}_r$. It coincides with the degree of the dual variety of the Zariski closure $\bar{\mathcal{U}}_r$, whenever this dual variety is a hypersurface (cf. [**38**, Theorem 13]).

Notice that the formulas of algebraic degree depend on genericity assumptions in many of the cases of polynomial optimization discussed in this section.

In the Lagrange multiplier cases, the map α is assumed to be regular on the variety $\bar{Y}$. Often the set of points $x \in \mathbb{P}^m_{\mathbb{C}}$ where the map α is regular is not compact. This leads to problems of residual or excess intersection, cf. Section 2.1.4.

In the examples of discrete and Gaussian statistical models similar problems are even more explicit. In the case of discrete models the Zariski closure $\bar{Y}$ intersects the locus $\mathcal{H}$, leading to possible false solutions in computations of critical loci. In the case of Gaussian models and in semidefinite programming, the Zariski closure $\bar{\mathcal{U}}_r$ intersects the locus of matrices of rank $< r$, which again may lead to problems of residual or excess intersection.

In the next section will both analyze and in some cases find solutions to these problems when interpreted by Cremona transformations.

4. Cremona transformations and more degree formulas

In a mathematical model, the choice of coordinates or parameters are crucial, both in the analysis of the model, and in accessing effective computations. An algebraic choice of coordinates is a homomorphism

$$\varphi_{\text{aff}} : \mathbb{C}[x_1, \ldots, x_m] \to \mathbb{C}(x_1, \ldots, x_m) \quad x_i \mapsto f_i(x_1, \ldots, x_m)$$

where the f_i are rational functions $f_i = g_i/h_i$, with $g_i, h_i \in \mathbb{C}[x_1, \ldots, x_m]$. Let h_0 be the least common multiple of the h_i, and let d be the largest degree of the polynomials $h_0 f_i$. Then the above homomorphism extends in the projective context to a homomorphism

$$\varphi : \mathbb{C}[x_0, \ldots, x_m] \to \mathbb{C}[x_0, \ldots, x_m] \quad x \mapsto F_i(x_0, \ldots, x_m)$$

where $F_i = x_0^d h_0(x_i/x_0) f_i(x_i/x_0)$.

A reasonable requirement for φ_{aff} to be called a choice of coordinates is that the values $f_1(a), \ldots, f_n(a)$ distinguish the points $a \in \mathbb{C}^n$, i.e., that the map

$$\varphi_{\text{aff}} : \mathbb{C}^n \to \mathbb{C}^n; \quad a \mapsto (f_1(a), \ldots, f_n(a))$$

is injective on some open dense set that contains the set of points $a \in \mathbb{C}^n$ that are relevant for our model. The homomorphism φ is called *a Cremona transformation* if φ_{aff} is injective on some open dense set.

Choosing the forms F_i as coordinates on a variety $Y \subset \mathbb{P}^n_{\mathbb{C}}$ may be interpreted as viewing the image

$$\varphi(Y) \subset \mathbb{P}^n_{\mathbb{C}}$$

with standard coordinates. When the forms F_i are not linear, then the degree computations in Y and on $\varphi(Y)$ may differ.

In fact, let $Z(\varphi) \subset \mathbb{P}^n_{\mathbb{C}}$ be the set of common zeros of the forms F_i. This set $Z(\varphi)$ is called the base locus or fundamental locus of the Cremona transformation. Let $G_j(x_0, \dots, x_m), j = 1, \dots, r$ be a collection of homogeneous forms in the x_i. Then

$$Z(\varphi) \subset \cap_{j=1}^r \{G_j(F_0, \dots, F_m) = 0\}.$$

Let us assume that the common zero locus of the G_j is a variety W. When $Z := Z(\varphi) \cap Y \neq \emptyset$, the degrees of $\varphi^{-1}(W) \cdot Y$ in coordinates F_i and $W \cdot \varphi(Y)$ in coordinates x_i differ.

The general procedure to compute the difference is to modify Y by blowing up the locus Z of common zeros of the F_i on Y. The blow up is a variety $\tilde{Y}$ and a map $\pi : \tilde{Y} \to Y$ such that $\pi^{-1}(Z) = E_Z \subset \tilde{Y}$ is a divisor while π is an isomorphism outside the preimage of Z. Since E_Z is a divisor, it is defined locally by a single form e_Z. The F_i are forms on $\tilde{Y}$ via π, that vanishes on E_Z. If the general form in $\langle F_0, \dots, F_m \rangle$ has multiplicity m along E_Z, then the rational functions $\tilde{F}_i = F_i / e_Z^m$ do not vanish on E_Z. So we replace Y by $\tilde{Y}$ and the forms F_i by the rational functions $\tilde{F}_i$ to avoid the zero locus Z in the degree computations. If the functions $\tilde{F}_i$ have common zeros on $\tilde{Y}$, the modification continues by further blowups. It is a famous result by Hironaka [**26**], that the procedure ends with rational functions without common zeros on a modified Y after a finite number of blowups.

Computing common zeros of some of the forms $\tilde{F}_i$ on $\tilde{Y}$ depends clearly on the set Z. Although modern intersection theory may give a general recipe, explicit computations depend very much on Y and the zero locus Z. The general recipe typically applies to the case when Z and Y are both nonsingular projective varieties. When Y or Z or both are singular, ad hoc methods are often necessary.

We consider two important examples, the classical Cremona transformation and the symmetric Cremona transformation, and show how they appear naturally in modeling.

4.1. Classical Cremona. The affine map

$$Cr_{\text{aff}} : (x_1, \dots, x_n) \mapsto (1/x_1, \dots, 1/x_n)$$

is injective off the union of coordinate hyperplanes. It extends to a map

$$Cr : \mathbb{P}^n \dashrightarrow \mathbb{P}^n; \quad a \mapsto \phi_i(a),$$

where $\phi_i = (x_0 x_1 \cdots x_n / x_i)$ are forms of degree n.

A slight variation of this map is used in discrete statistical models. Fix $u = (u_1, \dots, u_n) \in (\mathbb{C}^*)^n$ with

$$Cr_u : (x_1, \dots, x_n) \mapsto (u_1/x_1, \dots, u_n/x_n), \quad u_i \in \mathbb{C}^*.$$

On the one hand the map Cr_u is certainly bijective off all the coordinate hyperplanes, but on the other hand it is not defined along the intersections of two or more coordinate hyperplanes.

EXAMPLE 4.1 ([**15**, Theorem 7.2.4]). A k-dimensional linear space $L \subset \mathbb{P}^n$ that intersects the intersection of any r coordinate hyperplanes in dimension $k - r$ for any $r = 1, \ldots, k$, is mapped by Cr_u to a k-dimensional variety $Cr_u(L) \subset \mathbb{P}^n$ of degree $\binom{n}{k}$.

EXAMPLE 4.2. The computation of the ML-degree for the discrete statistical model in example 3.3 involves a Cremona transformation: Let $Y_0 \subset \Delta_n \subset \mathbb{R}^{n+1}$ be an r-dimensional nonsingular variety with Zariski closure $\bar{Y} \subset \mathbb{P}^n_{\mathbb{C}}$. Consider the likelihood function

$$\varphi_u = \Pi_{i=0}^n x_i^{u_i}$$

with $u = (u_0, \ldots, u_n) \in \mathbb{Z}^n$. The Zariski closure $\mathrm{ML}_u(\bar{Y})$ in $\bar{Y}$ of the likelihood locus in Y_0 is a Lagrangian locus: Let

$$\alpha : \bar{Y} \to \mathrm{Grass}(n - r, (\mathbb{C}^{n+1})^*)$$

be the Gauss map, and let

$$\Lambda_u : U \to \mathrm{Grass}(2, n+1), \quad x \mapsto (Cr_u(x), e),$$

where $e = (1, \ldots, 1) \in \mathbb{C}^{n+1}$. Then $\mathrm{ML}_u(\bar{Y})$
is the Lagrangian locus $X_{\Lambda_u, \alpha}(\bar{Y})$ on $\bar{Y}$ of Λ with respect to α.

If $\bar{Y}$ is a complete intersection $V(F_1, \ldots, F_r)$ of forms of degree $\deg F_i = d_i$, then this Lagrangian locus is the locus where the following matrix has rank at most $n - r + 1$:

$$M_{\bar{Y},u}(x) = \begin{pmatrix} u_0/x_0 & u_1/x_1 & \ldots & u_n/x_n \\ 1 & 1 & \ldots & 1 \\ \partial F_1/\partial x_0 & \partial F_1/\partial x_1 & \ldots & \partial F_1/\partial x_n \\ \partial F_2/\partial x_0 & \partial F_2/\partial x_1 & \ldots & \partial F_2/\partial x_n \\ \vdots & \vdots & \vdots & \vdots \\ \partial F_r/\partial x_0 & \partial F_r/\partial x_1 & \ldots & \partial F_r/\partial x_n \end{pmatrix}.$$

It is not defined on coordinate hyperplanes, but off these hyperplanes the rank is the same after multiplying each column i by x_i, after which the degree of the Lagrangian locus and hence the ML-degree was computed in the Example 3.3 above. Of course the degree makes sense only if the Lagrangian locus does not intersect the coordinate hyperplanes.

EXAMPLE 4.3. (Complex hyperplane arrangements, cf. [**27, 40, 45**]) We have seen in §2.2.2 how critical loci of a function may be modeled with tangencies in a linear one-parameter family, or pencil, of hypersurfaces. In fact, the critical locus may sometimes be computed by comparing the Euler characteristic of Y with the Euler characteristic of hypersurface sections in a general pencil. It requires, however the computation of the Euler characteristic of all hypersurface sections in the pencil. When a Cremona transformation is used, an obstacle is often the base locus.

A very interesting example where this is avoided is the case of very affine varieties. A very affine variety is isomorphic to a variety in $(\mathbb{C}^*)^n$. Huh, following arguments of Orlik and Terao, applies the Euler characteristic method to get a formula for the ML-degree of a variety that lies entirely in the complement of a finite number of hypersurfaces.

Let $Y \subset \mathbb{R}^m$ be a nonsingular very affine variety with an embedding

$$f : Y \to (\mathbb{C}^*)^{n+1}; \quad y \mapsto (f_0(y), \ldots, f_n(y)).$$

Consider the function

$$\varphi_u = \Pi_{i=0}^n f_i^{u_i},$$

with $u = (u_0, \ldots, u_n) \in \mathbb{Z}^{n+1}$. We call φ_u the master function for f on Y.

In a discrete statistical model with $n+1$ observables, the f_i is the probability of the ith event, while the data comes as a set of positive integers where u_i is the number of times the ith event was observed. The maximal likelihood degree MLdegree(Y) of Y_0 is the number of critical points of the master function φ_u on Y_0 for general $u = (u_0, \ldots, u_n)$.

The following result, when the f_i are linear, was first conjectured by Varchenko and proved by Varchenko and by Orlik and Terao; see [**40, 45**].

Huh proved it in this more general form.

THEOREM 4.4 ([**27**, Theorem 1]). *Let Y be an r-dimensional affine variety with an embedding $(f_1, \ldots, f_n) : Y \to (\mathbb{C}^*)^n$, then*

$$\mathrm{MLdegree}(Y) = (-1)^r \chi(Y)$$

where $\chi(Y)$ is the signed Euler characteristic.

Notice how the MLdegree of Y does not depend on the embedding!

4.2. Symmetric Cremona. Let S^n be the $N = \binom{n+1}{2}$-dimensional space of symmetric $(n \times n)$-matrices. Then

$$\sigma : S^n \to S^n; \quad M \mapsto M^{-1}$$

is a Cremona transformation used in Gaussian statistical models.

The map σ is certainly bijective on the set of nonsingular matrices M, and it is homogeneous, so it naturally defines a Cremona transformation

$$\sigma : \mathbb{P}(S^n) \dashrightarrow \mathbb{P}(S^n); \quad M \mapsto \mathrm{Adj}(M)$$

where $\mathrm{Adj}(M)$ is the adjoint matrix; i.e., $\mathrm{Adj}(M) = det(M) \cdot M^{-1}$.

Recall the trace inner product on S^n:

$$\langle M, N \rangle = \mathrm{trace}(M \cdot N).$$

Let $L \subset S^n$ be a linear space. Then

$$\langle -, N \rangle : L \to \mathbb{C}$$

is a linear form on L for any $N \in S^n$. Let $L^\perp = \{N \mid \langle -, N \rangle_L = 0\}$. Then the dual space L^* is naturally isomorphic to the quotient $S^n/L^\perp$.

Now, let $\mathbb{P}(L) \subset \mathbb{P}(S^n)$ be a d-dimensional linear subspace. For a Gaussian statistical model of distributions on $\mathbb{R}^n$, the subspace $\mathbb{P}(L) \subset \mathbb{P}(S^n)$ is the complexified space of covariance matrices of the distributions, while $\sigma(\mathbb{P}(L))$ is the corresponding variety of concentration matrices.

The ML-degree of the Gaussian statistical model is the degree of the map

$$\pi_{L^\perp} : \sigma(\mathbb{P}(L)) \dashrightarrow \mathbb{P}(L^*),$$

where the $\mathbb{P}(L^\perp) \subset \mathbb{P}(S^n)$ is a subspace of the concentration matrices and $L^* = S^n/L^\perp$, whenever this map is finite. In particular, if

$$\mathbb{P}(L^\perp) \cap \sigma(\mathbb{P}(L)) = \emptyset,$$

then the ML-degree equals the degree of $\sigma(\mathbb{P}(L))$.

Let $D^r_{S^n}$ be the rank $\leq r$ locus

$$D^r_{S^n} = \{[M] \mid \mathrm{rank}\, M \leq r\} \subset \mathbb{P}(S^n).$$

It is a variety of codimension $\binom{n-r+1}{2}$ and degree

$$\deg(D^r_{S^n}) = \prod_{j=0}^{n-r-1} \frac{\binom{n+j}{n-r-j}}{\binom{2j+1}{j}}.$$

Let $\mathbb{P}(L) \subset \mathbb{P}(S^n)$ be a d-dimensional linear subspace that intersects the rank r-locus in codimension $\binom{n-r+1}{2}$ for each r. Then we set

$$\phi(n,d) := \text{degree}(\sigma(\mathbb{P}(L))). \tag{4.1}$$

Let $Y_d^{r,n} = \mathbb{P}(L) \cap D^r_{S^n}$. For each $r < n$, the variety $Y_d^{n,r}$ is a proper subvariety of $\mathbb{P}(L)$ and has a dual variety in $\mathbb{P}(L^*)$. Whenever this dual variety is a hypersurface we denote its degree:

$$\delta(d,n,r) := \text{degree}(Y_d^{n,r})^*$$

Recall that the degree of the dual of $Y_d^{n,r}$ maybe interpreted as the algebraic degree in semidefinite programming; see Section 3.5.

Consider, in the product

$$\mathbb{P}(S^2(V)) \times \mathbb{P}(S^2(V \wedge V)) \cdots \times \mathbb{P}(S^2(\wedge^{n-1} V)),$$

the closure CQ of the graph of the map

$$M \mapsto (M, \wedge^2 M, \ldots, \wedge^{n-1} M)$$

defined on invertible matrices M. Then CQ is called the variety of complete quadrics. The closure in $\mathbb{P}(S^2(V)) \times \mathbb{P}(S^2(\wedge^{n-1} V)) = \mathbb{P}(S^2 V) \times \mathbb{P}(S^2 V^*)$ of the map σ is the image of CQ by the projection to the first and last factor.

EXAMPLE 4.5. Enumerative problems in the space of nonsingular quadrics are often best handled in the variety of complete quadrics. An example is the problem of finding the cardinality of the set of nonsingular quadrics in $\mathbb{P}^n$ that pass through $\binom{n+1}{2} - d$ general points and are tangent to $d-1$ general hyperplanes. This number in fact coincides with the number $\phi(n,d)$ defined in (4.1) (see [**34**]).

A relation between the degree $\phi(n,d)$ and the degrees $\delta(d,n,r)$ now follows from a fundamental relation between divisors in this space of complete quadrics.

PROPOSITION 4.6. [**34**, Cor. 3.6]

$$\phi(n,d) = \frac{1}{n} \sum_{\binom{s+1}{2} < d} s \cdot \delta(d,n,n-s).$$

The formula for the degree of $\delta(d,n,r)$ has been given by the Lascoux-coefficients.

These coefficients depend on strict partitions of nonnegative integers. A sequence $I = (i_1 < i_2 < \cdots < i_r)) \subset \{0, \ldots, n-1\}$ is called a strict partition of $[n] := \{0, \ldots, n-1\}$ of cardinality r, and $\sum I := i_1 + \cdots + i_r$ denotes the sum of its elements. Consider the Pascal triangle extended by zeroes in each row,

$$PT = \begin{pmatrix} 1 & 0 & 0 & 0 & 0 & 0 & 0\ldots \\ 1 & 1 & 0 & 0 & 0 & 0 & 0\ldots \\ 1 & 2 & 1 & 0 & 0 & 0 & 0\ldots \\ 1 & 3 & 3 & 1 & 0 & 0 & 0\ldots \\ 1 & 4 & 6 & 4 & 1 & 0 & 0\ldots \\ 1 & 5 & 10 & 10 & 5 & 1 & 0\ldots \\ \vdots & \vdots & \vdots & \vdots & \vdots & \vdots & \vdots \end{pmatrix}.$$

The Lascoux coefficient ψ_I is the sum of the $(r \times r)$-minors of the submatrix PT_I of PT consisting of the rows $i_1+1, \ldots, i_r+1$. Alternatively, these coefficients may be defined as follows:

Let

$$\psi_i = 2^i, \quad \psi_{i,j} = \sum_{k=i+1}^{j} \binom{i+j}{k} \quad \text{when} \quad i < j,$$

and

$$\psi_{i_1,\ldots,i_r} = \mathrm{Pf}(\psi_{i_k,i_l})_{0<k<l\leq r} \quad \text{if } r \text{ is even},$$

$$\psi_{i_1,\ldots,i_r} = \mathrm{Pf}(\psi_{i_k,i_l})_{0\leq k<l\leq r} \quad \text{if } r \text{ is odd},$$

where $\psi_{i_0,i_k} := \psi_{i_k}$ and Pf denotes the Pfaffian (i.e., the square root of the determinant) of a skew-symmetric matrix of even size.

Consider also the Schur Q-function $Q_I(x_1,\ldots,x_n)$, where I is a strict partition of $[n]$ of cardinality $s \leq n$ (cf. [**32**, §III.8]). $Q_I(x_1,\ldots,x_n)$ is the coefficient of $t_1^{i_1}\cdots t_s^{i_s}$ in

$$Q(t_1,\ldots,t_s) = \prod_{i=1}^{n}\left(\frac{1+t_ix_i}{1-t_ix_i}\right)\prod_{i<j}\left(\frac{1-t_j/t_i}{1+t_j/t_i}\right).$$

The formula for $\delta(m,n,r)$ is expressed in terms of Lascoux coefficients ψ_I and values of Schur Q-functions Q_{I+1_s}. We set $b_I(n) = Q_{I+1_s}(\frac{1}{2},\frac{1}{2},\ldots,\frac{1}{2})$, where $\frac{1}{2}$ appears n times in the argument. Manivel, Michałek, Monin, Seynnaeve, and Vodǐcka prove the following formula.

THEOREM 4.7 ([**34**, Theorem 1.6]). *If $\binom{s+1}{2} \leq m < \binom{s+2}{2}$ then*

$$\delta(m,n,n-s) = \sum_{\sum I\leq m-s} (-1)^{m-s-\sum I}\cdot\psi_I\cdot b_I(n)\cdot\binom{m-1}{m-s-\sum I},$$

where the sum is over all strict partitions I of $[n]$ of cardinality s.

Here $m-\binom{s+1}{2}$ is the dimension of $Y_m^{n-s,n}$, and the formula was first proved in the case when $Y_m^{n-s,n}$ is smooth, by Nie, Ranestad, and Sturmfels in [**38**]. A general formula was first given in [**23**].

THEOREM 4.8 ([**23**, Theorem 1.1]). *Let $\binom{n-r+1}{2} \leq m \leq \binom{n+1}{2}-\binom{r+1}{2}$. Then*

$$\delta(m,n,r) = \sum_I \psi_I\psi_{I^c},$$

where the sum runs over all strict partitions I of $[n]$ with $\sum I = i_1+\cdots+i_{n-r} = m-n+r$, and the partition I^c is the complement $[n]\setminus I$.

A corollary of Theorem 4.7 is therefore the identity between the two formulas.

THEOREM 4.9 ([**34**, Theorems 1.3 and 1.5]). *The degree $\delta(m,n,n-s)$ is a polynomial in n, given m and s. Similarly $\phi(n,d)$ is polynomial in n given d.*

In [**34**], Manivel, Michałek, Monin, Seynnaeve, and Vodǐcka extend their results to square matrices and skew symmetric matrices.

5. Closing remarks

These notes do not claim to cover more than some basic concepts and a small sampling of examples that illustrates how algebraic degree appears in algebraic models and what is involved in computing this degree.

Paul Breiding and Kathlén Kohn, together with Bernd Sturmfels, co-authored the recent book *Metric Algebraic Geometry* [**5**] that serves as a nice guide to a wide range of examples and research activities on nonlinear algebraic models. It covers a proliferation of algebraic models and methods where the computation of algebraic degree is key to a complexity assessment, and it manifests the key role that Bernd Sturmfels has played in showing the importance of algebraic degree in algebraic models.

References

[1] P. Aluffi and C. Harris, *The Euclidean distance degree of smooth complex projective varieties*, Algebra Number Theory **12** (2018), no. 8, 2005–2032, DOI 10.2140/ant.2018.12.2005. MR3892971

[2] C. Améndola, L. Gustafsson, K. Kohn, O. Marigliano, and A. Seigal, *The maximum likelihood degree of linear spaces of symmetric matrices*, Matematiche (Catania) **76** (2021), no. 2, 535–557, DOI 10.4418/2021.76.2.15. MR4334903

[3] C. Améndola and J. L. Rodriguez, *Likelihood geometry*, Combinatorial, Computational, and Applied Algebraaic Geometry: A Tribute to Bernd Sturmfels, Proc. Sympos. Pure Math., vol. 111, Amer. Math. Soc., ©2025.

[4] M. Belotti, M. Joswig, and M. Panizzut, *Algebraic degrees of 3-dimensional polytopes*, Vietnam J. Math. **50** (2022), no. 3, 581–597, DOI 10.1007/s10013-022-00559-2. MR4447397

[5] P. Breiding, K. Kohn, and B. Sturmfels, *Metric algebraic geometry*, Oberwolfach Seminars, vol. 53, Birkhäuser/Springer, Cham, [2024] ©2024, DOI 10.1007/978-3-031-51462-3. MR4738534

[6] P. Breiding, F. Sottile, and J. Woodcock, *Euclidean distance degree and mixed volume*, Found. Comput. Math. **22** (2022), no. 6, 1743–1765, DOI 10.1007/s10208-021-09534-8. MR4509115

[7] F. Catanese, S. Hoşten, A. Khetan, and B. Sturmfels, *The maximum likelihood degree*, Amer. J. Math. **128** (2006), no. 3, 671–697. MR2230921

[8] T. Ö. Çelik, A. Jamneshan, G. Montúfar, B. Sturmfels, and L. Venturello, *Wasserstein distance to independence models*, J. Symbolic Comput. **104** (2021), 855–873, DOI 10.1016/j.jsc.2020.10.005. MR4180150

[9] D. Cifuentes, K. Ranestad, B. Sturmfels, and M. Weinstein, *Voronoi cells of varieties*, J. Symbolic Comput. **109** (2022), 351–366, DOI 10.1016/j.jsc.2020.07.009. MR4316044

[10] O. Clarke, S. Hoşten, N. Kushnerchuk, and J. Oldekop, *Matroid stratification of ML degrees of independence models*, Algebr. Stat. **15** (2024), no. 2, 199–223, DOI 10.2140/astat.2024.15.199. MR4836464

[11] D. T. Hiep, *A formula for the algebraic degree in semidefinite programming*, Kodai Math. J. **39** (2016), no. 3, 484–488, DOI 10.2996/kmj/1478073765. MR3567226

[12] D. T. Hiep, N. T. N. Giao, and N. T. M. Van, *A characterization of the algebraic degree in semidefinite programming*, Collect. Math. **74** (2023), no. 2, 443–455, DOI 10.1007/s13348-022-00358-5. MR4576958

[13] G. DePaul, S. Hoşten, N. Metya, and I. Nometa, *Degrees of the Wasserstein distance to small toric models*, Algebr. Stat. **15** (2024), no. 2, 249–267, DOI 10.2140/astat.2024.15.249. MR4836466

[14] S. Di Rocco, D. Eklund, and M. Weinstein, *The bottleneck degree of algebraic varieties*, SIAM J. Appl. Algebra Geom. **4** (2020), no. 1, 227–253, DOI 10.1137/19M1265776. MR4078795

[15] I. V. Dolgachev, *Classical algebraic geometry: A modern view*, Cambridge University Press, Cambridge, 2012, DOI 10.1017/CBO9781139084437. MR2964027

[16] J. Draisma, E. Horobeţ, G. Ottaviani, B. Sturmfels, and R. R. Thomas, *The Euclidean distance degree of an algebraic variety*, Found. Comput. Math. **16** (2016), no. 1, 99–149, DOI 10.1007/s10208-014-9240-x. MR3451425

[17] D. Eisenbud and J. Harris, *3264 and all that—A second course in algebraic geometry*, Cambridge University Press, Cambridge, 2016, DOI 10.1017/CBO9781139062046. MR3617981
[18] C. Eur, T. Fife, J. A. Samper, and T. Seynnaeve, *Reciprocal maximum likelihood degrees of diagonal linear concentration models*, Matematiche (Catania) **76** (2021), no. 2, 447–459, DOI 10.4418/2021.76.2.10. MR4334898
[19] W. Fulton, *Intersection theory*, Ergebnisse der Mathematik und ihrer Grenzgebiete (3) [Results in Mathematics and Related Areas (3)], vol. 2, Springer-Verlag, Berlin, 1984, DOI 10.1007/978-3-662-02421-8. MR732620
[20] W. Fulton, *Introduction to toric varieties*, Annals of Mathematics Studies, vol. 131, Princeton University Press, Princeton, NJ, 1993. The William H. Roever Lectures in Geometry, DOI 10.1515/9781400882526. MR1234037
[21] M. Gałązka, *Initial values of ML-degree polynomials*, Adv. in Appl. Math. **164** (2025), Paper No. 102831, 8, DOI 10.1016/j.aam.2024.102831. MR4839475
[22] L. D. García Puente, M. Garrote-López, and E. Shehu, *Computing algebraic degrees of phylogenetic varieties*, Algebr. Stat. **14** (2023), no. 2, 215–231, DOI 10.2140/astat.2023.14.215. MR4746404
[23] H.-C. Graf von Bothmer and K. Ranestad, *A general formula for the algebraic degree in semidefinite programming*, Bull. Lond. Math. Soc. **41** (2009), no. 2, 193–197, DOI 10.1112/blms/bdn114. MR2496496
[24] E. Gross, M. Drton, and S. Petrović, *Maximum likelihood degree of variance component models*, Electron. J. Stat. **6** (2012), 993–1016, DOI 10.1214/12-EJS702. MR2988436
[25] J. Harris and L. W. Tu, *On symmetric and skew-symmetric determinantal varieties*, Topology **23** (1984), no. 1, 71–84, DOI 10.1016/0040-9383(84)90026-0. MR721453
[26] H. Hironaka, *Resolution of singularities of an algebraic variety over a field of characteristic zero. I, II*, Ann. of Math. (2) **79** (1964), 109–203; 79 (1964), 205–326, DOI 10.2307/1970547. MR199184
[27] J. Huh, *The maximum likelihood degree of a very affine variety*, Compos. Math. **149** (2013), no. 8, 1245–1266, DOI 10.1112/S0010437X13007057. MR3103064
[28] Y. Jiang, K. Kohn, and R. Winter, *Linear spaces of symmetric matrices with non-maximal maximum likelihood degree*, Matematiche (Catania) **76** (2021), no. 2, 461–481, DOI 10.4418/2021.76.2.11. MR4334899
[29] K. Kubjas, O. Kuznetsova, and L. Sodomaco, *Algebraic degree of optimization over a variety with an application to p-norm distance degree*, Preprint, `arXiv:2105.07785`, 2021.
[30] J. Lindberg, L. Monin, K. Rose, *The algebraic degree of sparse polynomial optimization*, Preprint, `arXiv:2308.07765`, (2023).
[31] J. Lindberg, N. Nicholson, J. I. Rodriguez, and Z. Wang, *The maximum likelihood degree of sparse polynomial systems*, SIAM J. Appl. Algebra Geom. **7** (2023), no. 1, 159–171, DOI 10.1137/21M1422550. MR4566812
[32] I. G. Macdonald, *Symmetric functions and Hall polynomials*, 2nd ed., Oxford Mathematical Monographs, The Clarendon Press, Oxford University Press, New York, 1995. With contributions by A. Zelevinsky; Oxford Science Publications. MR1354144
[33] L. Manivel, *Proof of a conjecture of Sturmfels, Timme, and Zwiernik*, Preprint, `arXiv:2108.11085`, (2021).
[34] L. Manivel, M. Michałek, L. Monin, T. Seynnaeve, and M. Vodička, *Complete quadrics: Schubert calculus for Gaussian models and semidefinite programming*, J. Eur. Math. Soc. (JEMS) **26** (2024), no. 8, 3091–3135, DOI 10.4171/jems/1330. MR4756951
[35] L. G. Maxim, J. I. Rodriguez, and B. Wang, *Defect of Euclidean distance degree*, Adv. in Appl. Math. **121** (2020), 102101, 22, DOI 10.1016/j.aam.2020.102101. MR4136171
[36] M. Michałek and B. Sturmfels, *Invitation to nonlinear algebra*, Graduate Studies in Mathematics, vol. 211, American Mathematical Society, Providence, RI, [2021] ©2021, DOI 10.1090/gsm/211. MR4423369
[37] C. Meroni, B. Reinke, and K. Wang, *The algebraic degree of the Wasserstein distance*, Preprint, `arXiv:2401.12735`, (2024).
[38] J. Nie, K. Ranestad, and B. Sturmfels, *The algebraic degree of semidefinite programming*, Math. Program. **122** (2010), no. 2, Ser. A, 379–405, DOI 10.1007/s10107-008-0253-6. MR2546336
[39] J. Nie, K. Ranestad, and T. Xindong, *Algebraic Degrees of Generalized Nash Equilibrium Problems*, Preprint, `arXiv:2208.00357`, (2024).

[40] P. Orlik and H. Terao, *The number of critical points of a product of powers of linear functions*, Invent. Math. **120** (1995), no. 1, 1–14, DOI 10.1007/BF01241120. MR1323980

[41] G. Ottaviani, L. Sodomaco, and E. Ventura, *Asymptotics of degrees and ED degrees of Segre products*, Adv. in Appl. Math. **130** (2021), Paper No. 102242, 36, DOI 10.1016/j.aam.2021.102242. MR4277048

[42] B. Sturmfels and S. Telen, *Likelihood equations and scattering amplitudes*, Algebr. Stat. **12** (2021), no. 2, 167–186, DOI 10.2140/astat.2021.12.167. MR4350875

[43] B. Sturmfels, *Beyond linear algebra*, ICM—International Congress of Mathematicians. Vol. 6. Sections 12–14, EMS Press, Berlin, [2023] ©2023, pp. 4820–4840. MR4680425

[44] N. T. Thang and P. T. Thuy, *Euclidean distance degree of complete intersections via Newton polytopes*, Preprint, arXiv:2404.17237, (2024).

[45] A. Varchenko, *Critical points of the product of powers of linear functions and families of bases of singular vectors*, Compositio Math. **97** (1995), no. 3, 385–401. MR1353281

Matematisk Institutt, Universitetet i Oslo, Oslo, Norway
Email address: ranestad@math.uio.no

Proceedings of Symposia in Pure Mathematics
Volume **111**, 2025
https://doi.org/10.1090/pspum/111/02049

Tensors: From complex problems to real applications

Mateusz Michałek

Dedicated to Bernd Sturmfels
on the occasion of his sixtieth birthday.

Abstract. Tensors are among the most important notions in nonlinear algebra, generalizing such fundamental objects as a vector, a matrix and a polynomial. They play a prominent role in abstract algebraic geometry and are used in practice in computer science, statistics, artificial intelligence, physics and many more. In this expository article we give a glimpse at Bernd Sturmfels's remarkable contributions to this theory. We focus on interplays of geometry and the theory of tensors, as well as applications of complex numbers to study real objects.

1. Introduction

Tensors appear in almost every branch of modern mathematics. So does Bernd, hence it should not be surprising that their paths intersected. Here, we describe just a few of those amazing encounters.

Tensors could be regarded as generalisations of matrices to higher dimensions. Thus, as an $a_1 \times a_2$ matrix is defined as a family of numbers (a_{ij}) indexed by two integers $1 \leq i \leq a_1$ and $1 \leq j \leq a_2$, a tensor of format $a_1 \times \cdots \times a_k$ is a family of numbers $t_{i_1,\ldots,i_k}$ indexed by integers $1 \leq i_s \leq a_s$ for each $s = 1, \ldots, k$.

Tensors are considered to be quite complicated objects with properties very different from those of matrices. Still, following Bernd's approach, tensors are not really different, but rather more interesting. Their versatile nature allows for many groundbreaking applications. Two of such we describe in Sections 2.1 and 2.2. In the first one we discuss how the results about nonnegative rank of a tensor can help answering fundamental questions in statistics. One of the final theorems we present is [**3**, Theorem 1.1] where a complete description of tensors of nonnegative rank two is provided.

In Section 2.2 we briefly present main ideas in algebraic phylogenetics. Here, a statistical model, say describing a history of evolution of certain species, is identified with the image of a polynomial map. We focus on so-called tree models, showing how they may encode the history of evolution. Finally, referring to [**30**], we emphasize how application of mathematical methods, like discrete Fourier transform, may allow to answer questions that seem intractable from a computational perspective. This topic brings us closer to toric varieties, which are images of monomial maps.

2020 *Mathematics Subject Classification.* Primary 15A69, 13P25, 14N07.

The author is funded by the Deutsche Forschungsgemeinschaft - Projektnummer 467575307.

Next, in Section 2.3 we present results how application of toric methods could be extremely useful to study classical varieties in tensor spaces, like the secant variety and the tangential variety of the Segre variety.

As the author learned from Bernd, when approached from a correct perspective, a lot of notions typical for matrices could be generalized to tensors in a natural way. Among those are eigenvalues and eigenvectors. These are presented in Section 3.1. Here, we will describe a remarkable result about the number of eigenvalues and eigenvectors of a generic symmetric tensor [**9**, **10**], see Theorem 3.3. We also present a very important result on eigendiscriminant of the tensor in Theorem 3.4, based on [**1**].

Finally, in Section 3.2 we describe interactions among real and complex geometry, focusing on real Waring rank. Here, the key problem is to present a homogeneous polynomial as a sum of powers of linear forms. As the reader will see, this is related to a lot of beautiful algebraic and geometric constructions. The real rank boundary is the algebraic hypersurface that separates real polynomials that allow, in a sense minimal, presentation as sums of powers from those that do not allow such a representation. We present several explicit examples of such hypersurfaces in Theorems 3.10 and 3.11. We also describe the *space of real sums of powers*—a semialgebraic set, serving as a moduli space for the real Waring decompositions.

We recall a quote by Hilbert:

> *"He who seeks for methods without having a definite problem in mind seeks in the most part in vain."*

Bernd's unique ability to discuss sophisticated problems with scientists from various disciplines, combined with his great expertise in mathematics allows him to not only find new methods and concepts that are fundamental in applications, but to create whole new subfields of mathematics, that turn out to be utterly fascinating.

Notation. We denote by $[n]$ the set $\{1, \dots, n\}$. The group of permutations of $[n]$ is denoted by $\mathfrak{S}_n$. K will be always a field of real or complex numbers. A tensor $T \in K^{a_1} \otimes \cdots \otimes K^{a_k}$ is called a tensor of format $a_1 \times \cdots \times a_k$. For a vector $v \in K^a$ we denote by $(v)_i \in K$ its ith coordinate.

2. Theory and applications

2.1. Nonnegative rank. One of the principal applications of tensors is data science and statistics. An example is a group of k people, where we can ask for hair color (dark, later corresponding to 0 on the first coordinate of the tensor, or blond later corresponding to 1), height (short or tall), and weight (low or high). The statistical information about the group naturally forms a tensor T of format $2 \times 2 \times 2$, where on first coordinate we put information about hair color, on second about height and on third about weight. Thus one entry, say, $T_{0,1,1}$ is the number of people with dark hair that are tall and have high weight. We could ask fundamental questions in statistics, like "is the hair color independent of height?" or "given dark hair color is the height independent of weight?" Such questions translate to basic algebraic properties of a tensor.

REMARK 2.1. Here and in what follows we have to remember that the entries of the tensor, like $T_{0,1,1}$, represent data. Thus, $T_{0,1,1}/k$ approximates the probability of a person having dark hair and being tall and heavy. However, we cannot expect it will be exactly equal to that probability, even when the sample size k is large. In particular, the algebraic relations we refer to, in practice, will only hold up to a small error. This is another, very interesting field of study, where we can ask for example how far is the data tensor T from the locus of tensors that have certain property. In this article, we will not address this problem, just pointing out its existence.

For example, assume we would like to exclude the fact that hair color, height and weight are independent. If they were, then we could write, abusing notation, also d for the probability of a person having dark hair, b for the probability of having blond hair, t of being tall etc. We expect that $T_{0,1,1}$ is $(dth)k$, $T_{1,1,0}$ is $(bts)k$ etc. as we assume the properties are independent. However, note that not all tensors are of this form! Indeed, such tensors are *rank one* tensors, i.e., they are of the form $v_1 \otimes v_2 \otimes v_3$, and satisfy certain algebraic relations. One relation we may check is:

$$T_{0,1,1}T_{1,1,0} = T_{1,1,1}T_{0,1,0}, \text{ or equivalenty } T_{0,1,1}T_{1,1,0} - T_{1,1,1}T_{0,1,0} = 0.$$

We encourage the reader to think about other algebraic relations—one may prove that the ideal of all possible relations is generated by relations similar to the one above. We make this statement more precise we introduce the following definition.

DEFINITION 2.2. A tensor T of format $a_1 \times \cdots \times a_k$ has *rank one* if there exist vectors $v_i \in K^{a_i}$ such that $T = v_1 \otimes \cdots \otimes v_k$. In other words for every entry of the tensor we have the equality:

$$T_{i_1,\ldots,i_k} = (v_1)_{i_1} \cdots (v_k)_{i_k}.$$

The locus of rank one tensors is known as the *Segre variety*. We know that to check if a tensor has rank one it is enough to verify a set of degree-two equations, similar to the example above. These are generalizations of 2×2 minors of a matrix, if we imagine T as a multidimensional array of numbers.

As a next step we imagine our tensor is not of rank one. The next question we could ask is if there is a hidden factor, i.e., that maybe our group of people could be somehow subdivided into two groups, such that in each group the properties we ask for are independent. Note that we have no idea what this hidden factor could be, but still we would like to know if it is likely that it exists. In such a case, for each group we would get new tensors T_1 and T_2, each one of rank one, such that $T = T_1 + T_2$.

In classical algebraic geometry, one defines the *rank* of a tensor T as a minimum r such that T is a sum of r rank one tensors. Thus, it seems that our statistical question translates to asking if T has rank two. We do know the polynomial equations that vanish on tensors of rank at most two[1]—in fact the ideal is generated by 3×3 minors of so-called flattening matrices [**17**, **26**]. Unfortunately, almost every tensor of format $2 \times 2 \times 2$ has rank two, thus it seems we cannot do anything to exclude such a situation. Fortunately, Bernd can!

The trick is to realize that T_1 and T_2 are not just arbitrary rank one tensors, but they come from probability distributions, that have to be nonnegative.

[1]As we will see below this is not a closed set, still the description of the Zariski closure is known.

DEFINITION 2.3. A tensor $T \neq 0$ of format $a_1 \times \cdots \times a_k$ has *nonnegative rank one* if there exist vectors $v_i \in \mathbb{R}^{a_i}$ with all coordinates that are nonnegative and $T = v_1 \otimes \cdots \otimes v_k$. In other words for every entry of the tensor we have the equality

$$T_{i_1,\ldots,i_k} = (v_1)_{i_1} \cdots (v_k)_{i_k}.$$

LEMMA 2.4. *A tensor has nonnegative rank one if and only if it has rank one and all its entries are nonnegative.*

PROOF. If the tensor has nonnegative rank one, it has rank one and its entries are nonnegative. We prove the other implication.

Fix T of rank one with nonnegative entries. We must have $T = v_1 \otimes \cdots \otimes v_k$. Let $\tilde{v}_i$ be obtained from v_i by taking absolute values of the coordinates. We claim that $T = \tilde{v}_1 \otimes \cdots \otimes \tilde{v}_k$. Clearly, we did not change entries of T that are equal to zero. For other entries, we know that $0 < T_{i_1,\ldots,i_k} = (v_1)_{i_1} \cdots (v_k)_{i_k}$. Thus, there must be an even number of negative numbers among $(v_1)_{i_1}, \ldots, (v_k)_{i_k}$. Hence, $(v_1)_{i_1} \cdots (v_k)_{i_k} = (\tilde{v}_1)_{i_1} \cdots (\tilde{v}_k)_{i_k}$, which finishes the proof. □

COROLLARY 2.5. *The set of tensors of nonnegative rank one is Euclidean closed.*

PROOF. The set of tensors of rank one is closed and so is the set of tensors with nonnegative entries. □

The nonnegative rank of a tensor T is the smallest r such that T is a sum of r tensors of nonnegative rank one. We highlight the main differences among rank and nonnegative rank:

(1) the rank of a tensor is finite for any tensor, while nonnegative rank is finite only for tensors with nonnegative entries;
(2) the locus of tensors of rank at most r, in general is not Euclidean closed. The most classical example is the tensor $e_2 \otimes e_1 \otimes e_1 + e_1 \otimes e_2 \otimes e_1 + e_1 \otimes e_1 \otimes e_2$. It has rank three, although it is a limit of rank-two tensors
$$\frac{1}{\epsilon}\left((e_1 + \epsilon e_2) \otimes (e_1 + \epsilon e_2) \otimes (e_1 + \epsilon e_2) - e_1 \otimes e_1 \otimes e_1\right)$$
as $\epsilon \to 0$. However, for nonnegative rank we have the following result.

PROPOSITION 2.6. *The set of tensors of nonnegative rank at most r is closed.*

PROOF. Consider a sequence $T_{n,1} + \cdots + T_{n,r}$ of sums of r nonnegative rank-one tensors $T_{n,i}$ converging to a tensor T as $n \to \infty$. Note that each entry of each $T_{n,i}$ is nonnegative. As the sum converges to T entrywise, the entries of $T_{n,i}$ must be uniformly bounded. Thus, we can assume that $T_{n,i}$ are chosen from a compact set. Passing to a subsequence we may assume that for every i the sequence $T_{n,i}$ converges to T_i. By Corollary 2.5 each $T_i \neq 0$ has nonnegative rank one.

Thus, as $T = T_1 + \cdots + T_r$, it must have nonnegative rank at most r. Hence, the locus of tensors of nonnegative rank at most r is Euclidean closed. □

The above proof fails for rank, as in the sum we may have cancellations and the entries of some $T_{n,i}$'s may diverge to infinity. This leads to numerical problems while computing ranks and motivated the study of secant varieties and border rank, cf. [**22**, Chapter 9.2], [**16**] for more information.

How can we decide if a tensor has nonnegative rank two? Clearly, it needs to have rank at most two. The other condition is *supermodularity*.

DEFINITION 2.7. A tensor $T \in \mathbb{R}^{a_1} \otimes \cdots \otimes \mathbb{R}^{a_k}$ is called *supermodular* if there exist permutations π_1 of $[a_1]$, π_2 of $[a_2]$, ..., π_k of $[a_k]$ such that

$$T_{i_1,\ldots,i_k} T_{j_1,\ldots,j_k} \leq T_{l_1,\ldots,l_k} T_{p_1,\ldots,p_k}$$

whenever for any $1 \leq s \leq k$ we have $\{i_s, j_s\} = \{l_s, p_s\}$ and $\pi_s(l_s) \leq \pi_s(p_s)$.

Below we have a remarkable result describing all polynomial inequalities needed to verify if a tensor has nonnegative rank two. Recall that 3×3 minors of flattening matrices generate the ideal of polynomials vanishing on rank three tensors [**26**].

THEOREM 2.8 ([**3**, Theorem 1.1]). *A nonnegative tensor has nonnegative rank at most two if and only if it is supermodular and the* 3×3 *minors of flattening matrices vanish also on the given tensor.*

2.2. Algebraic phylogenetics. Another topic in algebraic statistics, particularly dear to the author, is algebraic phylogenetics [**24, 32**]. Here Bernd is considered not only one of the world leading experts but also a pioneer. Our datum will be now a sequence of letters A, C, G, T representing the DNA. Such sequences are typically considered to be very long, individual for each object and measuring how much the two individuals differ. Thus two humans will have distinct DNA, but much more alike than that of a cow.

We fix representative DNA sequences $S_1, \ldots, S_k$ of k species. Here k is usually assumed to be small—we may even imagine $k = 4$. If we look at first positions in the DNA, we obtain a sequence of k letters $S_1(1), \ldots, S_k(1)$. Going through the whole DNA we can count how many times we obtained any given sequence of k letters. This provides us with a probability distribution P on the set of sequences of length k of letters.

REMARK 2.9. Our model below is based on evolution of a given site, i.e. position in the sequence. We use the whole DNA string only to obtain probability distribution on sequences of k letters. Note that, we would obtain the same result if we acted on all k sequences with the same permuation. To make this biologically relevant, one needs to make several assumptions, for example that sites evolve independently.

Let us identify A, C, G, T with four basis vectors of $\mathbb{R}^4$. In this way we may identify sequences of length k with the basis vectors of $(\mathbb{R}^4)^{\otimes k}$ and the probability distribution P with a tensor in $(\mathbb{R}^4)^{\otimes k}$. Even for small k, this is a high dimensional vector space!

Our aim will be to try to reconstruct the history of evolution of the k given species. According to Bernd a good example is worth much more than pure formalism. We thus fix $k = 4$. To simplify notation, assume we only allow two letters A, B. We consider two models of evolution:

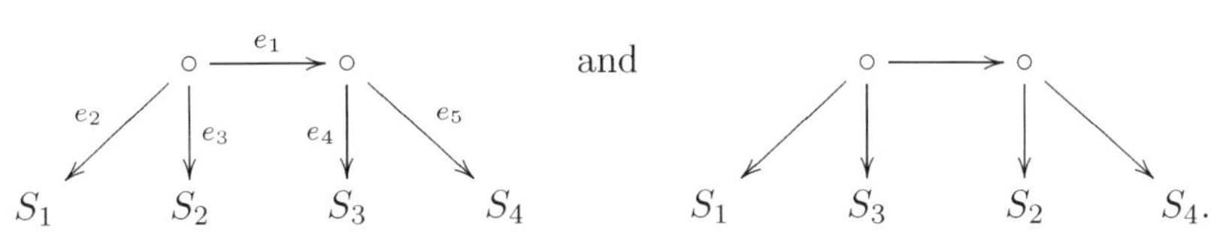

The first one describes the following situation. There was a common ancestor, that evolved into species S_1, S_2 and the ancestor of species S_3 and S_4. Note that in this model we expect S_3 and S_4 to be closer related than S_2 and S_4, while in the other model we have the opposite situation. We make a few further simplifying assumptions. We assume that each position in DNA was evolving independently and could either stay the same (with some probability) or change. In other words, over every edge e of our graph we have a parameter q_e denoting the probability of mutation. Let $p_e := 1 - q_e$ be the probability that mutation did not happen. Also, as we do not know anything about the DNA of the common ancestor, we assume that letters A, B appear uniformly at random, that is each one with probability $\frac{1}{2}$. We focus on the first model, where we have six parameters corresponding to edges.

What is the probability of observing the sequence $AAAA$ at the four leaves? We have to consider first all possibilities for the state of the common ancestor, then the possible states of its unknown neighbour. We obtain the sum of four probabilities:

(1) for common ancestor A and neighbor vertex A we obtain:
$$\frac{1}{2}p_{e_1}p_{e_2}p_{e_3}p_{e_4}p_{e_5},$$
(2) for common ancestor A and neighbor vertex B we obtain:
$$\frac{1}{2}q_{e_1}p_{e_2}p_{e_3}q_{e_4}q_{e_5},$$
(3) for common ancestor B and neighbor vertex A we obtain:
$$\frac{1}{2}q_{e_1}q_{e_2}q_{e_3}p_{e_4}p_{e_5},$$
(4) for common ancestor B and neighbor vertex B we obtain:
$$\frac{1}{2}p_{e_1}q_{e_2}q_{e_3}q_{e_4}q_{e_5}.$$

The sum gives a quite complicated polynomial and we have just computed one coordinate out of $2^4 = 16$ after making many simplifying assumptions! From experiments, we obtain a probability distribution in the 16-dimensional space, but how could we decide if this point is in the image of the map from the parameter space for the first model, or maybe for the second one? And what to do when the trees and models will be more complicated?

In their groundbreaking work [**30**], basing on the earlier work in theoretical biology, Sturmfels and Sullivant provide a unified approach leading to beautiful, efficient answers. The trick is based on discrete Fourier transform. This is nothing else then a linear change of coordinates. After performing this change of coordinates the map from parameters to the space with probability distributions is given by monomials!

As demonstrated in the same article, the theorem has tremendous consequences, as the (Zariski closure of the) set of probabilities allowed by our model is a *toric variety*. This is one of the best understood class of varieties. This approach had consequences not only for applications, but also inspired many algebraic geometers to study the geometry of phylogenetic models, cf. [**8**].

2.3. Toric methods for secants and tangents. The example from the previous section is by far not the only one showing how toric methods can simplify a very difficult problem. In fact, detecting toric structure and using it to obtain results that were impossible before is one of "Bernd's spécialités de la maison." Below we present one more case when such application was particularly useful for tensors.

Recall that the locus of rank one tensors in $\mathbb{P}(V_1 \otimes \cdots \otimes V_k)$ is the Segre variety $X := \mathbb{P}(V_1) \times \cdots \times \mathbb{P}(V_k)$. It is classically well known that it is a smooth, toric algebraic variety. For $k = 2$ the locus of rank at most two matrices is also closed, however as we have seen above this fails for $k > 2$. For the time being let us fix the field $K = \mathbb{C}$.

DEFINITION 2.10. The rth *secant variety* $\sigma_r(X)$ of the Segre variety is the closure of the locus of tensors of rank at most r.

The tangential variety $\tau(X)$ of the Segre variety is the union of all tangent spaces to X.

Note that tensors of rank at most r are simply tensors lying on a space spanned by r points of X. We call such planes secant r-planes. Thus $\sigma_r(X)$ is simply the closure of the union of all secant r-planes. In this way we may extend the definition of the secant and the tangential variety to an arbitrary projective or affine, smooth[2] variety X.

We see that $X \subset \tau(X) \subset \sigma_2(X)$, as any tangent line can be approximated by secant lines. The relations between $\tau(Y)$ and $\sigma_2(Y)$ for arbitrary variety Y have been classically studied. An example of a beautiful theorem due to Zak is presented below.

THEOREM 2.11 ([**33**, Simplified Theorem 1.4]). *For any smooth, projective variety Y of dimension d either:*

- *the second secant and tangential varieties have maximal possible dimension, i.e., respectively, $2d + 1$ and $2d$, or*
- $\tau(Y) = \sigma_2(Y)$.

In the case $k = 2$ of matrices much is known about $\sigma_r(X)$. These are simply the varieties defined by $(r + 1) \times (r + 1)$ minors. We always have $\tau(X) = \sigma_2(X)$. Further, the singular locus of $\sigma_{r+1}(X)$ is $\sigma_r(X)$, as long as the secant variety does not fill the ambient space. However, for $k > 2$ similar results are very hard to find. In particular, we lack the description (or even examples of elements!) of the ideal of $\sigma_r(X)$ for large r. We mention an interesting problem, related to phylogenetics.

The general Markov model is a phylogenetic model that does not distinguish any specific symmetries. Consider the simplest tree, that is a tree with one inner vertex and three leaves. As in Section 2.2, from this data we obtain a parametrization of probability distributions. The Zariski closure of all probabilities allowed by the model is $\sigma_4(\mathbb{P}^3 \times \mathbb{P}^3 \times \mathbb{P}^3)$, that is the locus of tensors of format $4 \times 4 \times 4$ and border rank at most four. The complete description of the ideal of this variety is unknown. Elizabeth Allman offered a prize of an Alaskan salmon to anyone who would solve this challenge. For more information we refer to [**5, 11, 18**].

Further, in general, the description of the singular locus of $\sigma_{r+1}(X)$ is also challenging. In particular, it is not equal to $\sigma_r(X)$, although $\sigma_r(X)$ is always contained in the singular locus of $\sigma_{r+1}(X)$, as long as the latter is not equal to the whole ambient space.

The amazing work by Sturmfels and Zwiernik shows that $\tau(X)$ and $\sigma_2(X)$ are much more friendly [**31**]. This is one of very few works, where methods from statistics found new, unknown applications in classical algebraic geometry.

[2]When X is not smooth one has to be more careful with the definition of the tangential variety.

THEOREM 2.12 ([**21**, **31**]). *The tangential variety $\tau(X)$ and the secant variety $\sigma_2(X)$ of the Segre variety are covered by open affine toric varieties.*

The proof goes in the following steps:

(1) consider a hyperplane $H \subset \mathbb{P}(V_1 \otimes \cdots \otimes V_k)$ and the varieties $Y_H := \sigma_2(X) \setminus H$ and $Y'_H := \tau(X) \setminus H$. Here, H must come from a linear form corresponding to a rank one tensor in the dual space.
(2) Clearly, the varieties Y_H and Y'_H are affine and cover respectively $\sigma_2(X)$ and $\tau(X)$ as we change H.
(3) The key step is to apply automorphisms of the affine space $\mathbb{P}(V_1 \otimes \cdots \otimes V_k) \setminus H$ inspired by relations among moments and cumulants in statistics. This leads to transforming the natural parametrisations of Y_H and Y'_H into monomial parametrisations, exhibiting their toric structures.

The result again had enormous consequences leading, e.g., to better understanding of the singularities of $\sigma_2(X)$ and $\tau(X)$, in particular classifications of normal, Cohen–Macaulay and Gorenstein singularities [**14**, **28**].

3. Invariants of tensors

3.1. Tensors and their eigenvalues. Eigenvectors and eigenvalues are among the most useful notions in algebra. As a square matrix A may be identified with a linear endomorphism f_A of a vector space, the usual definition of an eigenvector v is that f_A rescales v. As tensors are more general objects we must be careful when generalizing this notion. Our exposition follows [**22**, Chapter 9]. For more information about eigenvectors and eigenvalues of tensors we refer to [**25**, **29**].

We start with the case of symmetric matrices. Every symmetric $n \times n$ matrix $A \in K^n \otimes K^n$ is naturally identified with a homogeneous degree-two polynomial P_A in n variables $x_1, \ldots, x_n$. Let $x = (x_1, \ldots, x_n)$ be the vector of the variables. We have:[3]

$$P_A(x) := xAx^t,$$

where x^t denotes the transpose of x. Let

$$\nabla P_A := \left(\frac{\partial P_A}{\partial x_1}, \frac{\partial P_A}{\partial x_2}, \ldots, \frac{\partial P_A}{\partial x_n} \right)$$

be the gradient of P_A.

We note that, as linear operators, $\nabla P_A = 2A$. As partial derivatives of a homogeneous polynomial are homogeneous and of the same degree (if nonzero), we may also, abusing notation, regard ∇P_A as a rational map $\mathbb{P}^{n-1} \dashrightarrow \mathbb{P}^{n-1}$. Thus we make an equivalent definition of eigenvectors and eigenvalues.

DEFINITION 3.1. An eigenvector of the homogeneous polynomial P_A is a vector $v \in K^n \setminus \{0\}$ such that either $\nabla P_A([v]) = [v] \in \mathbb{P}^n$ or the rational map ∇P_A is not defined on $[v]$. Equivalently, if there exists $\lambda \in K$ such that $\nabla P_A(v) = \lambda v \in K^n$.

Miraculously, the definition above can be taken verbatim for any homogeneous polynomial, even if it has higher degree! The only difference is that now ∇P_A is a polynomial map, possibly not linear. Just as we identified symmetric matrices with

[3] A nonsymmetric matrix A also defines a quadratic polynomial xAx^t. However, this is the same polynomial as $x(\frac{1}{2}(A + A^t))x^t$ and $\frac{1}{2}(A + A^t)$ is symmetric. Thus, it is natural to identify quadratic polynomials and symmetric matrices.

degree-two homogeneous polynomials, we may identify higher degree homogeneous polynomials with symmetric tensors.

DEFINITION 3.2. A tensor $T \in (K^n)^{\otimes k}$ is *symmetric* if for any permutation $\sigma \in \mathfrak{S}_k$ and any indices $1 \leq i_1, \ldots, i_k \leq n$ we have:

$$T_{i_1,\ldots,i_k} = T_{i_{\sigma(1)},\ldots,i_{\sigma(k)}}.$$

We identify such a symmetric tensor with a degree-k homogeneous polynomial P_T in n variables by:

$$P_T(x) = \sum_{1\leq i_1,\ldots,i_k\leq n} T_{i_1,\ldots,i_k} x_{i_1} \cdots x_{i_k}.$$

This extends the well-known correspondence between a symmetric matrix A and a quadratic polynomial P_A.

Thus Definition 3.1 provides us with the notion of an eigenvector of a symmetric tensor a.k.a. homogeneous polynomial. We know from linear algebra, that a general symmetric $n \times n$ matrix has, up to scaling, precisely n distinct eigenvectors. What about tensors? The theorem below provides an exact answer for arbitrary n and k.

THEOREM 3.3 ([**9**, Simplification of Theorem 1.2], [**10**]). *The number of eigenvectors of a general symmetric tensor $T \in (\mathbb{C}^n)^{\otimes k}$ equals:*

$$\sum_{i=0}^{n-1} (k-1)^i.$$

For $k > 2$ this equals $\frac{(k-1)^n - 1}{k-2}$.

This is a beautiful extension of the case of matrices, where clearly we have $n = \sum_{i=0}^{n-1} 1$.

We note that the theorem holds over an algebraically closed field. In general, even for real symmetric tensors the eigenvectors can be complex—recall that for a real symmetric matrix there exists a basis of the vector space consisting of real, orthogonal eigenvectors. In fact, for tensors it was not known if it is possible for all eigenvectors to be real, for all k and n, until a proof by Kozhasov [**15**], confirming a conjecture by Abo, Seigal, and Sturmfels [**1**].

The eigenvalues are distinct for a general tensor, however they may coincide for special tensors. The polynomial, in the entries of the tensor, that vanishes when two eigenvalues coincide is known as the *eigendiscriminant*. Computing is, in general, hard. However, we do know the general formula for its degree.

THEOREM 3.4 ([**1**]). *The eigendiscriminant is an irreducible homogeneous polynomial of degree $n(n-1)(k-1)^{n-1}$.*

Apart from the fact that the formula is very nice, it could also be used in practice to obtain explicit formulas for the eigendiscriminant. Indeed, one general approach to obtain such formulas is via elimination theory. The software to obtain a polynomial after elimination runs usually faster when given degree bounds.

Next we turn our attention to arbitrary matrices $B \in K^n \otimes K^m$. In this case we may identify B with a bilinear map $P_B : K^n \times K^m \to K$. Fixing variables $x_1, \ldots, x_n$ for K^n and $y_1, \ldots, y_m$ for K^m the bilinear form P_B is given by:

$$P_B((x_1,\ldots,x_n),(y_1,\ldots,y_m)) = \sum_{1\leq i\leq n, 1\leq j\leq m} B_{ij} x_i y_j = xBy^t.$$

As before, we may consider the gradient

$$\nabla P_B = ((\frac{\partial P_B}{x_1}, \dots, \frac{\partial P_B}{x_n}), (\frac{\partial P_B}{y_1}, \dots, \frac{\partial P_B}{y_m}))$$

of P_B:

$$\nabla P_B : K^n \times K^m \to K^n \times K^m.$$

We note that the first part of the gradient corresponds to multiplication of matrix B from right, while the second part corresponds to multiplication from left. Hence, trying to mimic the approach where we look at the fixed points of the gradient map, after projectivizing we obtain the following definition.

DEFINITION 3.5. A pair $x \in K^n$ and $y \in K^m$ of nonzero vectors is called a pair of *singular vectors* if there exists a scalar $\lambda \in K$ such that

$$By^t = \lambda x^t \text{ and } xB = \lambda y.$$

In particular, λ^2 is a common eigenvalue of the matrices BB^t and B^tB, as $B^tBy^t = \lambda B^t x^t = \lambda^2 y^t$ and $BB^t x^t = \lambda B y^t = \lambda^2 x^t$.

In such a case, we call y the *right singular vector*, x the *left singular vector* and λ a *singular value*. Here for simplicity let us assume $\lambda \neq 0$, for the general case referring to [**25**].

We note that the definition above implies that singular vector pairs (with nonzero singular value) correspond to fixed points of the map between products of projective spaces:

$$\nabla P_B : \mathbb{P}(K^n) \times \mathbb{P}(K^m) \dashrightarrow \mathbb{P}(K^n) \times \mathbb{P}(K^m).$$

This motivates the following definition.

DEFINITION 3.6. Let $T \in \mathbb{C}^{a_1} \otimes \cdots \otimes \mathbb{C}^{a_k}$. To T we naturally associate the k-linear map

$$P_T(x_{1,1}, \dots, x_{1,a_1}, \dots, x_{k,a_k}) := \sum_{1 \le i_1 \le a_1, \dots, 1 \le i_k \le a_k} T_{i_1,\dots,i_k} x_{1,i_1} \cdots x_{k,i_k}.$$

The gradient of this map

$$\nabla P_T : \mathbb{P}(\mathbb{C}^{a_1}) \times \cdots \times \mathbb{P}(\mathbb{C}^{a_k}) \dashrightarrow \mathbb{P}(\mathbb{C}^{a_1}) \times \cdots \times \mathbb{P}(\mathbb{C}^{a_k})$$

is a rational map of projective spaces. We call the fixed points $(v_1, \dots, v_k)$ of this map *k-tuples of singular vectors*.

The number of singular vector tuples for a generic tensor is known.

THEOREM 3.7 ([**12**]). *For a general tensor $T \in \mathbb{C}^{a_1} \otimes \cdots \otimes \mathbb{C}^{a_k}$, the number of singular vector tuples is the coefficient of the monomial $z_1^{a_1-1} \cdots z_k^{a_k-1}$ in the polynomial*

$$\prod_{i=1}^{k} \frac{\hat{z}_i^{a_i} - z_i^{a_i}}{\hat{z}_i - z_i},$$

where $\hat{z}_i = z_1 + \cdots + z_{i-1} + z_{i+1} + \cdots + z_k$.

3.2. Complex and real geometry of tensors. Often mathematicians perceive real and complex algebraic geometry as two distinct fields. As the author learned from Bernd, the fact that these two branches exhibit different methods, should rather be considered as an opportunity than difficulty. Indeed, one may apply methods from complex algebraic geometry to provide new approaches to problems coming from real algebraic geometry. The latter often have a more direct connection to other sciences, including statistics (as we have seen), biology, physics, computer vision, and many others. It is also an opportunity to find challenging problems coming from real geometry. In this section we present results concerning real Waring rank of polynomials.

DEFINITION 3.8. The complex Waring rank of a homogeneous polynomial P of degree k with complex coefficients is the minimal number r, such that

$$P(x) = \sum_{i=1}^{r} l_i(x)^k,$$

where l_i are linear forms.

The real Waring rank of a real homogeneous polynomial P of degree k is the minimal number r, such that

$$P(x) = \sum_{i=1}^{r} \lambda_i l_i(x)^k,$$

for some $\lambda_i \in \mathbb{R}$ and real linear forms l_i.

Here, we note that in the complex case or in the real case when k is odd, the definition does not change if we remove the λ_i's. However, for k even, without the λ_i's the right hand side would be a sum of squares, in particular a positive polynomial, thus it would not be possible to find such a representation for certain polynomials.

The definition of Waring rank is very close to the definition of tensor rank, considering P as a symmetric tensor. However, they are not the same. The reason is that in the definition of Waring rank, we only allow powers of linear forms and these are rank one *symmetric tensors*.

Consider the vector space $S^k K^n$ of degree k polynomials in n variables. When $K = \mathbb{C}$ there is a Zariski open (nonempty) subset of this space of polynomials of constant rank $r(k,n)$. The number $r(k,n)$ is known as *the generic rank*. Switching to $K = \mathbb{R}$ the polynomials of real rank $r(k,n)$ form a full-dimensional semialgebraic set that in general does not contain a Zariski open, dense set. If for certain r the set of polynomials of real Waring rank r contains a Euclidean open set, we call such r a *typical rank*. Note that $r(k,n)$ is the smallest typical rank. The exact value of $r(k,n)$ is known due to Alexander–Hirshowitz theorem [**2**]. For example, $r(k,3) = \lceil \frac{(k+2)(k+1)}{6} \rceil$, for $k \neq 2,4$ and $r(2,3) = 3$, $r(4,3) = 6$. It is known that for fixed k and n, if $r_1 \leq r_2$ are typical ranks, then so are all integers greater than r_1 and smaller than r_2 [**6**]. Thus typical ranks are integers from an interval that starts with $r(k,n)$. Still, determining all typical ranks for all k and n seems a very hard task.

Natural questions that arise are as follows:

(1) Is it true that for most k, n the integer $r(k,n)+1$ is a typical rank? For $k \geq 4$ and $n = 3$ this is conjectured in [**20**].

(2) If $r(k,n)+1$ is a typical rank, what is the Zariski closure of the boundary of the locus of real tensors of real rank $r(k,n)$?

One of the easiest cases is $k = 2$, i.e. quadrics. Here a full rank real quadric (considered as a symmetric matrix) has, up to change of coordinates, a unique presentation as $f = x_1^2 + \cdots + x_p^2 - x_{p+1}^2 - \cdots - x_n^2$. Thus we have only one typical rank.

Let us now study in detail the case of **ternary forms**, i.e., $n = 3$. For cubic forms $k = 3$ we also have only one typical rank, namely $r(3,3) = 4$, cf. [**4**]. For $k = 4$ we know that we have more typical ranks.

DEFINITION 3.9. The *real rank boundary*, denoted by $\partial_{alg}(R_{k,n})$, is defined as the Zariski closure of the topological boundary of forms of typical rank $r(k,n)$.

THEOREM 3.10 ([**20**, Theorem 4.1]). *The algebraic boundary $\partial_{alg}(R_{4,3})$ is a reducible hypersurface in the space of quartics. One of its irreducible components has degree* 51.

The proof of the theorem provides an explicit formula for the degree-51 polynomial, known as the *Blekherman discriminant.* The fact that the boundary is reducible is proved by presenting an explicit interval, that does not intersect the zero-locus of the Blekherman discriminant, but joins points of minimal and non-minimal real rank. An explicit conjectural description of the boundary is given in [**20**, Conjecture 4.3].

A very beautiful geometric situation happens for quintics $k = 5$. Here, by classical results due to Hilbert, Richmond, and Palatini, a general ternary quintic has a *unique* representation as a sum of seven fifth powers of linear forms. Note that this is a stronger statement than $r(5,3) = 7$. To a general quintic P we may associate the seven linear forms $\{l_1, \ldots, l_7\}$. Even if P is real, the linear forms do not have to be. Clearly they are real precisely when the real Waring rank coincides with the complex one. However, if the linear forms are not real, the set of those forms must be closed with respect to conjugation. Indeed, if $P = \sum l_i^5$, then $P = \overline{P} = \sum \overline{l_i}^5$, but the decomposition must be unique.

To understand the algebraic boundary $\partial_{alg}(R_{5,3})$ we have to ask the question: how is it possible that a real decomposition changes to a complex, self-conjugate decomposition, as P changes? This seems impossible, as if any single l_i becomes not real, it will be "missing" its conjugate. We see that the only possibility is that some of the l_i's come together to single l and then change simultaneously into complex linear forms. In our case, generically two l_i's come together and then form a conjugate pair. The formal proof is given in [**20**, Theorem 5.3].

THEOREM 3.11. *The algebraic boundary $\partial_{alg}(R_{5,3})$ is an irreducible hypersurface of degree* 168. *It is the join of the fifth secant variety of the Veronese variety and the tangential variety of the Veronese variety.*

For further results, where some components of the algebraic boundary are identified, for sextics and septics we refer to [**20**, Proposition 5.6 and Theorem 6.1]. The case of binary forms and different ranks is very nicely discussed in [**7**].

We have seen that the fact that for ternary quintics we only have one decomposition as a sum of powers plays an important role. Indeed, for ternary forms such a situation happens only when $k = 5$, cf. [**19**]. Still, we have finitely many decompositions, unless $k = 2, 4$ or k is divisible by three. The geometry of all decompositions is governed by the *variety of sums of powers*.

DEFINITION 3.12. Let P be a homogeneous polynomial of degree k, in n variables and of Waring rank r. The *variety of sums of powers* $VSP(P)$ is the closure, in the Hilbert scheme, of the locus consisting of r-tuples of linear forms $l_i \in (\mathbb{P}^{n-1})^\vee$ such that $P = \sum l_i^k$.

For example, when P is a general ternary quintic, then $VSP(P)$ is one point. Inside $VSP(P)$ one may define a semialgebraic subset corresponding to those decompositions that are real. It is known as the *space of real sums of powers* and denoted by $SSP(P)$. Note that $SSP(P)$ is nonempty if and only if the real Waring rank of P coincides with the complex Waring rank. For experts, we point out that $SSP(P)$ is contained in the locus of real points of $VSP(P)$, but the containment is in general strict.

Let us explain the difference between VSP and SSP on the example of quadrics. Let P be a real quadric in n variables, of rank n. Recall, that a basic invariant of a real quadric is its *signature* $(p, n-p)$. Precisely, up to change of coordinates we have

$$P(x_1, \dots, x_n) = x_1^2 + \dots + x_p^2 - x_{p+1}^2 - \dots - x_n^2.$$

First, suppose $p, n-p \neq n$. We note that in this case $SSP(P)$ is a proper subset of the real part of $VSP(P)$. Indeed, we have

$$P(x_1, \dots, x_n) = x_1^2 - x_{p+1}^2 + \dots = (\frac{1}{\sqrt{2}}(x_1 + ix_{p+1}))^2 + (\frac{1}{\sqrt{2}}(x_1 - ix_{p+1}))^2 + \dots,$$

where the second decomposition is not real, i.e., is an example of a point not belonging to $SSP(P)$. When the quadric is definite, i.e., $p = n$ or $p = 0$, this phenomenon does not happen [**20**, Theorem 2.1].

The geometry of the VSP, even in case of ternary quadrics is very interesting. By the results of Mukai [**23**] and Jelisiejew, Ranestad, and Schreyer [**13**, **27**] the VSP in this case is a linear, codimension three, cut of the Grassmannian $G(3, 5)$ under the Plücker embedding. This is a degree five Fano threefold, known as the *quintic del Pezzo threefold*. The explicit formula for the hypersurface that is the boundary of the $SSP(P)$ for $P = x_1^2 + x_2^2 - x_3^2$ is derived in [**20**, Example 2.5] using hyperdeterminants of tensors.

Acknowledgment

We are grateful to Rodica Dinu, Vincenzo Galgano, and Anatoli Shatsila for useful comments.

References

[1] H. Abo, A. Seigal, and B. Sturmfels, *Eigenconfigurations of tensors*, Algebraic and geometric methods in discrete mathematics, Contemp. Math., vol. 685, Amer. Math. Soc., Providence, RI, 2017, pp. 1–25. MR3625569

[2] J. Alexander and A. Hirschowitz, *Polynomial interpolation in several variables*, J. Algebraic Geom. **4** (1995), no. 2, 201–222. MR1311347

[3] E. S. Allman, J. A. Rhodes, B. Sturmfels, and P. Zwiernik, *Tensors of nonnegative rank two*, Linear Algebra Appl. **473** (2015), 37–53, DOI 10.1016/j.laa.2013.10.046. MR3338324

[4] M. Banchi, *Rank and border rank of real ternary cubics*, Boll. Unione Mat. Ital. **8** (2015), no. 1, 65–80, DOI 10.1007/s40574-015-0027-z. MR3376916

[5] D. J. Bates and L. Oeding, *Toward a salmon conjecture*, Exp. Math. **20** (2011), no. 3, 358–370, DOI 10.1080/10586458.2011.576539. MR2836258

[6] A. Bernardi, G. Blekherman, and G. Ottaviani, *On real typical ranks*, Boll. Unione Mat. Ital. **11** (2018), no. 3, 293–307, DOI 10.1007/s40574-017-0134-0. MR3855760

[7] M. C. Brambilla and G. Staglianò, *Algebraic boundaries among typical ranks for real binary forms of arbitrary degree*, Found. Comput. Math. **21** (2021), no. 4, 1003–1022, DOI 10.1007/s10208-020-09474-9. MR4298238

[8] W. Buczyńska and J. A. Wiśniewski, *On geometry of binary symmetric models of phylogenetic trees*, J. Eur. Math. Soc. (JEMS) **9** (2007), no. 3, 609–635, DOI 10.4171/JEMS/90. MR2314109

[9] D. Cartwright and B. Sturmfels, *The number of eigenvalues of a tensor*, Linear Algebra Appl. **438** (2013), no. 2, 942–952, DOI 10.1016/j.laa.2011.05.040. MR2996375

[10] J. E. Fornaess and N. Sibony, *Complex dynamics in higher dimension. II*, Modern methods in complex analysis (Princeton, NJ, 1992), Ann. of Math. Stud., vol. 137, Princeton Univ. Press, Princeton, NJ, 1995, pp. 135–182. MR1369137

[11] S. Friedland and E. Gross, *A proof of the set-theoretic version of the salmon conjecture*, J. Algebra **356** (2012), 374–379, DOI 10.1016/j.jalgebra.2012.01.017. MR2891138

[12] S. Friedland and G. Ottaviani, *The number of singular vector tuples and uniqueness of best rank-one approximation of tensors*, Found. Comput. Math. **14** (2014), no. 6, 1209–1242, DOI 10.1007/s10208-014-9194-z. MR3273677

[13] J. Jelisiejew, K. Ranestad, and F.-O. Schreyer, *The variety of polar simplices II*, Preprint, arXiv:2304.00533, (2023).

[14] M. A. Khadam and M. Vodička, *Cohen–Macaulay and Gorenstein tangential varieties of the Segre–Veronese varieties*, Bull. Lond. Math. Soc. **54** (2022), no. 5, 1721–1740, DOI 10.1112/blms.12652. MR4505729

[15] K. Kozhasov, *On fully real eigenconfigurations of tensors*, SIAM J. Appl. Algebra Geom. **2** (2018), no. 2, 339–347, DOI 10.1137/17M1145902. MR3814009

[16] J. M. Landsberg, *Tensors: Geometry and applications*, Graduate Studies in Mathematics, vol. 128, American Mathematical Society, Providence, RI, 2012, DOI 10.1090/gsm/128. MR2865915

[17] J. M. Landsberg and L. Manivel, *On the ideals of secant varieties of Segre varieties*, Found. Comput. Math. **4** (2004), no. 4, 397–422, DOI 10.1007/s10208-003-0115-9. MR2097214

[18] J. M. Landsberg and L. Manivel, *Generalizations of Strassen's equations for secant varieties of Segre varieties*, Comm. Algebra **36** (2008), no. 2, 405–422, DOI 10.1080/00927870701715746. MR2387532

[19] M. Mella, *Base loci of linear systems and the Waring problem*, Proc. Amer. Math. Soc. **137** (2009), no. 1, 91–98, DOI 10.1090/S0002-9939-08-09545-2. MR2439429

[20] M. Michałek, H. Moon, B. Sturmfels, and E. Ventura, *Real rank geometry of ternary forms*, Ann. Mat. Pura Appl. (4) **196** (2017), no. 3, 1025–1054, DOI 10.1007/s10231-016-0606-3. MR3654943

[21] M. Michałek, L. Oeding, and P. Zwiernik, *Secant cumulants and toric geometry*, Int. Math. Res. Not. IMRN **12** (2015), 4019–4063, DOI 10.1093/imrn/rnu056. MR3356747

[22] M. Michałek and B. Sturmfels, *Invitation to nonlinear algebra*, Graduate Studies in Mathematics, vol. 211, American Mathematical Society, Providence, RI, [2021] ©2021, DOI 10.1090/gsm/211. MR4423369

[23] S. Mukai, *Fano 3-folds*, Complex projective geometry (Trieste, 1989/Bergen, 1989), London Math. Soc. Lecture Note Ser., vol. 179, Cambridge Univ. Press, Cambridge, 1992, pp. 255–263, DOI 10.1017/CBO9780511662652.018. MR1201387

[24] L. Pachter and B. Sturmfels, *Statistics*, Algebraic statistics for computational biology, Cambridge Univ. Press, New York, 2005, pp. 3–42, DOI 10.1017/CBO9780511610684.004. MR2205866

[25] L. Qi, *Eigenvalues and invariants of tensors*, J. Math. Anal. Appl. **325** (2007), no. 2, 1363–1377, DOI 10.1016/j.jmaa.2006.02.071. MR2270090

[26] C. Raicu, *Secant varieties of Segre–Veronese varieties*, Algebra Number Theory **6** (2012), no. 8, 1817–1868, DOI 10.2140/ant.2012.6.1817. MR3033528

[27] K. Ranestad and F.-O. Schreyer, *Varieties of sums of powers*, J. Reine Angew. Math. **525** (2000), 147–181, DOI 10.1515/crll.2000.064. MR1780430

[28] K. Rose, *Toric non-gorenstein loci*, Preprint, arXiv:2204.01317, (2022).

[29] B. Sturmfels, *Tensors and their eigenvectors*, Notices Amer. Math. Soc. **63** (2016), no. 6, 604–606, DOI 10.1090/noti1389. MR3495220

[30] B. Sturmfels and S. Sullivant, *Toric ideals of phylogenetic invariants*, J. Comp. Biol. **12** (2005), no. 4, 457–481.

[31] B. Sturmfels and P. Zwiernik, *Binary cumulant varieties*, Ann. Comb. **17** (2013), no. 1, 229–250, DOI 10.1007/s00026-012-0174-1. MR3027579

[32] S. Sullivant, *Algebraic statistics*, Graduate Studies in Mathematics, vol. 194, American Mathematical Society, Providence, RI, 2018, DOI 10.1090/gsm/194. MR3838364

[33] F. L. Zak, *Tangents and secants of algebraic varieties*, Translations of Mathematical Monographs, vol. 127, American Mathematical Society, Providence, RI, 1993. Translated from the Russian manuscript by the author, DOI 10.1090/mmono/127. MR1234494

University of Konstanz, Germany, Fachbereich Mathematik und Statistik, Fach D 197 D-78457 Konstanz, Germany

Email address: mateusz.michalek@uni-konstanz.de

Proceedings of Symposia in Pure Mathematics
Volume **111**, 2025
https://doi.org/10.1090/pspum/111/02050

Snapshot of algebraic vision

Joe Kileel and Kathlén Kohn

Abstract. In this survey article, we present interactions between algebraic geometry and computer vision, which have recently come under the header of algebraic vision. The subject has given new insights in multiple view geometry and its application to 3D scene reconstruction and carried a host of novel problems and ideas back into algebraic geometry.

Computer vision is the research field that studies how computers can gain understanding from 2D images and videos, similar to human cognitive abilities. Typical computer vision tasks include the automatic recognition of objects in images, the detection of events in videos, and the reconstruction of 3D scenes from many given 2D images. A general overview of computer vision is presented in textbook form in [**174**]. The subject is a pillar in the AI revolution.

Algebraic vision is the symbiosis of computer vision and algebraic geometry. Motivated by Chris Aholt's PhD thesis titled *Polynomials in Multiview Geometry* [**9**] and earlier works, the term "algebraic vision" was coined during a particular lunch held at a Seattle office of Google in early spring 2014, attended by Sameer Agarwal, Chris Aholt, Joe Kileel, Hon-Leung Lee, Max Lieblich, Bernd Sturmfels, and Rekha Thomas. The intent was to encourage interactions between the applied algebraic geometry community and the 3D reconstruction community in computer vision. A short discussion of algebraic vision can be found in the review [**34**] on nonlinear algebra and its applications.

Historically, computer vision made substantial use of projective geometry and computational algebra in parts of its foundations. Specifically *multiple view geometry*, as described in the textbook [**94**] of Hartley and Zisserman, is modeled on projective three-space and two-space and group-equivariant (multi-)linear transformations between these. Similar algebraic treatments of the subject are the textbooks [**135**] and [**139**]. Previously, this connection was not well-appreciated by the computational algebra geometry community. However, in the last decade, algebro-geometric papers and workshops on 3D reconstruction have been appearing, leading

2020 *Mathematics Subject Classification.* Primary 68T45, 14Q20, 13P25; Secondary 13P15, 65H14, 13P10.

The first author was supported in part by NSF awards DMS-2309782 and IIS-2312746, and start-up grants from the Department of Mathematics and Oden Institute at UT Austin.

The second author was supported in part by the Wallenberg AI, Autonomous Systems and Software Program (WASP) funded by the Knut and Alice Wallenberg Foundation.

to novel results in multiple view geometry while motivating developments in applied algebraic geometry.

The present article provides a survey of algebraic vision. No previous knowledge of computer vision is assumed, and the prerequisites for computational algebraic geometry are kept mostly to the level of undergraduate texts [**51**]. Due to space limitations, the article makes no attempt to be comprehensive in any way, but instead it focuses narrowly on the role of projective varieties and systems of polynomial equations in 3D vision. An outline of the sections is as follows:

- In Section 1, we introduce the problem of 3D scene reconstruction from unkown cameras and its algebro-geometric nature.
- In Section 2, we discuss a variety of usual models for cameras.
- In Section 3, we study *multiview varieties* which characterize feasible images of points under fixed cameras. Their defining equations play a key role in 3D reconstruction algorithms, and their Euclidean distance degrees measure the intrinsic complexity of noisy triangulation (i.e., the task of recovering the 3D coordinates of a point observed by known cameras).
- In Section 4, we consider the space of all cameras. We explain how tuples of cameras—up to changes of world coordinates—can be encoded via *multifocal tensors* [**94**].
- In Section 5, we overview the most popular algorithmic pipeline to solve 3D scene reconstruction, highlighting *minimal problems* that are the algebro-geometric heart of the pipeline.
- In Section 6, we describe polynomial solvers for minimal problems, focusing on Gröbner basis methods using *elimination templates* and homotopy continuation. Those method applies to zero-dimensional parameterized polynomial systems in general.
- In Section 7, we discuss algebro-geometric approaches to understand degenerate world scenes and image data, where uniqueness of reconstruction breaks down and algorithms can encounter difficulty.

After reading Sections 1 and 2, the other sections are essentially independent; only Section 6 builds on Section 5. We provide specific pointers to earlier sections in case of partial dependencies.

Some important topics in algebraic vision that are omitted include group synchronization (e.g., [**129, 163**]), uses of polynomial optimization (e.g., [**1, 44, 50, 104, 192**]), and approaches based on differential invariants (e.g., [**30, 42**]). Readers may consult [**149**] for a survey that covers numerical and large-scale optimization aspects in 3D reconstruction.

1. Computer vision through the algebraic lens

One of the main challenges in computer vision is the *structure-from-motion* (SfM) problem: given many 2D images, the task is to reconstruct the 3D scene and also the positions of the cameras that took the pictures. This has many applications such as 3D mapping from images taken by drones [**160**], to localize and navigate autonomous cars and robots in a 3D world [**83**], or in the movie industry to reconstruct 3D backgrounds [**109**], for photo tourism [**3**], and for combining real and virtual worlds [**60**].

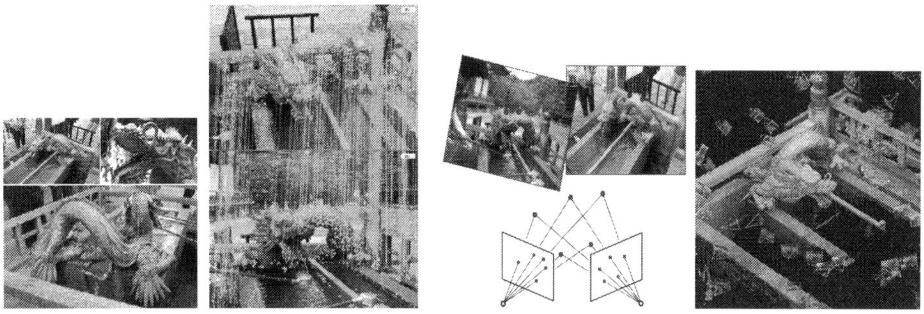

(A) Input images (B) Image matching (C) Reconstruct cameras and 3D points (D) Output

FIGURE 1. 3D reconstruction pipeline (courtesy of Tomas Pajdla).

The structure-from-motion problem is typically solved using the 3D reconstruction pipeline. We will now sketch a highly simplified version of that pipeline, illustrated in Figure 1. We provide more details in Section 5.1. Given a set of 2D images, the first step in the pipeline is to take a few of the given images and identify geometric features, such as points or lines, that they have in common. In Figure 1(B), a detection algorithm has been used that only identifies points. In the second step of the pipeline, we forget the original images and only keep the geometric features we have identified. We reconstruct the 3D coordinates of those features and also the camera poses, that is, the locations and orientations of the cameras. In Figure 1(C), five common points were identified on two images, so we aim to reconstruct the five points in 3-space and the two cameras. Finally, we repeat this process several times until we have recovered all cameras and also enough geometric features to approximate the 3D scene.

As the second step of the pipeline forgets the pictures and only works with algebro-geometric features, such as points or lines, the reconstruction problem becomes purely algebraic. More specifically, we aim to compute a fiber of the *joint camera map*

$$\Phi : \mathcal{X} \times \mathcal{C}_m \dashrightarrow \mathcal{Y}, \tag{1.1}$$

that maps an arrangement $X \in \mathcal{X}$ of 3D features and a tuple $(C_1, \ldots, C_m) \in \mathcal{C}_m$ of cameras to the m 2D images of X taken by the cameras. For instance in Figure 1(C), the joint camera map becomes

$$\Phi : \left(\mathbb{R}^3\right)^5 \times \mathcal{C}_2 \dashrightarrow \left(\mathbb{R}^2\right)^5 \times \left(\mathbb{R}^2\right)^5. \tag{1.2}$$

A full specification of the joint camera map requires a choice of camera model. The simplest model is a *pinhole camera*; see Figure 2. Such a camera simply takes a picture of a point in space by projecting it onto a plane. A pinhole camera in standard position is typically assumed to be centered at the origin such that its image plane is $H = \{(x, y, z) \in \mathbb{R}^2 \mid z = 1\}$. In these coordinates, the pinhole camera is the map

$$\mathbb{R}^3 \dashrightarrow H, \quad (x, y, z) \longmapsto \left(\tfrac{x}{z}, \tfrac{y}{z}, 1\right).$$

Often homogeneous coordinates are used to model cameras. This means that each point in the image plane is identified with the light ray passing through the

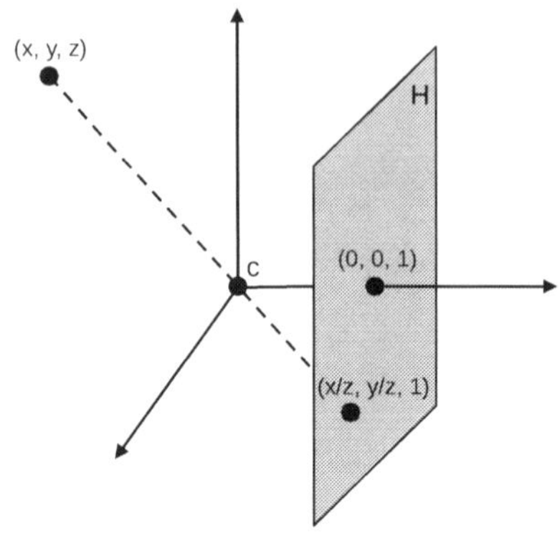

FIGURE 2. A pinhole camera in standard position is centered at $c = (0,0,0)$ and maps world points (x,y,z) to image points $(\frac{x}{z}, \frac{y}{z}, 1)$ on the image plane H.

point and the origin. In homogeneous coordinates, the standard pinhole camera in Figure 2 becomes

$$\mathbb{P}^3_{\mathbb{R}} \dashrightarrow \mathbb{P}^2_{\mathbb{R}}, \quad [x:y:z:w] \longmapsto [x:y:z].$$

This map is defined everywhere except at the camera center $[0:0:0:1]$, i.e., the origin in the affine chart where $w = 1$.

The projective geometry approach in modeling cameras is thoroughly explained in the textbook [**94**]. That book laid many foundations and conventions used in modern computer vision and offers a great entry point for the algebraic community into the field of computer vision. The main focus of [**94**] is multiview geometry, where a 3D object is viewed by several cameras, such as in Figure 1. In that setting, we cannot assume that all cameras are in standard position as described above. Instead, a pinhole camera is more generally given by a 3×4 matrix A of rank three. The corresponding camera map

$$\mathbb{P}^3_{\mathbb{R}} \dashrightarrow \mathbb{P}^2_{\mathbb{R}}, \quad X \longmapsto AX.$$

is defined everywhere except at the camera center that is given by the kernel of A. The standard camera in Figure 2 corresponds to the matrix

$$\begin{bmatrix} 1 & 0 & 0 & 0 \\ 0 & 1 & 0 & 0 \\ 0 & 0 & 1 & 0 \end{bmatrix}.$$

Hence, when using pinhole cameras and homogeneous coordinates, the *camera variety* $\mathcal{C}_m$ in (1.1) that describes all m-tuples of such cameras is

$$\mathcal{C}_m = (\mathbb{P}\,\mathrm{Mat}_3^{3\times 4})^m,$$

where $\mathrm{Mat}_3^{3\times 4} \subset \mathbb{R}^{3\times 4}$ denotes the set of 3×4 matrices of rank three. For instance, the joint camera map in (1.2) becomes

$$\begin{aligned} \Phi : \left(\mathbb{P}^3_{\mathbb{R}}\right)^5 \times (\mathbb{P}\,\mathrm{Mat}_3^{3\times 4})^2 &\dashrightarrow \left(\mathbb{P}^2_{\mathbb{R}}\right)^5 \times \left(\mathbb{P}^2_{\mathbb{R}}\right)^5, \\ (X_1, \ldots, X_5, A_1, A_2) &\longmapsto (A_1X_1, \ldots, A_1X_5, A_2X_1, \ldots, A_2X_5). \end{aligned}$$

In the next section, we review common camera models and highlight algebraic vision articles studying camera geometry. In the remaining sections, our focus returns to the joint camera map in (1.1): We will see that many computer vision problems can be formulated using the joint camera map—such as understanding the image of a shape in space or reconstructing a 3D shape from several images—and are thus natural to study through the algebraic lens. The recent paper [**2**] gives a similar such unifying algebro-geometric framework for computer vision problems.

2. Camera models

Calibrated cameras. The camera model described in the previous section is known as the *projective/uncalibrated pinhole camera*. The *calibrated pinhole camera* model assumes that every camera is obtained from the standard pinhole camera in Figure 2 by translation and rotation. This means that every camera matrix A is of the form $[R \mid t]$ where $R \in \mathrm{SO}(3)$ is the relative rotation from the standard pinhole camera to the camera with matrix A and the relative translation can be read off from the vector $t \in \mathbb{R}^3$: the camera center c, which is the origin in Figure 2, is now $c = -R^\top t$ (note that the vector $(c,1)^\top \in \mathbb{R}^4$ spans the kernel of the camera matrix $[R \mid t]$). In particular, every calibrated pinhole camera has 6 degrees of freedom (3 for R and 3 for t), whereas a projective pinhole camera has 11 degrees of freedom.

Calibrated pinhole cameras are a commonly used model in applications, corresponding to the case when the internal parameters of the cameras are known (such as from meta data stored inside the image file). There is also a variety of *partly calibrated pinhole cameras*, e.g., a camera with unknown focal length, that have less strict structural assumptions on the 3×4 camera matrices than the fully calibrated model described above. Partly calibrated pinhole cameras are modeled as $K[R \mid t]$ where K is a 3×3 upper triangular *calibration matrix* whose entries are partially known [**94**, Chapter 6].

Distortion. In practice, cameras are not as ideal as in the calibrated model. As seen in Figure 2, the pinhole cameras described so far assume that the world point, the camera center, and the image point are collinear. This assumption does not hold for real-life camera lenses, because they are affected by various kinds of distortion. The main factor of deviation from the idealistic pinhole camera model is typically radial distortion; see Figure 3.

Often, calibrated cameras are a sufficient approximation of real-life cameras. However, sometimes the impact of radial distortion is too big, e.g., for fisheye cameras. One approach to address radial distortion is to make the camera model more complicated by adding distortion parameters that have to be estimated during 3D reconstruction (see [**94**, Chapter 7.4] for an overview and [**108**] for an algebraic treatment of distortion varieties). Another approach is to simplify the camera model by not estimating the radial distortion at all: Once the center of radial distortion on a given image is determined, we know for every 3D point onto which line through the distortion center it gets mapped by the camera (see Figure 3), although we do not know its exact point in the image. Thus, the camera simplifies to a map $\mathbb{P}^3 \dashrightarrow \mathbb{P}^1$ that sends 3D points to radial lines through the distortion center in the image. This is known as the *1D radial camera model* [**126, 175**].

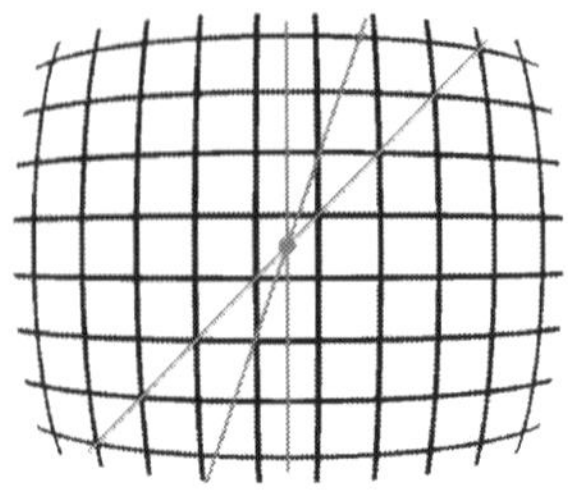

FIGURE 3. A grid affected by radial distortion. Radial lines pass through the distortion center.

Rolling shutter and pushbroom cameras. Many real-world cameras do not function like pinhole cameras that were described so far. For instance, smartphones, that produce a massive and yet growing amount of picture data, use *rolling shutters* (see, e.g., [**53**]). When taking a picture, these cameras scan across the scene, capturing the resulting image row by row. For static cameras taking pictures of still scenes, there is no difference between rolling shutters and standard pinhole cameras that capture an entire image at the same instant (also known as global shutter). However, for moving cameras or objects, in particular videos, global shutter pinhole cameras are not a good approximation of rolling shutters. For example, when a rolling shutter camera moves with constant speed along a line, the picture of any other line in three-space would typically be a conic in the image plane, whereas a pinhole camera maps lines to lines.

A related camera model is the *pushbroom camera*, which is common in satellites [**94**, Chapter 6.4.1]. Such a camera considers only a plane in the 3D world that passes through the camera center and then projects that plane onto a single line. Hence, a nonmoving pushbroom camera produces only an image line, not a plane. Two-dimensional images arise by moving the camera.

Rolling shutter and pushbroom cameras have received little attention from the algebraic community so far. In particular, the theory described in Sections 3–7 is largely undeveloped for such cameras.

Congruences. Ponce, Sturmfels, and Trager [**156**] propose a general framework to study camera geometry that includes all camera models without distortion discussed above. Their notion of a camera is an abstraction from the classical way of modeling a camera by a map $\mathbb{P}^3 \dashrightarrow \mathbb{P}^2$. Instead of identifying a concrete image plane with coordinates, they define a *rational geometric camera* to be a map $\mathbb{P}^3 \dashrightarrow \mathrm{Gr}(1, \mathbb{P}^3)$ from three-space into the Grassmannian of lines. The fiber over any image point under a classical camera $\mathbb{P}^3 \dashrightarrow \mathbb{P}^2$ without distortion is a line, namely a *viewing ray* of the camera. The associated rational geometric camera maps each world point directly to its corresponding viewing ray. For instance, a geometric pinhole camera assigns to every world point distinct from the camera center the unique line spanned by the world point and the camera center.

The image of a rational geometric camera is a *congruence*, i.e., a two-dimensional family of lines. To obtain an actual photographic camera, one composes the geometric camera map with a map from its image congruence to $\mathbb{P}^2$ that models the image formation process via mapping viewing rays to image coordinates.

These physical realizations of rational geometric cameras are the focus of Trager, Sturmfels, Canny, Hebert, and Ponce in [**179**].

In [**156**], the authors suggest an even more general camera definition by saying that a *geometric camera* is an arbitrary congruence in $\mathrm{Gr}(1, \mathbb{P}^3)$. The *order* of a congruence is the number of lines in the congruence that pass through a general point in $\mathbb{P}^3$. Hence, congruences of order one are exactly those that are the images of rational geometric cameras $\mathbb{P}^3 \dashrightarrow \mathrm{Gr}(1, \mathbb{P}^3)$. Congruences of higher order are geometric cameras such that a world point typically appears several times on each image taken by the camera. This happens for instance for many moving rolling shutter cameras, or for *catadioptric cameras* which are cameras that use curved mirrors. The idea of modeling cameras by congruences had been expressed by computer vision researchers before [**151, 154, 171**].

Chirality. In real life, a 3D point has to lie in front of a camera in order to be seen by it. Imposing this semialgebraic constraint on a camera model is referred to as *chirality*. The theoretical investigation of chirality was initiated by Hartley [**88**]; see also [**94**, Chapter 21]. Further studies of this concept were for instance undertaken by Laveau and Faugeras [**127**], Werner and Pajdla [**187–190**], and Agarwal, Pryhuber, Sinn, and Thomas [**5, 157**].

3. Multiview varieties

Given m fixed pinhole cameras $C = (C_1, \dots, C_m)$, where $C_i : \mathbb{P}^3_{\mathbb{R}} \dashrightarrow \mathbb{P}^2_{\mathbb{R}}$, we can specialize the joint camera map Φ from (1.1) to

$$\Phi_C : \mathbb{P}^3_{\mathbb{R}} \dashrightarrow (\mathbb{P}^2_{\mathbb{R}})^m, \tag{3.1}$$

sending a 3D point to its m images under the cameras. The *joint image* of the cameras, first introduced by Triggs [**180**], is $\Phi_C(\mathbb{P}^3_{\mathbb{R}})$. The first algebraic study of the joint image was done by Heyden and Åström [**96**]. They called it the *natural descriptor* and observed that it is not Zariski closed. The Zariski closure $\mathcal{M}_C := \overline{\Phi_C(\mathbb{P}^3_{\mathbb{R}})}$ of the joint image is called the *joint image variety* [**177**] or *multiview variety* [**8**]. It coincides with the Euclidean closure of the joint image [**5**, Theorem 4].

3.1. Multiview constraints.

Many computer vision works have studied systems of polynomials that vanish on the joint image, often referred to as *multiview constraints*; see, e.g., [**6, 8, 73, 74, 96, 135, 177**]. Their main motivation is that the joint image completely determines the cameras $C_1, \dots, C_m$, up to changes of coordinates in $\mathbb{P}^3$. So one way of thinking about the 3D reconstruction pipeline in Figure 1 is to first obtain point correspondences between the given images (i.e., (noisy) samples on the joint image) and then estimate sufficiently many multiview constraints from the correspondences so as to be able to reconstruct the cameras, then the 3D points.

We now describe the multiview constraints studied by Heyden and Åström [**96**]. Let $A_1, \dots, A_m$ be 3×4 matrices of rank three that define the cameras, i.e., $C_i : X \mapsto A_i X$. We write $x_i = (x_{i1} : x_{i2} : x_{i3})$ for the homogeneous coordinates on the i-th image $\mathbb{P}^2_{\mathbb{R}}$, and form the $3m \times (m+4)$ matrix

$$M_A = M_A(x_1, \ldots, m) := \begin{bmatrix} A_1 & x_1 & 0 & \cdots & 0 \\ A_2 & 0 & x_2 & \cdots & 0 \\ \vdots & \vdots & \vdots & \ddots & \vdots \\ A_m & 0 & 0 & \cdots & x_m \end{bmatrix}.$$

A tuple of image points $(x_1, \ldots, x_m) \in (\mathbb{P}^2_{\mathbb{R}})^m$ lies in the joint image if and only if there is some 3D point X and nonzero scalars λ_i such that $A_i X = \lambda_i x_i$ for all $i = 1, \ldots, m$. This implies that $[X, -\lambda_1, \ldots, -\lambda_m]^\top$ is in the kernel of M_A, so the matrix M_A must be rank deficient. In fact, the multiview variety $\mathcal{M}_C$ is cut out by the maximal minors of M_A:

$$\mathcal{M}_C = \{(x_1, \ldots, x_m) \in (\mathbb{P}^2_{\mathbb{R}})^m \mid \operatorname{rk} M_A(x_1, \ldots, x_m) < m+4\}. \tag{3.2}$$

These maximal minors are multilinear, i.e., they are linear in every triplet x_i of variables.

EXAMPLE 3.1. For $m = 2$ cameras, M_A is a 6×6 matrix. Its determinant defines the multiview hypersurface $\mathcal{M}_C$ in $\mathbb{P}^2_{\mathbb{R}} \times \mathbb{P}^2_{\mathbb{R}}$. This equation is a bilinear form in x_1, x_2 represented by a matrix $F \in \mathbb{R}^{3\times 3}$, i.e., $\det(M_A) = x_2^\top F x_1$. Here F is called the *fundamental matrix* for projective pinhole cameras, and the *essential matrix* for calibrated pinhole cameras. Longuet-Higgins's seminal work [**133**] introduced the essential matrix to the computer vision community along with a first 3D reconstruction algorithm. We will discuss essential matrices, fundamental matrices, and their generalizations to more than two cameras in Section 4.

For every subset $J \subseteq \{1, \ldots, m\}$ of cardinality k, we consider the $3k \times (k+4)$ submatrix $M_{A,J}$ of M_A whose rows and columns correspond to the cameras C_j for $j \in J$. As the maximal minors of $M_{A,J}$ are multilinear constraints in the k triplets of variables $\{x_j \mid j \in J\}$, we refer to the set of all maximal minors of all $M_{A,J}$ with $|J| = k$ as the *k-linearities*. As seen in (3.2), the multiview variety $\mathcal{M}_C$ of m cameras is cut out by the m-linearities. Heyden and Åström [**96**] showed that $\mathcal{M}_C$ is in fact cut out by the 2-linearities if the m camera centers are not coplanar.

An ideal-theoretic treatment of the multiview variety was first done by Aholt, Sturmfels, and Thomas [**8**], and later continued by Agarwal, Pryhuber, and Thomas [**6**]. They study the vanishing ideal of $\mathcal{M}_C$, called the *multiview ideal*.

THEOREM 3.2.

(1) *If the m camera centers are pairwise distinct, the multiview ideal is generated by the 2- and 3-linearities* [**6**, Theorem 3.7].
(2) *If all 4×4 minors of the $4 \times 3m$ matrix $[A_1^\top \; A_2^\top \; \cdots \; A_m^\top]$ are nonzero, then the 2-, 3-, and 4-linearities form a universal Gröbner basis of the multiview ideal* [**8**, Theorem 2.1].

A nice overview of the joint image and its Zariski closure (equivalently, Euclidean closure) is presented in *the joint image handbook* by Trager, Hebert, and Ponce [**177**]. They provide an explicit description of the actual joint image (instead of its closure) as a constructible set, and discuss sets of k-linearities that cut out the multiview variety as well as sets of k-linearities that are not sufficient to define $\mathcal{M}_C$ but uniquely determine the cameras $C_1, \ldots, C_m$, up to coordinates changes in $\mathbb{P}^3_{\mathbb{R}}$.

3.2. Euclidean distance degree and triangulation. *Triangulation* refers to the problem of reconstructing the 3D coordinates of a point from 2D images when the cameras are known. For pinhole cameras $C = (C_1, \dots, C_m)$, this means to compute the fiber of the specialized joint camera map Φ_C in (3.1). In theory, triangulation is trivial to solve for $m \geq 2$ cameras, since the fiber under each camera $C_i : \mathbb{P}^3_{\mathbb{R}} \dashrightarrow \mathbb{P}^2_{\mathbb{R}}$ is a line and so the desired 3D point is the intersection of these m lines in $\mathbb{P}^3_{\mathbb{R}}$. However, in practice, the image measurements are noisy, which means that the given image points $(x_1, \dots, x_m) \in (\mathbb{P}^2_{\mathbb{R}})^m$ do not lie on the multiview variety $\mathcal{M}_C$. Thus, to solve triangulation, one first needs to *denoise* the given data, for instance by finding a nearby point on the multiview variety $\mathcal{M}_C$ to the given data point $x = (x_1, \dots, x_m) \in (\mathbb{P}^2_{\mathbb{R}})^m$.

In practical applications, the given image points are finite, which means that after a choice of affine chart $\mathbb{R}^2$ of each $\mathbb{P}^2_{\mathbb{R}}$ we may assume that $x \in \mathbb{R}^{2m}$. In this setting, we want to find the closest point x on the restriction of $\mathcal{M}_C$ to the chosen affine chart, that is, on the *affine multiview variety*:

$$\mathcal{M}_C^{\alpha} = \mathcal{M}_C \cap \mathbb{R}^{2m}.$$

Typically, distance is measured via the squared Euclidean distance on $\mathbb{R}^{2m}$. Regarding $\mathcal{M}_C^{\alpha}$ as a complex variety in $\mathbb{C}^{2m}$, for almost all data points $x \in \mathbb{C}^{2m}$, this optimization problem has a constant number of complex critical points [**61**]. That number is known as the *Euclidean distance degree* (*ED degree* for short) of the variety $\mathcal{M}_C^{\alpha}$. ED degree measures the algebraic complexity of the triangulation problem, in the sense that it provides an upper bound for the number of real critical points for generic data points $x \in \mathbb{R}^{2m}$.

Hartley and Sturm [**92**] showed that the ED degree of the affine multiview variety $\mathcal{M}^{\alpha}_{(C_1,C_2)}$ for two cameras is 6. Stewénius and Nistér computed the ED degree of $\mathcal{M}_C^{\alpha}$ for $m \leq 7$ cameras and observed that it grows cubically [**90**]. The problem of determining the ED degree of the affine multiview variety actually motivated the general introduction of Euclidean distance degrees of algebraic varieties by Draisma, Horobeţ, Ottaviani, Sturmfels, and Thomas [**61**, Example 3.3], as well as follow-up work on orthogonally invariant matrix sets [**62**, **63**]. In [**61**, Conjecture 3.4], a precise cubic formula for the ED degree of $\mathcal{M}_C^{\alpha}$ for an arbitrary number of cameras was conjectured. Harris and Lowengrub [**86**] provided an upper bound for the ED degree, showing that it is indeed bounded by some cubic polynomial in the number of cameras, before [**61**, Conjecture 3.4] was finally proven by Maxim, Rodriguez, and Wang.

THEOREM 3.3 ([**138**]). *When the $m \geq 2$ cameras are in general position, the Euclidean distance degree of the affine multiview variety $\mathcal{M}_C^{\alpha}$ is*

$$\frac{9}{2}m^3 - \frac{21}{2}m^2 + 8m - 4.$$

Informally, the theorem means triangulation with a large number of cameras may be quite difficult: the number of critical points grows cubically with m.

3.3. Extensions. There are several extensions of multiview varieties which incorporate additional constraints or apply to different models. We discuss some recent extensions below.

Chirality. The discussion of multiview varieties and triangulation above ignores the constraint of chirality, i.e., that the 3D point must lie in front of each camera in order to be seen by it. Agarwal, Pryhuber, Sinn, and Thomas [**5**] provide semialgebraic descriptions (using polynomial equalities and inequalities) of both the *chiral domain*, that is the set of points in $\mathbb{P}^3_{\mathbb{R}}$ that lie in front of each of the m pinhole cameras C_i, and the *chiral joint image*, i.e., the image of the chiral domain under the specialized joint camera map Φ_C in (3.1). The chiral joint image is a subset of Triggs's joint image $\Phi_C(\mathbb{P}^3_{\mathbb{R}})$ [**180**] and describes the true image of the 3D world as seen by the cameras $C = (C_1, \ldots, C_m)$. Hence, the semialgebraic description of the chiral joint image from [**5**] is a refinement of the multiview constraints discussed above.

Congruences. Multiview varieties can be naturally generalized to other camera models (besides pinhole cameras as described above). Recall that a rational geometric camera is a map

$$\mathbb{P}^3 \dashrightarrow \Sigma \subset \mathrm{Gr}(1, \mathbb{P}^3)$$

that maps a world point X to a line passing through X. The Zariski closure of the image of the map is a congruence Σ, i.e., a surface in the Grassmannian $\mathrm{Gr}(1, \mathbb{P}^3)$ of lines. The map is defined everywhere except at its *focal locus*, that is, the set of points $X \in \mathbb{P}^3$ that belong to more than one line on Σ.

Given m rational geometric cameras $C = (C_1, \ldots, C_m)$, $C_i : \mathbb{P}^3 \dashrightarrow \Sigma_i \subset \mathrm{Gr}(1, \mathbb{P}^3)$, the *multi-image variety* $\mathcal{M}_C$ is the Zariski closure of the image of

$$\Phi_C : \mathbb{P}^3 \dashrightarrow \Sigma_1 \times \ldots \times \Sigma_m \subset \mathrm{Gr}(1, \mathbb{P}^3)^m, \quad X \longmapsto (C_1(X), \ldots, C_m(X)).$$

Ponce, Sturmfels, and Trager [**156**] showed that the multi-image variety often has an easy description in terms of the *concurrent lines variety* $V_m \subset \mathrm{Gr}(1, \mathbb{P}^3)^m$ that consists of all m-tuples of lines that meet in a common point.

THEOREM 3.4 ([**156**, Theorem 5.1]). *If the m focal loci of the cameras $C_1, \ldots, C_m$ are pairwise disjoint, then $\mathcal{M}_C = V_m \cap (\Sigma_1 \times \ldots \times \Sigma_m)$.*

The authors also describe a minimal set of generators and a reduced Gröbner bases for the prime ideal of the concurrent lines variety V_m in [**156**, Theorem 3.1]. Moreover, using the Plücker embedding $\mathrm{Gr}(1, \mathbb{P}^3) \subset \mathbb{P}^5$, they conjectured a formula for the multidegree of V_m in $(\mathbb{P}^5)^m$, which was proven by Escobar and Knutson [**68**, Theorem 2.4]. The latter theorem also provides the multidegree of the multi-image variety $\mathcal{M}_C$ in $(\mathbb{P}^5)^m$ under the assumption that the m focal loci of the cameras $C_1, \ldots, C_m$ are pairwise disjoint.

3D data. Another way to generalize multiview varieties is by not only considering the image of a single 3D point under the given cameras, but rather instead considering the image of more complex 3D objects. The *line multiview variety*, i.e., the Zariski closure of the image of the map $\mathrm{Gr}(1, \mathbb{P}^3) \dashrightarrow (\mathrm{Gr}(1, \mathbb{P}^2))^m$ that sends a 3D line to its images under m pinhole cameras, is described in [**35**, **74**]. Multiview varieties for more than one point are studied in [**102**]. For instance, the authors there consider the set of point pairs in $\mathbb{R}^3 \times \mathbb{R}^3$ that are a Euclidean distance of 1 apart from each other. Let $\mathcal{X} \subset \mathbb{P}^3_{\mathbb{R}} \times \mathbb{P}^3_{\mathbb{R}}$ be the Zariski closure of said set. Given m pinhole cameras $C = (C_1, \ldots, C_m)$, the authors restrict the joint camera map from (1.1) to:

$$\mathcal{X} \dashrightarrow (\mathbb{P}^2_{\mathbb{R}})^m \times (\mathbb{P}^2_{\mathbb{R}})^m, \quad (X_1, X_2) \longmapsto (\Phi_C(X_1), \Phi_C(X_2)), \tag{3.3}$$

and define the *rigid multiview variety* as the Zariski closure of the image of this map. They derive a set of polynomial equations cutting out the rigid multiview variety and give a conjecture for a minimal generating set of its prime ideal. The authors of [**102**] also describe defining polynomial equations for other variations of multiview varieties, e.g., when $\mathcal{X}$ in (3.3) is instead taken to be an arbitrary hypersurface in $\mathbb{P}^3_{\mathbb{R}} \times \mathbb{P}^3_{\mathbb{R}}$, or when $\mathcal{X}$ is in $(\mathbb{P}^3_{\mathbb{R}})^3$ and consists of triples of points with fixed pairwise Euclidean distances, or when $\mathcal{X}$ is in $(\mathbb{P}^3_{\mathbb{R}})^4$ and consists of all 4-tuples of points that lie on a common plane in $\mathbb{P}^3_{\mathbb{R}}$.

Higher dimensions. Several works have studied higher-dimensional generalizations of the specialized joint camera map in (3.1), i.e., maps from a projective space onto a product of projective spaces, all of arbitrary dimension, e.g., [**49, 100, 130**]. The study of higher-dimensional pinhole cameras is motivated by capturing *dynamic scenes*. We will sketch this application and some higher-dimensional extensions of multiview varieties in Section 4.3.5.

4. Understanding camera tuples

4.1. Symmetry of joint camera map. The joint camera map $\Phi : \mathcal{X} \times \mathcal{C}_m \dashrightarrow \mathcal{Y}$ often carries a symmetry: depending on the specific camera models and 3D object types, there exists a group G acting on the fibers of Φ. This symmetry arises by simultaneously acting on the cameras and the 3D objects via 3D change of coordinates which preserve the camera model and 3D objects. This has the important implication that in 3D reconstruction (such as structure-from-motion) we cannot hope to precisely recover a world scene $(X, (C_1, \ldots, C_m))$ from its 2D image data $\Phi(X, (C_1, \ldots, C_m))$. Instead the best we can reconstruct is the world scene up the action of G.

For example, if $\mathcal{C}_m = (\mathbb{P}\,\mathrm{Mat}_3^{3\times 4})^m$ consists of tuples of projective pinhole cameras and $\mathcal{X} = (\mathbb{P}^3_{\mathbb{R}})^n$ consists of tuples of points in 3D projective space, then the group G is the projective general linear group $G = \mathrm{PGL}(4, \mathbb{R})$. Indeed, note

$$Ax = (Ag)(g^{-1}x)$$

for each $A \in \mathrm{Mat}_3^{3\times 4}$, $x \in \mathbb{P}^3$ and $g \in \mathrm{PGL}(4, \mathbb{R})$. In this case, we can only aim to recover the world scene up to a projective transformation.

As another example, suppose that that $\mathcal{C}_m$ consists of tuples of calibrated cameras $([R_1|t_1], \ldots, [R_m|t_m])$ where $R_i \in \mathrm{SO}(3)$ and $t_i \in \mathbb{R}^3$. Take $\mathcal{X} = (\mathbb{P}^3)^n$ again to consist of tuples of 3D projective points. Now the relevant group is

(4.1)
$$G = \left\{ g \in \mathrm{GL}(4, \mathbb{R}) : g = \begin{bmatrix} R & t \\ 0 & \lambda \end{bmatrix} \text{ for some } R \in \mathrm{SO}(3), t \in \mathbb{R}^3, \lambda \in \mathbb{R} \setminus \{0\} \right\},$$

i.e., the scaled special Euclidean group of $\mathbb{R}^3$ (also known as the similarity group). This is because G is the largest subgroup of 3D changes of coordinates which maps calibrated cameras to calibrated cameras. the world scene is ambiguous up to a proper rigid motion R, and a central scale λ, t So when we know the cameras are calibrated, the world scene is ambiguous up to a proper rigid motion (R, t) and a central scale (λ).

A novel approach to detect all symmetries in a given 3D reconstruction problem has been recently developed by Duff, Korotynskiy, Pajdla, and Regan [**66**]. Their approach is based on based on *Galois / monodromy groups*, and uses numerical homotopy continuation computations (cf. Section 6.2).

4.2. Camera configurations. To cope with such ambiguities, it has been useful in computer vision to consider *camera tuples modulo the relevant symmetries*. Working out camera tuples modulo symmetries can be viewed as one of the accomplishments of multiview geometry [**94**]. From a mathematical standpoint, Aholt and Oeding observed in [**7**] that formally we can regard the space of possible camera configurations as the GIT quotient:

$$\mathcal{C}_m//G. \tag{4.2}$$

Here GIT stands for geometric invariant theory, developed by Mumford [**143**]. The GIT quotient $\mathcal{C}_m//G$ is an abstract, nonembedded scheme or stack. While this GIT description can be made precise to the abstract algebraic geometer, it would not be useable in 3D reconstruction problems. The perspective does tell us that multiview geometry found certain explicit birational models for these GIT quotients, which are useful for 3D reconstruction.

4.3. Multifocal tensors. Camera configurations are represented by *multifocal tensors* in multiview geometry. We present them for configurations of two, three and four cameras, and then give a higher-dimensional construction called *Grassmann tensors* [**91**] motivated by dynamic scenes.

4.3.1. *Two projective cameras: Fundamental matrices.* Fix two projective pinhole cameras $C_1, C_2 : \mathbb{P}^3 \dashrightarrow \mathbb{P}^2$ represented by matrices A_1, A_2. Consider the set of corresponding point pairs defined as the multiview variety:

$$\mathcal{M}_C = \left\{(x_1, x_2) \in \mathbb{P}^2 \times \mathbb{P}^2 : \exists X \in \mathbb{P}^3 \text{ s.t. } C_1(X) = x_1, C_2(X) = x_2\right\}.$$

This set is invariant under the action of $G = \mathrm{PGL}(4, \mathbb{R})$ on $C = (C_1, C_2)$, i.e., it depends only on the configuration C modulo G. On the other hand, $\mathcal{M}_C$ is generically a hypersurface of bidegree $(1,1)$ as explained in Example 3.1:

$$\mathcal{M}_C = \left\{(x_1, x_2) : \exists X \in \mathbb{P}^3\, \exists \lambda_1, \lambda_2 \in \mathbb{R} \text{ s.t. } \begin{bmatrix} A_1 & x_1 & 0 \\ A_2 & 0 & x_2 \end{bmatrix} \begin{bmatrix} X \\ -\lambda_1 \\ -\lambda_2 \end{bmatrix} = 0\right\}$$
$$= \left\{(x_1, x_2) : \det \begin{bmatrix} A_1 & x_1 & 0 \\ A_2 & 0 & x_2 \end{bmatrix} = 0\right\}.$$

This equation for $\mathcal{M}_C$ may be written as $x_2^\top F x_1 = 0$ where for $1 \leq i, j \leq 3$,

$$F_{ij} = (-1)^{i+j} \det \begin{bmatrix} A_{1,\hat{j}} \\ A_{2,\hat{i}} \end{bmatrix},$$

where $A_{1,\hat{j}}$ denotes A_1 with the jth row dropped and likewise for $A_{2,\hat{i}}$. One regards F as in $\mathbb{P}(\mathrm{Mat}^{3\times 3})$, that is, defined only up to nonzero scale. In multiview geometry, F is called the *fundamental matrix* for the pair of cameras C. It may be shown that F generically determines C modulo G [**94**], thus F furnishes an explicit model for the configuration. The collection of all F arising in this way is exactly the set of rank 2 matrices, $\mathbb{P}(\mathrm{Mat}_2^{3\times 3})$, with Zariski closure given by the determinantal variety $\{F \in \mathbb{P}(\mathrm{Mat}^{3\times 3}) : \det(F) = 0\}$. The construction is practically useful in 3D reconstruction because point pairs in $\mathcal{M}_C$ can be detected in real-life 2D images.

4.3.2. *Two calibrated cameras: Essential matrices.* We can restrict the fundamental matrix construction to the case of two calibrated cameras. The restriction yields a subset of the fundamental matrices, whose points are called *essential matrices* and customarily denoted by E in computer vision. Demazure [**57**] determined that the resulting subvariety of the determinantal variety is 5 dimensional, degree 10, with ideal minimally generated by the following ten cubic equations:

$$\det(E) = 0 \qquad \text{and} \qquad EE^\top E - \frac{1}{2}\operatorname{trace}(EE^\top)E = 0. \tag{4.3}$$

Another description is that essential matrices $E \in \mathbb{P}(\mathrm{Mat}^{3\times 3})$ are those with rank 2 whose top singular values are equal, i.e., $\sigma_1(E) = \sigma_2(E) > \sigma_3(E) = 0$. Yet another description is that, after taking the *complex* Zariski closure, the variety of essential matrices can be realized as a complex hyperplane section of the determinantal variety of rank ≤ 2 symmetric 4×4 matrices [**77**].

Here E only depends on C modulo the relevant group, i.e., the scaled special Euclidean group G (4.1). In this case however, the association of a calibrated configuration to an essential matrix is generically two-to-one (cf. *twisted pairs* in [**94**]). Thus the GIT quotient (4.2) gives a degree-2 cover of the variety of essential matrices. The advantage of essential matrices is that they capture corresponding image point pairs which can be detected in real-life images.

4.3.3. *Three cameras: Trifocal tensors.* Given three projective cameras C_1, C_2, $C_3 : \mathbb{P}^3 \dashrightarrow \mathbb{P}^2$ represented by matrices A_1, A_2, A_3, consider the set of corresponding point-line-line triples:

$$\Psi_C = \big\{(x_1, \ell_2, \ell_3) \in \mathbb{P}^2 \times (\mathbb{P}^2)^* \times (\mathbb{P}^2)^* : \\ \exists X \in \mathbb{P}^3 \text{ s.t. } C_1(X) = x_1, C_2(X) \in \ell_2, C_3(X) \in \ell_3\big\},$$

where $(\mathbb{P}^2)^*$ is the dual projective plane consisting of lines in $\mathbb{P}^2$. It then holds

$$\Psi_C = \left\{(x_1, \ell_2, \ell_3) : \exists X \in \mathbb{P}^3\, \exists \lambda_1 \in \mathbb{R} \text{ s.t. } \begin{bmatrix} A_1 & x_1 \\ \ell_2^\top A_2 & 0 \\ \ell_3^\top A_3 & 0 \end{bmatrix} \begin{bmatrix} X \\ -\lambda_1 \end{bmatrix} = 0\right\}$$

$$= \left\{(x_1, \ell_2, \ell_3) : \det \begin{bmatrix} A_1 & x_1 \\ \ell_2^\top A_2 & 0 \\ \ell_3^\top A_3 & 0 \end{bmatrix} = 0\right\},$$

where we have identified ℓ_i with its normal vector of 3 homogeneous coordinates. The equation for Ψ_C may be written using tensor contraction as $T(x_1, \ell_2, \ell_3) = 0$ where $T \in \mathbb{P}(\mathbb{R}^{3\times 3\times 3})$ is given by

$$T_{ijk} \;=\; (-1)^{i+j+k} \det \begin{bmatrix} A_{1,i} \\ A_{2,j} \\ A_{3,\hat{k}} \end{bmatrix}$$

for $1 \leq i, j, k \leq 3$. Here $A_{1,i}$ denotes the ith row of A_1, likewise for $A_{2,j}$, and $A_{3,\hat{k}}$ denotes A_3 with the kth row removed. In multiview geometry T is known as the *trifocal tensor* associated with the camera triple C. The map from (C_1, C_2, C_3) modulo $\mathrm{PGL}(4, \mathbb{R})$ to T is generically one-to-one, so the Zariski closure of the set of trifocal tensors is a birational model for configurations of three (projective) cameras. As proven in [**7**], the trifocal tensors variety has dimension 18, degree 297 and an ideal minimally generated by 10 polynomials in degree 3, 81 polynomials in degree 5, and 1980 polynomials in degree 6. Trifocal tensors are relevant to 3D reconstruction, because of their relation to image correspondence data. They capture point-line-line correspondences by definition via tensor-vector contraction. However it is also true that trifocal tensors capture other correspondence types (e.g., point-point-point) via other multilinear algebraic operations, see [**94**, Part III] for details.

If we restrict (C_1, C_2, C_3) to calibrated pinhole cameras, then we obtain only certain trifocal tensors and a subvariety called the *calibrated trifocal variety* in [**107**]. The calibrated trifocal variety is a birational model for the GIT quotient of triples of calibrated cameras modulo the scaled special Euclidean group G (4.1). Currently the ideal defining the variety is partially understood [**137**].

4.3.4. *Four cameras: Quadrifocal tensors.* A similar construction can be carried out for four projective pinhole cameras C_1, C_2, C_3, C_4 based on correspondences between quadruples of image lines:

$$\big\{(\ell_1,\ell_2,\ell_3,\ell_4) \in (\mathbb{P}^2)^* \times (\mathbb{P}^2)^* \times (\mathbb{P}^2)^* \times (\mathbb{P}^2)^* : \\ \exists X \in \mathbb{P}^3 \text{ s.t. } C_1(X) \in \ell_1, C_2(X) \in \ell_2, C_3(X) \in \ell_3, C_4(X) \in \ell_4\big\}.$$

Similar to the above, this is a hypersurface defined by $Q(\ell_1,\ell_2,\ell_3,\ell_4) = 0$ where $Q \in \mathbb{P}(\mathbb{R}^{3\times3\times3\times3})$ is given by

$$Q_{ijkl} \;=\; (-1)^{i+j+k+l} \det \begin{bmatrix} A_{1,i} \\ A_{2,j} \\ A_{3,k} \\ A_{4,l} \end{bmatrix}.$$

The tensor Q is known as the *quadrifocal tensor* in multiple view geometry. The ideal of the Zariski closure of quadrifocal tensors was studied up to degree 9 in [**147**]. Quadrifocal tensors for calibrated cameras have not been investigated.

4.3.5. *Dynamic scenes.* The fundamental matrix and the trifocal and quadrifocal tensors (and their calibrated counterparts) are used for the reconstruction of static scenes. There are many approaches for the reconstruction of dynamic scenes, i.e., where the 3D points undergo different motions. Here we highlight two algebraic ideas.

Wolf and Shashua [**191**] and Huang, Fossum, and Ma [**99**] argue that many motions of 3D points can be modeled by embedding them into a larger dimensional space such that the camera map becomes a projection of the form $\mathbb{P}^N \dashrightarrow \mathbb{P}^2$; see also the general framework presented by Kaminski and Teicher in [**106**]. This motivated Hartley and Schaffalitzky [**91**] to construct the Grassmann tensor, which we explain in Section 4.3.6. For instance, a 3D point $X = (X_1, X_2, X_3, 1)^\top$ that moves linearly in an affine chart of $\mathbb{P}^3$ with constant velocity in the direction $V = (V_1, V_2, V_3, 0)^\top$ can be embedded into $\mathbb{P}^6$ as follows: $\bar{X} = (V_1 : V_2 : V_3 : X_1 : X_2 : X_3 : 1)^\top$. If A_i denotes the 3×4 matrix representing a moving pinhole camera at time i, the image taken of the moving point X at time i is $A_i(X + iV)$. Writing A_i' for the left-most 3×3 submatrix of A_i and $\bar{A}_i = [iA_i' \mid A_i] \in \mathbb{R}^{3\times7}$, we can express the image as $A_i(X + iV) = \bar{A}_i\bar{X}$. Hence, the camera map is a projection $\mathbb{P}^6 \dashrightarrow \mathbb{P}^2$.

Another algebraic approach to dynamic scenes was suggested by Vidal and coauthors [**72, 186**]. Suppose that a scene observed by a pinhole camera has n independently moving objects and that we take two pictures of the scene at two different times. For each of the n objects, we have a fundamental matrix F_i describing the relative pose of the object at the two instants of time. Any pair of image points (x_1, x_2) coming from a common 3D point must lie on one of the n objects and thus satisfies the equation $(x_2^\top F_1 x_1)\cdots(x_2^\top F_n x_1) = 0$. As this equation is homogeneous of degree n in both x_1 and x_2, we can apply the Veronese embedding $\nu_n : \mathbb{P}^2 \to \mathbb{P}^N, N = \binom{n+2}{2} - 1$, to obtain the following bilinear equation in $\nu(x_1)$ and $\nu(x_2)$: $\nu(x_2)^\top F \nu(x_1) = 0$, where $F \in \mathbb{R}^{N\times N}$ is called the *multibody fundamental matrix*. This construction has been generalized, e.g., to *multibody trifocal tensors* [**93**] and other motion models [**184**]. Finally, we note that 3D motion segmentation from m views, i.e., the problem to determine the number n of moving objects and which given m-tuples of image points belong to which object, can also be solved with *generalized principal component analysis (GPCA)*. GPCA is a method considered by Vidal, Ma, and Sastry to fit an unknown number of linear subspaces of unknown dimensions to given data points in a real vector space [**185**]. Other algebraic methods for GPCA include [**182**].

4.3.6. *Grassmann tensors.* Here we review the Grassmann tensor construction of Hartley and Schaffalitzky [**91**] that unifies multifocal tensors with higher-dimensional generalizations. Consider $\mathbb{P}^N$, noting that $N > 3$ can model dynamic scenes as in Section 4.3.5. Let $C_i : \mathbb{P}^N \dashrightarrow \mathbb{P}^{n_i}$, $X \mapsto A_i X$ be m surjective linear projections. Assume $m \le N + 1 \le \sum_{i=1}^m n_i$. Then we can fix integers $c_1, \ldots, c_m$ satisfying $1 \le c_i \le n_i$ and $c_1 + \ldots + c_m = N + 1$. The special cases $n_i = 2$ and $n_i = 1$ appear in pinhole cameras and 1D radial cameras [**126, 175**], respectively.

Write $\mathrm{Gr}(n - c, \mathbb{P}^n)$ for the Grassmannian of linear subspaces of $\mathbb{P}^n$ of codimension c. The main object is the set of "corresponding subspace tuples":

$$\begin{aligned}(4.4)\quad \{(L_1, \ldots, L_m) \in \mathrm{Gr}(n_1 - c_1, \mathbb{P}^{n_1}) \times \ldots \times \mathrm{Gr}(n_m - c_m, \mathbb{P}^{n_m}) :\\ \exists X \in \mathbb{P}^N \text{ s.t. } C_1(X) \in L_1, \ldots, C_m(X) \in L_m\}.\end{aligned}$$

The authors of [**91**] proved that (4.4) is an algebraic hypersurface; moreover its defining equation is multilinear in the Plücker coordinates for the Grassmannians. Thus, the coefficients of defining equation form an m-way tensor

$$\mathbf{T} = \mathbf{T}_{(C_1,\ldots,C_m),(c_1,\ldots,c_m)}$$

of size $\binom{n_1+1}{c_1} \times \ldots \times \binom{n_m+1}{c_m}$. One calls $\mathbf{T}$ the *Grassmann tensor* for the projections $(C_1, \ldots, C_m)$ with respect to the codimensions $(c_1, \ldots, c_m)$ [**91**]. The authors proved that $\mathbf{T}$ depends only on $(C_1, \ldots, C_m)$ modulo $\mathrm{PGL}(N + 1, \mathbb{R})$. Further, if at least n_i strictly exceeds 1 then $(C_1, \ldots, C_m)$ modulo $\mathrm{PGL}(N + 1)$ is generically uniquely determined by $\mathbf{T}$. Fundamental matrices, trifocal tensors and quadrifocal tensors correspond to $N = 3$ and $(c_1, c_2) = (2, 2)$, $(c_1, c_2, c_3) = (2, 1, 1)$ and $(c_1, c_2, c_3, c_4) = (1, 1, 1, 1)$, respectively.

The construction of $\mathbf{T}$ can be derived here. For $L_i \in \mathrm{Gr}(n_i - c_i, \mathbb{P}^{n_i})$ we choose a basis of the dual space $L_i^* \subseteq (\mathbb{P}^{n_i})^*$ comprised of the hyperplanes containing L_i; let the basis be

$$\mathcal{V}_i = \{v_{i1}, \ldots, v_{ic_i}\} \subseteq (\mathbb{P}^{n_i})^*.$$

The dual linear map $C_i^* : (\mathbb{P}^{n_i})^* \to (\mathbb{P}^N)^*$ pulls $\mathcal{V}_i$ back to $(\mathbb{P}^N)^*$ to give

$$A_i^\top \mathcal{V}_i = \{A_i^\top v_{i1}, \ldots, A_i^\top v_{ic_i}\} \subseteq (\mathbb{P}^N)^*.$$

The requirement on $X \in \mathbb{P}^N$ in (4.4) is $\langle A_i X, v_{ij}\rangle = 0$ for each i, j, or equivalently $\langle X, A_i^\top v_{ij}\rangle = 0$. Hence we require $A_1^\top \mathcal{V}_1 \cup \ldots \cup A_m^\top \mathcal{V}_m$ to not span $(\mathbb{P}^N)^*$, i.e.,

$$(4.5)\quad \begin{bmatrix} A_1^\top v_{11} & \ldots & A_1^\top v_{1c_1} & \ldots & \ldots & A_m^\top v_{m1} & \ldots & A_m^\top v_{mc_m}\end{bmatrix} \in \mathbb{R}^{(N+1)\times(N+1)}.$$

is rank-deficient. The matrix (4.5) factors as

$$\underbrace{\begin{bmatrix} A_1^\top & A_2^\top & \dots & A_m^\top \end{bmatrix}}_{=A^\top} \underbrace{\begin{bmatrix} \mathcal{V}_1 & 0 & \dots & 0 \\ 0 & \mathcal{V}_2 & \dots & 0 \\ \vdots & \vdots & \ddots & \vdots \\ 0 & 0 & \dots & \mathcal{V}_m \end{bmatrix}}_{=\mathcal{V}}. \tag{4.6}$$

We require the determinant of (4.6) to vanish, which, by the Cauchy–Binet formula, is

$$\sum_{\substack{\mathcal{I} \subseteq [(n_1+1)+\ldots+(n_m+1)] \\ \#\mathcal{I}=N+1}} \det\big(A^\top[:,\mathcal{I}]\big)\ \det(\mathcal{V}[\mathcal{I},:]). \tag{4.7}$$

Here $A^\top[:,\mathcal{I}]$ denotes the $(N+1)\times(N+1)$ submatrix consisting of the columns indexed by $\mathcal{I}$, etc. Recalling that the rows in $\mathcal{V}$ containing $\mathcal{V}_i \in \mathbb{R}^{(n_i+1)\times c_i}$ have rank c_i and that $c_1+\ldots+c_m = N+1$, we can restrict the sum in (4.7) as follows:

$$\sum_{\substack{\mathcal{I}_1 \subseteq [n_1+1] \\ \#\mathcal{I}_1 = c_1}} \cdots \sum_{\substack{\mathcal{I}_m \subseteq [n_m+1] \\ \#\mathcal{I}_m = c_m}} \det\left(A_1^\top[:,\mathcal{I}_1] \ \ldots\ A_m^\top[:,\mathcal{I}_m]\right) \prod_{i=1}^m \det\left(\mathcal{V}_i[\mathcal{I}_i,:]\right).$$

At this point, note $\det(\mathcal{V}_i[\mathcal{I}_i,:])$ for $\mathcal{I}_i \subseteq [n_i+1]$ are the dual Plücker coordinates of $L_i \in \mathrm{Gr}(n_i - c_i, \mathbb{P}^{n_i})$. These depend on the choice of basis $\mathcal{V}_i$ only through a global nonzero scale, and (after sign flips) equal the primal Plücker coordinates of L_i up to scale. That is, letting $\mathcal{U}_i = \{u_{i1}, \ldots, u_{i,n_i-c_i+1}\} \subseteq \mathbb{P}^{n_i}$ be a basis for L_i one has

$$\begin{aligned} &\Big(\det(\mathcal{V}_i[\mathcal{I}_i,:]) \ \mid\ \mathcal{I}_i \subseteq [n_i+1], \#\mathcal{I}_i = c_i\Big) \\ &\qquad \propto\ \Big((-1)^{c_i+\sum\mathcal{I}_i} \det(\mathcal{U}_i[\mathcal{I}_i^c,:]) \ \mid\ \mathcal{I}_i \subseteq [n_i+1], \#\mathcal{I}_i = c_i\Big), \end{aligned} \tag{4.8}$$

where $\mathcal{I}_i^c = [n_i+1] \setminus \mathcal{I}_i$ and $\sum \mathcal{I}_i = \sum_{\ell\in\mathcal{I}_i} \ell$.

In view of this discussion, we define the entries of the Grassmann tensor as

$$\mathbf{T}_{\mathcal{I}_1,\ldots,\mathcal{I}_m} = (-1)^{N+1+\sum\mathcal{I}_1+\ldots+\sum\mathcal{I}_m} \det\big(A_1^\top[:,\mathcal{I}_1]\ \ldots\ A_m^\top[:,\mathcal{I}_m]\big), \tag{4.9}$$

and conclude the following result.

THEOREM 4.1 ([**91**]). *Given surjective linear projections* $C_i : \mathbb{P}^N \dashrightarrow \mathbb{P}^{n_i}$ *for* $i = 1,\ldots,m$ *and* $c_i \in \mathbb{Z}$ *such that* $1 \le c_i \le n_i$ *and* $c_1+\ldots+c_m = N+1$, *define the set of corresponding subspaces by* (4.4). *Up to Zariski closure, this is an algebraic hypersurface cut out in primal Plücker coordinates by* $\mathbf{T}$ *in* (4.9).

4.3.7. *Beyond four cameras.* When $N = 3$ and there are $m > 4$ pinhole cameras, a tensor-based model for camera configurations is no longer possible since there is no suitable tuple of codimensions $(c_1,\ldots,c_m)$ for the Grassmann tensor construction.

The issue is that various embedded subvarieties that one may associate with the cameras are not hypersurfaces, and thus not specified by a single equation. However models for camera configurations still exist inside appropriate *Hilbert schemes.* Roughly speaking, a Hilbert scheme is the set of ideals of a fixed polynomial ring with a prescribed number of equations in each degree. In particular, one can associate $(C_1,\ldots,C_m)$ modulo G to the ideal of its multiview variety $\mathcal{M}_C$, viewed as a

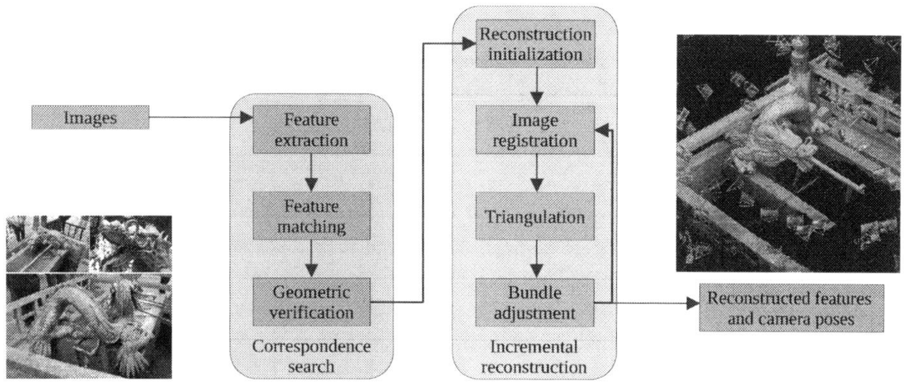

FIGURE 4. Incremental structure-from-motion pipeline.

point in a suitable Hilbert scheme [**8**]. Aholt, Thomas and Sturmfels [**8**] proved that this association is generically one-to-one if $m > 2$. The Hilbert scheme approach was recently extended by Lieblich and Van Meter [**132**] via functorial methods. Their work has the benefit of incorporating calibrated cameras as well.

A rather different line of work encodes the configuration of m cameras through their *viewing graph*. The graph has m vertices, one for each camera. There is an edge between two vertices if the relative position of the two corresponding cameras is known. The edge is then labeled with the fundamental matrix of the camera pair. In most practical scenarios with many cameras, the viewing graph is incomplete. The structure of the graph then determines if the given fundamental matrices are generically sufficient to recover the global camera configuration [**11, 178**].

5. 3D reconstruction

In this section, we discuss the role that multiview geometry plays in 3D reconstruction algorithms and how zero-dimensional polynomial systems arise. The next section will focus on methods that are used to solve these polynomials with sufficient speed.

5.1. Structure-from-motion pipeline. We have introduced the structure-from-motion problem already in Section 1. Here, we give more detailed algorithmic steps to solve this 3D reconstruction problem. Figure 4 overviews the usual *incremental structure-from-motion pipeline*; see [**26**] for a full description.

Feature extraction. The first step in the pipeline is to detect distinguished points or lines (or possibly other information such as conics) in each given image. A standard algorithm is the scale-invariant feature transform (SIFT) due to Lowe [**134**].

Feature matching. Next, the detected features are matched across different images; cf. Figure 1(B). In particular, feature matching identifies which images show a common part of the scene.

Geometric verification. Typically, not all feature matches from the feature matching step will be correct; many will be outliers. For two (or more) images with matched features, the geometric verification step tries to compute a geometric transformation that correctly maps sufficiently many features between the images. If such a transformation can be found, the pair (or tuple) of images and their common features that could be mapped are considered to be geometrically verified. For instance, the fundamental (see Section 4.3.1) or essential matrix (see Section 4.3.2) encode the relative pose of two pinhole cameras and thus provide the desired transformation between their two images. The output of the geometric verification step is the viewing graph whose nodes are images and whose (hyper-)edges connect geometrically verified image pairs (or tuples). The (hyper-)edges are labeled with the computed geometric transformations. See Section 4.3.7 for more on the viewing graph.

Since geometric verification forgets the original images and only works with the feature matches from feature matching (see Figure 1(C)), its core is an algebro-geometric problem that amounts to solving systems of polynomial equations. However, due to the presence of outliers, this step must use robust estimation techniques, usually RANSAC (random sample consensus). More details are in Section 5.2.

Reconstruction initialization. This step picks a starting pair (or tuple) of geometrically verified images (typically, in a dense area of the viewing graph), marks them as *registered*, and reconstructs the 3D coordinates of their matched features (as shown in Figure 1(C)). Note that the relative pose of the cameras was already precomputed in the previous step.

Image registration. This is the first step of the incremental reconstruction phase. It registers one new image that is connected with previously registered images in the viewing graph by (re-)calculating its camera pose relative to the reconstructed camera poses of the registered images. This can be either done from matched 2D features (as in the geometric verification step) or from 2D-3D correspondences if some of the already reconstructed 3D features are detected 2D features on the new image. Both ways solve a system of polynomial equations, but the latter problem—known as *perspective-n-point (PnP)* if all detected features are points—can typically be solved faster.

Triangulation. This step reconstructs the 3D coordinates of all features that the newly registered image has in common with any of the previously registered images (except, of course, for those features whose 3D coordinates were reconstructed before); see Section 3.2.

Bundle adjustment. This step refines the 3D features and camera poses reconstructed so far to minimize the accumulation of errors in the incremental reconstruction phase; see [**181**] for more details. Afterward, a new image will be registered and the incremental reconstruction phase repeats until (almost) all cameras and sufficiently many 3D features are reconstructed to approximate the 3D scene. Bundle adjustment is a very important step in the pipeline, but a fuller account falls outside our scope. See [**149**] for another survey article that emphasizes numerical optimization aspects in structure-from-motion more ours does.

5.2. Geometric verification. The algebro-geometric problems within the structure-from-motion pipeline aim to reconstruct the relative camera positions locally. That is, we seek the relative pose for only a few (e.g., two) of the given images at a time, based on feature matches between the images. A salient issue is that some of the purported feature matches will be *completely erroneous.* Indeed feature extraction/matching methods applied to real images invariably produce a nontrivial fraction of incorrect matches; and the problem becomes severe for scenes with repeated textures (e.g., a brick wall), among others. Therefore reconstruction methods must be robust to outliers. A general paradigm for estimation in the presence of outliers was introduced by Fischler and Bolles [**76**], who were motivated by the computer vision setting. Their *(RANSAC)* and its subsequent variants remain popular today for computing local reconstructions.

To illustrate RANSAC and show how it motivates solving zero-dimensional polynomial systems in vision, we discuss essential matrix estimation. In this example, data consists of several purported point matches

$$(x_1, y_1), \dots, (x_n, y_n) \in \mathbb{P}^2_{\mathbb{R}} \times \mathbb{P}^2_{\mathbb{R}}.$$

Let $\mathcal{E} \subset \mathbb{P}(\mathrm{Mat}^{3\times 3})$ be the variety of essential matrices defined by (4.3). We seek $E \in \mathcal{E}$ which best fits the veridical point pairs amongst the data. By construction, if (x_i, y_i) is a correct (and noiseless) image point pair and E is the ground-truth essential matrix, then

$$y_i^\top E x_i = 0.$$

Notice that this constraint on E is linear and of codimension 1. Thus given $\dim(\mathcal{E}) = 5$ such constraints, we expect up to $\deg(\mathcal{E}) = 10$ real solutions for E. However it is unknown which of the purported points pairs are veridical. RANSAC achieves robust estimation by randomly sampling a subset $\mathcal{I} \subseteq [n] = \{1, \dots, n\}$ of five point pairs and fitting to them exactly by solving the polynomial system:

$$(5.1) \quad \det(E) = 0, \qquad EE^\top E - \frac{1}{2}\operatorname{trace}(EE^\top)E = 0, \qquad y_i^\top E x_i = 0 \ \ (i \in \mathcal{I}).$$

Letting $E_1, \dots, E_s$ be the real solutions, RANSAC checks if there exists a consensus amongst the other $n - 5$ point pairs as to whether any of the solutions agree with most of the other point pairs. RANSAC re-samples $\mathcal{I}$ until a consensus is found. We note that this involves a choice of error metric to measure the agreement between E_k and (x_i, y_i) for $i \in \mathcal{I}$. A simple choice is $(y_i^\top E_k x_i)^2$ often called algebraic error, but often the geometric reprojection error or its first-order approximation called Sampson error are preferred [**94**]. We also remark that popular variants of RANSAC, like LO-RANSAC [**48**], polish the winning E_k by using all point pairs in $[n]$ that agree with E_k to form a nonlinear least squares cost which is minimized starting from E_k.

5.3. Minimal problems. The polynomial system (5.1) is called a minimal problem in the vision literature, reflecting the fact that (5.1) uses the minimal amount of data such that the solution is uniquely determined up to finitely many solutions. More generally, minimal problems are 3D reconstruction problems such that random input instances have a finite positive number of solutions. Formally, a structure-from-motion problem is *minimal* if the associated joint camera map $\Phi : \mathcal{X} \times \mathcal{C}_m \dashrightarrow \mathcal{Y}$ over the complex numbers has generically finite nonempty fibers, modulo the group G that acts on the fibers. In other words, for a generic (complex) $y \in \mathcal{Y}$, the fiber $\Phi^{-1}(y)$ is nonempty and consists of finitely many G-orbits.

By choosing appropriate coordinates for camera configurations modulo the group G (for example, by using the multifocal tensors in Section 4.3), a minimal problem can be cast as a zero-dimensional parameterized polynomial system. Thus one expects that a minimal problem can be solved efficiently, as long as its generic number of complex solutions is sufficiently small. This generic number of complex solutions—known as the *algebraic degree* of the minimal problem—is a measure of the minimal problem's intrinsic difficulty.

EXAMPLE 5.1. Reconstructing five points in general position observed by two calibrated pinhole cameras (see Figure 1(C)) is a minimal problem. This is because, generically, the system (5.1) has 10 complex solutions (i.e., essential matrices), so 20 complex pairs of cameras (modulo the scaled special Euclidean group), and for each camera pair there are five unique 3D points mapping to the given image points. Note that the number of *real* solutions to (5.1) depends on the image data. The real count is generically amongst $\{0, 2, 4, 6, 8, 10\}$, but varies with the specific five points. In particular, understanding when there exists at least a real solution is subtle [**4**]. An even more difficult problem is to determine when there is at least one chiral reconstruction, i.e., a real solution where the 3D points lie *in front* of the cameras [**157**]. If we are given six points in two calibrated views, the system (5.1) no longer corresponds to a minimal problem, as generically there is no complex solution. We need the image data to lie on the *Chow hypersurface* in order for (5.1) to be soluble; see [**77**] for an explicit Pfaffian formula defining the Chow hypersurface obtained using Ulrich sheaves, or [**29**] for a simpler construction based on Jordan algebras.

The literature on minimal problems is vast. Many minimal problems have been described, solvers have been developed for them, and new minimal problems are constantly appearing; see, e.g., [**13, 14, 43, 67, 69, 101, 105, 117, 119–121, 128, 141, 146, 148, 158, 165, 169, 183, 194**].

There are also many variations of the above definition. For instance, when solving 3D reconstruction by RANSAC as described in Section 5.2, one often aims to recover the camera parameters only. Hence, often a structure-from-motion problem is referred to as minimal if for a generic image tuple $y \in \mathcal{Y}$, the projection of the fiber $\Phi^{-1}(y) \subset X \times \mathcal{C}_m$ onto the space $\mathcal{C}_m$ of cameras is nonempty and finite modulo G. In the following, we call such problems *camera-minimal* and reserve the term *minimal* for when we aim to recover the 3D structure in $\mathcal{X}$ as well (as in the definition before Example 5.1). Note that every minimal problem is camera-minimal but not vice versa.

EXAMPLE 5.2. The following problem is camera-minimal (shown in [**107**] where it is called 3PPP + 1PPL): Reconstruct four points $X_1, X_2, X_3, X_4 \in \mathbb{P}^3$ and a line $L \subset \mathbb{P}^3$ incident to X_4 that are observed by three calibrated pinhole cameras such that the first two cameras see only the four points and the third camera only sees X_1, X_2, X_3 and the line L; see Figure 5. In fact, it typically has 272 complex solutions in terms of camera parameters. However, this problem is *not* minimal since the line L cannot be recovered uniquely when the three cameras are known: There is a one-dimensional family of lines that yield the same image line in the third view.

In standard structure-from-motion reconstructions, (camera-)minimal problems need to be solved many times to ensure RANSAC finds a solution that is not corrupted by outliers. Thus, minimal solvers need to be fast. Furthermore,

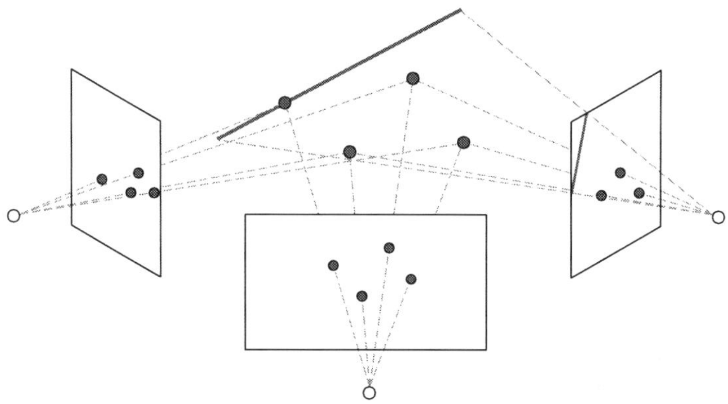

FIGURE 5. A camera-minimal problem with algebraic degree 272.

they should be accurate, since numerical errors could jeopardize RANSAC's success as well. We also would like to handle various nearly-degenerate geometries, such as near coplanarities amongst the five point pairs in (5.1), as such situations do arise in natural images. In Section 6, we review the symbolic and numerical minimal solvers that are used to meet these needs.

The large zoo of minimal problems in the literature is motivated by the quest to find reconstruction problems that (1) admit a feasible minimal solver as described above; (2) their necessary input data can be detected by state-of-the-art feature extraction methods; and (3) cover the wide variety of reconstruction applications in computer vision. The latter includes variations of the camera model (see Section 2), of the parameters to be reconstructed, or of the input data (also beyond the structure-from-motion problem we focus on here, e.g., camera pose estimation from known 3D *and* 2D data). We list some of the most classical minimal problems in Table 1.

Efforts have been undertaken to provide complete catalogs of some classes of minimal problems. Kileel [**107**] describes all 66 camera-minimal problems (incl. their algebraic degrees) for three calibrated pinhole cameras that can be formulated using linear constraints on the trifocal tensor. Duff, Kohn, Leykin, and Pajdla [**64**] characterize all 30 minimal problems (including their algebraic degrees) of reconstructing points, lines and their incidences for an arbitrary number of calibrated pinhole cameras under the assumption that all points and lines are observed by every camera. In a follow-up work allowing partial visibility [**65**], they classify all camera-minimal and minimal problems of three calibrated pinhole cameras observing point-line arrangements such that each line is incident to at most one point. In particular, there are 759 (equivalence classes of) camera-minimal problems with algebraic degree less than 300.

6. Polynomial solvers for minimal problems

In this section, we discuss symbolic and numerical methods to construct solvers for minimal problems in computer vision, e.g., the minimal problems in Table 1.

TABLE 1. Some representative minimal problems. Several of these minimal problems are classical and integrated in vision pipelines.

to be reconstructed	minimal data	degree	references
essential matrix	5 point pairs	10	[**57**, **145**]
fundamental matrix	7 point pairs	3	[**87**] already known in 19th century [**95**, **172**]
relative pose of 2 calibrated cameras with unknown common focal length	6 point pairs	15	[**169**]
absolute pose of 1 calibrated camera (P3P, image registration)	3 world-image point pairs	4	[**85**] already known in 19th century [**84**]
planar homography	4 point pairs	1	
calibrated planar homography	4 point pairs	12	[**64**, **75**]
trifocal tensor	9 line triples	36	[**124**]
calibrated trifocal tensor	3 point triples +1 line triple	216	[**107**]
relative pose of 2 projective cameras with unkown radial lens distortion	8 point pairs	16	[**121**]
world point under noise (triangulation, reprojection error)	known cameras with: • 1 point pair • 1 point triple	• 6 • 47	[**92**] [**170**]

Algebraically, we model a minimal problem as a *zero-dimensional parameterized polynomial system:*

$$
\begin{cases} f_1(x_1, \dots, x_n; \theta_1, \dots, \theta_t) = 0 \\ \qquad \vdots \\ f_k(x_1, \dots, x_n; \theta_1, \dots, \theta_t) = 0. \end{cases} \tag{6.1}
$$

Here $x = (x_1, \dots, x_n)$ are variables, and $\theta = (\theta_1, \dots, \theta_t)$ are parameters (e.g., coming from image point correspondences). The equations f_i are polynomials in both x and θ, typically with integer coefficients (e.g., the epipolar constraints (5.1)). Write $\mathcal{I}$ for the ideal generated by the equations

$$
\mathcal{I} = \{p_1 f_1 + \dots + p_k f_k \mid p_i \in \mathbb{R}[x_1, \dots, x_n; \theta_1, \dots, \theta_t]\}.
$$

Let $|_\theta$ denotes specialization to concrete parameter values $\theta \in \mathbb{R}^t$. Minimality of the problem translates to the following two assumptions:

(A1) For generic θ, the ideal $\mathcal{I}|_\theta$ is zero-dimensional (i.e., has finitely many roots);

(A2) For generic θ, the ideal $\mathcal{I}|_\theta$ is radical (i.e., has no repeated roots)

The conditions imply the existence of a positive integer $N_{\mathbb{C}} = N_{\mathbb{C}}(\mathcal{I})$ such that for generic parameters θ, the number of $\mathbb{C}$-solutions to (6.1) is $N_{\mathbb{C}}$. Furthermore, each root has multiplicity 1. Note that $N_{\mathbb{C}}$ is the algebraic degree of the minimal problem discussed in Section 5.3.

By definition, a *solver* is meant to receive values for the parameter $\theta \in \mathbb{R}^t$, and then output the set of corresponding real solutions $V_{\mathbb{R}}(\mathcal{I}|_\theta) \subset \mathbb{R}^n$ to (6.1). A salient issue to bear in mind is the *online vs. offline* distinction in computer vision: When given a specific parameter value θ, for many vision tasks we must run the solver in real-time. Therefore, online computations must be fast. On the other hand, when developing the solver, we are free to dedicate far more time and computational resources offline.

6.1. Symbolic method: Elimination templates. Currently a symbolic method is the most popular one to build minimal solvers in vision. At the heart of the method is a classical fact that zero-dimensional polynomial systems can be converted to eigenvector problems.[1]

LEMMA 6.1. *Let $\theta \in \mathbb{R}^t$, and the corresponding solutions to (6.1) be $V_{\mathbb{C}}(\mathcal{I}|_\theta) = \{v_1, \ldots, v_{N_{\mathbb{C}}}\} \subset \mathbb{C}^n$. Let $P \in \mathbb{R}[x_1, \ldots, x_n]$ be a polynomial separating the solutions, that is, $P(v_i) \neq P(v_j)$ for all $i \neq j$. Let $\mathcal{B} = \{b_1, \ldots, b_{N_{\mathbb{C}}}\} \subseteq \mathbb{R}[x_1, \ldots, x_n]$ be such that its reduction modulo $\mathcal{I}|_\theta$ is a vector space basis $\overline{\mathcal{B}} = \{\overline{b}_1, \ldots, \overline{b}_{N_{\mathbb{C}}}\}$ of the quotient ring $\mathbb{R}[x_1, \ldots, x_n]/\mathcal{I}|_\theta$. Multiplication by P is a linear endomorphism on $\mathbb{R}[x_1, \ldots, x_n]/\mathcal{I}|_\theta$, represented with respect to $\overline{\mathcal{B}}$ by a matrix*

$$(6.2) \qquad [\mathrm{mult}(P)|_\theta]_{\overline{\mathcal{B}}} \in \mathbb{R}^{N_{\mathbb{C}} \times N_{\mathbb{C}}}, \qquad \overline{Pb_j} := \sum_{i=1}^{N_{\mathbb{C}}} \left([\mathrm{mult}(P)|_\theta]_{\overline{\mathcal{B}}}\right)_{ij} \overline{b}_i.$$

Then $[\mathrm{mult}(P)|_\theta]_{\overline{\mathcal{B}}}$ is diagonalizable with eigenvalue/left eigenvectors pairs given by $(P(v_i), \mathcal{B}(v_i)) \in \mathbb{C} \times \mathbb{C}^{N_{\mathbb{C}}}$ for $i = 1, \ldots, N_{\mathbb{C}}$.

In [**52**], Cox gives an interesting history of Lemma 6.1, which is often (wrongly) attributed to Stickelberger. We call P the *action polynomial* and $[\mathrm{mult}(P)|_\theta]_{\overline{\mathcal{B}}}$ the *action matrix*. Notice that if the action matrix is available, then Lemma 6.1 shows how the solutions v_i to (6.1) may be computed: from the eigenvectors w_i of the action matrix we require $\mathcal{B}(v_i)$ to be proportional to w_i. To make the recovery of v_i easy, typically $\mathcal{B}$ is chosen to consist of monomials.

Thus the main challenge is to obtain the action matrix online. In principle, it requires computing identities of the form

$$(6.3) \qquad Pb_j = \sum_{i=1}^{N_{\mathbb{C}}} \left([\mathrm{mult}(P)|_\theta]_{\overline{\mathcal{B}}}\right)_{ij} b_i \; + \; \sum_{i=1}^{k} q_{ij} f_i|_\theta \; \in \; \mathbb{R}[x_1, \ldots, x_n],$$

for $j = 1, \ldots, N_{\mathbb{C}}$ where $q_{ij} \in \mathbb{R}[x_1, \ldots, x_n]$ and $\left([\mathrm{mult}(P)|_\theta]_{\overline{\mathcal{B}}}\right)_{ij} \in \mathbb{R}$. Gröbner basis computations are too slow to be performed online, and not viable in floating point arithmetic anyways. Instead, the main tool is *elimination templates*.

Fix choices of $\mathcal{B} \subset \mathbb{R}[x_1, \ldots, x_n]$ and $P \in \mathbb{R}[x_1, \ldots, x_n]$ such that:

(A2) For generic θ, $\overline{\mathcal{B}} \subset \mathbb{R}[x_1, \ldots, x_n]/\mathcal{I}|_\theta$ is a vector space basis;
(A3) For generic θ, P separates $V_{\mathbb{C}}(\mathcal{I}|_\theta)$.

In the present setting, elimination templates are defined as follows.

DEFINITION 6.2. A set $\mathcal{E} = \{e_1, \ldots, e_T\} \subset \mathcal{I} \subseteq \mathbb{R}[x_1, \ldots, x_n; \theta_1, \ldots, \theta_t]$ is an *elimination template* if **A5)** For generic θ, we have $P \cdot \mathcal{B} \subset \mathrm{span}(\mathcal{B} \cup \mathcal{E}|_\theta)$.

[1] A different symbolic approach, based instead on *resultants*, is being developed by Bhayani, Heikkilä, and Kukelova [**24, 25**].

Often $\mathcal{E}$ is chosen to be x-monomials multiplied by $f_1, \ldots, f_k$. By Definition 6.2,

$$Pb_j = \sum_{i=1}^{N_{\mathbb{C}}} ([\mathrm{mult}(P)|_\theta]_{\overline{\mathcal{B}}})_{ij}\, b_i \; + \; \sum_{\ell=1}^{T} \star\, e_\ell|_\theta \;\in\; \mathbb{R}[x_1, \ldots, x_n] \tag{6.4}$$

for $j = 1, \ldots, N_{\mathbb{C}}$ and scalars $\star$. Computing the action matrix thus becomes a linear system. Specifically, fix $\mathcal{C} = \{c_1, \ldots, c_M\} \subset \mathbb{R}[x_1, \ldots, x_n]$ such that

(A5) For generic θ, $\mathcal{E}|_\theta \subset \mathrm{span}(\mathcal{B} \cup \mathcal{C})$.

Often $\mathcal{C}$ is all x-monomials appearing in any $e_\ell \in \mathcal{E}$ with nonzero coefficient in $\mathbb{R}[\theta_1, \ldots, \theta_t]$, not already in $\mathcal{B}$. Form a matrix called the *elimination template matrix*, denoted $[\mathrm{elim}(P)|_\theta]_{\mathcal{B},\mathcal{E},\mathcal{C}} \in \mathbb{R}^{(T+N_{\mathbb{C}})\times(N_{\mathbb{C}}+M)}$, with rows indexed by $\mathcal{E}|_\theta \cup P \cdot \mathcal{B}$ and columns indexed by $\mathcal{B} \cup \mathcal{C}$:

$$\begin{array}{c} \\ e_1|_\theta \\ \vdots \\ e_T|_\theta \\ Pb_1 \\ \vdots \\ Pb_{N_{\mathbb{C}}} \end{array} \begin{array}{c} \begin{array}{cc} b_1 \;\ldots\; b_{N_{\mathbb{C}}} & c_1 \;\ldots\; c_M \end{array} \\ \left(\begin{array}{c|c} \bigstar & \bigstar \\ \hline \bigstar & \bigstar \end{array}\right) \end{array}, \tag{6.5}$$

where each entry is the coefficient of the column in the row. Thanks to (6.4), elementary row operations can reduce the elimination template matrix to the form

$$\begin{array}{c} \\ e_1|_\theta \\ \vdots \\ e_T|_\theta \\ Pb_1 \\ \vdots \\ Pb_{N_{\mathbb{C}}} \end{array} \begin{array}{c} \begin{array}{cc} b_1 \;\ldots\; b_{N_{\mathbb{C}}} & c_1 \;\ldots\; c_M \end{array} \\ \left(\begin{array}{c|c} \bigstar & \bigstar \\ \hline [\mathrm{mult}(P)|_\theta]_{\overline{\mathcal{B}}} & 0 \end{array}\right) \end{array}. \tag{6.6}$$

The bottom-left block of (6.6) is the desired action matrix.

The eigenvectors and elimination templates method is summarized in Algorithm 1. In the online part, the most costly step is computing the pseudo-inverse of the top-right block of the elimination template matrix. In the offline phase, we seek $\mathcal{B}, P, \mathcal{E}, \mathcal{C}$ such that the size of the elimination template matrix is as small as possible.

For the offline procedure, full details are beyond our scope but we mention some of the highlights. A breakthrough was obtained by Kukelova, Bujnak, and Pajdla in 2008, who showed how find elimination templates automatically, as opposed to by hand [**118**]. Behind their work, a principle is that "shape" of symbolic calculations should be the same at generic parameters θ, and also preserved if we reduce (6.1) modulo a prime. This enables traditional Gröbner basis computations in the offline phase. In [**124**] Larsson, Åström, and Oskarsson observed that q_{ij} in (6.3) are uniquely defined only up to the first syzygy module of $(f_1|_\theta, \ldots, f_k|_\theta)$. Syzygy-reduction [**124**, Proposition 1] was shown to result in choices of $\mathcal{E}$ with

Algorithm 1 Offline/online solver for the parameterized system (6.1)

1: **Offline Input**

$\langle f_1, \ldots, f_k \rangle \subseteq \mathbb{R}[x_1, \ldots, x_n; \theta_1, \ldots, \theta_t]$ satisfying **A1-A2**.

2: **Offline Precomputations**

Find $\mathcal{B} \subseteq \mathbb{R}[x_1, \ldots, x_n]$ satisfying **A3**,
$P \in \mathbb{R}[x_1, \ldots, x_n]$ satisfying **A4**,
$\mathcal{E} \subseteq \mathbb{R}[x_1, \ldots, x_n; \theta_1, \ldots, \theta_t]$ satisfying **A5**,
$\mathcal{C} \subseteq \mathbb{R}[x_1, \ldots, x_n]$ satisfying **A6**.

Compute the bottom $N_{\mathbb{C}}$ rows of (6.5), and denote these by $\begin{bmatrix} E_{21} & E_{22} \end{bmatrix} \in \mathbb{R}^{N_{\mathbb{C}} \times (N_{\mathbb{C}}+M)}$.

3: **Online Input**

Generic $\theta \in \mathbb{R}^t$.

4: **Online Computations**

Evaluate the top T rows of (6.5) at θ, and denote these by $\begin{bmatrix} E_{11} & E_{12} \end{bmatrix} \in \mathbb{R}^{T \times (N_{\mathbb{C}}+M)}$.

Thus $[\text{elim}(P)|_\theta]_{\mathcal{B},\mathcal{E},\mathcal{C}} = \begin{bmatrix} E_{11} & E_{12} \\ E_{21} & E_{22} \end{bmatrix}$.

Compute $E_{21} - E_{22}E_{12}^{\dagger}E_{11} \in \mathbb{R}^{N_{\mathbb{C}} \times N_{\mathbb{C}}}$.
This equals $[\text{mult}(P)|_\theta]_{\overline{\mathcal{B}}}$.

Compute all real eigenvalue/eigenvector pairs $(\lambda_1, w_1), \ldots, (\lambda_r, w_r) \in \mathbb{R} \times \mathbb{R}^{N_{\mathbb{C}}}$ of $[\text{mult}(P)|_\theta]_{\overline{\mathcal{B}}}$.

Determine $v_1, \ldots, v_r \in \mathbb{C}^n$ such that $P(v_i) = \lambda_i$ and $\mathcal{B}(v_i) \propto w_i$.

5: **Online Output**

$V_{\mathbb{R}}(\mathcal{I}|_\theta) = \{v_1, \ldots, v_r\} \cap \mathbb{R}^n$.

substantially smaller elimination template matrices for many minimal problems. In [**125**], Larsson, Oskarsson, Åström, Wallis, Kukelova, and Pajdla leveraged the fact that the basis $\mathcal{B}$ need not be standard monomials with respect to some Gröbner basis for $\mathcal{I}|_\theta$, only condition **A3** is required (see [**142**]). The authors chose $\mathcal{B}$ by a random monomial-sampling strategy, and showed the strategy can sometimes produce smaller elimination template matrices than achievable anywhere on the Gröbner fan. Martyushev, Vrablikova and Pajdla further improved the state-of-the-art with a greedy search strategy to reduce the size of $\mathcal{C}$ in [**136**]. Integrating many of these ideas, Li and Larsson released code for building solvers in [**131**].

6.2. Numerical method: Homotopy continuation. An alternative method for building minimal problem solver is homotopy continuation. Homotopy continuation [**168**] is a path-following, ODE-based numerical method to solve parametrized zero-dimensional polynomial systems of a significantly different nature than the eigenvectors and elimination templates method described above. It uses only floating-point arithmetic operations, and exploits the offline vs. online paradigm in a different way.

In the offline phase, homotopy continuation solves one generic instance of the minimal problem (i.e., the polynomial system $f(x;\theta_*) = 0$ for a fixed choice of parameters). The offline solve is done effectively from scratch. Specifically, homotopy continuation chooses a *start system* $g(x) = 0$ of polynomials with the same variables and degree structure as $f(x;\theta_*) = 0$ that is trivial to solve (e.g., the start solutions are roots of unity). Then homotopy continuation interpolates from the start system to the minimal problem instance via linear interpolation:

$$h(x;t) = (1-t)f(x;\theta_*) + tg(x),$$

as t goes from 1 to 0. During the interpolation, we numerically track how the solutions to $g(x) = 0$ evolve [**55**, **56**]. In general, many solution paths diverge. However the paths with finite limits give all solutions to $f(x;\theta_*) = 0$ [**168**, Theorem 7.1.1]. The offline procedure is costly: its complexity is driven by the number of solutions to $g(x) = 0$, often exponential in the number of variables.

In the online phase, homotopy continuation also interpolates between polynomial systems. But now it tracks solutions of the precomputed minimal problem instance $f(x;\theta_*) = 0$ to the solutions of the desired instance of the minimal problem (i.e., $f(x;\theta) = 0$ for a different choice of parameters). The complexity of the online phase is driven by the number of solutions to $f(x;\theta_0) = 0$, rather than the number of solutions to $g(x) = 0$. Consequently the online solve is much faster than the offline phase. Indeed, in the online solve we track only $N_{\mathbb{C}}$ many paths, which is the intrinsic algebraic degree of the minimal problem. General-purpose software for homotopy continuation includes Bertini [**15**], HomotopyContinuation.jl [**36**], and NumericalImplicitization.m2 [**45**], among others.

In computer vision, homotopy continuation has been used to determine the algebraic degrees of (camera-)minimal problems [**64**, **65**, **107**]. Homotopy continuation was also previously suggested as a solver for such problems [**97**, **116**], however for many years it was not seen to be competitive with the elimination templates approach. Recently, homotopy continuation yielded the fastest solver for the minimal structure-from-motion problem shown in Figure 6. For this problem no elimination template matrix of reasonable size could be found by the symbolic method of Section 6.1. However, with homotopy continuation a numerically stable solver was built that runs on the order of a second or less [**69**]. The minimal problem in Figure 6 has algebraic degree $N_{\mathbb{C}} = 312$, so homotopy continuation tracks 312 paths per online solve. The solver was validated on real-data scenarios where the state-of-the-art pipeline COLMAP [**166**] failed to find enough point-based features, which also highlights the importance of investigating (camera-)minimal problems that involve not only points but also lines.

More recently, homotopy continuation has been combined with machine learning to build competitive solvers for minimal problems of large algebraic degree: Hruby, Duff, Leykin, and Pajdla [**98**] achieve fast computational speeds by first using machine learning to pick a single start solution and then tracking that solution to a single solution of the desired target system. The start solution does not come from a fixed start system, but rather determines the start system used for homotopy continuation, meaning that the start system is also learned. Another

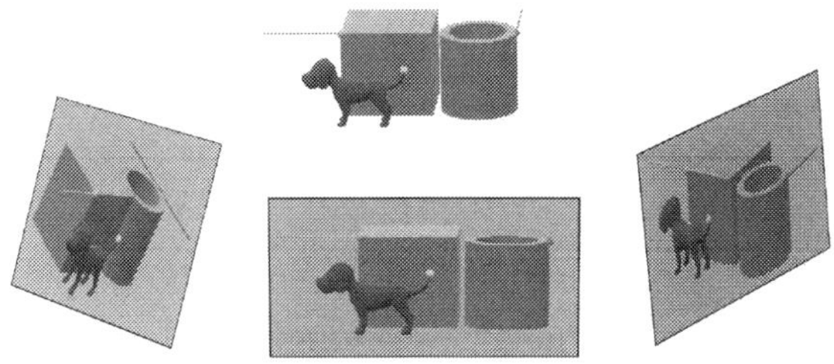

FIGURE 6. A minimal problem with algebraic degree 312: three calibrated pinhole cameras observe three points, two of which have an incident line. Homotopy continuation was used to build a solver for this minimal problem taking 1 second or less, whereas symbolic methods have not yielded a practical solver [**69**].

important advance has been to combine GPU computing with homotopy continuation. This was recently carried out by Chien, Fan, Abdelfattah, Tsigaridas, Tomov, and Kimia [**46**] on various benchmark minimal problems.

7. Degeneracies and discriminants

This section is devoted to degenerate configurations of 3D scenes and cameras. Section 7.1 focuses on configurations that do not admit unique recovery from image data. Degenerate views of 3D curves and surfaces are discussed in Section 7.2. Finally, Section 7.3 outlines the connection between degenerate instances of minimal problems and condition numbers that measure how sensitive numerical reconstruction algorithms are to errors in the input.

7.1. Critical loci. A *critical configuration* in a structure-from-motion problem is a pair $(X, C) \in \mathcal{X} \times \mathcal{C}_m$ consisting of an ordered subset X of $\mathbb{P}^3_{\mathbb{R}}$ and an m-tuple of cameras C such that the resulting m images $\Phi(X, C)$ do not have a unique reconstruction modulo the group G that acts on the fibers of the joint camera map Φ. In other words, a configuration of 3D points $X \in \mathcal{X}$ and cameras $C \in \mathcal{C}_m$ is critical if there is another configuration $(X', C') \in \mathcal{X} \times \mathcal{C}_m$ that is not in the same G-orbit as (X, C) and produces the same images, i.e., $\Phi(X, C) = \Phi(X', C')$.

The study of critical configurations for two pinhole cameras goes back to the German-Austrian literature on photogrammetry, where critical loci where known as *gefährliche Örter*: Krames [**114**] showed that an ordered set of 3D points and two pinhole cameras form a critical configuration only if the points and the two camera centers lie on a real ruled quadric surface; see also Maybank [**139**] for a modern reference. Recall that a quadric surface in $\mathbb{P}^3_{\mathbb{R}}$ is said to be *ruled* if it contains a line. There are four projectively nonequivalent ruled quadrics. They can be distinguished based on the rank of a symmetric matrix $S \in \mathbb{R}^{4\times 4}$ giving a defining equation $X^\top S X = 0$ of the quadric: smooth ruled quadric ($\operatorname{rank} S = 4$), cone ($\operatorname{rank} S = 3$), two planes ($\operatorname{rank} S = 2$), and double plane ($\operatorname{rank} S = 1$); see Figure 7.

A first characterization of critical configurations of 3D points and an *arbitrary* number of projective pinhole cameras modulo the group $\mathrm{PGL}(4, \mathbb{R})$ was provided by Hartley and Kahl [**89**]. In particular, for $m = 2$, they derive.

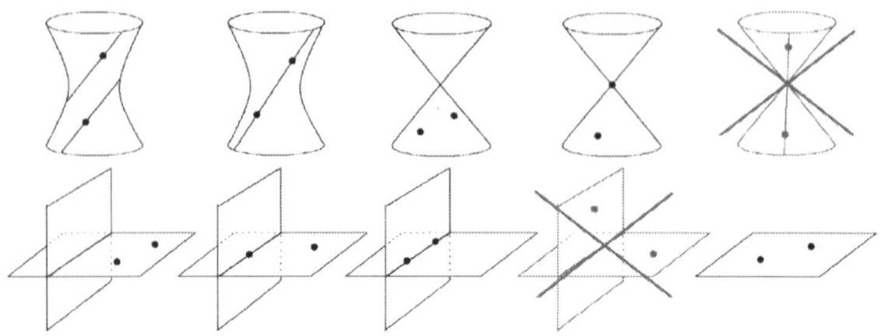

FIGURE 7. The eight critical quadrics for 3D points and two projective pinhole cameras. The marked points are the camera centers. The exceptions in Theorem 7.1 are marked with a cross. (Courtesy of Martin Bråtelund, [**33**].)

THEOREM 7.1. *An ordered set of 3D points and two pinhole cameras form a critical configuration modulo* $\mathrm{PGL}(4, \mathbb{R})$ *if and only if the points and the two camera centers lie on a real ruled quadric surface* Q*, with the following two exceptions: 1)* Q *is a cone and both camera centers lie on the same line on* Q *but not on its vertex, and 2)* Q *is two planes and the camera centers do not lie on the same plane. Hence, up to projective equivalence, there are eight critical quadrics with two marked camera centers, shown in Figure* 7.

The classification in [**89**] has some mistakes for three or more cameras. Bråtelund developed new algebraic techniques [**33**] and corrected the classification for three cameras [**32**].

Kahl and Hartley [**103**] also studied the critical configurations of 3D points and calibrated pinhole cameras, modulo the scaled special Euclidean group of $\mathbb{R}^3$. They provide a full characterization for two cameras and show the existence of critical configurations for an arbitrary number of points and cameras. It is still an open problem to derive the complete catalog of all critical configurations consisting of 3D points and more than two calibrated pinhole cameras.

REMARK 7.2. Another natural notion of critical locus that is not as prominent in the vision literature as the critical configurations described above is the ramification locus of the joint camera map. For instance, for a minimal problem of algebraic degree larger one, almost all configurations $(X, C) \in \mathcal{X} \times \mathcal{C}_m$ are critical (over $\mathbb{C}$). However, the ramification locus of the joint camera map of such a minimal problem carries crucial meaning: configurations close to the ramification locus often correspond to ill-conditioned problem instances [**41, 58, 59**], which has negative algorithmic consequences such as numerical instability; see also Section 7.3.

Critical configurations are also understood in a variety of other settings. Critical loci for 3D lines and three pinhole cameras are for instance investigated by Buchanan [**38, 39**], Maybank [**140**], Navab and Faugeras [**144**], and Zhao and Chung [**195**]. They are congruences in $\mathrm{Gr}(1, \mathbb{P}^3)$ that can be parametrized by Bordiga surfaces in $\mathbb{P}^4$. It would be interesting to generalize those results to more cameras, which is currently open.

Buchanan [**37**] also described the critical configurations for a single pinhole camera, i.e., for the problem of *camera calibration* where both 3D and image points

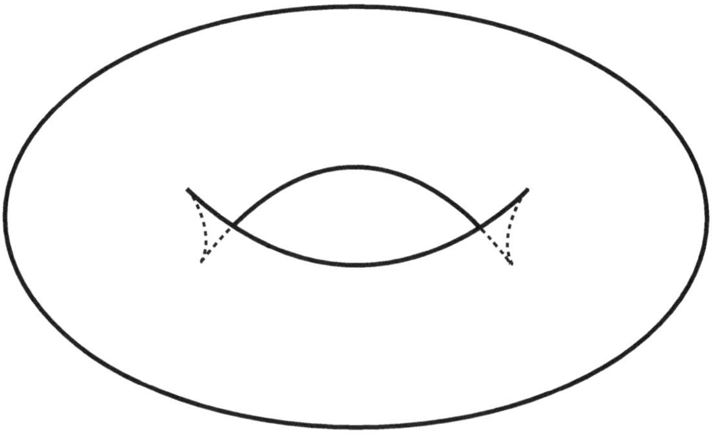

FIGURE 8. Image contour of a torus.

are known and only the camera parameters are to be recovered. Here the critical 3D points and the camera center lie either on a (possibly reducible but connected) twisted cubic or on the union of a line and a plane such that the camera center lies on the line.

Åström and Kahl [**12**] classified the critical loci for 1D structure-from-motion problems where 2D points are projected via pinhole cameras of the form $\mathbb{P}^2 \dashrightarrow \mathbb{P}^1$. Critical configurations of 2D points and camera centers lie on cubic curves.

Bertolini, Turrini, and coauthors generalized several of the results above to higher dimensions, i.e., where the classical pinhole cameras $\mathbb{P}^3 \dashrightarrow \mathbb{P}^2$ are replaced by projections $\mathbb{P}^N \dashrightarrow \mathbb{P}^n, N > n$; see, e.g., [**16, 18, 19, 21, 22**]. As explained at the end of Section 4.3, the study of general projections is motivated by the reconstruction of dynamic scenes. Several of their works also perform simulated experiments to test numerical instability phenomena for problem instances near critical loci; see in addition [**17, 20, 23**].

Despite the literature on critical configurations outlined above, critical loci of vision problems are still unexplored in many settings, e.g., for 3D point-line arrangements or for other camera models besides pinhole cameras.

7.2. Aspect graphs and visual events. In this section, we focus on taking pictures of an algebraic curve or surface X in $\mathbb{P}^3$ with a pinhole camera $C : \mathbb{P}^3 \dashrightarrow \mathbb{P}^2$. Clearly, if X is a curve, its image is typically a plane curve and we write $Y_C(X)$ for its Zariski closure in $\mathbb{P}^2$. If X is a surface, we denote by $Y_C(X)$ its *image contour*, also known as *silhouette* or *outline curve*, in the image plane $\mathbb{P}^2$. This is the curve that bounds the two-dimensional region that is the image of the surface X taken by the camera C. In other words, it is the natural sketch one might use to depict the surface; see Figure 8 for an image contour of a torus. Formally, $Y_C(X)$ is the branch locus of the map C restricted to X, i.e., it is the projection of the critical points where viewing lines through the camera center are tangent to the surface X.

The reconstruction of an algebraic surface from its image contour has been studied intensively, e.g., by Zariski [**193**]; Segre [**167**]; Biggiogero [**27, 28**]; Chisini

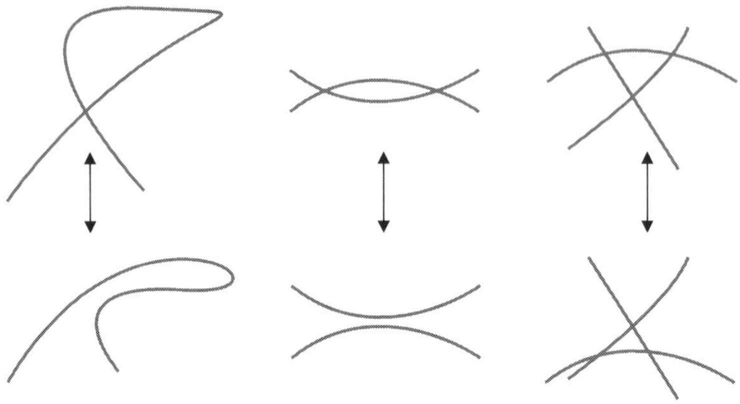

FIGURE 9. The three types of visual events of a space curve.

[**47**], who conjectured that a surface can be reconstructed from its image contour; Kulikov [**122, 123**], who solved Chisini's conjecture; and Forsyth [**79**], who solved the special case of a smooth surface and a general camera. Algorithmic solutions are for instance provided in [**54, 80, 81, 115**].

The *aspects* of a curve or surface X are the stable views when the camera moves, i.e., where the topology and singularities of the plane curve $Y_C(X)$ do not change under perturbations of the camera center. There is a finite number of aspects and they are the nodes of the *aspect graph*. Its edges are the *visual events* that transition between different aspects. For instance, Figure 8 shows an aspect of a torus, and a visual event occurs when we rotate the torus until the real singularities disappear. The aspect graph was first introduced by Koenderink and van Doorn [**111**] under the name of *visual potential*. A detailed discussion of image contours and their visual events can be found in Koenderink's book [**110**].

Although there has not been much use of aspect graphs in real-life applications, they have been an active research topic in the computer vision community; see, e.g., Bowyer and Dyer's survey [**31**] or Forsyth and Ponce's book [**78**, Chapter 13]. In particular, several algorithms for the computation of aspect graphs of algebraic surfaces were proposed, using both symbolic and numerical methods [**150, 153, 155, 161, 162**].

From the algebraic perspective, visual events have been studied by Petitjean [**152**] and Kohn, Sturmfels, and Trager [**113**]. The *visual event surface* $\mathcal{V}(X)$ of the curve or surface X is the Zariski closure in $\mathbb{P}^3$ of the set of camera centers where a visual event occurs. For a curve X in $\mathbb{P}^3$, the visual event surface typically has three irreducible components, each representing a different type of visual event. On the image curve $Y_C(X)$, these types correspond to the three Reidemeister moves from knot theory [**159**]; see Figure 9.

If X is a surface, its visual event surface $\mathcal{V}(X)$ typically has five irreducible components that arise from six different types of visual events (here two types of events cannot be distinguished algebraically, i.e., they lie on the same component of $\mathcal{V}(X)$, but they represent two distinct behaviors over the real numbers). These six types of events correspond to the nongeneric singularities from catastrophe theory [**10, 176**].

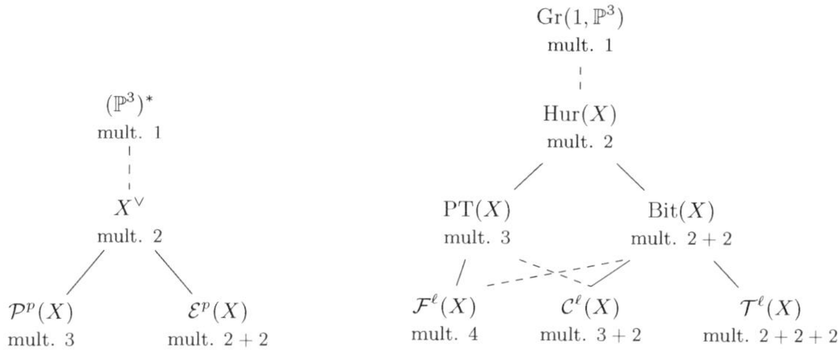

FIGURE 10. Loci of planes and lines that meet a surface X with assigned multiplicities. For instance, "mult. $2 + 2 + 2$" on the right-hand side refers to the locus of tritangent lines. A dashed (resp. solid) edge means that the lower variety is contained in (the singular locus of) the upper one [**113**].

In [**113**], the components of the visual event surface $\mathcal{V}(X)$ of a curve or surface X are alternatively characterized via the iterated singular loci of the *higher associated hypersurfaces* [**82**]/*coisotropic hypersurfaces* [**112**] of X. For a surface X, the coisotropic hypersurfaces are the *dual surface* $X^\vee$ in $(\mathbb{P}^3)^*$ and the *Hurwitz threefold* $\mathrm{Hur}(X)$ in $\mathrm{Gr}(1, \mathbb{P}^3)$, i.e., the Zariski closure of the set of planes (respectively lines) tangent to X. Figure 10 shows the iterated singular loci of these hypersurfaces till the level of curves. The five components of $\mathcal{V}(X)$ are encoded by the curves in the last row of Figure 10: two components are the dual surfaces to the curves on the left. The remaining three components are ruled by the lines on the curves shown in the right diagram.

EXAMPLE 7.3. Consider the curve $\mathcal{F}^\ell(X) \subset \mathrm{Gr}(1, \mathbb{P}^3)$ associated with a surface $X \subset \mathbb{P}^3$ of degree at least four. The curve $\mathcal{F}^\ell(X)$ is the Zariski closure of the set of all lines that intersect the surface X at some point with multiplicity four. Such lines are called *flecnodal*. The union of all flecnodal lines is the *flecnodal surface* of X; it is one of the irreducible components of the visual event surface $\mathcal{V}(X)$. The visual event corresponding to the flecnodal surface is called *swallowtail event*.

For an explicit example, let us consider the quartic surface parametrized by $(s,t) \mapsto (s, t, s^2 + 3st + t^4)$. The surface is shown in yellow in Figure 11. In Figure 11(A), the surface's image contour is smooth. In Figure 11(B), the image contour has two cusps and a node, as illustrated in the torus example in Figure 8. The transition occurs when the camera center is located on a flecnodal line. On a general surface, there is a one-dimensional family of points such that one of the tangent lines at that point is flecnodal. The family makes up the curve shown in red in Figure 11. For the particular surface there, every red point has two flecnodal lines. The lines are shown for a particular point in Figures 11(A) and 11(B), and for many points in Figure 11(C).

Similarly, if X is a curve its coisotropic hypersurfaces are the dual surface $X^\vee$ in $(\mathbb{P}^3)^*$ and the *Chow threefold* in $\mathrm{Gr}(1, \mathbb{P}^3)$, that is, the set of lines meeting X. Their (iterated) singular curves give rise to the three components of $\mathcal{V}(X)$ analogously.

The degrees of the components of $\mathcal{V}(X)$ for a general curve or surface X of fixed degree (and genus in the curve case) are classically known; see [**113**, Theorems 3.1

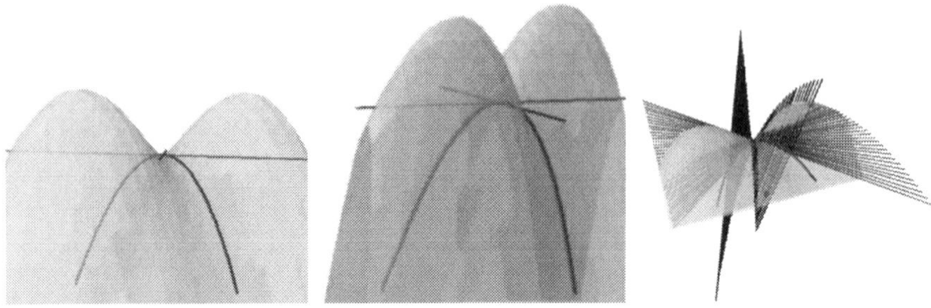

(A) Smooth image contour (B) Singular image contour (C) Flecnodal surface

FIGURE 11. Quartic surface parametrized by $(s, t, s^2 + 3st + t^4)$ with flecnodal lines that meet the surface with multiplicity 4 at one of the marked points. A swallowtail event happens when the 2 cusps and the node in Figure 11(B) come together to a higher order singularity.

and 4.1]. Petitjean provided degree formulas for the five components of $\mathcal{V}(X)$ in the surface case [**152**], but they appeared already in paragraphs 597, 598, 599, 608, and 613 of Salmon's 1882 book [**164**].

7.3. Hurwitz hypersurfaces and condition numbers. The minimal problem (5.1) of computing essential matrices consistent with five point pairs admits the following geometric interpretation: We are computing the intersection of the essential variety $\mathcal{E} \subset \mathbb{P}(\mathbb{R}^{3\times 3})$ with a data-dependent linear subspace $L \subset \mathbb{P}(\mathbb{R}^{3\times 3})$ of complementary dimension, namely $L = \{x_i y_i^\top : i \in \mathcal{I}\}^\perp$. In numerical algebraic geometry language, we are computing a *witness set* of $\mathcal{E}$ [**168**, Chapter 13.3]. Several, though not all, minimal problems in computer vision usefully admit such a interpretation. For instance, Kileel [**107**] describes all 66 minimal problems for three calibrated pinhole cameras that arise from slicing the calibrated trifocal variety $\mathcal{T}_{\text{cal}}$ with linear spaces of complementary dimension.

In the applied algebra community, this description has motivated theoretical works on intersecting a fixed variety $Z \subset \mathbb{P}^n$ with varying linear subspaces of complementary dimension. This includes the detailed study of the *Hurwitz hypersurface* $\text{Hur}(Z)$ by Sturmfels [**173**]. Writing c for the codimension of Z in $\mathbb{P}^n$ and d for its degree, $\text{Hur}(Z)$ is the Zariski closure in $\text{Gr}(c, \mathbb{P}^n)$ of the set of all c-dimensional projective subspaces L of $\mathbb{P}^n$ such that the intersection $Z \cap L$ does not consist of d reduced points. The Hurwitz hypersurface is intimately linked with the *condition number* [**41**] of the algebraic function $L \mapsto Z \cap L$. That condition number measures how much the intersection $Z \cap L$ changes when L gets perturbed, which is crucial for the understanding of how much numerical computations such as homotopy continuation are affected by errors. Bürgisser showed that the set of *ill-posed* L, i.e., those with infinite condition, is essentially the Hurwitz hypersurface $\text{Hur}(Z)$ and that (almost) all L with large condition are contained in a small tube around $\text{Hur}(Z)$ [**40**, Corollary 1.10]. Bürgisser's work follows the general idea that condition numbers are given by the inverse distance to ill-posedness from numerical analysis [**58, 59**].

Recently, there has been work on condition numbers of minimal problems in the more practical setting, where the dependence is in terms of the perturbation of the image data (instead of perturbation of L in the Grassmannian) by Fan, Kileel, and Kimia [**70**, **71**]. Their work drew distinctions between criticality (Section 7.1 above) and ill-posedness (when the condition number is infinite), see [**71**, Section 5]. For the 5- and 7-point problems, the authors derived condition number formulas in terms of certain Jacobian matrices, and they characterized ill-posed world scenes and ill-posed image point pairs geometrically. Furthermore, real-data experiments showed that poor conditioning can actually plague some computer vision data sets.

Acknowledgment

We thank Sameer Agarwal, Paul Breiding, Luca Carlone, Tim Duff, Hongyi Fan, Fredrik Kahl, Anton Leykin, Tomas Pajdla, Jean Ponce, Kristian Ranestad, Felix Rydell, Elima Shehu, Rekha Thomas, Matthew Trager, and Uli Walther for their comments on earlier versions of the manuscript.

References

[1] S. Agarwal, C. Aholt, and R. Thomas, *A QCQP approach to triangulation*, European Conference on Computer Vision, Springer, 2012, pp. 654–667.

[2] S. Agarwal, T. Duff, M. Lieblich, and R. R. Thomas, *An atlas for the pinhole camera*, Found. Comput. Math. **24** (2024), no. 1, 227–277, DOI 10.1007/s10208-022-09592-6. MR4707063

[3] S. Agarwal, Y. Furukawa, N. Snavely, I. Simon, B. Curless, S. M. Seitz, and R. Szeliski, *Building Rome in a day*, Communications of the ACM **54** (2011), no. 10, 105–112.

[4] S. Agarwal, H.-L. Lee, B. Sturmfels, and R. R. Thomas, *On the existence of epipolar matrices*, Int. J. Comput. Vis. **121** (2017), no. 3, 403–415, DOI 10.1007/s11263-016-0949-7. MR3606541

[5] S. Agarwal, A. Pryhuber, R. Sinn, and R. R. Thomas, *The chiral domain of a camera arrangement*, J. Math. Imaging Vision **64** (2022), no. 9, 948–967, DOI 10.1007/s10851-022-01101-2. MR4503458

[6] S. Agarwal, A. Pryhuber, and R. Thomas, *Ideals of the multiview variety*, IEEE Transactions on Pattern Analysis and Machine Intelligence **43** (2019), no. 4, 1279–1292.

[7] C. Aholt and L. Oeding, *The ideal of the trifocal variety*, Math. Comp. **83** (2014), no. 289, 2553–2574, DOI 10.1090/S0025-5718-2014-02842-1. MR3223346

[8] C. Aholt, B. Sturmfels, and R. Thomas, *A Hilbert scheme in computer vision*, Canad. J. Math. **65** (2013), no. 5, 961–988, DOI 10.4153/CJM-2012-023-2. MR3095002

[9] C. Aholt, *Polynomials in Multiview Geometry*, ProQuest LLC, Ann Arbor, MI, 2012. Thesis (Ph.D.)–University of Washington. MR3152447

[10] V. I. Arnol′d, *Catastrophe theory*, Springer-Verlag, Berlin, 1984. Translated from the Russian by R. K. Thomas, DOI 10.1007/978-3-642-96799-3. MR727639

[11] F. Arrigoni, A. Fusiello, E. Ricci, and T. Pajdla, *Viewing graph solvability via cycle consistency*, Proceedings of the IEEE/CVF International Conference on Computer Vision, 2021, pp. 5540–5549.

[12] K. Åström and F. Kahl, *Ambiguous configurations for the 1D structure and motion problem*, J. Math. Imaging Vision **18** (2003), no. 2, 191–203, DOI 10.1023/A:1022120702016. MR1965320

[13] D. Barath, *Five-point fundamental matrix estimation for uncalibrated cameras*, Proceedings of the IEEE Conference on Computer Vision and Pattern Recognition, 2018, pp. 235–243.

[14] D. Barath, T. Toth, and L. Hajder, *A minimal solution for two-view focal-length estimation using two affine correspondences*, Proceedings of the IEEE Conference on Computer Vision and Pattern Recognition, 2017, pp. 6003–6011.

[15] D. J. Bates, J. D. Hauenstein, A. J. Sommese, and C. W. Wampler, *Numerically solving polynomial systems with Bertini*, Software, Environments, and Tools, vol. 25, Society for Industrial and Applied Mathematics (SIAM), Philadelphia, PA, 2013. MR3155500

[16] M. Bertolini, G. M. Besana, R. Notari, and C. Turrini, *Critical loci in computer vision and matrices dropping rank in codimension one*, J. Pure Appl. Algebra **224** (2020), no. 12, 106439, 27, DOI 10.1016/j.jpaa.2020.106439. MR4110277

[17] M. Bertolini, G. M. Besana, and C. Turrini, *Instability of projective reconstruction from 1-view near critical configurations in higher dimensions*, Algebra, geometry and their interactions, Contemp. Math., vol. 448, Amer. Math. Soc., Providence, RI, 2007, pp. 1–12, DOI 10.1090/conm/448/08654. MR2389231

[18] M. Bertolini, G. Besana, and C. Turrini, *Critical loci for projective reconstruction from multiple views in higher dimension: A comprehensive theoretical approach*, Linear Algebra Appl. **469** (2015), 335–363, DOI 10.1016/j.laa.2014.11.021. MR3299067

[19] M. Bertolini and L. Magri, *Critical hypersurfaces and instability for reconstruction of scenes in high dimensional projective spaces*, Machine Graphics and Vision **29** (2020), no. 1/4, 3–20.

[20] M. Bertolini, L. Magri, and C. Turrini, *Critical loci for two views reconstruction as quadratic transformations between images*, J. Math. Imaging Vision **61** (2019), no. 9, 1322–1328, DOI 10.1007/s10851-019-00908-w. MR4017694

[21] M. Bertolini, R. Notari, and C. Turrini, *Smooth determinantal varieties and critical loci in multiview geometry*, Collect. Math. **73** (2022), no. 3, 457–475, DOI 10.1007/s13348-021-00329-2. MR4467909

[22] M. Bertolini and C. Turrini, *Critical configurations for 1-view in projections from* $\mathbb{P}^k \to \mathbb{P}^2$, J. Math. Imaging Vision **27** (2007), no. 3, 227–287, DOI 10.1007/s10851-007-0649-6. MR2325854

[23] M. Bertolini, C. Turrini, and G. M. Besana, *Instability of projective reconstruction of dynamic scenes near critical configurations*, Proceedings of the IEEE International Conference on Computer Vision, 2007, pp. 1–7.

[24] S. Bhayani, J. Heikkilä, and Z. Kukelova, *Sparse resultant-based minimal solvers in computer vision and their connection with the action matrix*, J. Math. Imaging Vision **66** (2024), no. 3, 335–360, DOI 10.1007/s10851-024-01182-1. MR4747079

[25] S. Bhayani, Z. Kukelova, and J. Heikkilä, *A sparse resultant based method for efficient minimal solvers*, Proceedings of the IEEE/CVF Conference on Computer Vision and Pattern Recognition, 2020, pp. 1770–1779.

[26] S. Bianco, G. Ciocca, and D. Marelli, *Evaluating the performance of structure from motion pipelines*, Journal of Imaging **4** (2018), no. 8, 98.

[27] G. Masotti Biggiogero, *La caratterizzazione della curva di diramazione dei piani tripli, ottenuta mediante sistemi di curve pluritangenti* (Italian), Ist. Lombardo Sci. Lett. Rend. Cl. Sci. Mat. Nat. (3) **11(80)** (1947), 151–160 (1949). MR35471

[28] G. Masotti Biggiogero, *Sulla caratterizzazione della curva di diramazione dei piani quadrupli generali*, Ist. Lombardo Sci. Lett. Rend Cl. Sci. Mat. Nat **80** (1947), 269–280.

[29] A. Bik, H. Eisenmann, and B. Sturmfels, *Jordan algebras of symmetric matrices*, Matematiche (Catania) **76** (2021), no. 2, 337–353, DOI 10.4418/2021.76.2.3. MR4334891

[30] P.-L. Bazin and M. Boutin, *Structure from motion: A new look from the point of view of invariant theory*, SIAM J. Appl. Math. **64** (2004), no. 4, 1156–1174, DOI 10.1137/S003613990340246X. MR2068664

[31] K. W. Bowyer and C. R. Dyer, *Aspect graphs: An introduction and survey of recent results*, International Journal of Imaging Systems and Technology **2** (1990), no. 4, 315–328.

[32] M. Bråtelund, *Critical configurations for three projective views*, Math. Scand. **129** (2023), no. 3, 401–444. MR4667876

[33] M. Bråtelund, *Critical configurations for two projective views, a new approach*, J. Symbolic Comput. **120** (2024), Paper No. 102226, 22, DOI 10.1016/j.jsc.2023.102226. MR4583113

[34] P. Breiding, T. Ö. Çelik, T. Duff, A. Heaton, A. Maraj, A.-L. Sattelberger, L. Venturello, and O. Yürük, *Nonlinear algebra and applications*, Numer. Algebra Control Optim. **13** (2023), no. 1, 81–116, DOI 10.3934/naco.2021045. MR4520402

[35] P. Breiding, F. Rydell, E. Shehu, and A. Torres, *Line multiview varieties*, SIAM J. Appl. Algebra Geom. **7** (2023), no. 2, 470–504, DOI 10.1137/22M1482263. MR4598813

[36] P. Breiding and S. Timme, *HomotopyContinuation.jl: A package for homotopy continuation in Julia*, International Congress on Mathematical Software, Springer, 2018, pp. 458–465.

[37] T. Buchanan, *The twisted cubic and camera calibration*, Comput. Vision Graphics Image Process. **42** (1988), no. 1, 130–132, DOI 10.1016/0734-189X(88)90146-6. MR1255702

[38] T. Buchanan, *Critical sets for 3D reconstruction using lines*, European Conference on Computer Vision, Springer, 1992, pp. 730–738.
[39] T. Buchanan, *On the critical set for photogrammetric reconstruction using line tokens in* $P_3(\mathbf{C})$, Geom. Dedicata **44** (1992), no. 2, 223–232, DOI 10.1007/BF00182950. MR1187642
[40] P. Bürgisser, *Condition of intersecting a projective variety with a varying linear subspace*, SIAM J. Appl. Algebra Geom. **1** (2017), no. 1, 111–125, DOI 10.1137/16M1077222. MR3633770
[41] P. Bürgisser and F. Cucker, *Condition: The geometry of numerical algorithms*, Grundlehren der mathematischen Wissenschaften [Fundamental Principles of Mathematical Sciences], vol. 349, Springer, Heidelberg, 2013, DOI 10.1007/978-3-642-38896-5. MR3098452
[42] E. Calabi, P. J. Olver, C. Shakiban, A. Tannenbaum, and S. Haker, *Differential and numerically invariant signature curves applied to object recognition*, International Journal of Computer Vision **26** (1998), 107–135.
[43] F. Camposeco, T. Sattler, and M. Pollefeys, *Minimal solvers for generalized pose and scale estimation from two rays and one point*, European Conference on Computer Vision, Springer, 2016, pp. 202–218.
[44] L. Carlone, *Estimation contracts for outlier-robust geometric perception*, Foundations and Trends in Robotics **11** (2023), no. 2-3, 90–224.
[45] J. Chen and J. Kileel, *Numerical implicitization*, J. Softw. Algebra Geom. **9** (2019), no. 1, 55–63, DOI 10.2140/jsag.2019.9.55. MR3988096
[46] C.-H. Chien, H. Fan, A. Abdelfattah, E. Tsigaridas, S. Tomov, and B. Kimia, *GPU-based homotopy continuation for minimal problems in computer vision*, Proceedings of the IEEE/CVF Conference on Computer Vision and Pattern Recognition, 2022, pp. 15765–15776.
[47] O. Chisini, *Sulla identità birazionale di due funzioni algebriche di più variabili, dotate di una medesima varietà di diramazione* (Italian), Ist. Lombardo Sci. Lett. Rend. Cl. Sci. Mat. Nat. (3) **11(80)** (1947), 3–6 (1949). MR34054
[48] O. Chum, J. Matas, and J. Kittler, *Locally optimized RANSAC*, Joint Pattern Recognition Symposium, Springer, 2003, pp. 236–243.
[49] Y. Cid-Ruiz, O. Clarke, and F. Mohammadi, *A study of nonlinear multiview varieties*, J. Algebra **620** (2023), 363–391, DOI 10.1016/j.jalgebra.2022.12.036. MR4536152
[50] D. Cifuentes, S. Agarwal, P. A. Parrilo, and R. R. Thomas, *On the local stability of semidefinite relaxations*, Math. Program. **193** (2022), no. 2, Ser. B, 629–663, DOI 10.1007/s10107-021-01696-1. MR4438380
[51] D. A. Cox, J. Little, and D. O'Shea, *Ideals, varieties, and algorithms: An introduction to computational algebraic geometry and commutative algebra*, 4th ed., Undergraduate Texts in Mathematics, Springer, Cham, 2015, DOI 10.1007/978-3-319-16721-3. MR3330490
[52] D. A. Cox, *Stickelberger and the eigenvalue theorem*, Commutative algebra, Springer, Cham, [2021] ©2021, pp. 283–298, DOI 10.1007/978-3-030-89694-2_8. MR4394411
[53] Y. Dai, H. Li, and L. Kneip, *Rolling shutter camera relative pose: Generalized epipolar geometry*, Proceedings of the IEEE Conference on Computer Vision and Pattern Recognition, 2016, pp. 4132–4140.
[54] J. d'Almeida, *Courbe de ramification de la projection sur* $\mathbf{P}^2$ *d'une surface de* $\mathbf{P}^3$ (French), Duke Math. J. **65** (1992), no. 2, 229–233, DOI 10.1215/S0012-7094-92-06509-4. MR1150585
[55] D. F. Davidenko, *On a new method of numerical solution of systems of nonlinear equations* (Russian), Doklady Akad. Nauk SSSR (N.S.) **88** (1953), 601–602. MR54339
[56] D. F. Davidenko, *On approximate solution of systems of nonlinear equations* (Russian), Ukrain. Mat. Žurnal **5** (1953), 196–206. MR57029
[57] M. Demazure, *Sur deux problemes de reconstruction*, Tech. Report 882, *INRIA*, 1988.
[58] J. W. Demmel, *On condition numbers and the distance to the nearest ill-posed problem*, Numer. Math. **51** (1987), no. 3, 251–289, DOI 10.1007/BF01400115. MR895087
[59] J. W. Demmel, *The geometry of ill-conditioning*, J. Complexity **3** (1987), no. 2, 201–229, DOI 10.1016/0885-064X(87)90027-6. MR907197
[60] T. Dobbert, *Matchmoving: The invisible art of camera tracking*, Sybex, 2005.
[61] J. Draisma, E. Horobeţ, G. Ottaviani, B. Sturmfels, and R. R. Thomas, *The Euclidean distance degree of an algebraic variety*, Found. Comput. Math. **16** (2016), no. 1, 99–149, DOI 10.1007/s10208-014-9240-x. MR3451425

[62] D. Drusvyatskiy, H.-L. Lee, G. Ottaviani, and R. R. Thomas, *The Euclidean distance degree of orthogonally invariant matrix varieties*, Israel J. Math. **221** (2017), no. 1, 291–316, DOI 10.1007/s11856-017-1545-4. MR3705855

[63] D. Drusvyatskiy, H.-L. Lee, and R. R. Thomas, *Counting real critical points of the distance to orthogonally invariant matrix sets*, SIAM J. Matrix Anal. Appl. **36** (2015), no. 3, 1360–1380, DOI 10.1137/15M1007835. MR3402784

[64] T. Duff, K. Kohn, A. Leykin, and T. Pajdla, *PLMP – Point-line minimal problems in complete multi-view visibility*, Proceedings of the IEEE/CVF International Conference on Computer Vision, 2019, pp. 1675–1684.

[65] T. Duff, K. Kohn, A. Leykin, and T. Pajdla, *PL_1P – Point-line minimal problems under partial visibility in three views*, European Conference on Computer Vision, Springer, 2020, pp. 175–192.

[66] T. Duff, V. Korotynskiy, T. Pajdla, and M. H. Regan, *Galois/monodromy groups for decomposing minimal problems in 3D reconstruction*, SIAM J. Appl. Algebra Geom. **6** (2022), no. 4, 740–772, DOI 10.1137/21M1422872. MR4522866

[67] A. Elqursh and A. Elgammal, *Line-based relative pose estimation*, Proceedings of the IEEE Conference on Computer Vision and Pattern Recognition, 2011, pp. 3049–3056.

[68] L. Escobar and A. Knutson, *The multidegree of the multi-image variety*, Combinatorial algebraic geometry, Fields Inst. Commun., vol. 80, Fields Inst. Res. Math. Sci., Toronto, ON, 2017, pp. 283–296. MR3752504

[69] R. Fabbri, T. Duff, H. Fan, M. H. Regan, D. da Costa de Pinho, E. Tsigaridas, C. W. Wampler, J. D. Hauenstein, P. J. Giblin, B. Kimia, A. Leykin, and T. Pajdla, *TRPLP – Trifocal relative pose from lines at points*, Proceedings of the IEEE/CVF Conference on Computer Vision and Pattern Recognition, 2020, pp. 12073–12083.

[70] H. Fan, J. Kileel, and B. Kimia, *On the instability of relative pose estimation and RANSAC's role*, Proceedings of the IEEE/CVF Conference on Computer Vision and Pattern Recognition, 2022, pp. 8935–8943.

[71] H. Fan, J. Kileel, and B. Kimia, *Condition numbers in multiview geometry, instability in relative pose estimation, and RANSAC*, Preprint, arXiv:2310.02719, (2023).

[72] X. Fan and R. Vidal, *The space of multibody fundamental matrices: Rank, geometry and projection*, Dynamical Vision, Springer, 2006, pp. 1–17.

[73] O. Faugeras and Q.-T. Luong, *The geometry of multiple images*, MIT Press, Cambridge, MA, 2001. The laws that govern the formation of multiple images of a scene and some of their applications; With contributions from Théo Papadopoulo. MR1823958

[74] O. D. Faugeras and B. Mourrain, *On the geometry and algebra of the point and line correspondences between n images*, Proceedings of IEEE International Conference on Computer Vision, 1995, pp. 951–956.

[75] O. D. Faugeras and F. Lustman, *Motion and structure from motion in a piecewise planar environment*, International Journal of Pattern Recognition and Artificial Intelligence **2** (1988), no. 03, 48–508.

[76] M. A. Fischler and R. C. Bolles, *Random sample consensus: A paradigm for model fitting with applications to image analysis and automated cartography*, Comm. ACM **24** (1981), no. 6, 381–395, DOI 10.1145/358669.358692. MR618158

[77] G. Fløystad, J. Kileel, and G. Ottaviani, *The Chow form of the essential variety in computer vision*, J. Symbolic Comput. **86** (2018), 97–119, DOI 10.1016/j.jsc.2017.03.010. MR3725215

[78] D. A. Forsyth and J. Ponce, *Computer vision: A modern approach*, 2nd ed., Pearson, 2012.

[79] D. A. Forsyth, *Recognizing algebraic surfaces from their outlines*, International Journal of Computer Vision **18** (1996), no. 1, 21–40.

[80] M. Gallet, N. Lubbes, J. Schicho, and J. Vršek, *Reconstruction of surfaces with ordinary singularities from their silhouettes*, SIAM J. Appl. Algebra Geom. **3** (2019), no. 3, 472–506, DOI 10.1137/18M1220911. MR4000204

[81] M. Gallet, N. Lubbes, J. Schicho, and J. Vršek, *Reconstruction of rational ruled surfaces from their silhouettes*, J. Symbolic Comput. **104** (2021), 366–380, DOI 10.1016/j.jsc.2020.08.002. MR4180133

[82] I. M. Gel′fand, M. M. Kapranov, and A. V. Zelevinsky, *Discriminants, resultants, and multidimensional determinants*, Mathematics: Theory & Applications, Birkhäuser Boston, Inc., Boston, MA, 1994, DOI 10.1007/978-0-8176-4771-1. MR1264417

[83] G. Grisetti, R. Kümmerle, C. Stachniss, and W. Burgard, *A tutorial on graph-based SLAM*, IEEE Intelligent Transportation Systems Magazine **2** (2010), no. 4, 31–43.

[84] J. A. Grunert, *Das Pothenot'sche problem in erweiterter gestalt; nebst bemerkungen über seine anwendung in der geodäsie*, Grunerts Archiv fur Mathematik und Physik (1841), 238–248.

[85] B. M. Haralick, C.-N. Lee, K. Ottenberg, and M. Nölle, *Review and analysis of solutions of the three point perspective pose estimation problem*, International Journal of Computer Vision **13** (1994), no. 3, 331–356.

[86] C. Harris and D. Lowengrub, *The Chern-Mather class of the multiview variety*, Comm. Algebra **46** (2018), no. 6, 2488–2499, DOI 10.1080/00927872.2017.1392545. MR3778407

[87] R. Hartley, *Projective reconstruction and invariants from multiple images*, IEEE Transactions on Pattern Analysis and Machine Intelligence **16** (1994), no. 10, 1036–1041.

[88] R. Hartley, *Chirality*, International Journal of Computer Vision **26** (1998), 41–61.

[89] R. Hartley and F. Kahl, *Critical configurations for projective reconstruction from multiple views*, International Journal of Computer Vision **71** (2007), no. 1, 5–47.

[90] R. Hartley and F. Kahl, *Optimal algorithms in multiview geometry*, Asian Conference on Computer Vision, Springer, 2007, pp. 13–34.

[91] R. Hartley and F. Schaffalitzky, *Reconstruction from projections using Grassmann tensors*, International Journal of Computer Vision **83** (2009), no. 3, 274–293.

[92] R. Hartley and P. Sturm, *Triangulation*, Computer Vision and Image Understanding **68** (1997), no. 2, 146–157.

[93] R. Hartley and R. Vidal, *The multibody trifocal tensor: Motion segmentation from 3 perspective views*, Proceedings of the IEEE Conference on Computer Vision and Pattern Recognition, 2004.

[94] R. Hartley and A. Zisserman, *Multiple view geometry in computer vision*, Cambridge University Press, Cambridge, 2001. With a foreword by Olivier Faugeras. MR1823669

[95] O. Hesse, *Die cubische Gleichung, von welcher die Lösung des Problems der Homographie von M. Chasles abhängt* (German), J. Reine Angew. Math. **62** (1863), 188–192, DOI 10.1515/crll.1863.62.188. MR1579236

[96] A. Heyden and K. Åström, *Algebraic properties of multilinear constraints*, Math. Methods Appl. Sci. **20** (1997), no. 13, 1135–1162, DOI 10.1002/(SICI)1099-1476(19970910)20:13¡1135::AID-MMA908¿3.0.CO;2-9. MR1465398

[97] B. K. P. Horn, *Relative orientation revisited*, Journal of the Optical Society of America A **8** (1991), no. 10, 1630–1638.

[98] P. Hruby, T. Duff, A. Leykin, and T. Pajdla, *Learning to solve hard minimal problems*, Proceedings of the IEEE/CVF Conference on Computer Vision and Pattern Recognition, 2022, pp. 5532–5542.

[99] K. Huang, R. Fossum, and Y. Ma, *Generalized rank conditions in multiple view geometry with applications to dynamical scenes*, European Conference on Computer Vision, Springer, 2002, pp. 201–216

[100] A. Ito, M. Miura, and K. Ueda, *Projective reconstruction in algebraic vision*, Canad. Math. Bull. **63** (2020), no. 3, 592–609, DOI 10.4153/s0008439519000687. MR4148121

[101] B. Johansson, M. Oskarsson, and K. Åström, *Structure and motion estimation from complex features in three views*, Indian Conference on Computer Vision, Graphics and Image Processing, 2002.

[102] M. Joswig, J. Kileel, B. Sturmfels, and A. Wagner, *Rigid multiview varieties*, Internat. J. Algebra Comput. **26** (2016), no. 4, 775–788, DOI 10.1142/S021819671650034X. MR3513935

[103] F. Kahl and R. Hartley, *Critical curves and surfaces for Euclidean reconstruction*, European Conference on Computer Vision, Springer, 2002, pp. 447–462.

[104] F. Kahl and D. Henrion, *Globally optimal estimates for geometric reconstruction problems*, Int. J. Comput. Vis. **74** (2007), no. 1, 3–15.

[105] J. Y. Kaminski and A. Shashua, *Multiple-view geometry of general algebraic curves*, Int. J. Comput. Vis. **56** (2004), no. 3, 195–219.

[106] J. Y. Kaminski and M. Teicher, *A general framework for trajectory triangulation*, J. Math. Imaging Vision **21** (2004), no. 1, 27–41, DOI 10.1023/B:JMIV.0000026555.79056.b8. MR2075715

[107] J. Kileel, *Minimal problems for the calibrated trifocal variety*, SIAM J. Appl. Algebra Geom. **1** (2017), no. 1, 575–598, DOI 10.1137/16M1104482. MR3705779

[108] J. Kileel, Z. Kukelova, T. Pajdla, and B. Sturmfels, *Distortion varieties*, Found. Comput. Math. **18** (2018), no. 4, 1043–1071, DOI 10.1007/s10208-017-9361-0. MR3833649

[109] M. Kitagawa and B. Windsor, *MoCap for artists: Workflow and techniques for motion capture*, Focal Press, 2008.

[110] J. J. Koenderink, *Solid shape*, MIT Press Series in Artificial Intelligence, MIT Press, Cambridge, MA, 1990. MR1045203

[111] J. J. Koenderink and A. J. Van Doorn, *The internal representation of solid shape with respect to vision*, Biological Cybernetics **32** (1979), no. 4, 211–216.

[112] K. Kohn and J. C. Mathews Jr., *Isotropic and coisotropic subvarieties of Grassmannians*, Adv. Math. **377** (2021), Paper No. 107492, 37, DOI 10.1016/j.aim.2020.107492. MR4182913

[113] K. Kohn, B. Sturmfels, and M. Trager, *Changing views on curves and surfaces*, Acta Math. Vietnam. **43** (2018), no. 1, 1–29, DOI 10.1007/s40306-017-0240-1. MR3755795

[114] J. Krames, *Zur Ermittlung eines Objektes aus zwei Perspektiven. (Ein Beitrag zur Theorie der "gefährlichen Örter.")* (German), Monatsh. Math. Phys. **49** (1941), 327–354, DOI 10.1007/BF01707311. MR6413

[115] D. J. Kriegman and J. Ponce, *On recognizing and positioning curved 3-D objects from image contours*, IEEE Transactions on Pattern Analysis and Machine Intelligence **12** (1990), no. 12, 1127–1137.

[116] D. J. Kriegman and J. Ponce, *Geometric modeling for computer vision*, Curves and Surfaces in Computer Vision and Graphics II, vol. 1610, International Society for Optics and Photonics, 1992, pp. 250–260.

[117] Y. Kuang and K. Åström, *Pose estimation with unknown focal length using points, directions and lines*, Proceedings of the IEEE International Conference on Computer Vision, 2013, pp. 529–536.

[118] Z. Kukelova, M. Bujnak, and T. Pajdla, *Automatic generator of minimal problem solvers*, European Conference on Computer Vision, Springer, 2008, pp. 302–315.

[119] Z. Kukelova, J. Heller, M. Bujnak, A. Fitzgibbon, and T. Pajdla, *Efficient solution to the epipolar geometry for radially distorted cameras*, Proceedings of the IEEE International Conference on Computer Vision, 2015, pp. 2309–2317.

[120] Z. Kukelova, J. Kileel, B. Sturmfels, and T. Pajdla, *A clever elimination strategy for efficient minimal solvers*, Proceedings of the IEEE Conference on Computer Vision and Pattern Recognition, 2017, pp. 4912–4921.

[121] Z. Kukelova and T. Pajdla, *A minimal solution to the autocalibration of radial distortion*, Proceedings of the IEEE Conference on Computer Vision and Pattern Recognition, 2007, pp. 1–7.

[122] Vik. S. Kulikov, *On Chisini's conjecture* (Russian, with Russian summary), Izv. Ross. Akad. Nauk Ser. Mat. **63** (1999), no. 6, 83–116, DOI 10.1070/im1999v063n06ABEH000267; English transl., Izv. Math. **63** (1999), no. 6, 1139–1170. MR1748562

[123] Vik. S. Kulikov, *On Chisini's conjecture. II* (Russian, with Russian summary), Izv. Ross. Akad. Nauk Ser. Mat. **72** (2008), no. 5, 63–76, DOI 10.1070/IM2008v072n05ABEH002423; English transl., Izv. Math. **72** (2008), no. 5, 901–913. MR2473772

[124] V. Larsson, K. Åström, and M. Oskarsson, *Efficient solvers for minimal problems by syzygy-based reduction*, Proceedings of the IEEE Conference on Computer Vision and Pattern Recognition, 2017, pp. 820–829.

[125] V. Larsson, M. Oskarsson, K. Åström, A. Wallis, Z. Kukelova, and T. Pajdla, *Beyond Gröbner bases: Basis selection for minimal solvers*, Proceedings of the IEEE Conference on Computer Vision and Pattern Recognition, 2018, pp. 3945–3954.

[126] V. Larsson, N. Zobernig, K. Taskin, and M. Pollefeys, *Calibration-free structure-from-motion with calibrated radial trifocal tensors*, European Conference on Computer Vision, Springer, 2020, pp. 382–399.

[127] S. Laveau and O. Faugeras, *Oriented projective geometry for computer vision*, European Conference on Computer Vision, Springer, 1996, pp. 147–156.

[128] G. H. Lee, *A minimal solution for non-perspective pose estimation from line correspondences*, European Conference on Computer Vision, Springer, 2016, pp. 170–185.

[129] G. Lerman and Y. Shi, *Robust group synchronization via cycle-edge message passing*, Found. Comput. Math. **22** (2022), no. 6, 1665–1741, DOI 10.1007/s10208-021-09532-w. MR4509114

[130] B. Li, *Images of rational maps of projective spaces*, Int. Math. Res. Not. IMRN **13** (2018), 4190–4228, DOI 10.1093/imrn/rnx003. MR3829180

[131] B. Li and V. Larsson, *Gaps: Generator for automatic polynomial solvers*, Preprint, arXiv:2004.11765 (2020).

[132] M. Lieblich and L. Van Meter, *Two Hilbert schemes in computer vision*, SIAM J. Appl. Algebra Geom. **4** (2020), no. 2, 297–321, DOI 10.1137/18M1200117. MR4101360

[133] H. C. Longuet-Higgins, *A computer algorithm for reconstructing a scene from two projections*, Nature **293** (1981), no. 5828, 133–135.

[134] D. G. Lowe, *Distinctive image features from scale-invariant keypoints*, International Journal of Computer Vision **60** (2004), no. 2, 91–110.

[135] Y. Ma, S. Soatto, J. Košecká, and S. S. Sastry, *An invitation to 3-D vision*, Interdisciplinary Applied Mathematics, vol. 26, Springer-Verlag, New York, 2004. From images to geometric models, DOI 10.1007/978-0-387-21779-6. MR2015496

[136] E. Martyushev, J. Vrablikova, and T. Pajdla, *Optimizing elimination templates by greedy parameter search*, Proceedings of the IEEE/CVF Conference on Computer Vision and Pattern Recognition, 2022, pp. 15754–15764.

[137] E. V. Martyushev, *On some properties of calibrated trifocal tensors*, J. Math. Imaging Vision **58** (2017), no. 2, 321–332, DOI 10.1007/s10851-017-0712-x. MR3636592

[138] L. G. Maxim, J. I. Rodriguez, and B. Wang, *Euclidean distance degree of the multiview variety*, SIAM J. Appl. Algebra Geom. **4** (2020), no. 1, 28–48, DOI 10.1137/18M1233406. MR4048615

[139] S. J. Maybank, *Theory of reconstruction from image motion*, Springer Series in Information Sciences, vol. 28, Springer-Verlag, Berlin, 1993, DOI 10.1007/978-3-642-77557-4. MR1226451

[140] S. J. Maybank, *The critical line congruence for reconstruction from three images*, Appl. Algebra Engrg. Comm. Comput. **6** (1995), no. 2, 89–113, DOI 10.1007/BF01225646. MR1317188

[141] P. Miraldo, T. Dias, and S. Ramalingam, *A minimal closed-form solution for multi-perspective pose estimation using points and lines*, European Conference on Computer Vision, Springer, 2018, pp. 474–490.

[142] B. Mourrain, *A new criterion for normal form algorithms*, Applied algebra, algebraic algorithms and error-correcting codes (Honolulu, HI, 1999), Lecture Notes in Comput. Sci., vol. 1719, Springer, Berlin, 1999, pp. 430–443, DOI 10.1007/3-540-46796-3_41. MR1846517

[143] D. Mumford, J. Fogarty, and F. Kirwan, *Geometric invariant theory*, 3rd ed., Ergebnisse der Mathematik und ihrer Grenzgebiete (2) [Results in Mathematics and Related Areas (2)], vol. 34, Springer-Verlag, Berlin, 1994. MR1304906

[144] N. Navab and O. Faugeras, *The critical sets of lines for camera displacement estimation: A mixed Euclidean-projective and constructive approach*, International Journal of Computer Vision **23** (1997), no. 1, 17–44.

[145] D. Nistér, *An efficient solution to the five-point relative pose problem*, IEEE Transactions on Pattern Analysis and Machine Intelligence **26** (2004), no. 6, 756–770.

[146] D. Nistér and H. Stewénius, *A minimal solution to the generalised 3-point pose problem*, J. Math. Imaging Vision **27** (2007), no. 1, 67–79, DOI 10.1007/s10851-006-0450-y. MR2297999

[147] L. Oeding, *The quadrifocal variety*, Linear Algebra Appl. **512** (2017), 306–330, DOI 10.1016/j.laa.2016.09.034. MR3567527

[148] M. Oskarsson, A. Zisserman, and K. Åström, *Minimal projective reconstruction for combinations of points and lines in three views*, Image and Vision Computing **22** (2004), no. 10, 777–785.

[149] O. Özyeşil, V. Voroninski, R. Basri, and A. Singer, *A survey of structure from motion*, Acta Numer. **26** (2017), 305–364, DOI 10.1017/S096249291700006X. MR3653853

[150] S.-I. Pae and J. Ponce, *On computing structural changes in evolving surfaces and their appearance*, International Journal of Computer Vision **43** (2001), no. 2, 113–131.

[151] T. Pajdla, *Stereo with oblique cameras*, International Journal of Computer Vision **47** (2002), no. 1, 161–170.

[152] S. Petitjean, *The complexity and enumerative geometry of aspect graphs of smooth surfaces*, Algorithms in algebraic geometry and applications (Santander, 1994), Progr. Math., vol. 143, Birkhäuser, Basel, 1996, pp. 317–352. MR1414457

[153] S. Petitjean, J. Ponce, and D. J. Kriegman, *Computing exact aspect graphs of curved objects: Algebraic surfaces*, International Journal of Computer Vision **9** (1992), no. 3, 231–255.

[154] J. Ponce, *What is a camera?*, Proceedings of the IEEE Conference on Computer Vision and Pattern Recognition, 2009, pp. 1526–1533.

[155] J. Ponce and D. J. Kriegman, *Computing exact aspect graphs of curved objects: Parametric surfaces*, no. 1579, Department of Computer Science, University of Illinois at Urbana-Champaign, 1990.

[156] J. Ponce, B. Sturmfels, and M. Trager, *Congruences and concurrent lines in multi-view geometry*, Adv. in Appl. Math. **88** (2017), 62–91, DOI 10.1016/j.aam.2017.01.001. MR3641809

[157] A. Pryhuber, R. Sinn, and R. R. Thomas, *Existence of two view chiral reconstructions*, SIAM J. Appl. Algebra Geom. **6** (2022), no. 1, 41–76, DOI 10.1137/20M1381848. MR4387191

[158] S. Ramalingam and P. Sturm, *Minimal solutions for generic imaging models*, Proceedings of the IEEE Conference on Computer Vision and Pattern Recognition, 2008, pp. 1–8.

[159] K. Reidemeister, *Knotentheorie* (German), Springer-Verlag, Berlin-New York, 1974. Reprint. MR345089

[160] F. Remondino, L. Barazzetti, F. Nex, M. Scaioni, and D. Sarazzi, *UAV photogrammetry for mapping and 3D modeling – Current status and future perspectives*, The International Archives of the Photogrammetry, Remote Sensing and Spatial Information Sciences **38** (2012), 25–31.

[161] J. H. Rieger, *Global bifurcation sets and stable projections of nonsingular algebraic surfaces*, International Journal of Computer Vision **7** (1992), no. 3, 171–194.

[162] J. H. Rieger, *Computing view graphs of algebraic surfaces*, J. Symbolic Comput. **16** (1993), no. 3, 259–272, DOI 10.1006/jsco.1993.1044. MR1259673

[163] D. M. Rosen, L. Carlone, A. S. Bandeira, and J. J. Leonard, *SE-Sync: A certifiably correct algorithm for synchronization over the special Euclidean group*, The International Journal of Robotics Research **38** (2019), no. 2-3, 95–125.

[164] G. Salmon, *A treatise on the analytic geometry of three dimensions*, Chelsea Publishing Co., New York, 1958. Revised by R. A. P. Rogers; 7th ed. Vol. 1; Edited by C. H. Rowe. MR94753

[165] O. Saurer, M. Pollefeys, and G. H. Lee, *A minimal solution to the rolling shutter pose estimation problem*, 2015 IEEE/RSJ International Conference on Intelligent Robots and Systems, 2015, pp. 1328–1334.

[166] J. L. Schonberger and J.-M. Frahm, *Structure-from-motion revisited*, Proceedings of the IEEE Conference on Computer Vision and Pattern Recognition, 2016, pp. 4104–4113.

[167] B. Segre, *Sulla caratterizzazione delle curve di diramazione dei piani multipli generali*, Memorie Della R. Accad. d'Italia **1** (1930), 5–31.

[168] A. J. Sommese and C. W. Wampler II, *The numerical solution of systems of polynomials*, World Scientific Publishing Co. Pte. Ltd., Hackensack, NJ, 2005. Arising in engineering and science, DOI 10.1142/9789812567727. MR2160078

[169] H. Stewénius, D. Nistér, F. Kahl, and F. Schaffalitzky, *A minimal solution for relative pose with unknown focal length*, Image and Vision Computing **26** (2008), no. 7, 871–877.

[170] H. Stewénius, F. Schaffalitzky, and D. Nistér, *How hard is 3-view triangulation really?*, Proceedings of the IEEE International Conference on Computer Vision, 2005, pp. 686–693.

[171] P. Sturm, *Multi-view geometry for general camera models*, Proceedings of the IEEE Conference on Computer Vision and Pattern Recognition, 2005, pp. 206–212.

[172] R. Sturm, *Das Problem der Projectivität und seine Anwendung auf die Flächen zweiten Grades* (German), Math. Ann. **1** (1869), no. 4, 533–574, DOI 10.1007/BF01448082. MR1509639

[173] B. Sturmfels, *The Hurwitz form of a projective variety*. part 1, J. Symbolic Comput. **79** (2017), no. part 1, 186–196, DOI 10.1016/j.jsc.2016.08.012. MR3550361

[174] R. Szeliski, *Computer vision: Algorithms and applications*, Springer Science & Business Media, 2010.

[175] S. Thirthala and M. Pollefeys, *Radial multi-focal tensors: applications to omnidirectional camera calibration*, Int. J. Comput. Vis. **96** (2012), no. 2, 195–211, DOI 10.1007/s11263-011-0463-x. MR2874567

[176] R. Thom, *Structural stability and morphogenesis*, W.A. Benjamin, Inc., 1972.

[177] M. Trager, M. Hebert, and J. Ponce, *The joint image handbook*, Proceedings of the IEEE International Conference on Computer Vision, 2015, pp. 909–917.

[178] M. Trager, B. Osserman, and J. Ponce, *On the solvability of viewing graphs*, European Conference on Computer Vision, Springer, 2018, pp. 321–335.

[179] M. Trager, B. Sturmfels, J. Canny, M. Hebert, and J. Ponce, *General models for rational cameras and the case of two-slit projections*, Proceedings of the IEEE Conference on Computer Vision and Pattern Recognition, 2017, pp. 1935–1943.

[180] B. Triggs, *Matching constraints and the joint image*, Proceedings of the IEEE International Conference on Computer Vision, 1995, pp. 338–343.

[181] B. Triggs, P. McLauchlan, R. Hartley, and A. Fitzgibbon, *Bundle adjustment – a modern synthesis*, International Workshop on Vision Algorithms, Springer, 1999, pp. 298–372.

[182] M. C. Tsakiris and R. Vidal, *Filtrated algebraic subspace clustering*, SIAM J. Imaging Sci. **10** (2017), no. 1, 372–415, DOI 10.1137/16M1083451. MR3623692

[183] J. Ventura, C. Arth, and V. Lepetit, *An efficient minimal solution for multi-camera motion*, Proceedings of the IEEE International Conference on Computer Vision, 2015, pp. 747–755.

[184] R. Vidal and Y. Ma, *A unified algebraic approach to 2-D and 3-D motion segmentation and estimation*, J. Math. Imaging Vision **25** (2006), no. 3, 403–421, DOI 10.1007/s10851-006-8286-z. MR2283615

[185] R. Vidal, Y. Ma, and S. S. Sastry, *Generalized principal component analysis*, Interdisciplinary Applied Mathematics, vol. 40, Springer, New York, 2016, DOI 10.1007/978-0-387-87811-9. MR3468872

[186] R. Vidal, Y. Ma, S. Soatto, and S. Sastry, *Two-view multibody structure from motion*, International Journal of Computer Vision **68** (2006), no. 1, 7–25.

[187] T. Werner, *Combinatorial constraints on multiple projections of a set of points.*, Proceedings of IEEE International Conference on Computer Vision, 2003, pp. 1011–1016.

[188] T. Werner, *Constraint on five points in two images*, Proceedings of the IEEE Conference on Computer Vision and Pattern Recognition, vol. 2, 2003, pp. II–203.

[189] T. Werner and T. Pajdla, *Cheirality in epipolar geometry*, Proceedings of IEEE International Conference on Computer Vision, vol. 1, 2001, pp. 548–553.

[190] T. Werner and T. Pajdla, *Oriented matching constraints.*, British Machine Vision Conference, 2001, pp. 1–10.

[191] L. Wolf and A. Shashua, *On projection matrices $\mathbb{P}^k \to \mathbb{P}^2, k = 3, \ldots, 6$, and their applications in computer vision*, International Journal of Computer Vision **48** (2002), no. 1, 53–67.

[192] H. Yang and L. Carlone, *Certifiably optimal outlier-robust geometric perception: Semidefinite relaxations and scalable global optimization*, IEEE Transactions on Pattern Analysis and Machine Intelligence **45** (2022), no. 3, 2816–2834.

[193] O. Zariski, *On the Problem of Existence of Algebraic Functions of Two Variables Possessing a Given Branch Curve*, Amer. J. Math. **51** (1929), no. 2, 305–328, DOI 10.2307/2370712. MR1506719

[194] J. Zhao, L. Kneip, Y. He, and J. Ma, *Minimal case relative pose computation using ray-point-ray features*, IEEE Transactions on Pattern Analysis and Machine Intelligence **42** (2019), no. 5, 1176–1190.

[195] M. Zhao and R. Chung, *Critical configurations of lines to geometry determination of three cameras*, 19th International Conference on Pattern Recognition, IEEE, 2008, pp. 1–5.

Department of Mathematics and Oden Insitute, University of Texas at Austin
Email address: jkileel@math.utexas.edu

Institutionen för Matematik, KTH Royal Institute of Technology
Email address: kathlen@kth.se

Selected Published Titles in This Series

111 **Serkan Hoşten, Diane Maclagan, and Frank Sottile, Editors,** Combinatorial, Computational, and Applied Algebraic Geometry, 2025

110 **Christine Berkesch, Benjamin Brubaker, Gregg Musiker, Pavlo Pylyavskyy, and Victor Reiner, Editors,** Open Problems in Algebraic Combinatorics, 2024

109 **Paul M. N. Feehan, Lenhard L. Ng, and Peter S. Ozsváth, Editors,** Frontiers in Geometry and Topology, 2024

108 **Pramod N. Achar, Kailash C. Misra, and Daniel K. Nakano, Editors,** Categorical, Combinatorial and Geometric Representation Theory and Related Topics, 2024

107 **Ron Donagi, Adrian Langer, Piotr Sułkowski, and Katrin Wendland, Editors,** String-Math 2022, 2024

106 **Gilles Godefroy, Mohammad Sal Moslehian, and Juan Benigno Seoane-Sepúlveda, Editors,** Geometry of Banach Spaces and Related Fields, 2024

105 **A. Connes, C. Consani, B. I. Dundas, M. Khalkhali, and H. Moscovici, Editors,** Cyclic Cohomology at 40: Achievements and Future Prospects, 2023

104 **A. Kechris, N. Makarov, D. Ramakrishnan, and X. Zhu, Editors,** Nine Mathematical Challenges, 2021

103 **Sergey Novikov, Igor Krichever, Oleg Ogievetsky, and Senya Shlosman, Editors,** Integrability, Quantization, and Geometry, 2021

102 **David T. Gay and Weiwei Wu, Editors,** Breadth in Contemporary Topology, 2019

101 **Avraham Aizenbud, Dmitry Gourevitch, David Kazhdan, and Erez M. Lapid, Editors,** Representations of Reductive Groups, 2019

100 **Chiu-Chu Melissa Liu and Motohico Mulase, Editors,** Topological Recursion and its Influence in Analysis, Geometry, and Topology, 2018

99 **Vicente Muñoz, Ivan Smith, and Richard P. Thomas, Editors,** Modern Geometry, 2018

98 **Amir-Kian Kashani-Poor, Ruben Minasian, Nikita Nekrasov, and Boris Pioline, Editors,** String-Math 2016, 2018

97 **Tommaso de Fernex, Brendan Hassett, Mircea Mustaţă, Martin Olsson, Mihnea Popa, and Richard Thomas, Editors,** Algebraic Geometry: Salt Lake City 2015 (Parts 1 and 2), 2018

96 **Si Li, Bong H. Lian, Wei Song, and Shing-Tung Yau, Editors,** String-Math 2015, 2017

95 **Izzet Coskun, Tommaso de Fernex, and Angela Gibney, Editors,** Surveys on Recent Developments in Algebraic Geometry, 2017

94 **Mahir Bilen Can, Editor,** Algebraic Groups: Structure and Actions, 2017

93 **Vincent Bouchard, Charles Doran, Stefan Méndez-Diez, and Callum Quigley, Editors,** String-Math 2014, 2016

92 **Kailash C. Misra, Daniel K. Nakano, and Brian J. Parshall, Editors,** Lie Algebras, Lie Superalgebras, Vertex Algebras and Related Topics, 2016

91 **V. Sidoravicius and S. Smirnov, Editors,** Probability and Statistical Physics in St. Petersburg, 2016

90 **Ron Donagi, Sheldon Katz, Albrecht Klemm, and David R. Morrison, Editors,** String-Math 2012, 2015

89 **D. Dolgopyat, Y. Pesin, M. Pollicott, and L. Stoyanov, Editors,** Hyperbolic Dynamics, Fluctuations and Large Deviations, 2015

88 **Ron Donagi, Michael R. Douglas, Ljudmila Kamenova, and Martin Rocek, Editors,** String-Math 2013, 2014

For a complete list of titles in this series, visit the
AMS Bookstore at **www.ams.org/bookstore/pspumseries/**.

FSC
www.fsc.org
MIX
Papier | Fördert
gute Waldnutzung
FSC® C083411

Kolibri 360 GmbH
Ferdinand-Jühlke-Straße 7
99095 Erfurt, Deutschland
produktsicherheit@kolibri360.de